D0782373

Thermodynamics

Fundamentals and Engineering Applications

This concise text provides an essential treatment of thermodynamics and a discussion of the basic principles built on an intuitive description of the microscopic behavior of matter. Aimed at a range of courses in mechanical and aerospace engineering, the presentation explains the foundations valid at the macroscopic level in relation to what happens at the microscopic level, relying on intuitive and visual explanations which are presented with engaging cases. With ad hoc, real-world examples related also to current and future renewable energy conversion technologies and two well-known programs used for thermodynamic calculations, FLUIDPROP and STANJAN, this text provides students with a rich and engaging learning experience.

William C. Reynolds (1933–2004) was a renowned and exceptionally creative scientist who specialized in turbulent flow and computational fluid dynamics. However, his competence spanned many areas of fluid mechanics and of mechanical engineering in general. He completed his bachelor's (1954), master's (1955), and doctoral (1957) degrees at Stanford. Reynolds chaired the Department of Mechanical Engineering from 1972 to 1982 and from 1989 to 1992. Pioneer in large eddy simulation for fluid modeling, he was elected to the National Academy of Engineering in 1979. He won the Fluid Engineering Award of the American Society of Mechanical Engineers in 1989 and the Otto Laporte Award by the American Physical Society in 1992. He was universally recognized as a gifted and original educator. He wrote several books, some on thermodynamics.

Piero Colonna is Professor and Chair of Propulsion and Power at Delft University of Technology, where he has been teaching *Thermodynamics* and *Modelling and Simulation of Energy Conversion Systems* since 2002. He completed his master's degree (1991) in aerospace engineering at Politecnico di Milano, obtained a master's degree in mechanical engineering from Stanford University (1995) and a doctoral degree again from Politecnico di Milano (1996). In 2005, he became the recipient of the VIDI personal grant of the Dutch Science Foundation (NWO) for his research on the gas dynamics of dense vapours and supercritical fluids. He is also a recognized world expert of Organic Rankine Cycle and Supercritical CO_2 power systems (renewable thermal energy conversion), and a pioneer of Non Ideal Compressible Fluid Dynamics (NICFD). He served as the Chairman of the Board of the International Gas Turbine Institute (2015-2017), and is advisor to the Board of the Global Power and Propulsion Society. He was also Associate Editor of the ASME *Journal for Engineering of Gas Turbines and Power*, and is currently Associate Editor of the *Journal of the Global Power and Propulsion Society*. His talent as lecturer is testified by the *Best MSc. Lecturer Award in Sustainable Energy Technology and Fluid Mechanics* which he received twice in 2010 and 2012.

Building on Bill Reynolds' classic book, *Engineering Thermodynamics*, this textbook offers a unique approach to teaching thermodynamics based on the understanding of simple physical principles. It will be of great benefit to future engineers tackling tough problems in energy.

– Tim Lieuwen, *Georgia Tech.*

An excellent textbook for undergraduate and graduate students, and good reference for practicing mechanical engineers. The book's strength is in its simplification of the science of thermodynamics and the listing of its practical applications.

– Hany Moustapha, *University of Québec*

A textbook that could be entitled "Lean thermodynamics" because it explains the core of this science by employing an effective and essential language and providing the fundamental concepts as they are needed. It focuses on topics and applications that are indispensable in facing today's challenges in energy, mechanical and aerospace engineering.

– Gianluca Valenti, *Politecnico di Milano*

This book is a must-have for students, academics and engineers working with thermodynamics. The authors present the foundations of thermodynamics by linking the beautiful formality of primary laws to a sharp description of the underlying microscopic world. Thermodynamics professionals and enthusiasts will keep this book within reach as an insightful reference.

– Alberto Guardone, *Politecnico di Milano*

The authors have taken care in clearly articulating the requisite topics for engineering students in a very engaging manner. The book will be an important resource for students in the classroom as well as into the future as they become practicing engineers.

– Karen A. Thole, *Pennsylvania State University*

The authors of this superb text have set out to present "the very few concepts of classical thermodynamics, in a much more essential treatment". Not overburdened with the inclusion of numerous repetitive examples, this book presents the basic concepts of energy and entropy in a simple equation form. The text is eminently suited for engineering undergraduate and first-year graduate level courses.

– Lee S. Langston, *University of Connecticut*

This comprehensive textbook of thermodynamics is strongly backed by plentiful educational experiences of the authors. It realizes that the fundamental concepts naturally connect to the practical consideration and understanding of real thermal systems, including modern engineering applications.

– Toshinori Watanabe, *University of Tokyo*

Thermodynamics

Fundamentals and Engineering Applications

William C. Reynolds
Stanford University, USA

Piero Colonna
Delft University of Technology, The Netherlands

CAMBRIDGE
UNIVERSITY PRESS

CAMBRIDGE
UNIVERSITY PRESS

University Printing House, Cambridge CB2 8BS, United Kingdom

One Liberty Plaza, 20th Floor, New York, NY 10006, USA

477 Williamstown Road, Port Melbourne, VIC 3207, Australia

314–321, 3rd Floor, Plot 3, Splendor Forum, Jasola District Centre, New Delhi – 110025, India

79 Anson Road, #06–04/06, Singapore 079906

Cambridge University Press is part of the University of Cambridge.

It furthers the University's mission by disseminating knowledge in the pursuit of education, learning, and research at the highest international levels of excellence.

www.cambridge.org
Information on this title: www.cambridge.org/9780521862738
DOI: 10.1017/9781139050616

First published 2018

Printed and bound in Great Britain by Clays Ltd, Elcograf S.p.A.

A catalog record for this publication is available from the British Library.

Library of Congress Cataloging-in-Publication Data
Names: Reynolds, William C., 1933–2004, author. | Colonna, Piero, author.
Title: Thermodynamics. Fundamentals and engineering applications /
William Reynolds, Piero Colonna, Delft University of Technology
Description: New York, NY, USA : Cambridge University Press, [2017]
Identifiers: LCCN 2017025042 | ISBN 9780521862738 (hardback)
Subjects: LCSH: Thermodynamics. | Engineering mathematics.
Classification: LCC QC311 .R423 2017 | DDC 536/.7–dc23
LC record available at https://lccn.loc.gov/2017025042

ISBN 978-0-521-86273-8 Hardback

Additional resources for this publication at www.cambridge.org/Reynolds

In memory of
Professor William Craig Reynolds
and
Professor Gianfranco Angelino,
Masters

To Mirella, Flavia, and Giulia, the lights of my life.

P. C.

CONTENTS

APPENDICES

PREFACE

Inspiring Principles

Bill and I talked many times about his idea of writing a new book on thermodynamics for engineers. We observed that there are many books on this subject which have one thing in common: They are very thick and every chapter becomes thicker with every new edition, by adding ever more examples and auxiliary material. We believed instead that the explanation of the very few founding concepts of classical thermodynamics, beautiful and challenging to the mind and expressed in formally very simple equations, would benefit from a much more essential treatment, which always starts with the description of the physical world at microscopic level. Concepts like energy and entropy must first be understood in their essence from situations that can be visualized with a few carefully conceived examples. Once these notions are clear, they can then be used to solve the inexhaustible variety of problems involving thermodynamics by means of a systematic approach, which always begins with applying these founding concepts.

This method is also a key element of this textbook. Illustrating or trying to solve a multitude of problems does not help deep understanding: It is better that students approach few but longer and more challenging exercises, with the intent of solving them using founding concepts and a rigorous method. Repetitively solving many examples, hoping that the mechanistic application of mysterious formulas will lead to the correct solution, is to be avoided, and this tendency is often stimulated by the availability of too many examples and exercises. The complexity and variety of the reality we want to model with thermodynamics is just excessive for the empirical approach based on learning from examples. To this purpose we recall the words of Leonardo da Vinci:[1] *"Quelli che s'innamorano della pratica senza la scienza sono come i nocchieri ch'entrano in mare sopra nave senza timone o bussola, che mai non hanno certezza dove si vadino"*, or, *"Those who fall in love with practice without science are like helmsmen who come on board and sail the seas without helm or compass, and never gain certainty of where they are heading."*

Intended Audience and Content

This book covers topics aimed at a range of courses in mechanical and aerospace engineering, but can also serve as a reference for the practicing engineer. Given the variety of engineering curricula in academic institutions all over the world, it is difficult to suggest where the chapters for the Bachelor program end and where those for the Masters program begins, but the progression of chapters is clearly from Bachelor to Masters, not only because of the content, but also because of the writing style, which is targeted to audiences of increasing age and knowledge.

The passion for energy technologies that will emancipate us from the use of fossil fuels has been the underlying driver for my love of thermodynamics, and talking about its societal relevance has often made me successful in motivating students towards its study. For this reason, whenever possible, examples and exercises are about renewable or more sustainable propulsion and power systems, and we trust instructors will be able to use these examples to draw an ever increasing number of skilled students to participate in the global effort to solve the energy issue. Throughout the book we sometime suggest that students work out some interesting examples on their

[1] L. da Vinci, Codex Urbinas Latinus 1270. Biblioteca Apostolica Vaticana, vol. Trattato della Pittura, Chapter XXIII.

own, starting with the information they just learned, as an attempt to engage them immediately in the application of concepts.

This book contains several aspects that are arguably innovative, including its essence-driven approach to the subject matter, and the emphasis on a systematic method for the solution of engineering problems. The entropy concept is developed as a measure of the degree of disorder and randomness at microscopic level.

Thermodynamic properties of fluids – and in particular of mixtures – are covered in a somewhat more extensive way as compared with other thermodynamics textbooks for mechanical engineers. We have identified a clear need for this if we look at the trend of next-generation power and propulsion systems. We also included the treatment of the Element Potential method for chemical equilibrium analysis that Professor Reynolds had first documented in a widely circulated report published in 1986.

Pedagogical Features

Worked examples

An abundance of examples throughout the text will enable students to fully understand the fundamental concepts of thermodynamics and to successfully apply them to all kinds of energy engineering problems. A passion for energy technologies that will emancipate humans from the use of fossil fuels has been Professor Colonna's underlying driver for his love of thermodynamics. Thus, many examples related to current and future renewable energy conversion technologies is included throughout the text.

Exercises

At the end of each chapter, students can find carefully conceived problems to help them put into practice what they have learnt about the topics discussed in the chapter. Exercises with the computer icon can be solved using the academic versions of FLUIDPROP and STANJAN, two well-known programs developed by the authors and colleagues and downloadable for free from www.asimptote.nl/software. These programs run on the Windows, Mac OS and Linux operating systems. Complete solutions to all exercises are available online at www.cambridge.org/Reynolds.

Boxed paragraphs

Boxed paragraphs provide information about curious or historical anecdotes as well as details on additional topics accompanying or supporting those treated in the body of the text.

Figures and charts

Colors have been used in many figures in order to facilitate the understanding of complex relations between the visualized data, or to make the represented simplified objects more recognizable. Many of the colorful charts and tables throughout the book have been obtained by running FLUIDPROP and STANJAN, and can thus be reproduced by the reader.

Appendices

Appendices at the end of the book provide a wealth of thermodynamic data (e.g., Appendix A) and an easy reference to auxiliary information, like mathematical relations between partial derivatives (Appendix B), numerical schemes for the calculation of saturated thermodynamic states (Appendix C), values of chemical equilibrium constants (Appendix D), and a synopsis of the method of Lagrange multipliers (Appendix E).

SI units

The authors have consistently used SI units throughout the book.

Software

Several exercises in Chapter 6 to 10 can be solved using either FLUIDPROP or STANJAN. These are computer programs to which the authors contributed founding ideas and much development. The models that are implemented in these programs are treated in detail in the book.

FLUIDPROP

FLUIDPROP is a server application that allows one to compute a large number of primary and secondary thermodynamic and transport properties of pure fluids and mixtures according to a variety of models. The full version of FLUIDPROP includes the following sub-programs or libraries:

- GASMIX, implementing the ideal gas equation of state;
- IF97, implementing the Industrial Formulation 97 model for water issued by the International Association for the Properties of Water and Steam (IAPWS);
- TPSI, implementing several pressure-explicit multiparameter equations of state coded in a very efficient way;
- STANMIX, implementing the improved Peng-Robinson cubic equation of state, and its extension to mixtures according to the Wong Sandler mixing rules;
- PCP-SAFT, implementing molecular-based models for pure fluids and mixtures, and
- REFPROP, the well-known program developed and maintained by the National Institute of Standards and Technology.

The academic version of FLUIDPROP contains GASMIX, IF97, TPSI and a reduced version of STANMIX and REFPROP. The use of FLUIDPROP is introduced starting with Chapter 6, while previous chapters suggest the use of tables in the appendices, as a formative step in thermodynamics education.

STANJAN

STANJAN is also a server application and it implements the element potential method for the calculation of the equilibrium of reacting systems explained in Section 10.7.

Client programs that can make use of FLUIDPROP and STANJAN are EXCEL™, MATLAB™, MAPLE™, MATHEMATICA™, and many others. Each program comes with a detailed on-line help and many illustrative examples. Our experience shows that students learn to use them in a very short time, as they require only typing one simple formula in cells (e.g., EXCEL™), or in lines of code (e.g., MATLAB™). The calculated result (fluid property, mixture composition, etc.) can then be used for complex thermodynamic analyses in spreadsheets or programs.

Cycle-Tempo

Cycle-Tempo is a flow sheeting program for the thermodynamic analysis and optimization of energy conversion systems.

Energy system analysis and its extension with exergy analysis as illustrated in the examples of Chapter 7 and Chapter 9 can be reproduced with Cycle-Tempo. The process flow diagrams in some of the figures of Chapter 7 are taken from the graphical user interface of Cycle-Tempo.

The complete walkthrough and instructions for Cycle-Tempo are available to instructors as password-protected online resources. Instructors will also be able to obtain a six-months-long Cycle-Tempo license from Asimptote by contacting their sales rep).

Piero Colonna

REMEMBERING BILL REYNOLDS

I started to work at this book in a very emotional situation, and those strong feelings accompanied me during the time I dedicated to its writing since then.

Professor Bill Reynolds, whom I first met in 1994 as my mentor while pursuing a Masters degree at Stanford University, contacted me in June 2003 and told me of the new thermodynamics textbook he had started to write some time before. He mentioned that he had health problems. He then asked me if I wanted to visit him before the end of the year to talk about the book. I had been spending periods at Stanford working with him since I graduated there in 1995. As always, I enthusiastically accepted the invitation and arrived at Stanford in November of 2003. The moment I met Bill in his office I understood that something was not quite right.

After a very short preamble, he told me "You remember our plan to work together on a new thermo book and to add your material on properties of mixtures, energy systems, and the rest? Well, I would like you to pick up from where I am leaving this book and finish it, as I am going to die soon of a brain tumor." I was shocked and immensely sad, and at the same time I felt honored and committed. Those last days I spent with him were extremely intense and moving. We only managed to exchange ideas about the book very briefly, as his health was rapidly deteriorating.

I will never forget the moment I parted from him. I was standing next to his bed in his home, holding his hand, with my eyes full of tears, in silence, while I was mentally thanking him for all he had done for my education. He was looking back at me, with tenderness. It was December 26th, 2003. Bill died on January 3rd, 2004.

I had become involved with this book earlier, during my Masters studies at Stanford. I was taking ME270-Engineering Thermodynamics, taught by Bill. He was the most extraordinary lecturer, as anybody who has ever followed one of his classes can testify. Meanwhile, I was one of his research assistants, working on thermodynamic property models for multicomponent fluids. During one of the first lectures he told the students "I am not happy with the way we have been teaching thermodynamics of mixtures so far. This year we'll do it differently, and during the next lecture Piero will tell us what he has learned about it." I was in the audience, surprised and uncomfortable at first: What a joke (he liked jokes)!

I would have never imagined that I could give an academic lecture at that young age, but I discovered that I very much enjoyed it. It was one of the first times I realized the difference between learning for yourself and learning in such a way that you can explain something to others.

This book took many years to complete, and I very much enjoyed working on it, even if at times I felt uncomfortable when realizing that the end was much farther away than I anticipated. I regret very much I did not have the opportunity to discuss with Bill each and every sentence of this book, but such is life, and I appreciate the growth in professional maturity that the responsibility of making all the choices has given to me. I smile when I think that Bill, wherever he is now, might also be satisfied of the outcome, though I am sure he would propose many improvements.

I hope that you, student, instructor, or professional engineer, will like it, and that it will contribute to the development of a keen interest for this wonderful subject. I also hope that it will engage you even more into studying and applying solutions that will help solving all the relevant global energy issues.

Piero Colonna

ACKNOWLEDGMENTS

I am profoundly grateful to Teus van der Stelt for our deep friendship and his inspiring and constant support, without which this book would not have taken shape. He provided careful checks, feedback, made most of the figures, solved exercises and examples, and much more. He added much value to this book also because of the computer programs accompanying it, which are for a large part the fruit of his coding craftsmanship.

My gratitude goes also to the following colleagues for their editorial recommendations: Professor André Bardow, Professor Paolo Chiesa, Professor Richard Gaggioli, Professor Arvind Gangoli Rao, Professor Godfrey Mungal, Professor Alberto Guardone, and Professor Joachim Gross. Special thanks go to Professor Tom Bowman, who, in addition, hosted me many times at Stanford, allowing me to seclude myself over summer periods solely dedicated to the writing of this book.

I am happy to thank the coworkers and students of the *Propulsion and Power Group* who contributed to the Solutions Manual with much dedication: Adam, André, Antonio, Carlo, Emiliano, Fiona, Irene, Jan, Lucia, Salvo, Sebastian, Timo, and Vincent.

During all these years Janice Reynolds always sustained me: in the moments when I needed encouragement, she shared memories of Bill also struggling with the completion of books. That meant a lot to me.

Walt and Kay Hays were also a constant source of motivation and followed every step forward with the emotional level of participation that only a profound relation of mutual love can achieve.

Finally, no words are adequate to express my feelings for my wife Mirella in relation to the completion of the book: she has been as much my companion in this as in any other facet of my life.

1 Introduction

CONTENTS

1.1 What is Thermodynamics?

Thermodynamics is the science of energy and its transformations. Engineering thermodynamics is the application of this science in the creation of new technology. It must be understood very well by engineers of all varieties. A basic understanding of thermodynamics can also be an asset to people in fields where the technology is used, such as medicine, and to any lawyer, politician, or citizen who participates in decisions on the appropriate uses of technology. Thermodynamic analysis provides a good model for general analytical thinking.

Basic principles

Like any science, thermodynamics rests on a small number of very fundamental principles. The *First Law of Thermodynamics* is the idea that energy is conserved; this means that the total energy in any isolated region is always the same, although the form of the energy may change. It is one of the two great conservation principles on which all of modern science rests (the other is conservation of momentum). These conservation principles cannot be proven by experiment because the things one measures in an experiment are evaluated by assuming that the principles are true. But by using the basic principles, if necessary inventing new forms of energy to keep energy conserved, or new forces to keep momentum conserved, we can explain the workings of nature and create devices that behave as we predict, and that give support to the theory.

Microscopic and macroscopic views

Energy exists at all scales, from the smallest subatomic scale to the grandest scale of the universe. At human scales we sense that fast-moving objects have lots of *kinetic energy*, and that heavy objects up high, springs wound up, or charged capacitors have lots of *potential energy*. We also sense that a very hot object has much energy, in modern terminology referred to as *internal energy* (not *heat*, which in modern terminology refers to a mechanism of energy *transfer* and not to energy *content*). At *microscopic* scales, internal energy is nothing more than the kinetic and potential energy of the molecules, which we cannot see in detail at our human *macroscopic* scale. The energy in the electronic bonds that hold molecules together is another form of *internal energy*.

When a macroscopic object is moving quickly, there is a high degree of organization in the motion of the molecules. In contrast, the object could be motionless but contain the same energy, this time in randomly oriented, disorganized molecular motions (internal energy). Often the job of an engineer is to find some way to convert microscopically *disorganized* energy (internal energy) into microscopically *organized* energy (kinetic or potential energy) so that we can produce a macroscopic motion or effect. There is a limit on the ability of macroscopic systems to produce microscopic order. The *Second Law of Thermodynamics* is the great principle of science that reflects this limit.

Entropy

Just as the first law is based on a fundamental property of matter (energy) so is the second law, which rests on the concept of *entropy*. Entropy is a measure of the *amount of microscopic disorder* in a macroscopic system. The second law says that *entropy can be produced, but is never destroyed*. In other words, molecules left to themselves will not become organized; all natural processes produce entropy. This law has been supported by more than a century of accurate predictions for the behavior of macroscopic systems, and in more recent times by new understanding of the dynamics of nonlinear systems with many degrees of freedom, which are known to exhibit chaotic behavior.

These two fundamental principles (first and second laws) are mostly what thermodynamics is all about. The first law gives us valuable equations to use in predicting the behavior of something we might like to build. But it does not tell us whether or not the processes we assumed would occur will actually take place or could take place. This missing information is provided by the second law, which tells us which way a process must go. Used together in combination in a technique called *exergy (or availability) analysis*, they can tell us the minimum electrical power required to liquefy a given stream of natural gas, or the maximum shaft power we could get out of the chemical energy in a barrel of oil, *without any reference to the sort of hardware we might employ in these systems*. This gives us the "best performance" against which we can compare the systems we design to see how much margin there might be for improvement. Thus, thermodynamics is an extremely powerful tool and one that an engineer should not be without.

Our approach

There are many approaches to teaching thermodynamics. The biggest differences surround the way in which the second law is approached. There are some who prefer a logically elegant, purely macroscopic approach in which entropy is given no microscopic interpretation and hence is purely a mathematical abstraction. Beginning students usually find this nonphysical approach very difficult to comprehend, and this often results in thermodynamics courses being regarded as mystery hours. Instead we will use an approach in which almost all of our work is *macroscopic* but much of our thinking builds on our understanding of the *microscopic* nature of matter. We believe that our approach will enable you to grasp the basic principles of thermodynamics quickly, to use them effectively in engineering analysis, and to explain the results of your analysis.

1.2 Accounting for the Basic Quantities

Much of engineering analysis involves the application of a basic principle to a carefully defined system, which yields an equation that describes something fundamental about the behavior of the system. There are two distinct steps in such an analysis. The first step is simply an accounting of the flows of the quantity in the principle (mass, energy, momentum, or entropy). The second step is to invoke the physics of the principle.

Production accounting

A very nice way to express the accounting is in terms of *production*. The production of anything by a system (automobiles by a factory, money by a bank account, momentum by a jet engine, energy by a power station, or entropy by a chemical reaction) is given by

$$\textbf{production} = \textbf{output} - \textbf{input} + \textbf{accumulation}. \tag{1.1a}$$

The *accumulation* is the increase in the amount within the system, or equivalently the excess of the final amount over the initial amount within the system,

$$\textbf{accumulation} = \textbf{final} - \textbf{initial}. \tag{1.1b}$$

For your bank account, "output" is your withdrawals plus bank charges, "input" is your deposits plus bank interest payments, "accumulation" is the increase in your balance, and "production" is zero (unless your bank prints new money and gives you some), *all measured over the same time period. This is the most important equation to master for success in engineering thermodynamics.* If you can balance your checkbook, you should have no trouble with this basic accounting equation.

Rate-basis production accounting

Sometimes we make our balances over a definite time interval (as above), and other times we make our balances on a *rate basis*, therefore we have that

$$\textbf{rate of production} = \textbf{rate of output} - \textbf{rate of input} + \textbf{rate of accumulation}. \tag{1.1c}$$

Given that the application of the balance results in an ordinary differential equation, and that the time-derivative is commonly at the left-hand side of the equation, we can also write

$$\textbf{rate of accumulation} = \textbf{rate of input} - \textbf{rate of output} + \textbf{rate of production}. \tag{1.1d}$$

Rate-basis accounting is useful in deriving differential equations that govern the system, but it is also useful in *steady-state* problems, where the rate of accumulation is zero and all of the other rates are constants. Momentum analyses are almost always made on a rate basis, because forces are (by concept; see Section 1.3) rates of momentum transfer.

Basic principles

The production (or rate of production) of any conserved quantity must be zero. Thus, denoting production by $\mathcal{P}$, the principles of conservation of energy, and momentum can be expressed in the very simple and easily remembered forms,

$$\mathcal{P}_{\text{energy}} = 0, \tag{1.2a}$$

$$\mathcal{P}_{\text{momentum}} = 0. \tag{1.2b}$$

For the case of non-relativistic mechanics, which covers almost all that is of interest in engineering thermodynamics, mass is also conserved, so

$$\mathcal{P}_{\text{mass}} = 0. \tag{1.2c}$$

Unlike these conserved entities, entropy is produced by natural processes; in the limit of certain idealized processes (*reversible* processes) entropy is ideally conserved. So the Second Law of Thermodynamics can be expressed very neatly as

$$\mathcal{P}_{\text{entropy}} \geq 0. \tag{1.2d}$$

The four equations (1.2a)–(1.2d) concisely give the pertinent physics of engineering thermodynamics. Used in conjunction with production accounting as described above, they provide the key tools for engineering analysis.

Alternative balance equations

The accounting equation may be written in other forms equivalent to (1.1a):

$$\textbf{production} = \textbf{net output} + \textbf{accumulation}, \tag{1.3a}$$

$$\textbf{input} + \textbf{production} = \textbf{output} + \textbf{accumulation}, \tag{1.3b}$$

$$\textbf{input} + \textbf{initial} + \textbf{production} = \textbf{output} + \textbf{final}. \tag{1.3c}$$

Some students find it easier to remember one of these other forms. Production accounting, (1.1a) or (1.3a), is often preferred because the basic principles are expressed in terms of production. Equation (1.3a) is the most compact and is often preferred in advanced treatments. We like to use (1.3b) for conserved quantities (mass, energy, momentum). In (1.3c) the accumulation term is split into two pieces (*initial* and *final*), which can make some problems harder rather than easier. You should use whatever form seems most natural to you and is acceptable to your instructor.

1.3 Analysis Methodology

How to be systematic

It is important to develop a systematic methodology for doing analysis. Here are the steps that should be taken every time you do an engineering analysis:

1. Define the system under study by dotted lines on a sketch. Try to put the boundaries where you know something or need to know something, never where you don't know something and don't need to know something.
2. Indicate the reference frame of the observer if it is other than fixed with respect to the system dots. This is particularly important in momentum analyses.
3. List the simplifying assumptions that will be used in writing the balance.
4. Indicate the time basis for the analysis (a specific time period or a rate basis).
5. On the sketch, show all of the non-zero terms that will appear in your balance equation: include all transfers with arrows defining the direction of positive transfer, the accumulations, and any productions. This defines your nomenclature.
6. Write the basic balance for your system, in terms of the quantities defined on your sketch. There should be a one-to-one correspondence between the terms in your equation and those in the sketch. Check that only transfers across the system boundary appear in your balance equation.
7. After you have written all the pertinent balances, bring in other modeling information as necessary to bring the analysis to the point where you have the same number of equations as unknowns and can therefore (in principle) solve for the unknowns. Do this using symbols at first, then substitute numbers to get what you need.

Make your analysis readable!

A good analysis, like a good computer program, should have comments here and there to help the reader understand the various steps. Numbering the equations helps. If you learn to make your analysis "readable" with a few well-chosen words, those who read it will understand it more easily, you will save time trying to understand it when you read it in the future, and you will find it much easier to write it up for a report or journal article.

1.4 Concepts from Mechanics

We assume that the student already has a background in mechanics and has some notion of the concepts of momentum, mass, force, kinetic energy, and potential energy. These fundamentals will be important in our study of thermodynamics. In the next two sections we review these familiar ideas to set the stage for our introduction to the fundamentals of thermodynamics.

Conservation of momentum

The key fundamental principle in mechanics is the *conservation of momentum*, which incorporates the following ideas:

- Matter can be treated as having a property, called *momentum*, that is an *extensive, conserved, vector function of its velocity* measuring its tendency to keep moving in the same direction at the same speed when the matter is *not* acted upon by external agents.

The terms used above are very important and have the following meanings:

- *extensive* means that the momentum of an object is the sum of the momenta of its parts;
- *conserved* means that the momentum of an isolated system does not change;
- *vector function of its velocity* means that the momentum of a little piece of matter (a particle) is a vector that depends on its velocity vector.

These physical ideas are enough to determine the mathematical representation of momentum. A mathematical theorem (the *representation theorem*) says that the only vector function of another vector is a scalar times the vector itself. Therefore, denoting the velocity vector of a particle by $\vec{\mathcal{V}}$ and its momentum vector by $\vec{\mathcal{I}}$, we know that the most general relationship possible between the momentum and velocity of a particle is

$$\vec{\mathcal{I}} = M(\mathcal{V}^2)\vec{\mathcal{V}}. \tag{1.4}$$

Here M is a scalar function that can depend at most on the only scalar that can be formed from the velocity vector, namely its magnitude (or magnitude-squared $\mathcal{V}^2$). Throughout this book we use a special parenthetical notation to distinguish the arguments of a function from a multiplicative factor. For example, $M(\mathcal{V}^2)$ denotes that M is a function of $\mathcal{V}^2$. As you know, we call M the *mass* of the particle.

Mass

With the addition of the basic concept of relativity, namely that the speed of light c must be the same in all reference frames, the functional form of $M(\mathcal{V}^2)$ can be determined. One considers an isolated system, in which two identical particles collide, as viewed in two different reference frames moving with respect to one another. Invoking conservation of momentum as written by observers in each frame, and using symmetry arguments and the Lorentz transformation, one discovers that the only functional form that allows momentum conservation in all frames is[1]

$$M = \frac{M_0}{\sqrt{1 - \mathcal{V}^2/c^2}}, \tag{1.5}$$

where M_0 is a constant called the *rest mass* of the particle. Note that just the basic ideas underlying the concept of momentum have led us to a formula for evaluating momentum! It is the same with energy and entropy, where careful formulations of the concepts lead to ways to measure these human representations of nature.

[1] See, *e.g.*, Wiedner and Sells, *Elementary Physics*, Allyn and Bacon, 1975, Ch. 35.

For a particle moving slowly compared with the speed of light, $\mathcal{V}^2 \ll c^2$ and so $M \approx M_0$. This is the realm of *Newtonian mechanics*, where mass also can be treated as a conserved quantity. Almost everything we shall do in this book is treated adequately by Newtonian mechanics, so we will make frequent use of the approximation that the mass of a particle is constant or that *mass is conserved*.

Force

When two systems interact in isolation, their combined momentum is conserved, but the momentum of one can increase while that of the other decreases by the same amount (Figure 1.1a). Often we want to analyze one system without detailed consideration of the other. We do this by drawing an imaginary boundary around the system of interest and replacing its interaction with the other system in some appropriate manner. In the case of a momentum analysis, we need to account for the possibility of momentum transfer from one to the other. As you know, we attribute this momentum transfer to a *force* acting between the two systems.

You were probably introduced to the concept of force (a push or pull) at a very early age, well before encountering the concept of momentum. You may have already studied the analysis of trusses and other static structures, using forced-based analyses that do not even mention momentum. However, at this stage in your education it is important to appreciate that the concepts of force and momentum are intimately related, and that the fundamental forces in nature have been invented to explain observed changes in momentum.

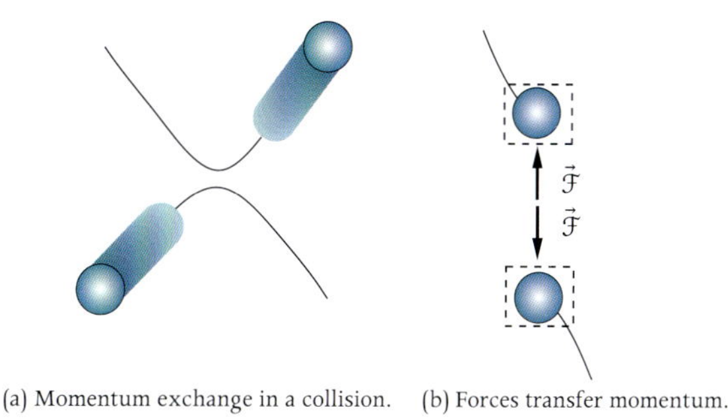

(a) Momentum exchange in a collision. (b) Forces transfer momentum.

Figure 1.1 Interactions.

We define the force acting *on* a system as the *rate of momentum transfer to the system*. Since the momentum transfer out of the first system is exactly the momentum transfer into the second, the force acting on the first system is always equal but opposite to the force acting on the second (Figure 1.1b).

Newton's law

With this understanding of the meaning of force, Newton's law can be interpreted as the *momentum balance* on a Newtonian particle. If $\vec{\mathcal{F}}$ is the force acting on a particle, and $\vec{\mathcal{V}}$ is its velocity, the momentum balance on the particle is

$$\underbrace{\vec{\mathcal{F}}}_{\substack{\text{rate of}\\ \text{momentum}\\ \text{input}}} = \underbrace{\frac{d(M\vec{\mathcal{V}})}{dt}}_{\substack{\text{rate of}\\ \text{momentum}\\ \text{accumulation}}} . \qquad (1.6)$$

If $\mathcal{V} \ll c$ then $M \approx M_0 = \text{constant}$; since the acceleration is $\vec{a} = d\vec{\mathcal{V}}/dt$, (1.6) then becomes $\vec{\mathcal{F}} = M\vec{a}$. Note that $\vec{\mathcal{F}}$, $\vec{\mathcal{V}}$, and $\vec{a}$ are all *vectors*, and *each component of the momentum* is conserved.

Gravitation

As viewed from the Sun, the Moon is constantly changing its direction (and hence its momentum vector) as it orbits the Earth. How can we explain this? The classical way is to ascribe the perceived momentum change to a *gravitational force* between the Earth and Moon, which causes both to orbit their common center of mass. Another way is to say that the Earth and Moon move freely, each with constant momentum, in a mass-distorted space–time frame (Einstein's theory). Both are legitimate views, and both invoke conservation of momentum. Since the classical view is conceptually simpler and works adequately for virtually every engineering task, we will use the gravitational force approach.

The gravitational force between two point masses M_1 and M_2 separated by a distance r can be reasoned to be given by

$$\mathcal{F} = k_G \frac{M_1 M_2}{r^2}, \qquad (1.7)$$

where $k_G = 6.67 \times 10^{-11}\ \text{m}^3/(\text{kg} \cdot \text{s}^2)$ is a physical constant. The force exerted by the Earth on an object (its *weight*, w) can be obtained by integrating (summing) the effects of all little pieces of the Earth and object, and is

$$w = Mg \qquad (1.8)$$

where M is the total mass of the object. Newton's law shows that the factor g is the acceleration of a freely-falling object, which is approximately $g = 9.8\ \text{m/s}^2$ on Earth.

Inertial frames

Momentum conservation analyses must be done in an *inertial reference frame*, a coordinate system in which a free particle would accelerate at the rate determined by the external gravitational field. For some analyses (aircraft dynamics) the inertial frame can be fixed to the Earth's surface; for other analyses (satellite launch) the inertial frame may be fixed at the Earth's center; still other analyses (interplanetary trajectories) require a Sun-based coordinate system; and so on.

Momentum analysis methodology

It is important that an engineer develops a good methodology for analysis. Meticulous attention to the following steps will greatly reduce the chance for an error in a momentum analysis:

1. Draw a diagram identifying the system to be analyzed by enclosing it within dotted lines.
2. Show the inertial reference frame to be used in writing the equations.
3. List any simplifying idealizations.
4. Show all forces (or momentum transfers) acting on (or transferred to/from) the system to define their positive directions.
5. Show the velocities to define their positive directions.
6. Write the momentum conservation equation for each component direction (x, y, z) of importance; there should be a one-to-one correspondence between the forces (or momentum transfers)

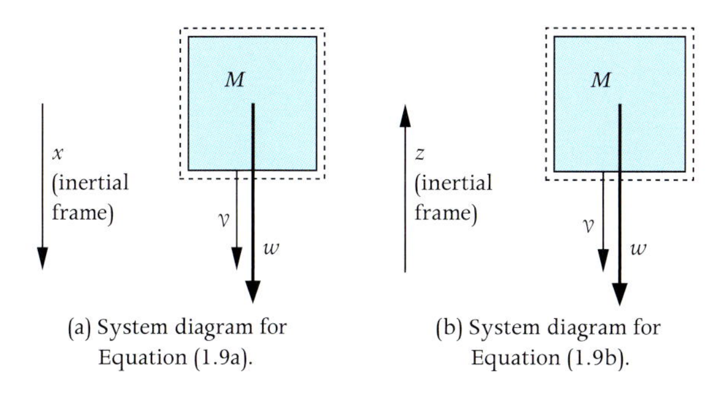

Figure 1.2 Momentum balance.

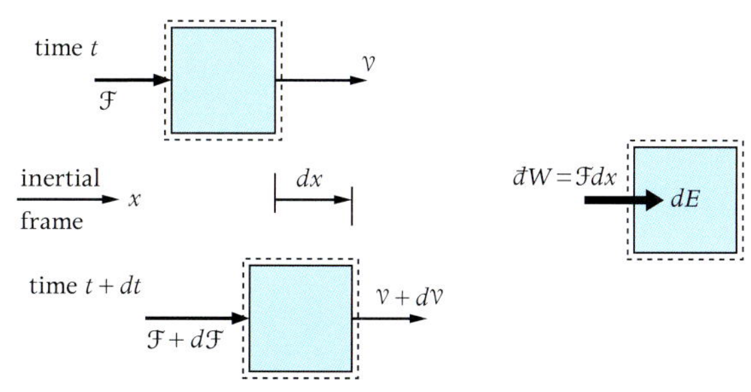

Figure 1.3 Definition of work.

on the diagram and those in the equations. The momentum balance is usually done on a rate basis.

7. Bring in other information as necessary to complete the analysis.

Example: being systematic. Let us consider a body of mass M falling freely towards the Earth (Figure 1.2a). It will accelerate as it falls due to the downward gravitational force w exerted on it by the Earth. Its momentum in the downward direction will increase as a result of the momentum transfer into the object by the downward force. We make the simplifying idealization that the resistance force exerted by the air is negligible. Then, denoting the downward velocity by $\mathcal{V}$, the balance of *downward* (x) momentum gives

$$\underbrace{w}_{\substack{\text{rate of}\\\text{momentum}\\\text{input}}} = \underbrace{\frac{d(M\mathcal{V})}{dt}}_{\substack{\text{rate of}\\\text{momentum}\\\text{accumulation}}}. \qquad (1.9a)$$

Had we chosen to write the momentum balance in the *upward* (z) coordinate system (Figure 1.2b) instead, the momentum balance of *upward* momentum would be

$$\underbrace{0}_{\substack{\text{rate of}\\\text{momentum}\\\text{input}}} = \underbrace{w}_{\substack{\text{rate of}\\\text{momentum}\\\text{output}}} + \underbrace{\frac{d(-M\mathcal{V})}{dt}}_{\substack{\text{rate of}\\\text{momentum}\\\text{accumulation}}}. \qquad (1.9b)$$

Note that the momentum in the z direction is $-M\mathcal{V}$, and the weight force w takes z momentum *out* of the system. The end result is the same, but using inconsistent frames for different terms would lead to errors. Being systematic helps avoid such errors.

1.5 Mechanical Concepts of Energy

Work

The concept of *work* is central in both mechanics and thermodynamics. Figure 1.3 shows an object that moves a small distance $d\vec{x}$ while force $\vec{\mathcal{F}}$ acts upon it. Note that we allow the velocity and force to change a little during this process. The work done *on* the object by the force, or the energy transfer as work *to* the object, is defined as

$$đW = \vec{\mathcal{F}} \cdot d\vec{x}. \qquad (1.10)$$

Here $đ$ is a special symbol that we use to denote *a small amount*, as opposed to the symbol d, which denotes *a small change*. Mathematically $đ$ denotes an *inexact differential*, meaning it is not the change of anything, while d denotes an exact differential, which is a change of something. Note that $\vec{\mathcal{F}}$ can vary during this process, and that $đW \neq d(\vec{\mathcal{F}} \cdot \vec{x})$. The $đ$ symbol reminds us that $đW$ is not the differential of W.

In thermodynamics, energy is conceived as a general conserved property of matter, and work as a form of energy transfer. Expressions for other forms of energy are then derived by making energy balances that relate these energy changes to work. We illustrate this in the following for two familiar forms of energy.

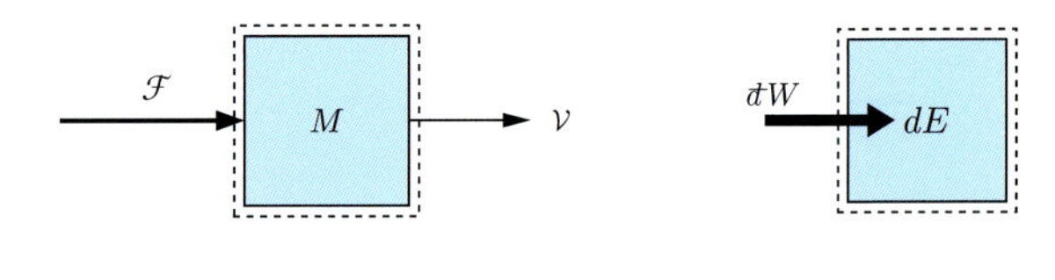

(a) Accelerating a mass increases its kinetic energy.

(b) Terms in the energy balance.

Figure 1.4 Kinetic energy balance.

Kinetic energy

Consider the acceleration of a body of fixed mass M acted on by force $\vec{\mathcal{F}}$ (Figure 1.4a). Newton's law relates $\vec{\mathcal{F}}$ to the acceleration,

$$\vec{\mathcal{F}} = M\frac{d\vec{\mathcal{V}}}{dt}. \tag{1.11}$$

Multiplying by $d\vec{x}$, and using $d\vec{x} = \vec{\mathcal{V}}dt$, we have

$$\begin{aligned} \text{đ}W &= \vec{\mathcal{F}} \cdot d\vec{x} = M\frac{d\vec{\mathcal{V}}}{dt} \cdot d\vec{x} = M\frac{d\vec{\mathcal{V}}}{dt} \cdot \vec{\mathcal{V}}dt \\ &= M\vec{\mathcal{V}} \cdot d\vec{\mathcal{V}} = d\left(\frac{1}{2}M\mathcal{V}^2\right), \end{aligned}$$

which we interpret as an *energy balance* (Figure 1.4b),

$$\underbrace{\text{đ}W}_{\substack{\text{energy input}\\ \text{as work}}} = \underbrace{d\left(\frac{1}{2}M\mathcal{V}^2\right)}_{\substack{\text{accumulation of}\\ \text{kinetic energy}}}. \tag{1.12}$$

We therefore *define* the kinetic energy E_k of a mass M moving at velocity $\mathcal{V}$ relative to the observer as

$$E_k \equiv \frac{1}{2}M\mathcal{V}^2. \tag{1.13}$$

Note that in this case $\text{đ}W = dE_k$ because there are no other energy changes or transfers to balance $\text{đ}W$; $\text{đ}W$ is always an inexact differential, but in this case it turns out to be balanced by the exact differential dE_k.

Potential energy

Next, consider a weight hanging from a rope in the Earth's gravitational field (Figure 1.5a). If we slowly raise the object by pulling on the rope, and neglect the slight extra force required to accelerate

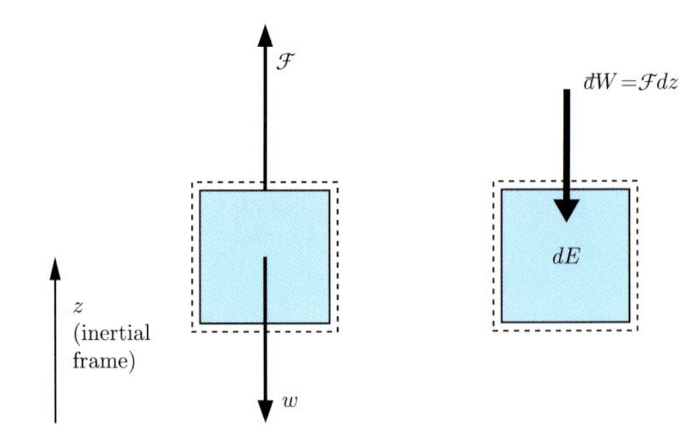

(a) Raising a weight increases its potential energy.

(b) Terms in the energy balance.

Figure 1.5 Potential energy balance.

the mass, then $\mathcal{F} = Mg$. If we treat g as constant, then the work done by the rope on the object when we increase its elevation by an infinitesimal amount dz is

$$\text{đ}W = Mg\,dz = d(Mgz),$$

which we interpret as an energy balance (Figure 1.5b),

$$\underbrace{\text{đ}W}_{\substack{\text{energy input}\\ \text{as work}}} = \underbrace{d(Mgz)}_{\substack{\text{accumulation of}\\ \text{potential energy}}}. \tag{1.14}$$

We therefore *define* the potential energy E_p of a mass M positioned a distance z above the (arbitrary) datum of the observer in a uniform gravitational field g as

$$E_p \equiv Mgz. \tag{1.15}$$

In this case $\text{đ}W = dE_p$ because there are no other energy changes or transfers to balance $\text{đ}W$; $\text{đ}W$ is always an inexact differential, but in this case it turns out to be balanced by the exact differential dE_p. *Note that we do include the work done by the gravitational force in the energy balance as this is accounted for by the potential energy change.*

Power

The *rate* of energy transfer is called *power*. In this book we use an overdot to denote a rate of *transfer* (but never a rate of *change*). For example, dividing (1.14) by dt, a small increment in time over which the

Table 1.1 SI mechanical units.

Primary quantity	SI unit	
Mass	kg (kilogram)	
Length	m (meter)	
Time	s (second)	
Secondary quantity	**SI units**	**Alias**
Velocity	m/s	
Acceleration	m/s^2	
Force	$kg \cdot m/s^2$	N (Newton)
Work, energy	$N \cdot m = kg \cdot m^2/s^2$	J (Joule)
Power	$J/s = kg \cdot m^2/s^3$	W (Watt)
Pressure	$N/m^2 = kg/\left(m \cdot s^2\right)$	Pa (Pascal)

change takes place, the energy balance of the system of Figure 1.5b, on a *rate basis*, is

$$\underbrace{\dot{W}}_{\text{rate of energy input as work}} = \underbrace{\frac{d(Mgz)}{dt}}_{\text{rate of accumulation of potential energy}}, \tag{1.16a}$$

where $\dot{W}$ denotes the *rate* of energy transfer as work,

$$\dot{W} \equiv \frac{đW}{dt}. \tag{1.16b}$$

Again we emphasize that $đW/dt$ is *not* the rate of *change* of W (this has no meaning), but is instead the rate of energy *transfer*.

1.6 Dimensions and Unit Systems

Eventually you will need numbers

In doing an analysis, it is a good idea to use symbols in order to keep the analysis general, and then insert specific numbers after the final equations have been developed symbolically. The quantitative results will be in the form of numerical values and units, and so engineers must be very comfortable working with numbers and various units of measure.

The SI unit system

This is a book for engineers of the future from all over the world, and for this reason we will work exclusively in the international standard system of units of measure known as SI (*Système Internationale*).

As for the USA, the transition to SI slowly began in the 1960s. SI is now the mandatory system for use in government laboratories, many technical journals, and a growing number of engineering firms (especially those that seek world markets). There are still vestiges of older measures which by the end of the twenty-first century are likely to be regarded as archaic. Probably by then we will no longer buy oil by the *barrel* (if there is any oil), or think of each barrel as containing about 6 million BTU (British Thermal Units) of energy. But the full conversion of US society to SI is likely to take at least another generation, so the modern engineer must be able to convert between these older measures and SI. This can be a very confusing task if one does not understand the basic ideas of units and dimensional systems. The purpose of this section is to help you develop this understanding.

Primary quantities

In any unit system there are a set of *primary quantities* for which one establishes some standards of measure. Table 1.1 lists the primary quantities and their units for the SI system. Note that the SI uses *mass, length, and time* as the primary mechanical quantities and hence is called an M-L-T system. The kg (kilogram), m (meter), and s (second) are the fundamental SI units of measure. Other SI primary quantities and corresponding units are: absolute temperature (Kelvin), electric current (Ampere), and luminous intensity (candela).

Standards for the primary quantities

The basic SI units will someday all be defined in terms of atomic standards easily reproduced anywhere with great precision. This is very nearly the case as of this writing. Prior to 1960 the second was defined as 1/86 400 of a mean solar day; then the definition was changed to 1/31 556 925.9747 of the tropical year 1900. Neither of these standards is very accessible, and so in 1964 the second was redefined in terms of the radiation frequency of a particular cesium spectral line, a standard accessible anywhere. The meter was redefined in 1960 as 1 650 763.73 times the wavelength of the orange-red line in the spectrum of krypton 86. Then, in 1986 the meter was redefined as the distance traveled through a vacuum by light in 1/299 792 458 s. Only the kilogram is still defined in terms of a particular artifact, namely a block of metal that is carefully maintained in Sèvres, France; someday the kilogram will be defined in terms of something more easily reproducible elsewhere, quite probably the rest mass of an electron.

The SI mass unit

Note that the *kilo*gram is *one* SI mass unit, and the *gram* is actually *one-thousandth* of the basic SI mass unit (1 gram = 10^{-3} kg = 1 mkg).[2] To maintain consistency in the nomenclature adopted by the SI system, a name for the mass unit without any prefix would seem more appropriate. A better choice might have been gram for the current kilogram and milligram for the current gram. These changes are favored by the authors, but not yet accepted by the science and engineering community.

[2] The prefix "kilo" is derived from the ancient Greek word for "one thousand", *i.e.*, *κιλιοι* (pronounced "kee-lee-oh-ee"). The prefix milli- is derived from the Latin word "mille" (pronounce "meel-leh") which means "one thousand".

Secondary quantities

Secondary quantities are defined in terms of primary quantities. The dimensions of secondary quantities, such as velocity, acceleration, force, and pressure, are determined by their definitions, and their units are combinations of the primary units. For convenience, a particular combination of the primary units is sometimes given an alias. Table 1.1 lists some important secondary quantities, their SI units, and the common alias used for the unit combinations.

Role of Newton's law

Newton's law, which relates the force exerted on a particle to its acceleration and mass, plays a crucial role in all unit systems. In general, Newton's law can be written

$$\vec{\mathcal{F}} = k_N M\vec{a}, \tag{1.17}$$

where k_N is a constant that depends on the unit system (Newton's constant). In the SI system, force is treated as a secondary quantity, and k_N is chosen to be unity and dimensionless. Thus, in SI, Newton's law sets the scale for force and makes the units of force $\text{kg} \cdot \text{m/s}^2$. This combination of primary units is given the *alias* Newton (N). It is rather appropriate that 1 N is about the weight of an apple... If you do not get the joke, Google it!

Non-uniqueness of SI

There is another physical law that relates force, mass, length, and time, namely the law for the gravitational attraction force between two point objects of mass M_1 and M_2, separated by a distance r,

$$\mathcal{F} = k_G \frac{M_1 M_2}{r^2}, \tag{1.18}$$

where k_G is the *universal gravitational constant*; in SI we have that $k_G = 6.67 \times 10^{-11}\ \text{m}^3/\text{kg} \cdot \text{s}^2$. It is pure chance that twentieth-century humans decided to make the constant in Newton's law unity when they set up the SI; they could have instead chosen

Table 1.2 US Customary System of mechanical units.

Primary quantity	USC unit	
Force	lb (pound)	
Length	ft (foot)	
Time	s (second)	
Secondary quantity	**USC units**	**Alias**
Mass	$\text{lb} \cdot \text{s}^2/\text{ft}$	slug
Velocity	ft/s	
Acceleration	ft/s^2	
Work, energy	$\text{ft} \cdot \text{lb}$	
Power	$\text{ft} \cdot \text{lb}/\text{s}$	
Pressure	lb/ft^2	psf

to make $k_G = 1$, which would have made force have the units of kg^2/m^2 and made the physical constant k_N have a weird value and units. It should be clear that humans and not nature gave Newton's law its simplest form, with $k_N = 1$.

US Customary FLT System

A great deal of twentieth-century engineering was done in a system based on English units, where, instead of *mass*, length, and time (M-L-T) being used for primary quantities as in the SI system, *force*, length, and time are the primary quantities. The unit of time is the same as in SI (second), the length unit is the foot (ft), and the force unit is the pound (lb). Then, the Newton's law constant is chosen to be unity, and hence mass, now a secondary quantity, comes out with the units of $\mathcal{F}/a$ or $\text{lb} \cdot \text{s}^2/\text{ft}$, which is given the alias *slug*. This system, summarized in Table 1.2, is now known officially as the US Customary System (the English have disowned it!). It is still in fairly wide use in the USA today, and it is the main system used in many American engineering textbooks in statics and dynamics.

USGC system

Other twentieth-century engineering was done in a different English-based system where mass, length, time, and force are all considered as primary quantities and given separate measurement standards, as shown in Table 1.3. The value and dimensions of k_N in such a system depend upon the standards chosen independently for force, mass, length, and time. We'll call this the US Grand Customary (USGC) system. In the USGC system, Newton's law is written as $\mathcal{F} = \frac{1}{g_c} M\vec{a}$, where $g_c = 32.2\ \text{ft} \cdot \text{lbm}/(\text{lbf} \cdot \text{s}^s)$ is a *physical constant* set by the independent standards for force, mass, length, and time. These standards are such that the weight of a 1 lbm mass is 1 lbf at a point where the acceleration of gravity g is 32.17 ft/s^2.

How to deal with with g_c

Although this book is in SI, and the world is converting to SI, important sectors of industry and society in the USA and elsewhere still employ the USGC system, and so you probably will need to work with it at some point. You may see books and journal articles in which Newton's law is written this way, or where kinetic energy is expressed as $\frac{1}{2g_c} M\mathcal{V}^2$ and potential energy as $\frac{1}{g_c} Mgz$. Put $g_c = 1$ to use this material in the SI system, or $g_c = 32.2\ \text{ft} \cdot \text{lbm}/\left(\text{lbf} \cdot \text{s}^2\right)$ for the USGC system. To use the USGC system with equations derived for SI in this text, insert $\frac{1}{g_c}$ in any term derived from Newton's law (kinetic energy, potential energy); a unit check will tell you clearly when this is necessary.

Alternative systems

We see that the number of primary quantities can be almost whatever we like. Some scientists have advocated use of a unit system in which *both* k_N and the gravitational constant k_G are unity, giving both laws their simplest possible form. In such a system there can be only two primary quantities, say length and time, so mass and force would both be secondary quantities. Even simpler is a system in which, in addition to $k_N = 1$ and $k_G = 1$, the speed of light $c = 1$. In this system there is only one primary quantity, which

Table 1.3 USGC mechanical units.

Primary quantity	USGC unit	
Force	lbf (pound force)	
Mass	lbm (pound mass)	
Length	ft (foot)	
Time	s (second)	
Secondary quantity	**USGC units**	**Alias**
Velocity	ft/s	
Acceleration	ft/s^2	
Work, energy	ft · lbf	
Power	ftlbf/s	
Pressure	lbf/ft^2	psf
Other common units and their equivalences in USGC and SI		
BTU (British Thermal Unit) 1 BTU = 778 ft · lbf = 1055 J		
HP (Horsepower) 1 HP = 550ft · lbf/s = 746 W		
psi (pound per square inch) 1 psi = 1/144 psf		

we might take as time. Working out the equivalents of the basic SI units in these systems is left as an exercise for the student who really wants to understand the fundamentals of unit systems.

Multiples and prefixes

Table 1.4 gives the standard prefixes and symbols for various multiplying powers of 10. Again we remind the student that the kg is *one* (*not* 1000!) basic SI mass unit. For example, 1 MW = 10^6 W (one megawatt).

Table 1.4 Multiples and prefixes for units.

Multiple	Prefix	Symbol
10^{12}	tera	T
10^9	giga	G
10^6	mega	M
10^3	kilo	k
10^2	deci	d
10^{-2}	centi	c
10^{-3}	milli	m
10^{-6}	micro	μ
10^{-9}	nano	n
10^{-12}	pico	p
10^{-15}	femto	f

Unit conversion

Although this book is in SI, often you will need to convert something from one set of units to another in order to use data or to compare with a published result. Such conversions are most easily accomplished by multiplying by unity, expressed in a suitable way, including

$$1 = 2.54\frac{\text{cm}}{\text{in}}, \qquad 1 = \frac{1}{2.54}\frac{\text{in}}{\text{cm}},$$
$$1 = 3600\frac{\text{s}}{\text{hr}}, \qquad 1 = 100\frac{\text{cm}}{\text{m}}.$$

These forms are obtained from *unit equivalents*, such as

$$1 \text{ in} = 2.54 \text{ cm}, \qquad 1 \text{ hr} = 3600 \text{ s},$$

which are available in various handbooks or textbook appendices, and also on many specialized websites on the internet. To make the unit conversion, one simply multiplies the number by unity (1) in forms that cancel out the unwanted units and replace them by the desired units.

Example: unit conversion. To find the speed in miles per hour of an airplane moving at 300 m/s, we multiply by some cleverly chosen forms of 1,

$$\mathcal{V} = 300\frac{\text{m}}{\text{s}} \times 3600\frac{\text{s}}{\text{hr}} \times 3.2808\frac{\text{ft}}{\text{m}} \times \frac{1}{5280}\frac{\text{mi}}{\text{ft}}$$
$$= 671 \text{ mi/hr}.$$

Example: determining unit equivalents. In order to determine an unknown dimensional equivalent, one must understand the bases of the two dimensional systems being employed. As an example, consider the dimensional system used in aeronautics in the early twentieth century. The primary quantities were force, length, and time. Force was measured in pounds (lbf), length in feet (ft), and time in seconds (s). The dimensions and units of mass were then determined by Newton's law, which was written as $\mathcal{F} = Ma$; so the dimensions of mass were those of $\mathcal{F}/a$, or force $\times$ time^2/length. The resulting mass unit, lbf $\cdot$ s^2/ft, was given the alias *slug*. In order to determine the number of slugs corresponding to one SI mass unit (1 kg), we consider what happens when a force of 1 N (1 kg $\cdot$ m/s^2) acts on a 1 kg mass, which we know accelerates the mass at the rate of 1 m/s^2. In the old aeronautics system the force and acceleration are

$$\mathcal{F} = 1\,\text{N} \times \frac{1}{4.448}\frac{\text{lbf}}{\text{N}} = 0.2248 \text{ lbf}$$

and

$$a = 1\frac{\text{m}}{\text{s}^2} \times 3.2808\frac{\text{ft}}{\text{m}} = 3.2808 \text{ ft/s}^2.$$

Since $\mathcal{F} = Ma$ in the old system, the mass in the old system is

$$M = \mathcal{F}/a = \frac{0.2248 \text{ lbf}}{3.2808 \text{ ft/s}^2} = 0.06852\frac{\text{lbf}\cdot\text{s}^2}{\text{ft}}$$
$$= 0.06852 \text{ slug}.$$

But since M is also 1 kg, we have the unit equivalence

$$1 \text{ kg} = 0.06852 \text{ slug} \qquad 1 \text{ slug} = 14.59 \text{ kg}.$$

If the early aerodynamicists had instead chosen to write Newton's law as $\mathcal{F} = \pi Ma$, which certainly was their option, the equivalences above would differ by a factor of π, but airplanes would fly no differently. This may sound silly, but exactly this sort of thing happened during the early development of electrical unit systems, where in some unit systems factors of π appeared in the equations and in other unit systems it did not. The dimensional system used by the extraterrestrials with which we may someday establish communication undoubtedly will be different from ours; analysis like this will be needed to develop dimensional equivalents between their systems and ours.

EXERCISES

1.1 An 80 kg student wishes to walk on snow without sinking and decides to make a pair of snow shoes. The student determined experimentally that the maximum pressure the snow can withstand is 0.6 kPa. What is the minimum surface required by one snow shoe? Would the student be able to sell snow shoes to his fellow classmates? Explain your answer.

1.2 Let $\mathcal{V}$ be the wind velocity approaching a windmill, D be the windmill rotor disk diameter, and ρ the air mass density. Show that the rate of kinetic energy flow through the windmill disk area is $\rho\pi D^2\mathcal{V}^3/8$. Defining η as the fraction of this energy that is converted into electrical energy (the *efficiency*), give the expression for the electrical power output. Apply your results to a windmill with $D = 5$ m, $\eta = 0.6$, using $\rho \approx 1$ kg/m^3, plotting the power in W versus the wind speed for velocities from 0 to 15 m/s. At 10 m/s, how many windmills are required by a city that needs 100 MW of electrical power? What is the power available to this city when the wind speed drops to 5 m/s?

1.3 Let $\mathcal{V}$ be the water velocity in a river of width w and depth d, flowing along at a height h above a hydraulic power station. Show that the maximum electrical power that could be obtained by converting the potential energy of the river water to electrical energy is $\rho \cdot \mathcal{V} \cdot w \cdot d \cdot h \cdot g$, where ρ is the water mass density. Calculate the maximum available power (MW) for $w = 20$ m, $d = 10$ m, $h = 100$ m, $\mathcal{V} = 2$ m/s, using $\rho = 1000$ kg/m^3.

1.4 Calculate the power (MW) that could be obtained by converting the *kinetic* energy of the river water in Exercise 1.3 to electricity at 100% efficiency.

1.5 Calculate the maximum energy (J) that could be obtained from 1 kg of water, assuming that the energy content of the mass, $E = Mc^2$, could be completely converted to useful energy in some sort of incredible reactor. Calculate the water consumption rate (kg/s) by such a reactor if it has the same power output as the maximum power output of the hydroelectric power plant of Exercise 1.3, and compare the water mass flow rates of the two systems.

1.6 Calculate the power requirements for lifting 40 kg sacks of cement up 1 m at the rate of 1 sack/second. If you had a team of superworkers who could do this all day long, how many would it take to equal the power available from the hydroelectric system of Exercise 1.3? Using a good current working wage, calculate the cost of the energy provided by these humans in, for example, \$/J and compare that to the current cost of electrical energy in your area. What does this lead you to conclude about society's need for energy conversion technology?

1.7 Make a table like Table 1.1 for a *Système Alternatif* in which $k_G = 1$ and the same primary quantities are used as in SI, choosing a suitable alias for each secondary quantity. What is k_N in this system? Hint: Consider two mutually gravitating 1 kg masses 1 m apart using both SI and SA.

1.8 Determine the equivalents of 1 kg, 1 N, 1 J, and 1 W in a *Système Deux* in which both k_N and k_G are unity and the meter and the second are the primary units of length and time. Choose a suitable alias for each secondary quantity. Make a table like Table 1.1 for this system. Hint: Consider two mutually gravitating 1 kg masses 1 m apart using both SI and SD.

1.9 Determine the equivalents of 1 m, 1 kg, 1 N, 1 J, and 1 W in a *Système Unis* in which both k_N and k_G are unity, the speed of light is unity, and the second is the only primary unit. Choose a suitable alias for each secondary quantity. Make a table like Table 1.1 for this system. Hint: Consider two mutually gravitating 1 kg masses 1 m apart using both SI and SU.

1.10 A home gas-fired heater has an output of 120 000 BTU/hr. The owner wants to exchange it with an electrical heater. What will be the equivalent electrical power in kW?

1.11 A 5 kW electrical heater is left on for an hour and a half. Calculate the amount of energy transfer as heat from the heater to the room in terms of the following units: a) unit for power in SI, b) ft · lbf, c) BTU.

1.12 An automobile engine has an output power of 120 HP. What is the equivalent motor power requirement in kW for an electric car?

1.13 A strong person might be able to lift a 110 lbf object up a distance of 5 ft in one second (at least once). What is the equivalent horsepower? What would be the average HP expenditure if the person could do it ten times per minute?

1.14 What is the kinetic energy in J of a 2000 lbm car moving at 60 mph? Work out the answer first by starting with the expression for kinetic energy in the SI system, converting the basic data into SI units in advance. Then, work out the answer starting with the expression for kinetic energy in the USGC system, and then converting the result into SI units. If you get two different results, explain.

1.15 An open office room holds 50 employees, each dissipating 430 kJ of energy as heat per hour. In addition, twenty 80 W light bulbs are all left on at all times. Calculate how many 6 kW air-conditioning units are required to keep the office room at a constant temperature of 22 °C, if:

a. The walls and windows are adiabatic;
b. The rate of energy transfer as heat into the office room through the walls and windows is 18 000 kJ/hr.

2 Energy

CONTENTS

In the first chapter we reviewed the concept of energy as it arises in mechanical systems. We interpreted work as a transfer of energy and used this idea to develop expressions for kinetic and potential energy. These were special instances of the general concept of energy, which is perhaps the most fundamental concept in all of science. In thermodynamics, another relevant mode of energy transfer is heat. Once all these concepts are clear, they can be used together with the first law to analyze and design an infinite variety of systems and devices, which are at the foundation of current and future societies.

2.1 Concept of Energy

The Energy Hypothesis

The ideas inherent in the general concept of energy are encapsulated in what we call the *Energy Hypothesis*:

> **Matter can be treated as having a property, called *energy*, that is an *extensive, conserved scalar* measuring its ability to cause change; *work* is a transfer of energy.**

The terms used above are very important:

- **extensive** means that the energy of a system is the sum of the energies of its parts;
- **conserved** means that the total amount of energy in an isolated system does not change;
- **scalar** means that energy is a quantity without directional (vector) character;
- **work is a transfer of energy;** it provides the key to the quantitative evaluation of the energy contained by matter.

In Section 1.4 we used these ideas to obtain expressions for the kinetic and potential energy of matter.

Other energy forms are evaluated in the same way. Since the energy hypothesis is itself used in the evaluation and measurement of the energy of matter, it is not possible to test the hypothesis of energy conservation by experimental measurements of energy. However, one can use the energy hypothesis to predict how matter will behave, and these predictions (when done correctly!) are always confirmed by experiment. Countless working devices, designed using energy analysis, provide enormous evidence of the validity and utility of the energy hypothesis. Believe in the concept of energy; it works!

2.2 Microscopic Energy Modes

A macroscopic quantity of matter contains a huge number of molecules, and its energy is the sum of the individual energies of all these molecules. Molecules and systems of molecules can have energy in a variety of different microscopic modes. Your understanding of the macroscopic changes that we observe will be enhanced by an understanding of the microscopic energy modes briefly discussed below.

Translational energy

The kinetic energy of molecules whizzing about in a gas or liquid is called *translational energy*. There are three degrees of freedom for translational motions, and all three components of the velocity of a molecule contribute to its translational energy. Since sound is propagated by molecular collisions within the fluid, it is not surprising that the average speed of molecules is approximately the speed of sound in the fluid, or about 300 m/s in air at room temperature. Translational energy increases with increasing temperature; for a low-density (ideal) gas, the translational energy is directly proportional to the absolute temperature (K) of the gas.

Rotational energy

The kinetic energy associated with the rotation of complex molecules about their centers of mass is called *rotational energy*. This energy is not very important for simple molecules with low moments of inertia, such as He, but it is very important for more complex molecules such as H_2O, CH_4OH, and complex organic molecules. As the temperature of a gas is increased, the rotational molecular motions become more intense, and so the rotational energy also increases with temperature.

Vibrational energy

Complex molecules are held together by the electrons surrounding their nuclei. These electronic bonds are springy, and with the heavy nuclei they act like spring-mass oscillators. Collisions between the molecules excite vibrations of the molecule, and the energy associated with these oscillations is called *vibrational energy*. The vibrations occur at a resonance frequency that depends on the molecular structure and the vibration mode. Quantum mechanics dictates that a molecule can have only certain discrete levels of vibrational energy. Each type of molecule has its own assortment of vibrational modes, which can be used in optical analysis to identify the molecule. Vibrational energy is not important at low temperatures because the translational velocities are so low that the vibrations excited by collisions are relatively weak. However, at high temperatures these vibrations become increasingly important. When the vibrations are sufficiently strong, the molecules can break apart (dissociate) to form simpler molecules or even single atoms.

Lattice energy

The atoms in a crystal lattice are held together by electrons that swirl around the atoms. These electronic bonds are springy, and so the atoms are constantly in a state of chaotic microscopic vibration during which kinetic energy is passed around from atom to atom. This chaotic microscopic energy increases as the temperature is raised, and vanishes as the temperature approaches absolute zero (0 K).

Electronic bonding energy

The electronic force fields that hold molecules together, or that bind atoms in a crystal, hold considerable *electronic bonding energy*. These bonds are broken during chemical reactions, which allows the atoms to rearrange themselves into different molecules, and the energy in the broken bonds is released. Usually this energy shows up as increased translational energy, macroscopically observed as an increase in temperature. The energy involved in making or breaking bonds is also called *chemical energy*.

Nuclear bonding energy

The force fields that hold the nuclei of atoms together are much stronger than electronic forces and hence there is much more energy associated with these bonds than with the electronic bonding of molecules. This energy is released in nuclear fission or fusion reactions, where it shows up in greatly increased microscopic kinetic energy.

2.3 Internal Energy

In a macroscopic system, such as the gas in an internal combustion engine or the silicon crystal in a semiconductor device, the sum of all the energy in the hidden microscopic modes (see Figure 2.1) is called the *internal energy*, U,

$$U = E_{\text{translation}} + E_{\text{vibration}} + E_{\text{rotation}} + E_{\text{electronic}} + \cdots \tag{2.1}$$

We emphasize that internal energy should not be called *heat*, a word used properly only for something else (see Section 2.6).

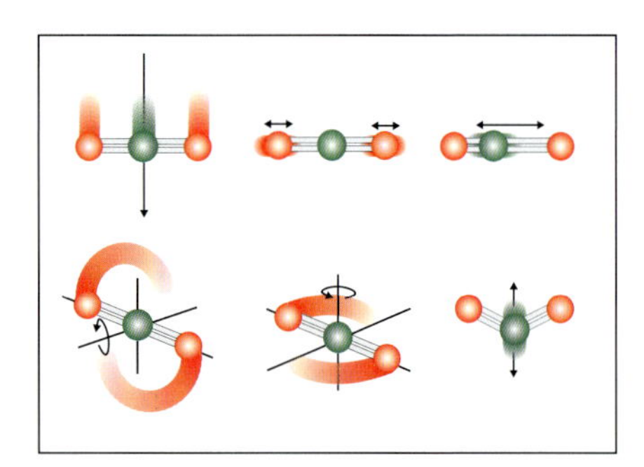

Figure 2.1 Internal energy is the energy of the hidden microscopic modes.

2.4 Total Energy

In general one can express the energy of matter as the sum of the internal energy arising from the hidden microscopic modes and the macroscopically observable forms of energy,

$$\begin{aligned} E &= U + E_{\text{kinetic}} + E_{\text{potential}} + \cdots \\ &= U + \frac{1}{2}M\mathcal{V}^2 + Mgz + \cdots \end{aligned} \tag{2.2}$$

Here $\mathcal{V}$ is the macroscopically observable velocity of the material relative to the coordinate frame, g is the gravitational acceleration, z is the height above the reference point for potential energy, and the dots represent other forms of macroscopically observable energy (electronic charge, magnetic dipoles, etc.). U includes the randomly oriented microscopic kinetic energy of the gas molecules due to motions relative to their common center of mass. Additional kinetic energy due to any coherent molecular motion that is revealed by motion of the center of mass is represented by the $\frac{1}{2}M\mathcal{V}^2$ term.

2.5 Energy Transfer as Work

In Chapter 1 we introduced the notion of work as energy transfer. Recall that the energy transferred as work by a force is defined as the product of the force times the displacement of the matter on which the force acts, as measured in the reference frame of the analysis *in the direction of the force*,

$$đW = \vec{\mathcal{F}} \cdot d\vec{x}. \tag{2.3}$$

We use the special symbol $đ$ to denote *a small amount* (here of work) whereas d denotes a small change (here in $\vec{x}$). The total amount of energy transferred as work to the system over the time period from t_1 to t_2 is

$$W_{12} = \int_1^2 \vec{\mathcal{F}} \cdot d\vec{x}. \tag{2.4}$$

Note that we do *not* write $W_2 - W_1$; this has no meaning, since work is associated with a *process over time* and not with the initial or final state. To carry out the integration one must know how $\vec{\mathcal{F}}$ varies with $\vec{x}$. Different functions $\vec{\mathcal{F}}(x)$ will give different results, which is why $đW$ is an *inexact* differential (see Section 1.3). Note that the integral is the area under a curve of $\vec{\mathfrak{F}}$ *vs.* $\vec{x}$; in engineering analysis this integral is sometimes determined analytically, sometimes by numerical integration, and sometimes by graphical integration, depending on what is most convenient.

Macroscopic work

At the microscopic level, work is the only mechanism by which energy can be transferred from one particle to another. When we do a macroscopic analysis we must somehow account for all of the energy transfer as work that occurs at the microscopic scale. If there is a macroscopically observable motion of the particles, we can account for some (perhaps all) of this energy transfer macroscopically as the work done by a macroscopic force pushing the matter through a macroscopically detectable distance.

In thermodynamic analysis, *energy transfer as work* refers to energy transfer (across a boundary) associated with microscopic motions which *are* observable macroscopically. The observability of the motion means that the motions of individual particles have some overall coherence. It is customary to represent an amount of energy transfer as work by W (or $đW$), and to indicate the direction of the transfer on an energy flow diagram.

Work done by an expanding gas

Consider the motion of a piston in an internal combustion engine cylinder under the influence of the force on the piston resulting from bombardment by molecules of the gas, as shown in Figure 2.2. The *pressure* P is the macroscopic representation of the average force per unit area exerted by the colliding molecules on the atoms in the piston. The macroscopically observable force in the direction of piston motion is therefore PA, where A is the piston area. If the piston is stationary, there is no macroscopically observable energy transfer as work. But if the piston atoms move an infinitesimal distance dx in the direction of the force, then the macroscopically observable energy transfer as work *from the gas to the piston* is

$$đW = PA\,dx.$$

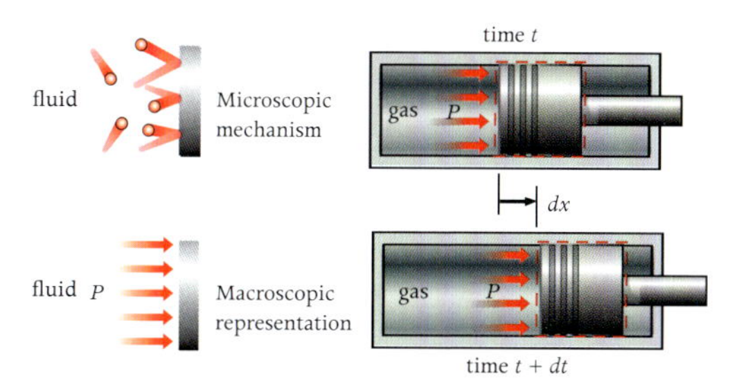

Figure 2.2 Pressure on a piston (a) and the macroscopic piston displacement (b).

Noting that the increase in volume of the gas is $dV = A\,dx$, we see that the energy transfer as work *from the gas* to the material that moves as the gas expands can be expressed as

$$đW_{\text{expansion}} = PdV. \tag{2.5}$$

Equation (2.5) holds for any expansion dV, regardless of the shape of the system, if the pressure P is everywhere the same on the system boundary.

Example: isobaric expansion in a piston–cylinder. Determine the work done on a piston by an expanding gas if the pressure is somehow maintained constant at P_1 as the gas expands from V_1 to V_2.

Solution: applying (2.5) gives

$$W_{12} = \int_1^2 đW = \int_1^2 PA\,dx = \int_1^2 P_1\,dV = P_1(V_2 - V_1).$$

Example: expansion in a piston–cylinder with prescribed pressure variation. Determine the work done on a piston by an expanding gas if the pressure is somehow made to vary inversely with the volume as the gas expands from V_1 to V_2.

Solution: the pressure as a function of volume $P(V)$ is given by $P = P_1V_1/V$, so (2.5) results in

$$W_{12} = \int_1^2 đW = \int_1^2 PA\,dx = \int_1^2 \left(P_1V_1\frac{1}{V}\right) dV$$
$$= P_1V_1 \ln(V_2/V_1).$$

Work for a polytropic process

The two previous examples illustrate the need for a *process model* in carrying out evaluations of work from the basic definition. A common model is the *polytropic process*,

$$PV^n = \text{constant}, \tag{2.6}$$

where n is the *polytropic exponent*. The case $n = 0$ is the first example above and the case $n = 1$ (corresponding to a constant temperature process for an ideal gas, as will be shown in the following) is the second. Measurements during the expansion stroke of typical internal combustion engines are fit rather well by values of n around 1.35. Using (2.6), the pressure as a function of volume is

$$P(V) = P_1\left(\frac{V_1}{V}\right)^n,$$

for which (for $n > 1$)

$$\begin{aligned} W_{12} &= \int_1^2 đW = \int_1^2 PA\,dx = \int_1^2 P_1\left(\frac{V_1}{V}\right)^n dV \\ &= P_1V_1^n\int_1^2 V^{-n}\,dV = P_1V_1^n\frac{1}{-\alpha+1}V^{-n+1}\Big|_1^2 \\ &= \frac{P_1V_1}{n-1}\left[1-\left(\frac{V_1}{V_2}\right)^{n-1}\right]. \end{aligned}$$

For example, if $n = 1.35$, $P_1 = 2 \times 10^6$ Pa (20 bar or about 20 atm), $V_2 = 10^{-3}$ m^3 (1 liter), and $V_2 = 10\,V_1$ (10:1 expansion ratio), the work done on the piston is

$$\begin{aligned} W_{12} = &\frac{1}{0.35} \times 2 \times 10^6\ \text{N/m}^2 \times 10^{-4}\ \text{m}^3 \\ &\times (1 - 0.1^{0.35}) = 316\ \text{Nm} = 316\ \text{J}. \end{aligned}$$

2.6 Energy Transfer as Heat

In the engine discussed previously the gas in the cylinder is typically hotter than the cylinder walls. When these energetic molecules bang into the more sluggish atoms of the cooler cylinder, they transfer energy to the wall atoms, which in turn become more active and then transfer energy to the cooling water molecules. At the microscopic level, this energy transfer takes place as work done by the gas molecules on the wall atoms, which in turn transfer energy to the cooling water molecules as depicted in Figure 2.3. The energy transfer results in microscopic vibrations of the wall atoms, but not in macroscopically observable motion, so we cannot compute this energy transfer as a force times an observable displacement. But we must account for this energy transfer in our macroscopic energy balance; we do this by including *energy transfer as heat*.

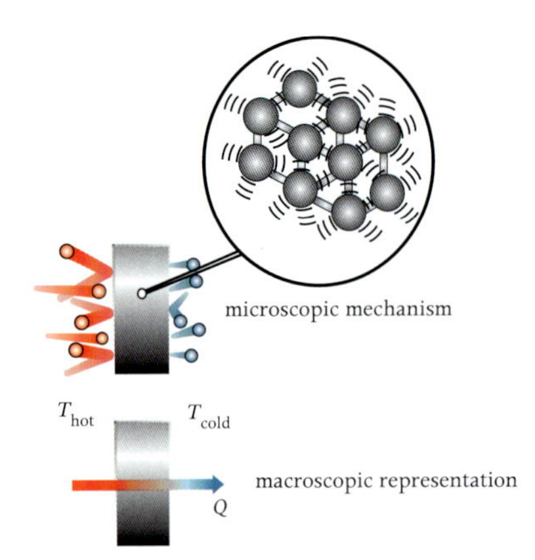

Figure 2.3 Energy transfer as heat.

In thermodynamic analysis, *energy transfer as heat* refers to that energy transfer, across a system boundary, associated with microscopic displacements that are *not* observable macroscopically. The motions are not macroscopically observable because they are microscopically chaotic (and not coherent). It is customary to represent an amount of energy transfer as heat by Q (or $đQ$), and to indicate the direction of the transfer on an energy flow diagram.

The following paragraphs introduce some additional ideas and terms related to energy transfer as heat. We will expand on these in subsequent chapters; the brief introductions below are intended to help you better understand the concepts now.

Heat and internal energy

We have cautioned against confusing internal energy and heat, and the importance of this cannot be overemphasized. *Internal energy* is the *stored* microscopically random energy of the matter; *heat* is a microscopically random *transfer* of energy. The internal energy can increase or decrease as energy is transferred to a system as heat, depending on what other energy transfers are involved. Without knowing what happened to a system in its past, it is impossible to tell whether it acquired its internal energy through energy transfer as heat, energy transfer as work, or both. There is absolutely no unique association between energy transferred as heat and a change in the internal energy.

Temperature

Energy transfer as heat takes place from material at high temperature to material at a lower temperature. Two systems at the same temperature are in *thermal equilibrium* and have no tendency to exchange energy as heat. These basic ideas, together with the concept of entropy, allow one to give a precise meaning to temperature and establish the absolute temperature scale. We will develop this fundamental basis for temperature in Chapter 5.

Heat transfer mechanisms

Energy is transferred as heat by two mechanisms, *conduction* and *radiation*. Energy is conducted in a fluid by molecular collisions, whereas conduction in solids is by lattice vibrations. Radiation is the transfer of energy as random electromagnetic radiation (photons), and can occur through a vacuum. One often hears of *convection* as a third mechanism of heat transfer. In convection, energy is transferred by a fluid that carries its *internal energy* from one place to another, where energy transfer as heat then occurs through a solid surface by conduction. Therefore, we regard convection as a means for *internal energy transport* and not as a fundamental mechanism for heat transfer.

Adiabatic boundaries

Often we idealize that a system has an *adiabatic boundary*, meaning it is impermeable to energy transfer as heat. Since a vacuum has no molecules to transfer energy, the adiabatic idealization is best for vacuum barriers that have been silvered to minimize radiative heat transfer (the laboratory Dewar flask used to hold cryogenic fluids such as liquid helium). Often a system can be treated as adiabatic if it is surrounded by a good thermal insulation so that the amount of energy transfer as heat is small compared with other energy transfers to or from the system. Even a highly conductive metal container can be treated as an adiabatic boundary for the gas inside if the processes of interest occur so quickly that there is no time for significant energy transfer as heat to occur.

Heat exchangers

Many systems require high rates of heat transfer in compact spaces. Examples include gas turbine engines, refrigeration systems, supercomputers, and artificial lungs. Special components, called *heat exchangers*, are used for this purpose. Most heat exchangers involve two flowing fluids, one hot and the other cold, separated by a thin solid surface through which energy is transferred by conduction. The automobile "radiator" is a familiar example, although radiation is not very important in this particular heat exchanger.

Heat, work, and entropy

Heat is a microscopically disorganized transfer of energy. We have briefly introduced the idea of entropy as a measure of microscopic disorganization. As we shall see later, entropy is also transferred whenever energy is transferred as heat. Work is microscopically organized transfer of energy (all the piston atoms move together) that takes place without a transport of entropy. These ideas are developed more fully in Chapter 5 as essential elements in the Second Law of Thermodynamics.

Vestiges of caloric theory

Nineteenth-century *caloric theory* treated "heat" as a conserved entity and used the term both for what we now call internal energy and what we now call heat; the theory was discarded because internal energy is just one of many forms of energy and is not itself conserved; there is no such thing as a "heat balance," although this term is still used. Our modern society depends critically on numerous inventions that convert disorganized molecular energy into useful macroscopically organized energy (jet engines, steam power plants, solar cells, etc.). These could not have been developed with caloric theory. You can help put caloric theory to rest (and do better in your work in thermodynamics) by being very precise in distinguishing between heat and internal energy.

2.7 Energy Balances

An energy balance is simply an equation that expresses the conservation of energy for a well-defined system. Energy balances are the primary tool for thermodynamic analysis, and you must be able to do them correctly. The basic balance is simply the bookkeeping statement described in Section 1.2:

energy input = energy output + energy accumulation.

If you have to memorize this, you are not thinking. If you are careful and systematic in your identification of the energy inputs, outputs, and energy storage modes, you should have no trouble in making correct energy balances.

Energy balance methodology

As with any analysis, a systematic approach is very helpful in eliminating errors. We recommend the following approach:

1. Draw a diagram identifying the system to be analyzed by enclosing it within dotted lines.
2. Show the forces, acting on the system, that you will use in the evaluation of energy transfer as work.
3. Identify the reference frame for evaluations of macroscopic kinetic and potential energy and work.
4. List all simplifying idealizations.
5. Show all significant energy transfers on your sketch to define the symbols, using arrows to indicate the directions chosen for *positive* energy transfer, and show the significant changes in energy storage.
6. State whether the energy balance is over some definite *time interval* or on a *rate basis*.
7. Write the energy balance: there should be a one-to-one correspondence between the terms in your equation and those on your diagram.
8. Bring in process, state, or other information as necessary to complete the analysis.

Importance of system boundaries

Since energy transfers between different pieces of matter within the system do not cross the system boundary, they are not involved in the energy balance of the system. That is why it is so important to have a clear definition of the system boundary. There are sometimes several options for choosing the boundary in an analysis, and great simplification can often be obtained by wise boundary choice. Since the only transfers that appear are those that cross the boundary, some good rules are as follows:

- Put the boundary where you know the transfer or want to know the transfer.
- Never put the boundary where you don't know the transfer and don't need to know the transfer!

Sign convention

Some textbooks use the convention that heat is always positive into a system and work always positive out; others treat both heat and work inputs as positive. We believe that these conventions lead to memorization of equations instead of thinking about the physics. Therefore, we always use a diagram to make a clear definition of positive energy transfers for each analysis.

Notation for energy accumulation

In energy analysis we often need to represent the increase in the energy within a system in an

Figure 2.4 Control mass and energy terms for first law analysis.

appropriate manner. We adopt the conventions of calculus, where the symbols Δ and d are understood to mean *final minus initial*, and denote the *increase* in the energy within the system (the accumulation of energy) by ΔE (if finite) or dE (if infinitesimal). Denoting the initial energy by E_i and the final by E_f,

$$\Delta E = E_\mathrm{f} - E_\mathrm{i}. \tag{2.7}$$

ΔE (or dE) will be positive if the energy increases, negative if it decreases, and 0 if it remains unchanged.

First Law of Thermodynamics

For the special case of a system of fixed mass, also called a *closed system* or *control mass*, there are only two macroscopic mechanisms for energy transfer to or from the system: *work* and *heat*. As already explained, both are actually work at the microscopic level, with heat being that part that is not associated with macroscopically observable displacements. Therefore, the basic energy balance for a control mass in which the only form of energy in the system is internal energy, and the energy transfers as work and heat are both defined to be positive *into* the system (see Figure 2.4), is

$$\underbrace{W}_{\substack{\text{energy}\\ \text{input}}} + \underbrace{Q}_{\substack{\text{energy}\\ \text{input}}} = \underbrace{\Delta U}_{\substack{\text{energy}\\ \text{accumulation}}}.$$

This is sometimes called *the First Law of Thermodynamics*. However, the real essence of the first law is simply the basic idea that energy is conserved. That is what you should remember and be able to apply to arbitrary systems using good energy bookkeeping. Students who instead operate by memorizing special formulae invariably get a sign wrong or use the formula when it does not apply; don't join that unfortunate group!

2.8 Examples

Gas compression

Consider the compression of gas in a cylinder, for example in an air compressor or during the compression stroke in an internal combustion engine (Figure 2.5). Note that here we define the work as positive *into* the gas and allow for energy transfer as heat from the gas. The energy balance is

$$\underbrace{đW_\mathrm{comp}}_{\substack{\text{energy}\\ \text{input}}} = \underbrace{đQ_\mathrm{comp}}_{\substack{\text{energy}\\ \text{output}}} + \underbrace{dU}_{\substack{\text{energy}\\ \text{accumulation}}}. \tag{2.8a}$$

Note that we use our special $đ$ symbol since $đW_\mathrm{comp}$ and $đQ_\mathrm{comp}$ represent infinitesimal *amounts* of energy transfer. If we integrate (2.8a) over the entire compression process, we get the energy balance for the entire compression process,

$$\underbrace{W_\mathrm{comp}}_{\substack{\text{energy}\\ \text{input}}} = \underbrace{Q_\mathrm{comp}}_{\substack{\text{energy}\\ \text{output}}} + \underbrace{\Delta U}_{\substack{\text{energy}\\ \text{accumulation}}}, \tag{2.8b}$$

where W_comp represents the total amount of energy transfer as work and Q_comp the total amount of energy transfer as heat. Note that if $Q_\mathrm{comp} > 0$ then ΔU could be positive or negative, depending on the relative magnitudes of W_comp and Q_comp. In a real engine, during the compression $\Delta U > 0$ because the temperature rises during compression, and $Q_\mathrm{comp} > 0$ because the warm compressed gas is cooled by heat transfer to the surrounding metal. Thus, there is an increase in temperature and a concurrent transfer of energy as heat out of the gas! This illustrates that there is no direct

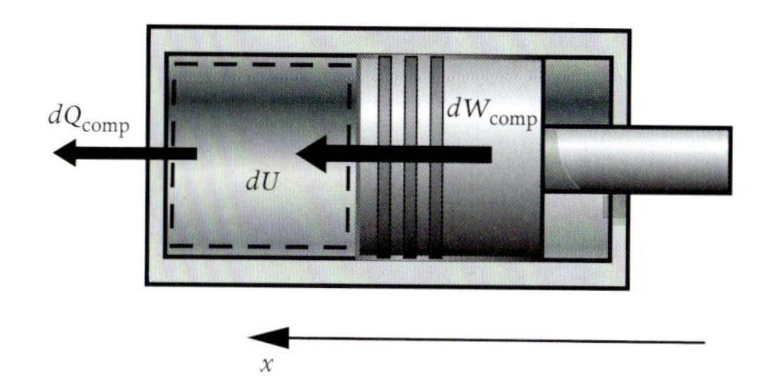

Figure 2.5 Control mass and energy terms for compression analysis.

relationship between the energy transfer as *heat* and the change in *internal energy* (except as given by the energy balance). That is why it is so important to distinguish between heat and internal energy.

The energy transfer as work from the piston to the gas (the work done by the piston on the gas) during the compression process can be computed using the work definition. In the macroscopic representation one sees the pressure P acting on the gas just inside the boundary adjacent to the piston, and this gas is moved a distance dx in the direction shown. Hence the work done on the gas is

$$W_{\text{comp}} = \int_i^f đW_{\text{comp}} = \int_i^f PA\,dx = -\int_i^f P\,dV,$$

where by the conventions of calculus $dV = -A\,dx$ denotes the *increase* in the gas volume (negative in this case). We would have to know how P varied with V during the compression to make this calculation. The integral is the area under a curve of P versus V, and could be determined analytically, by numerical integration, or graphically. Alternatively, if we knew Q and ΔU, we could use the energy balance to calculate W_{comp}.

Heat pump

A *heat pump* is a device used to transfer energy as heat from a cold region to a warmer region, for example to heat a home using energy taken from outdoors. A typical system is shown in Figure 2.6. Energy is taken from the outdoors using a heat exchanger through which a refrigerant fluid colder than the outdoor air is warmed by the outdoor air. This refrigerant is then compressed, which increases its energy and hence its temperature, to the point where it is warmer than the indoor air. Energy is then transferred as heat from the refrigerant to the cooler indoor air in a second heat exchanger. The refrigerant then flows through a tiny restriction (a *valve*) that causes a big drop in pressure. Refrigerants are fluids for which this pressure drop is accompanied by a temperature drop, so this cold fluid is then ready to make another pass through the outdoor heat exchanger.

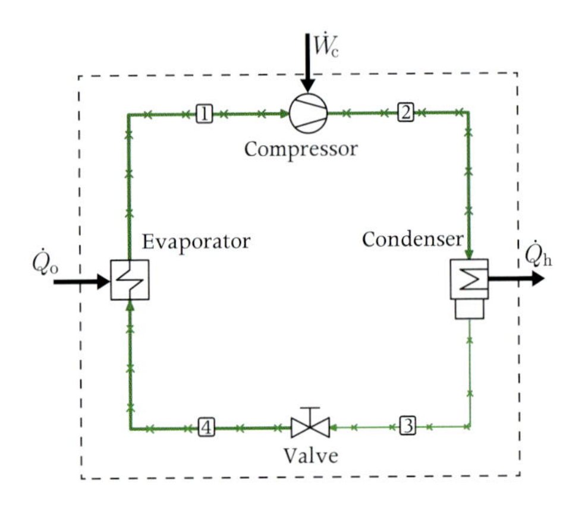

Figure 2.6 Heat pump system.

This is an example of the sort of thermodynamic system that you will soon be able to analyze in detail, component-by-component. Here we use the overall energy balance of the system indicated in Figure 2.6. We idealize the problem as *steady-state*, meaning that the start-up transients are over and it is now running without any changes in the temperature or pressure at any point in the system. Hence the energy accumulation in the system is zero, which greatly simplifies the analysis, since now only energy flows at the boundary appear in the energy balance. We make the energy balance on a *rate basis* (per unit time), and we use an overdot to denote a rate of energy *transfer* (not a rate of *change* as one might do in dynamics). Thus, $\dot{W}_{\text{comp}}$ is the compressor shaft power input (in Watt, W), and $\dot{Q}_o$ and $\dot{Q}_h$ are the rates of energy transfer (W) in the two heat exchangers. The energy balance is then

$$\underbrace{\dot{W}_{\text{comp}} + \dot{Q}_o}_{\substack{\text{rate of}\\ \text{energy input}}} = \underbrace{\dot{Q}_h}_{\substack{\text{rate of}\\ \text{energy}\\ \text{output}}} + \underbrace{0}_{\substack{\text{rate of}\\ \text{energy}\\ \text{accumulation}}}. \tag{2.9}$$

We see that $\dot{Q}_h > \dot{W}_{\text{comp}}$, so one gets more energy into the house than purchased from the utility company! The ratio $\dot{Q}_h/\dot{W}_{\text{comp}}$, the *coefficient of performance* (COP), is highest when the difference between the indoor and outdoor temperatures is not too great; in relatively mild climates the COP can be as much as 4 or 5! This leverage makes heat pumps attractive alternatives to other forms of heating, especially direct electrical heating.

Does the fact that the heat pump takes energy from the outdoors make things colder outside? To answer this one must realize that the heat pump is needed

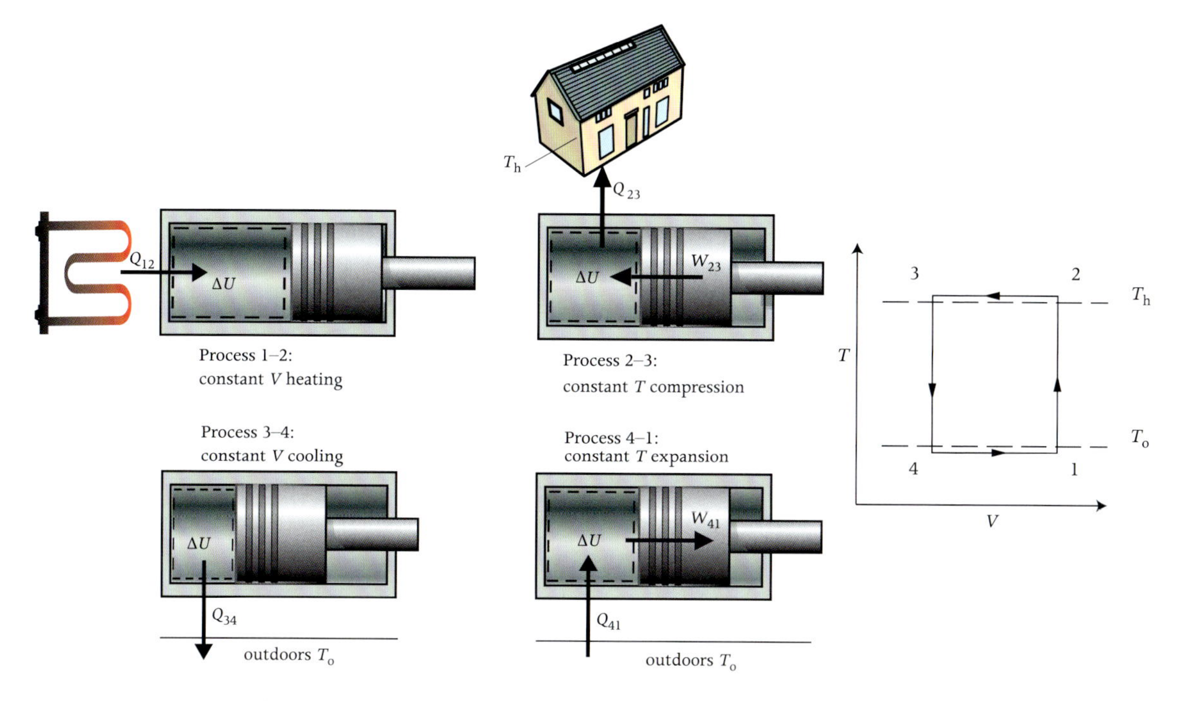

Figure 2.7 Inventor's heat pump proposal.

because the energy it puts into the house eventually leaks back to the outside. And since this is more than the energy originally taken from the outdoors, the outside actually gets warmer as the heat pump runs! The house and its immediate surroundings in essence are heated by the electrical energy that is brought in to run the heat pump.

Detailed heat pump cycle analysis

We will do one final example in which we look in more detail at the thermodynamic analysis of a proposed heat pump. This example will use most of the ideas introduced in this chapter and will give you a look at what one can do with simple thermodynamic analysis.

An inventor proposes to make a heat pump using a simple piston–cylinder system in which the contained gas is put through a repetitive series of four processes shown in Figure 2.7. We want to determine the performance of this design as a function of some key design parameters. The four processes are as follows:

1-2 We start with the cylinder at its largest volume and the gas inside at the temperature T_o of the cold outdoors (state 1). We turn on the electric heater; holding the volume fixed at V_1, we raise the gas temperature to just above the house temperature T_h (state 2).

2-3 At state 2 we bring the cylinder into thermal contact with the air inside the house, turn off the heater, and start compressing the gas in the cylinder. The compression will tend to raise the gas temperature above that of the house, so energy transfer as heat will take place from the warmer cylinder gas to the house air. To keep things simple, we will assume that the compression takes place with the cylinder gas temperature T maintained just above that of the house T_h. When we have reached the desired minimum volume (determined by the maximum pressure our device can tolerate), we stop the compression (state 3).

3-4 At state 3 we switch the device so that the cylinder is now in thermal contact with the cold outdoors at T_o. We now hold the volume fixed at V_3 and allow the gas in the cylinder to cool by energy transfer as heat to the environment, until the gas reaches the cold outdoor temperature T_o (state 4).

4-1 At state 4 we allow the gas in the cylinder to expand, doing work on the piston. This removes energy from the gas and tends to lower its temperature, so now the energy transfer as heat will be from the outdoor air to the (slightly) colder gas in the cylinder. We assume that the expansion takes place just below the outdoor temperature T_o. We continue this process until we again reach state 1, and then repeat the cycle.

The energy balances are as follows:

Process 1-2:

$$\underbrace{Q_{12}}_{\substack{\text{energy}\\\text{input}}} = \underbrace{0}_{\substack{\text{energy}\\\text{output}}} + \underbrace{U_2 - U_1}_{\substack{\text{energy}\\\text{accumulation}}}.$$

Process 2-3:

$$\underbrace{W_{23}}_{\substack{\text{energy}\\\text{input}}} = \underbrace{Q_{23}}_{\substack{\text{energy}\\\text{output}}} + \underbrace{U_3 - U_2}_{\substack{\text{energy}\\\text{accumulation}}}.$$

Process 3-4:

$$\underbrace{0}_{\substack{\text{energy}\\\text{input}}} = \underbrace{Q_{34}}_{\substack{\text{energy}\\\text{output}}} + \underbrace{U_4 - U_3}_{\substack{\text{energy}\\\text{accumulation}}}.$$

Process 4-1:

$$\underbrace{Q_{41}}_{\substack{\text{energy}\\\text{input}}} = \underbrace{W_{41}}_{\substack{\text{energy}\\\text{output}}} + \underbrace{U_1 - U_4}_{\substack{\text{energy}\\\text{accumulation}}}.$$

Note that if we add these four energy balances the energy accumulation terms all cancel, and so we obtain

$$\underbrace{Q_{12} + Q_{41} + W_{23}}_{\substack{\text{energy}\\\text{input}}} = \underbrace{Q_{23} + Q_{34} + W_{41}}_{\substack{\text{energy}\\\text{output}}} + \underbrace{0}_{\substack{\text{energy}\\\text{accumulation}}}.$$

This is the correct energy balance for the entire cycle, over which there is no energy accumulation. Assuming that we can save W_{41} in a flywheel and use this to help drive the next compression process, and that the net work is provided by an electric motor,

$$E_{\text{net}} = W_{23} - W_{41} + Q_{12}$$
$$Q_h = Q_{23}$$
$$Q_{o,\text{net}} = Q_{41} - Q_{34},$$

where E_{net} is the net electrical energy that we must purchase from the utility company, Q_h is the energy transfer as heat to the house air, and $Q_{o,\text{net}}$ is the net energy transfer as heat from the outdoor air (all per cycle of the device). Note that

$$Q_h = Q_{o,\text{net}} + E_{\text{net}},$$

which is consistent with the energy balance of the previous example.

We can carry the analysis much further if we bring in some more information about the gas. We will assume that it is an ideal gas (see Chapter 3), for which the pressure P, temperature T, mass M, and volume V are related by

$$PV = MRT,$$

and the internal energy is given by

$$U = Mc_vT,$$

where R and c_v are constants for the gas. Since PV and U both have dimensions of energy, R and c_v have the dimensions of energy/(mass · temperature). It can be proven that for an ideal gas $R/c_v = \gamma - 1$, where $\gamma = 5/3$ for a monatomic gas such as He and $7/5$ for a diatomic gas such as N_2.

Denoting the compression ratio of the device by $r_V = V_1/V_4$, and the temperature ratio by $r_T = T_2/T_1 = T_3/T_4$, use of the gas property relationships gives

$$U_2 - U_1 = Mc_vT_1(r_T - 1)$$
$$U_3 - U_2 = 0$$
$$U_4 - U_3 = Mc_vT_1(1 - r_T)$$
$$U_1 - U_4 = 0.$$

Carrying out the integrations to determine the two work terms gives

$$W_{23} = -\int_2^3 P\,dV = -MRT_2 \ln(V_3/V_2)$$
$$= MRT_1 r_T \ln r_V,$$
$$W_{41} = \int_4^1 P\,dV = MRT_1 \ln(V_1/V_4) = MRT_1 \ln r_V.$$

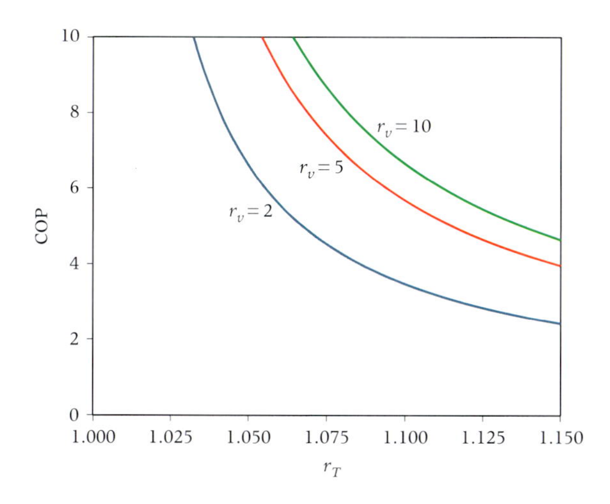

Figure 2.8 Performance of the idealized heat pump for $\gamma = 5/3$.

Substituting these results in the process energy balances allows us to determine the various energy transfers as heat as

$$Q_{12} = Mc_v T_1(r_T - 1)$$
$$Q_{23} = MRT_1 r_T \ln r_V$$
$$Q_{34} = Mc_v T_1(r_T - 1)$$
$$Q_{41} = MRT_1 \ln r_V.$$

Then, we can finally obtain

$$E_{\rm net} = MRT_1(r_T - 1)[\ln r_V + 1/(\gamma - 1)]$$
$$Q_{\rm h} = MRT_1 r_T \ln r_V$$
$$Q_{\rm o} = MRT_1[\ln r_V - (1 - r_T)/(1 - \gamma)].$$

The coefficient of performance is then

$$\mathrm{COP} = \frac{Q_{\rm h}}{E_{\rm net}} = \frac{r_T \ln r_V}{(r_T - 1)[\ln r_V + 1/(\gamma - 1)]}. \quad (2.10)$$

Figure 2.8 shows the COP as a function of r_T and r_V for $\gamma = 5/3$, as given by (2.10)

For a compression ratio of $r_V = 10$, (2.11) gives

$$\mathrm{COP} = \frac{0.6}{1 - T_1/T_2}, \quad (2.11)$$

so if $T_2 = 300$ K and $T_1 = 275$ K the COP would be 7.2 if He ($\gamma = 5/3$) were used as the working fluid. The energy gained by the house is predicted to be over seven times that purchased from the utility! This is a highly idealized analysis; we did not allow any temperature differences that are necessary to drive the heat transfer, we neglected friction, and so forth. A real device might have a COP of about 1/4 of that predicted here. After further study, you will be able to make more realistic analyses of systems like this.

EXERCISES

2.1 A person in an elevator moving upward at 10 m/s lifts a 60 kg mass 1.5 m off the elevator floor in 2 s. How much work did the person do relative to the elevator? How much relative to the ground? Why are these different, and what happens to the energy in each case?

2.2 How much energy transfer as work (J) is required to make fluid in an adiabatic container undergo a change $\Delta U = 200$ kJ/kg?

2.3 How much energy transfer as heat (J) is required to make fluid in a constant volume container undergo a change $\Delta U = 200$ kJ/kg?

2.4 The gas in an adiabatic piston–cylinder system undergoes a change of $\Delta U = -60$ kJ during a process. How much was the energy transfer as work to or from the gas, and in which direction?

2.5 The gas in a piston–cylinder system undergoes a change of $\Delta U = 30$ kJ during a process in which 40 kJ of energy as work are transferred from the gas to the piston. Was there energy transfer as heat involved in this process, and if so how much and in which direction?

2.6 In a constant-volume container, a fluid is to be heated so that its internal energy will change from 1200 kJ to 1600 kJ in 30 s. What is the power requirement (kW) of the electric heater?

2.7 The temperature of 10 kg of air inside a piston–cylinder is increased from 25 °C to 80 °C using an electrical heater, while the pressure inside the cylinder is maintained constant. During this process 50 kJ of energy is transferred as heat to the environment. Calculate the required energy transfer as heat into the system in kWh for cylinder pressures varying from 1 to 5 bar.

2.8 In an adiabatic constant-volume container, a fluid is to be stirred mechanically with a motor providing 40 W of shaft power. How long can this go on if the internal energy of the fluid is not allowed to rise more than 10 kJ?

2.9 Sketch on a P–v diagram the process that the air inside a piston–cylinder undergoes during an *isothermal* (constant-temperature) expansion reducing the pressure by 300 kPa. Energy is then transferred as heat either into or out of the system. Assume that air obeys the ideal gas law. Derive the expression for the net amount of energy transfer as heat either in terms of pressure and volume, or temperature, volumetric expansion ratio and determine its direction.

2.10 Consider the expansion of a gas at constant temperature in a water-cooled piston–cylinder system. The constant temperature is achieved by controlled input of energy as heat Q to the gas. Treating the gas as ideal, derive expressions for the energy output as work W and energy input as heat Q as a function of the expansion ratio V_2/V_1.

2.11 Consider the expansion of a gas at constant pressure in a water-cooled piston–cylinder system. The constant pressure is achieved by controlled input of energy as heat Q to the gas. Treating the gas as ideal (see Exercise 2.10), derive expressions for the energy output as work W and energy input as heat Q as a function of the expansion ratio V_2/V_1.

2.12 A water cannon for rock cutting uses compressed gas to accelerate a 70 kg metal piston to 65 m/s. The energy of the piston is transferred to 0.9 kg of water, which is fired out of a small nozzle. Assuming that all of the piston energy is transferred to the water, what is the water jet velocity in m/s?

2.13 Consider the system in the figure below. Initially the gas in the cylinder is at high pressure P_i and the piston held in place at volume V_i with a lock. At time zero the lock is released and the gas allowed to expand, accelerating the piston of mass M_p to the left. Friction is negligible. The expansion may be idealized as an adiabatic polytropic process with $\alpha = 1.4$, and there is negligible leakage. The pressure on the left side of the piston may be taken as a constant P_0 during the process.

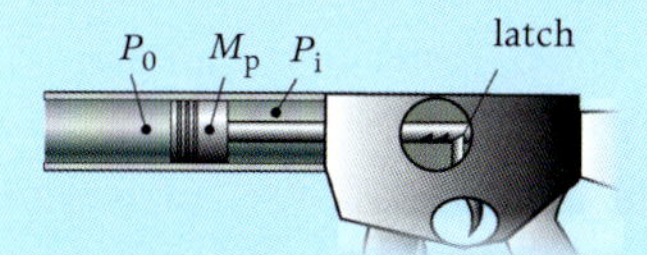

(a) Derive an expression for the velocity of the piston $\mathcal{V}_p$ as a function of the instantaneous gas volume V; the symbols defined above (and no others) will appear as parameters in your analysis.
(b) Plot $\mathcal{V}_p$ versus V and P versus V for the "can-thrower" case $M_p = 0.5$ kg, $P_0 = 10^5$ Pa, $P_i = 7 \times 10^5$ Pa, $V_i = 10^{-3}$ m^3, covering the range for which $P \geq P_0$.
(c) Specify the cylinder length required to produce 40 m/s when $P = P_0$ for the can-thrower, assuming the cylinder diameter is 8 cm.

2.14 Imagine turning the system of Exercise 2.13 vertically. Repeat the analysis, which now will involve the height of the piston. Using the same values as in Exercise 2.13(b), with $D = 8$ cm, plot the piston velocity as a function of its height z above its initial position. Neglecting air resistance, and assuming that $P = P_0$ at the moment of launch, how high would the mass rise?

2.15 An inventor claims to have a solar powered heat pump that receives energy as heat from the sun at the rate of 10 kW and extracts energy as heat from the environment at the rate of 7 kW. This system does not require any shaft or electrical power input. If you think this device is impossible, explain why, using basic principles to support your argument. If you think it might be possible, what would be the steady-state rate of transfer of energy as heat to the house?

2.16 A simple piston–cylinder engine operates using a sequence of three processes:

1-2 Compression at constant temperature, maintained by heat transfer to the environment.

2-3 Expansion at constant pressure, maintained by heating from flames.

3-1 Cooling at constant volume back to the initial state 1.

Using the ideal gas property information in (and following the format of) the *Detailed heat pump cycle analysis* example of Section 2.8 , derive expressions for the energy transfer as work and heat for each process. Express the efficiency of the cycle, $\eta \equiv W_{\text{net output}}/Q_{\text{input from flames}}$, as a function of the compression ratio $r_V = V_1/V_2$. Plot the efficiency over the r_V range 1–20, for $\gamma = 7/5$. Does this look like a promising engine cycle?

2.17 A simple piston–cylinder engine operates using a sequence of four processes:

1-2 Compression at constant temperature, maintained by cooling.
2-3 Heating at constant volume (by flames).
3-4 Expansion at constant temperature, maintained by heating (by flames).
4-1 Cooling at constant volume back to the initial state (1).

Using the ideal gas property information in (and following the format of) the *Detailed heat pump cycle analysis* example of Section 2.8, derive expressions for the energy transfer as work and heat for each process.

Express the efficiency of the cycle, $\eta \equiv W_{\text{net output}}/Q_{\text{input from flames}}$, as a function of the compression ratio $r_V = V_1/V_2$ and the temperature ratio $r_T = T_3/T_1$. Plot the efficiency over the r_V range 1–20 for $r_T = 2$ and 5, for $\gamma = 7/5$. Does this look like a promising engine cycle?

2.18 The motion actuator for a robotic manipulator consists of a simple piston–cylinder system where the pressure of the gas contained in the cylinder can be controlled by heating or cooling. The moving piston provides the actuation. In a particular application, the actuator starts from state 1 (T_1, P_1, V_1,...) and moves to state 2 (T_2, P_2, V_2,...). During this change in position, the heater is regulated such that the pressure increases linearly with the volume, $P = \frac{P_1}{V_1}V$. Derive an expression for the amount of energy transfer *to* the gas as heat, Q, for this process. Using the ideal gas property relations, express your result in terms of the pressure P and volume V at states 1 and 2 and constants for the gas.

2.19 A free-piston engine under development consists of a small piston in a cylinder. Each end of the cylinder behaves like a two-stroke engine, in which the following sequence of processes occurs repetitively:

1. A fresh charge of fuel–air mixture is blown in when the cylinder volume is approaching its maximum.
2. The fresh gas is compressed as the piston moves in one direction.
3. The mixture is ignited when the cylinder volume is at its minimum, producing a sudden rise in pressure.
4. The hot, high-pressure gas expands as the piston moves in the other direction.
5. The products of combustion are exhausted when the cylinder volume is near its maximum.

These processes happen in the two end volumes out of phase, with the gas at one end expanding while the gas at the other end is being compressed. There are no shafts in this engine. Instead, the piston is a lightweight ceramic magnet that moves back and forth inside a coil of wire surrounding the engine, generating AC electrical power directly.

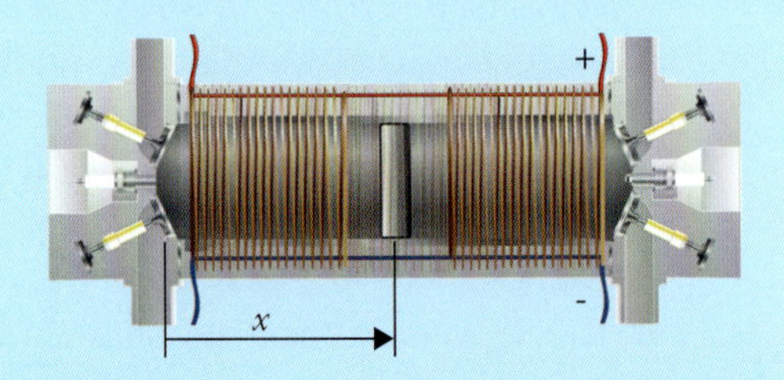

Energy transfer as work takes place between the piston and the electrons in the coil as a result of the force exerted by the electrons on the moving magnetic piston. Pressure measurements in the two cylinders as a function of the piston position x during the motion to the right are shown in the following figure. The piston cross-sectional area is 0.01 m^2.

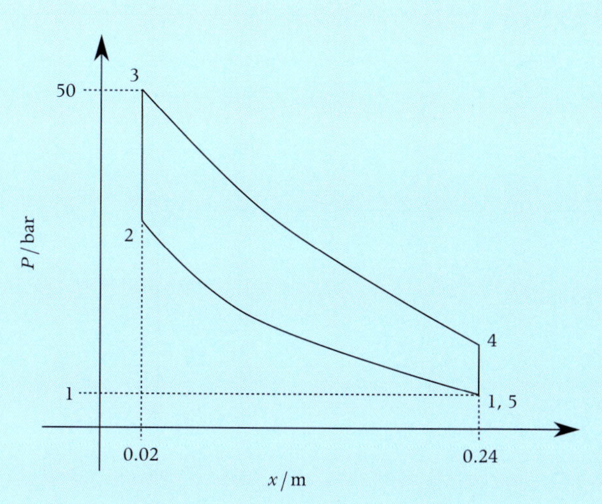

a. Calculate the work done (J) by the expanding left-side gas on the piston as it moves from one end to the other.

b. Calculate the work done (J) by the piston on the right-side gas.

c. Neglecting friction and electrical losses, calculate the power output of this engine (W), assuming that each one-way stroke takes 0.01 s.

d. The density of the fresh charge (at the start of the compression) is 1 kg/m^3. The measured temperature of the gas in the right side at the start of the compression is 300 K, and at the end of the compression, just before ignition, is 500 K. The internal energy of the fresh charge is given by

$$u = c_v T, \qquad c_v = 0.7\ \mathrm{kJ}/\left(\mathrm{kg}\cdot\mathrm{K}\right).$$

e. Calculate the energy transfer as heat (magnitude and direction) between the right-side gas and the surrounding solid during the compression process. Make a sketch of the system and of the energy transfers.

f. Suppose that the coil gets broken just as a stroke starts so that there is no force exerted on the piston by the electrons. Neglecting friction, assuming a piston mass of 0.8 kg, calculate the piston velocity at the end of the stroke, where normally the piston would be coming to rest at the other end. Make a sketch of the system and of the energy transfers.

3 Properties and States

CONTENTS

The examples of energy analysis in Chapter 2 show that, in order to obtain quantitative results from energy balances, we need to evaluate the thermodynamic properties of fluids at the relevant states of the processes. We used the ideal gas model with constant specific heat to calculate pressure, temperature, specific volume, and internal energy, an anticipation of what is explained and generalized in this chapter. In order to treat problems in which the working fluid can be a liquid, or a vapor, and in all cases in which the fluid does not obey the ideal gas law, we have to learn how we can obtain this information first from tables and charts, then using software. Before we master how to use these tools, we must learn in a more systematic way what thermodynamic properties and states of fluids are, understand the relations and transitions between the various possible states of matter, and how they can be visualized on appropriate diagrams, revealing their complex interaction.

3.1 Concepts of Property and State

Properties

A *property* is any characteristic or attribute of matter that can be quantitatively evaluated. Volume, mass, density, energy, charge, temperature, pressure, magnetic dipole moment, electric dipole moment, internal energy, momentum, surface tension, velocity, entropy, viscosity, and color are all properties. Work and heat are *not* properties because they are not something that matter *has*, but instead are amounts of energy transfer to or from matter.

States

The *state* of a system is its condition as described by the values of all of its properties. Usually we are content to describe a more limited aspect of the system. For example, the *geometric state* of a solid object can be described by the values of its various dimensions, which are the geometric properties. The dynamical state of a rigid body would be specified by its geometric state plus its three components of translational velocity and three angular rotation rates.

Intensive and extensive properties

An *extensive* property of a system is the sum of the values of that property for the parts of the system. Volume, mass, energy, charge, magnetic dipole moment, internal energy, momentum, and entropy are extensive properties; they depend on the *extent* of the system. Properties that are independent of the size of a system are called *intensive* properties. Temperature, pressure, density, viscosity, velocity, surface tension, and color are intensive properties. Extensive properties can be *intensified* by dividing by the amount of material present. For example, the *specific internal energy* $u \equiv U/M$ and the *specific volume* $v \equiv V/M$ are intensive properties (extensive property *per kilogram*). Data on the thermodynamic properties of substances are always given in terms of the *intensive* properties.

Thermodynamic properties and state

In thermodynamics we are concerned with energy and entropy. *Thermodynamic properties* are those *macroscopic properties* that are involved in energy and entropy analysis. For example, we have learned that the work done by an expanding fluid is $đW = PdV$, *independently of the shape* of the volume V. Therefore, the length, width, and breadth of a fluid element are not important in the energy analysis; only the volume V is relevant to the energy analysis and hence is a thermodynamic property. The *thermodynamic state* of a substance is its state as described by values of its *intensive* thermodynamic properties. The thermodynamic properties discussed in this chapter are listed in Table 3.1. We assume that you are familiar with some of these properties, and we will introduce you to the others.

Table 3.1 Thermodynamic properties.

Extensive	
V	volume
U	energy
H	enthalpy
S	entropy
Intensive	
T	absolute temperature
P	pressure
ρ	density, M/V
v	specific volume, V/M
u	specific internal energy, U/M
h	specific enthalpy, H/M
s	specific entropy, S/M
x	quality
c_V	specific heat at constant volume (isochoric)
c_P	specific heat at constant pressure (isobaric)
β	isothermal compressibility
κ	isobaric compressibility

Equilibrium states

Any well-defined system has well-defined values for its extensive properties (mass, energy volume, etc.). However, a piece of material that is hot on one side and cold on another does not have a well-defined temperature. But if we isolate the piece of material and allow its molecules to interact, the temperature variations will smooth out and the entire piece will end up at a single temperature. A state that is reached when a piece of material is isolated and allowed to adjust internally is called an *equilibrium state*. Most intensive properties are defined and evaluated only in equilibrium states. Therefore, thermodynamic properties are associated with equilibrium states.

Fixing a thermodynamic state

Any thermodynamic property of a substance is *fixed* (determined) when its thermodynamic state is fixed; but as we shall soon see, not all of these properties can be varied independently. Therefore, the thermodynamic state can be fixed by prescribing a small number of the thermodynamic properties, and all others are thereby determined. The *State Principle* tells us how many thermodynamic properties must be prescribed to fix the thermodynamic state. We will develop the State Principle after a brief review of the concepts and methods of measurement of two of the most important thermodynamic properties, pressure and temperature.

3.2 Pressure

Pressure at a solid boundary

In the two previous chapters, we used the idea that *pressure* is the macroscopic force per unit area that is exerted by a fluid on its container normal (perpendicular) to the surface. Alternatively, the pressure is the force per unit area exerted by the container on the fluid. The pressure results from collisions of the fluid molecules with those of the container.

Pressure within a fluid

We can also think of the pressure on an imaginary boundary drawn around fluid contained by more of the same fluid. When we mentally isolate a volume of fluid (to define a closed system) by removing the surrounding fluid outside the system boundary, we must replace the effect of the external fluid. Microscopically we have molecules crossing the boundary in both directions (Figure 3.1); if we think of the molecules that enter as being reflections of those that leave, the imaginary boundary acts like a solid boundary. Therefore, the pressure on an imaginary boundary *within* a fluid is the same as it would be if the boundaries were solid.

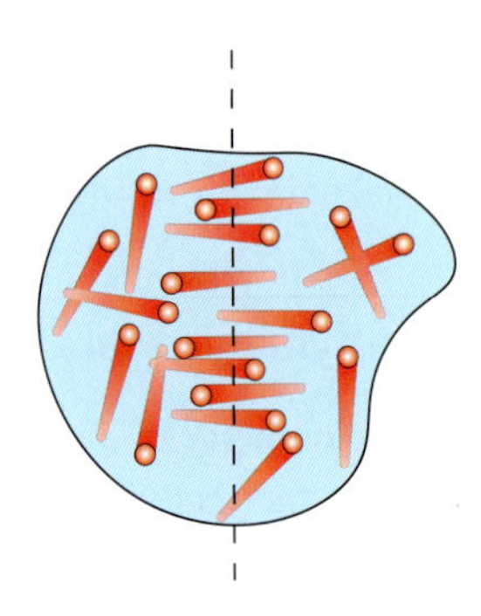

Figure 3.1 Pressure within a fluid is the same as it would be at a solid boundary.

Pressure is isotropic

At any given point in a fluid at rest the pressure is *isotropic*, meaning that it is the same regardless of the orientation of an infinitesimal area of the imaginary boundary. Isotropy occurs because molecules move randomly with equal probability in all directions. This means that the normal (pressure) force on every face of a little cube of fluid is the same (Figure 3.2), regardless of the orientation of the cube. Conversely, in a solid the normal forces need not be the same; for example, in a solid rod under tension the axial normal forces are much larger than the radial normal forces. In general, the *pressure* is defined to be the average of the magnitudes of the normal forces per unit area on faces of the cube, which *by definition* is isotropic.

Pressure in a fluid at rest is uniform in horizontal planes

Consider a thin horizontal column of fluid at rest (Figure 3.3). A horizontal force balance shows that the pressure is the same at both ends of the column. This is important in the measurement of pressure.

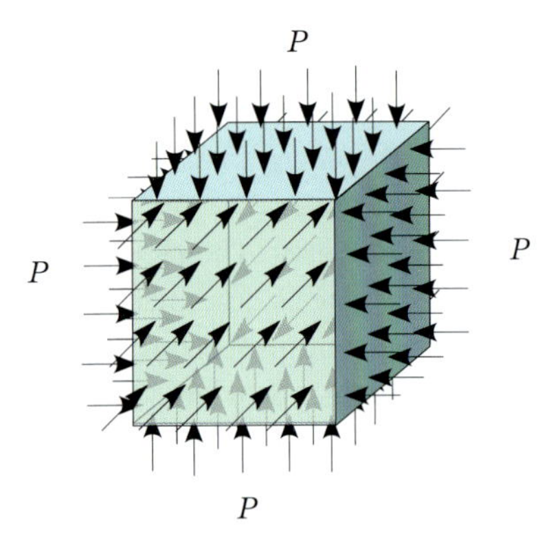

Figure 3.2 The pressure is the same on planes of all orientations.

Figure 3.3 The pressure is uniform in horizontal planes.

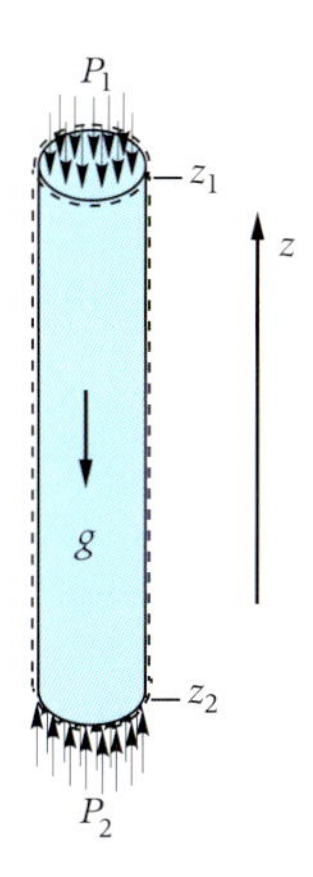

Figure 3.4 Hydrostatic pressure.

Hydrostatic pressure

Consider a vertical column of fluid at rest in a uniform gravitational field (Figure 3.4). A vertical force balance gives

$$P_2 - P_1 = \rho g(z_1 - z_2),$$

where g is the local acceleration of gravity. This is the basis for manometers used to measure pressure differences.

Atmospheric pressure

The pressure of the atmosphere can be measured using a special manometer called a barometer in which one end is exposed to the atmosphere and the other to a near vacuum. Mercury is an excellent liquid for this purpose since its vapor pressure is almost zero at room temperatures. One atmosphere is a unit of pressure defined as 1 atm ≡ 101 325 Pa, which is about the average found at sea level.

Gauge and absolute pressure

Pressures differences above atmospheric pressure are called *gauge pressures*, because that is what many mechanical pressure gauges sense. In this book P always means the *absolute* pressure.

3.3 Temperature

Temperature concept

The basic idea of temperature is that it is a property that is the same for two systems if there is no tendency for energy transfer as heat to occur between them. If two systems are at the same temperature, they are said to be in *equilibrium*. If two systems are in thermal equilibrium with a third system, all three have the same temperature, and hence are in thermal equilibrium with one another. Energy transfer as heat will take place between two systems if they are at different temperatures; the energy will flow from the hotter (higher temperature) system to the colder (lower temperature) system.

Empirical temperature scales

Early scientists used empirical temperature scales. They would use some thermometric device, such as a mercury-in-glass thermometer, which responds to temperature by expansion of the mercury in a constant-volume bulb, resulting in movement of the mercury interface in a connected capillary tube. First they would place the sensor in ice water, allow it to come to thermal equilibrium with the ice water, then mark the indicator position and label the mark 0 °C. Then they would place the sensor in boiling water (at 1 atmosphere), mark the indicator position, and label it 100 °C. One scientist might use a mercury-in-glass thermometer, another an alcohol thermometer, and another a bimetallic device that would bend differently at different temperatures. Of course all of these *two-point* temperature scales would agree at 0 °C and 100 °C, but when compared at intermediate temperatures they would not agree, because different thermometric media respond differently to temperature.

Constant-volume gas thermometer

This difficulty was solved by the invention of the *constant-volume gas thermometer*. It was discovered that the pressure of a gas confined at constant volume is a monotonically increasing function of the empirical temperatures. The pressure depends upon the amount of gas in the volume and the type of gas. But it was found that, in the limit as the gas mass approaches zero, the *ratio* of the pressure at any given temperature to the pressure at some calibration temperature, such as that provided by ice water, becomes independent of both the amount of gas in the volume and the nature of the gas, and is just a function of the test and reference temperatures T and T_0,

$$\lim_{M\to 0} \frac{P_{\mathrm{CVT}}(T)}{P_{\mathrm{CVT}}(T_0)} = F(T, T_0). \tag{3.1}$$

This limiting ratio is used to define the absolute temperature.

Absolute temperature

The absolute temperature is a *one-point* temperature scale that is *independent of the thermometric substance.* The choice for the function is not unique, and Earth's scientists chose to define the temperature T such that

$$\lim_{M\to 0} \frac{P_{\mathrm{CVT}}(T)}{P_{\mathrm{CVT}}(T_0)} = \frac{T}{T_0}. \tag{3.2}$$

The *SI temperature scale* was set by an international agreement in 1954 declaring that the triple point of water (a unique temperature and pressure where water liquid, vapor, and solid can exist together in equilibrium) occurs at $T_0 = 273.16$ K. This agreement established temperature as a primary quantity in the SI system, the Kelvin as its basic unit, and intimately linked the measurement of temperature to the behavior of gases at low density. In his early work, Kelvin himself explored the use of a logarithmic function for the temperature scale, but later switched to the linear function; it would be interesting to discover what choice was made on other planets!

In the US–English system, the absolute temperature is measured in Rankines, denoted by R. The Rankine scale is chosen such that the temperature in R is precisely 1.8 times the temperature in K. Hence, at the triple point of water, $T = 273.16\text{ K} = 491.69\text{ R}$.

The Celsius relative temperature (formerly centigrade), denoted by °C, is the difference between the temperature in Kelvin and the temperature at the point where ice forms at 1 atmosphere pressure (273.15 K), which therefore is the point of 0 °C. The difference between any two temperatures is the same on the Kelvin and Celsius scales.

The Fahrenheit relative temperature, denoted by °F, is the difference between the temperature in Rankines and the temperature at the point where ice forms at 1 atmosphere pressure (491.67 R), plus 32 °F, the (chosen) Fahrenheit temperature at the ice point. The difference between any two temperatures is the same on the Fahrenheit and Rankine scales.

3.4 The State Principle

Not all of the thermodynamic properties may be varied independently. For example, the *ideal* or *perfect gas,* which is a model for the gas used in constant-volume gas thermometers, is one for which P, ρ, and T are related by

$$P = \rho R T, \tag{3.3}$$

where R is a constant for the gas. Equation (3.3) illustrates the rule that *there are only two independent intensive thermodynamic properties for a simple fluid.* In this section we explain why this rule is true and give a more general rule applicable to other classes of substances.

Changing the thermodynamic state

Consider a system consisting of a fixed amount of a given substance. There are two basic ways that we can change its thermodynamic state; we can transfer energy in or out as work, or we can transfer energy in or out as heat. Since these are the only mechanisms for energy transfer, these are the only ways we can change the thermodynamic state. However, there may be more than one way to transfer energy as work to the substance. For example, we can transfer energy as work to ferrofluids either by compressing the fluid or by increasing the applied magnetic field, and there is one freely variable thermodynamic property associated with each of these work modes.

Reversible work modes

In Chapter 2 we learned that the energy transfer as work *from* a fluid undergoing an infinitesimal expansion can be represented by

$$đW = P\,dV. \tag{3.4a}$$

For magnetic materials the energy transfer *to* the material when its dipole moment is increased a small amount is[1]

[1] See, for example, D. Halliday and R. Resnick, *Physics*, 2nd edn., New York: John Wiley & Sons, 1962, Chapters 27, 28 and 33, 34.

$$đW = \mu_0 \vec{\mathcal{H}} \cdot d(V\vec{\mathcal{M}}), \tag{3.4b}$$

where $\vec{\mathcal{H}}$ is the applied magnetic field, $\vec{\mathcal{M}}$ is the magnetic dipole moment per unit volume, and μ_0 is the permeability of free space (a physical constant). For every work mode the expression for an infinitesimal transfer of energy as work is of the form

$$đW = F\,dX. \tag{3.5}$$

Here F denotes an intensive thermodynamic property and X an extensive thermodynamic property. For each such expression, the work is of equal but opposite sign if dX is replaced by $-dX$, and for that reason these work modes are called *reversible*. Each X is independently variable, so there are at least as many independent thermodynamic properties as reversible work modes. Chapter 5 discusses the relationship between reversible work and entropy and gives a deeper meaning to reversible work.

The State Principle

In addition to the reversible work modes, one can always transfer energy as heat to the substance, thereby changing the energy independently of any X changes. Thus, the State Principle is:

> **the number of independently variable thermodynamic properties for a specified amount of a given substance is the number of reversible work modes *plus one*.**

This is an important (but often overlooked) cornerstone of thermodynamics. It gives the *number* (N) of independent properties, but does not say that *any* set of N properties is independent. For example, for most fluids the internal energy and temperature are independent, but for the special case of an ideal gas they are not.

Application to a simple compressible substance

A *simple compressible substance (or fluid)* is one for which the only significant reversible work mode is *volume change*. Almost everything done in this book and in engineering thermodynamics involves only simple compressible substances.

For a specified mass M of a simple compressible substance, we can think of the two independent properties as V and U. The State Principle says that if we put a specified amount M of any such substance in a specified volume V and give it a specified energy U, *all other thermodynamic properties will be determined.* Once M, V, and U have been set, we are no longer free to vary the pressure, temperature, or any other thermodynamic property.

The intensive thermodynamic state of a specified simple compressible substance is therefore determined by specifying v and u; all other intensive thermodynamic properties must be functions of v and u, given by relationships of the form

$$T = T(v, u), \qquad P = P(v, u). \tag{3.6}$$

Imagine that we put the substance in a container of specified volume that we maintain at a specified temperature. The energy and pressure will adjust to equilibrium values (the thermodynamic state) corresponding to the specified temperature and volume. This suggests that we could invert the functional relationships (3.6) and instead consider T and v as the pair of independent properties,

$$u = u(T, v), \qquad P = P(T, v).$$

The State Principle says there are only two independently variable thermodynamic properties, but does not tell us that *any* two are independent. For example, v and ρ clearly are not independent because one is the reciprocal of the other.

With only two independent properties, one can build tables and charts of thermodynamic properties in which all states can be shown on a single plane. Several examples of such tables and charts can be found in Appendix A.

Application to a ferrofluid

A ferrofluid is a compressible fluid that can have a magnetic dipole moment. For a ferrofluid we can vary V, $\vec{\mathcal{M}}$, and U independently; there are *three independent properties*. It would be easier to vary V, $\vec{\mathcal{H}}$, and U independently, in which case $\vec{\mathcal{M}}$ would be determined by these three quantities. Or, if T and P are independent, we could take T, P, and $\vec{\mathcal{H}}$ as the independent properties, and express the intensive property relationships as

$$u = u(T, P, \vec{\mathcal{H}}), \qquad \vec{\mathcal{M}} = \vec{\mathcal{M}}(T, P, \vec{\mathcal{H}}). \qquad (3.7)$$

Graphing these functions would require three-dimensional charts, or sets of two-dimensional charts with one sheet for each pressure. This gets very complicated, and one seldom sees data for substances with three independent properties.

3.5 States of a Simple Compressible Substance

Liquid and vapor states

Suppose we place a known mass of a simple compressible substance in a vertical piston–cylinder system in which the pressure is fixed by the weight of the piston (Figure 3.5a). We can control the temperature by heating or cooling the container, and measure the resulting volume. This will allow us to trace out a line of constant pressure on a T–v plane. Figure 3.5b shows what would be obtained for methane (CH_4) at a pressure of 1 MPa. Imagine starting at state A on Figure 3.5b, where the substance is a liquid. Since in a liquid the molecules are relatively close together, the specific volume will be relatively low. For most substances, the volume will increase slightly as the temperature increases, as shown in Figure 3.5b (an important exception is H_2O below 4 °C, where the volume decreases with increasing temperature). This slight expansion will continue until the liquid starts to boil (point L on Figure 3.5b). Boiling occurs when molecules have enough energy to escape the electronic binding forces that hold them close to their neighbors in the liquid. The liquid is said to be *saturated*, meaning it has all of the energy it can hold at this pressure without evaporating. The gas that forms upon boiling is called *vapor*.

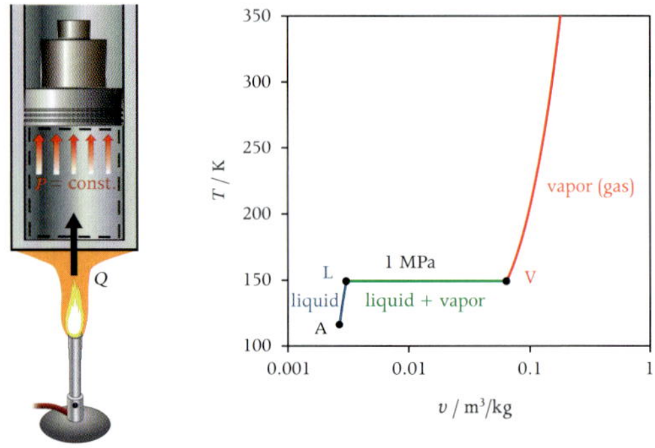

(a) Constant-pressure device. (b) A constant-pressure line for methane.

Figure 3.5 Isobaric heating of methane.

The liquid and vapor are two different *phases* of the substance that exist together in equilibrium under saturation conditions. The liquid is at state L in Figure 3.5b and the vapor is at state V. Since the molecules are much farther apart in the vapor than in the liquid, the specific volume of the vapor (v^V) is much higher than that of the liquid (v^L), and the specific volume of the mixture increases rapidly once boiling starts. As we continue to heat the system, more and more liquid is converted to vapor, but the temperature does not change as long as both liquid and vapor are present. During evaporation, every drop of liquid is at state L and every bubble of vapor is at state V, and the specific volume of the mixture is determined by v^L, v^V, and the mass fraction of the mixture that is vapor. The evaporation and expansion continue until all of the liquid has been vaporized and the system is entirely vapor at point V.

Continued heating at constant pressure will cause the gas to expand, with the molecules becoming farther and farther apart (on average) as the temperature increases. Eventually the density will be so low that forces between molecules will be negligible (except in molecular collisions), and so the gas will behave as an ideal gas, for which the v is proportional to T at constant P.

Saturation pressure and temperature, and normal boiling point

If we repeat the experiment at different pressures, different lines will be found. A line of constant pressure is called an *isobar*,[2] and several are shown on Figure 3.6 for methane. As the pressure is increased,

[2] From ancient Greek ἴσος (isos = equal) + βάρος (baros = weight).

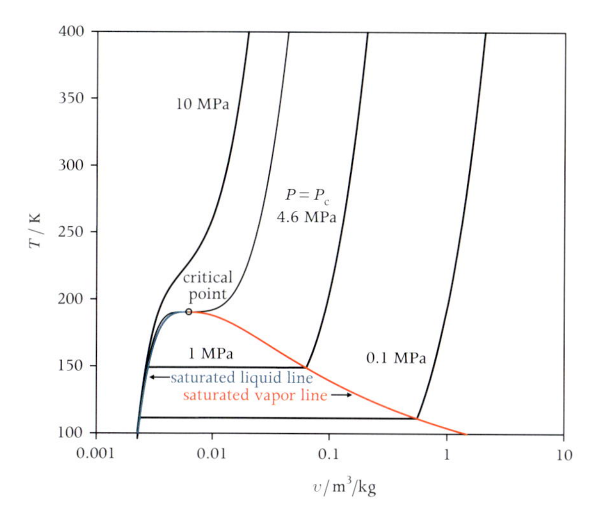

Figure 3.6 *T–v* **diagram for methane.**

boiling will occur at higher temperatures. At a given pressure, the temperature at which boiling occurs is called the *saturation temperature* T^{sat}. At a given temperature, the pressure at which boiling occurs is called the *saturation pressure* P^{sat}. Note that one is a function of the other: $P^{sat} = P^{sat}(T^{sat})$ and $T^{sat} = T^{sat}(P^{sat})$. The saturated state at which the pressure is equal to the atmospheric pressure at sea level (101.325 kPa) is called *normal boiling point*, thus $T^{NBP} = T^{sat}(P = 1 \text{ atm})$.

Critical point

As the pressure is increased, the volume difference between the saturated vapor (v^V) and saturated liquid (v^L) decreases. At the *critical pressure* the two saturation volumes are the same, and this point is called the *critical point*. The critical isobar is tangent to the critical-temperature line at the critical point, where it also has an inflection point. Above the critical pressure there is no two-phase region, and heating at constant pressure simply transforms the fluid slowly and continuously from something we would agree is liquid to something we would agree is gas.

Saturation lines, vapor dome

The locus of saturated liquid states is called the *saturated liquid line*, and states along this line are denoted with a superscript L (e.g., v^L). The locus of saturated vapor states is called the *saturated vapor line*, and saturated vapor states are denoted with the superscript V (*e.g.*, ρ^V). These two lines meet at the critical point. The region under these two lines is called the *vapor dome*.

Solid and liquid states

Suppose we start at a point corresponding to A in Figure 3.5b and cool the liquid at constant pressure (Figure 3.7a). At some lower temperature the liquid will start to freeze. The solid will be at state S, and will be in equilibrium with liquid at state L, and the specific volume of the mixture will be determined by v^S, v^L, and the fraction of the mixture that is liquid. Most fluids shrink upon freezing; water is unusual in that it expands when frozen (ice floats!). The freezing process defines a narrow region on the *T–v* diagram where two phases (solid and liquid) coexist, between the *solidus* and *liquidus* lines. Note that *T* and *P* lines coincide in the two-phase (liquid–solid) region, as in the vapor dome.

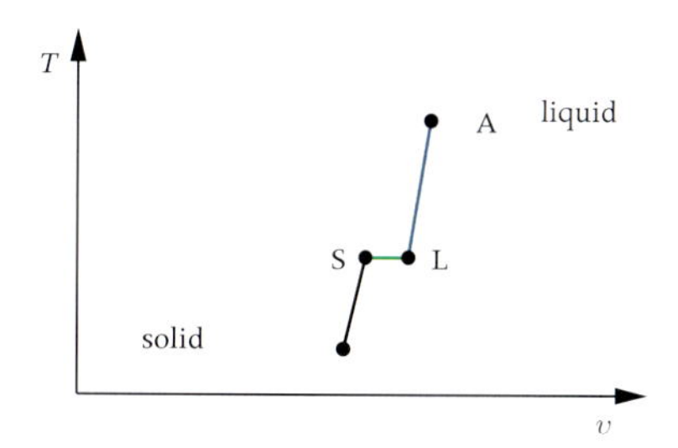

(a) *T–v* diagram for liquid–solid transition.

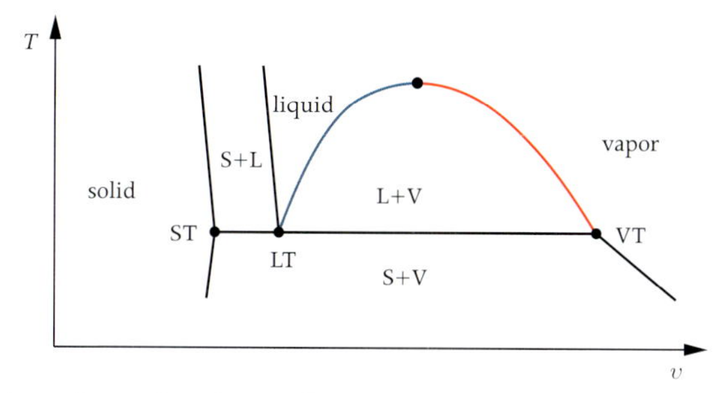

(b) *T–v* diagram for substance that expands upon melting (not to scale).

Figure 3.7 Phases of a simple compressible substance.

Triple point

As the pressure is lowered, the liquidus line and saturated liquid line come together at a unique temperature and pressure called the *triple point* (the horizontal line of Figure 3.7b). At this temperature and pressure three phases of the substance (solid, liquid, and gas) can coexist in equilibrium, with the liquid at state LT on Figure 3.7b, the gas at state VT, and the solid at state ST. Thermodynamic theory shows that the largest number of phases of a simple compressible substance that can exist together in equilibrium is three (see the *phase rule* in Section 8.5). At pressures below the triple point, the solid is converted into gas directly upon heating (*sublimation*); dry ice (solid CO_2) is a familiar example. Below the triple point the vapor transforms into solid, without passing through the liquid phase. This type of phase transition is called *desublimation* or *deposition*; snow is formed by this process.

$P-v-T$ surface

The relationships between pressure, specific volume, and temperature define a surface in the P–v–T space. Figure 3.8 shows the form of this surface for a substance that expands upon melting. A view of this surface perpendicular to the P axis gives the T–v plane discussed above, with isobars as contour lines formed by cuts at different P.

T–P phase diagram

A view of the P–v–T surface perpendicular to the v axis gives the T–P plane, shown for CO_2 in Figure 3.9.

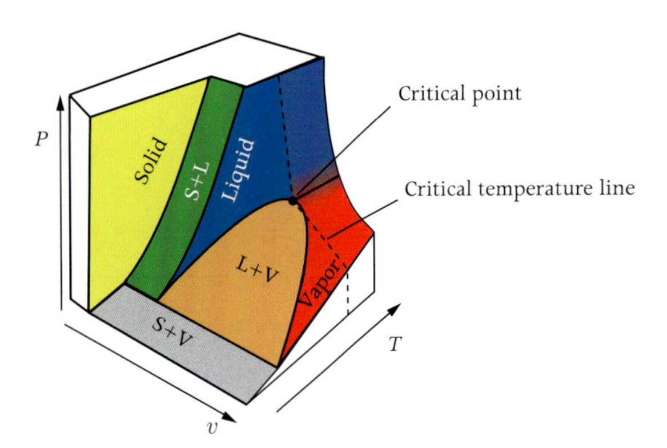

Figure 3.8 *P–v–T* surface for a substance that expands upon melting.

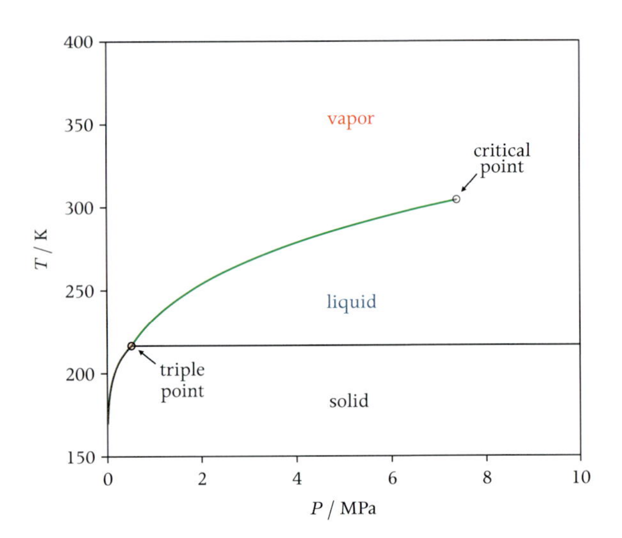

Figure 3.9 Temperature–pressure diagram for CO_2.

This diagram is especially useful to show the phases that exist at various temperatures and pressures.

Multiple solid phases

Some substances have more than one type of solid phase that depends upon the atomic organization in the crystal; the graphite and diamond forms of carbon are familiar examples. What phase is formed depends upon the pressure (or temperature) when the solid is formed. This is an interesting area for research in materials science, as relatively little is known about the phases and properties of most solids at ultra-high pressures.

3.6 Thermodynamic Property Data

Internal energy

In principle one could change the state of a simple compressible substance by transferring energy in or out as heat and/or work, carefully measure these energy transfers, and then use an energy balance to relate the internal energy change between the initial and final states to the measured energy transfers. If we pick some particular state as a reference point (the triple point is often used), we can measure the internal energy at all other states relative to that at the reference or *datum* state. Since only changes in the internal energy of the substance are involved in energy analysis, we never really need to know the internal energy at the reference state, and can take it as zero for simplicity in calculations. However, we have to be more careful if there are chemical reactions that change the substance into other species, as illustrated in Section 10.3. Thus, we can develop a map giving $u(T, v)$ as contour lines on the T–v plane. Alternatively, we can replot the data on a v–u plane, showing lines of constant pressure, as for example in the v–u diagram for H_2O, page 312 in Appendix A.3. A more sophisticated approach using equations developed by thermodynamic theory, measurements of P–v–T data, and c_v information at low density is actually used to determine the internal energy. This is discussed in Section 6.3.

Enthalpy

For reasons that will soon be very clear, a property related to the internal energy usually appears on thermodynamic property charts instead of u. The *enthalpy* is defined as the property combination

$$H \equiv U + PV. \tag{3.8}$$

Adding PV to U makes sense since both have units of energy ($đW = PdV$). The related intensive property is the *specific enthalpy* $h \equiv H/M$,

$$h \equiv u + Pv. \tag{3.9}$$

If one can find h, P, and v in a chart or table, u is easily determined, if needed, by subtracting Pv. Since h occurs more naturally in energy balances on flow systems (see Chapter 4), engineers decided a long time ago that it would be better to get or visualize h easily rather than u, and hence most tables and charts display h information but not u information.

Saturation tables

Thermodynamic properties along the saturated liquid and vapor lines can be compiled in a single-entry table with either T^{sat} or P^{sat} as the first column, with the corresponding value of the other in the second column. The properties of the saturated liquid (v^{L} and h^{L}) and the saturated vapor (v^{V} and h^{V}) are also included. For convenience in hand calculations, some tables also include the differences

$$v^{\text{LV}} \equiv v^{\text{V}} - v^{\text{L}}$$
$$h^{\text{LV}} \equiv h^{\text{V}} - h^{\text{L}}.$$

Properties in the vapor–liquid equilibrium region

The properties of a mixture of liquid and vapor are computed using the individual properties of the saturated liquid and vapor. For example, the total volume is

$$V = V^{\text{liquid}} + V^{\text{vapor}} = M^{\text{liquid}} v^{\text{L}} + M^{\text{vapor}} v^{\text{V}}.$$

Dividing by the total mass, and defining the *quality* or *vapor fraction* (x) of the mixture as the mass fraction of vapor,

$$x \equiv M^{\text{vapor}}/M, \tag{3.10}$$

then

$$M^{\text{vapor}}/M = x, \qquad M^{\text{liquid}}/M = (1 - x).$$

It follows that

$$v = (1 - x)v^{\text{L}} + xv^{\text{V}} = v^{\text{L}} + xv^{\text{LV}}. \tag{3.11a}$$

Similiarly,

$$u = (1 - x)u^{\text{L}} + xu^{\text{V}} = u^{\text{L}} + xu^{\text{LV}}, \tag{3.11b}$$

and, since both the liquid and vapor have the same pressure,

$$h = (1 - x)h^{\text{L}} + xh^{\text{V}} = h^{\text{L}} + xh^{\text{LV}}. \tag{3.11c}$$

Example: properties for a state in vapor–liquid equilibrium. Determine the pressure, specific volume, and internal energy of a mixture of water, liquid, and vapor at 400 K if the quality is 0.4. From the saturation tables (Table A.3) with T^{sat} as the entry column, $P^{\text{sat}} =$ 0.2458 MPa. The saturated liquid and saturated vapor specific volumes are given in the tables as

$v^L = 0.001067\ m^3/kg$, $v^V = 0.7303\ m^3/kg$, so for the mixture

$$v = (1 - 0.4) \times 0.001067 + 0.4 \times 0.7303 = 0.2928\ m^3/kg.$$

The (specific) enthalpy of the saturated liquid and vapor are read from the saturation table as $h^L = 532.9$ kJ/kg and $h^V = 2715.7$ kJ/kg, so the enthalpy of the mixture is

$$h = (1-0.4)\times 532.9 + 0.4\times 2715.7 = 1406.1\ kJ/kg.$$

So, the internal energy is

$$\begin{aligned} u &= h - Pv \\ &= 1406.1 \times 10^3\ J/kg - 0.2458 \times 10^6\ N/m^2 \\ &\quad \times 0.2928\ m^3/kg \\ &= 1334.1\ kJ/kg. \end{aligned}$$

Example: properties for a superheated state. In the vapor (or gas) region, historically called the *superheated* vapor region, double-entry tables are required. The usual tables have P and T as entry data, with values tabulated in the matrix cells. For example, for H_2O at 0.4 MPa and 600 K, Table A.5 gives

$$\begin{aligned} v &= 0.6867\ m^3/kg, \\ u &= 2847.6\ kJ/kg, \\ h &= 3122.3\ kJ/kg. \end{aligned}$$

The table also shows that the saturation temperature at a pressure of 0.4 MPa is 416.8 K, and the saturated vapor volume v^V is 0.4624 m^3/kg. The saturation data are useful if one is interpolating between the saturation data and the nearest entry in the superheated property tables.

Thermodynamic property charts

Graphical representations of thermodynamic property data are available for many substances, and several such charts are provided in Appendix A. An extensive collection of charts and tables can also be found in W. C. Reynolds, *Thermodynamic Properties in SI*, Dept. of Mechanical Engineering, Stanford University, 1979. Although charts are at a scale, they are suitable more for preliminary analysis and thinking rather than for accurate quantitative work. Several different thermodynamic planes are used in these charts. The T–s plane is widely used in textbooks, but it has certain drawbacks, chiefly in the liquid region, and it is something of a mystery for students who are not yet familiar with entropy (s). The P–h plane is very useful because the liquid region is nicely displayed, and the beginning student can ignore the lines of constant s. Therefore, at this point in your study we will have you work with P–h diagrams, and later we will introduce you to other thermodynamic planes that are especially useful for thinking about thermodynamic systems.

In using the P–h diagrams, you must remember that the isotherms coincide with isobars inside of the vapor dome. The isotherms are therefore not usually drawn inside the dome, although the graphs in Appendix A do have marks on either side of the dome to help you locate the isotherms. You should verify the calculations for H_2O of the previous examples using the P–h chart for H_2O, page 310 in Appendix A.3.

Properties software

In the past, double-entry interpolation in double-entry tables, and reading graphs with numerous confusing lines to low accuracy, has not helped making thermodynamics an enjoyable subject. But today virtually every new thermodynamics book comes with software that gives instant, precise thermodynamic properties information. The program FLUIDPROP (www.FluidProp.com) conceived and initially developed by the authors is available as freeware for academic use and runs on the Windows®, Linux®, and OS X® operating systems. FLUIDPROP can do all this for you and much more, allowing you to concentrate on the interesting aspects of engineering thermodynamics. We will introduce you to FLUIDPROP after you have become familiar with the tables and graphs through some relatively easy hand calculations like those above.

3.7 Derivative Properties

Derivatives of a thermodynamic property with respect to other thermodynamic properties, sometimes also called *secondary properties*, are themselves thermodynamic properties. For example, consider the specific volume v as a function of T and P, which we express as $v(T, P)$. This function has two partial derivatives,

$$\left(\frac{\partial v}{\partial T}\right)_P \quad \text{and} \quad \left(\frac{\partial v}{\partial P}\right)_T.$$

It is important to be very clear about what property is held constant when differentiating with respect to another, and the subscripts P and T above mean that P and T are held constant in these differentiations, respectively. These derivatives are thermodynamic properties. $(\partial v/\partial T)_P$ is the slope of a line of constant P on a v–T plane, and $(\partial v/\partial P)_T$ is the slope of a line of constant T on a v–P plane. For example, for an ideal gas where $v = RT/P$,

$$\left(\frac{\partial v}{\partial T}\right)_P = \frac{R}{P} \quad \text{and} \quad \left(\frac{\partial v}{\partial P}\right)_T = -\frac{RT}{P^2}. \tag{3.12}$$

The *total differential* of v is then (by the chain rule of calculus)

$$dv(T, P) = \left(\frac{\partial v}{\partial T}\right)_P dT + \left(\frac{\partial v}{\partial P}\right)_T dP.$$

Isobaric compressibility

The *isobaric compressibility* is defined as

$$\beta \equiv \frac{1}{v}\left(\frac{\partial v}{\partial T}\right)_P, \tag{3.13}$$

and measures the percentage volume increase per unit of temperature increase at constant pressure.

Isothermal compressibility

The *isothermal compressibility* is defined as

$$\kappa \equiv -\frac{1}{v}\left(\frac{\partial v}{\partial P}\right)_T, \tag{3.14}$$

and measures the percentage volume increase per unit of pressure increase at constant temperature.

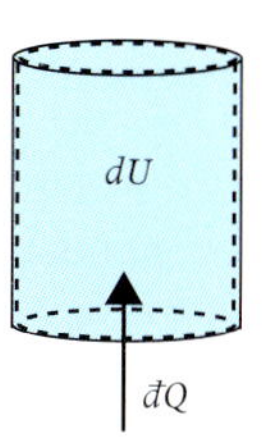

Figure 3.10 Heating at constant volume.

Specific heat at constant volume

Consider the heating of a substance maintained at constant volume by a rigid container (Figure 3.10). An energy balance on the substance gives

$$\underbrace{đQ}_{\text{energy input}} = \underbrace{d(Mu)}_{\text{energy accumulation}},$$

where $U = Mu$. Thinking of $u(T, v)$, the total differential du is

$$du = \left(\frac{\partial u}{\partial T}\right)_v dT + \left(\frac{\partial u}{\partial v}\right)_T dv. \tag{3.15}$$

The first of these derivatives is defined to be c_v,

$$c_v \equiv \left(\frac{\partial u}{\partial T}\right)_v. \tag{3.16}$$

Now, since $dv = 0$ for this system, and the mass M is constant, the energy balance for this problem gives

$$đQ = Mdu = Mc_v dT.$$

Note that c_v is the energy input as heat, per unit mass, per unit of temperature rise, for a constant-*volume* heating process; therefore, c_v is called the *specific heat at constant volume*, or *isochoric*[3] specific heat. In general, $c_v = (T, v)$, but we can often treat c_v as a constant in engineering problems, especially over limited ranges of T and v.

[3] From ancient Greek ισος (isos = equal) + Χορα (khora = space).

Specific heat at constant pressure

Consider the heating of a substance maintained at constant pressure in a container with a weighted piston on the top (Figure 3.11). An energy balance on the substance gives

$$\underbrace{\text{đ}Q}_{\text{energy input}} = \underbrace{\text{đ}W}_{\text{energy output}} + \underbrace{dU}_{\text{energy accumulation}} .$$

The work output due to expansion may be expressed as PdV. Since the pressure is constant for this problem, the energy transfer as heat may this time be expressed in terms of the enthalpy change,

$$\text{đ}Q = dU + PdV = dU + d(PV) = dH = Mdh.$$

Thinking of $h(T, P)$, the total differential dh is

$$dh = \left(\frac{\partial h}{\partial T}\right)_P dT + \left(\frac{\partial h}{\partial P}\right)_T dP.$$

The first of these derivatives is denoted as c_P,

$$c_P \equiv \left(\frac{\partial h}{\partial T}\right)_P . \tag{3.17}$$

Now, since $dP = 0$ for this system, the energy balance for this problem gives

$$\text{đ}Q = Mc_P\,dT.$$

Note that c_P is the energy transfer as heat, per unit mass, per unit of temperature rise, for a constant-*pressure* heating process; therefore, c_P is called the *specific heat at constant pressure*, or *isobaric* specific heat. In general, $c_P = f(T, P)$, but we can often treat c_P as a constant in engineering problems, especially over limited ranges of T and P. If you look at a P–h chart of a pure fluid, you see that inside of the vapor–liquid dome, T does not change at constant P, but h does, and hence c_P is infinite.

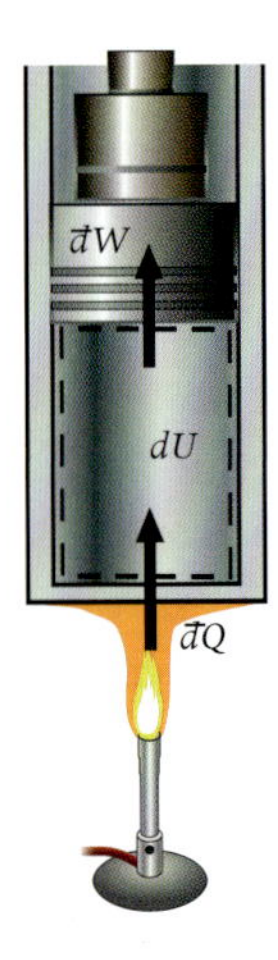

Figure 3.11 Heating at constant pressure.

Specific heat ratio

The ratio of the two specific heats is an important non-dimensional parameter, and is defined as

$$\gamma \equiv \frac{c_P}{c_v}. \tag{3.18}$$

For single phases of fluids, this ratio ranges from about 1 for very dense liquids to 5/3 for a monatomic ideal gas.

3.8 The Ideal (or Perfect) Gas

Definition

An *ideal (or perfect) gas* is any substance that obeys the equation of state

$$P = \rho RT. \tag{3.19a}$$

Here $R = \hat{R}/\hat{M}$, where $\hat{R} = 8.3143$ J/ (mol · K) is the *universal gas constant* and $\hat{M}$ is the molar mass (mass per mole) of the substance. Equivalent forms are

$$Pv = RT \tag{3.19b}$$

and

$$P\hat{v} = \hat{R}T, \tag{3.19c}$$

where $\hat{v}$ is the volume per mole. These are the intensive forms of the defining equation. For a sample of mass M containing N moles in a volume V, the extensive forms are

$$PV = MRT \tag{3.20a}$$

and

$$PV = N\hat{R}T. \tag{3.20b}$$

Conditions for ideal gas behavior

You can tell where the ideal gas model is a good approximation by looking at the thermodynamic property charts. For example, on Figure 3.6 the line for $P = 0.1$ MPa is very nearly straight throughout the gas region, and displaced in volume at any given temperature by a factor of 10 from the line for $P = 1$ MPa, indicating that the ideal gas model would be a good approximation in this region. You can see that the ideal gas model is best at very high volumes, or low densities, when the molecules are on average very far apart. In the limit of $\rho \to 0$, all substances behave like an ideal gas. For this reason, the ideal gas model is the most important properties model in thermodynamics; *but you must be careful not to use it where it does not apply!*

Energy of an ideal gas

The energy of the gas is the sum of the translational, rotational, and vibrational energy of each molecule; there is no significant energy associated with intermolecular forces because the molecules are on average very far apart. As a result, the internal energy of an ideal gas depends only upon temperature, which is a function of the average molecular energy:

$$u = u(T) \qquad \text{for an ideal gas.} \tag{3.21a}$$

If you examine the v–u diagram for H_2O, page 312 in Appendix A.3, you will see that at low pressures outside of the vapor dome the temperature lines and internal energy lines coincide, indicating that the internal energy is indeed independent of the density in the region where we expect ideal gas behavior.

Enthalpy for an ideal gas

Since $u = u(T)$ and $Pv = RT$, the enthalpy of an ideal gas $h = u + Pv$ is also only a function of temperature,

$$h = h(T) \qquad \text{for an ideal gas.} \tag{3.21b}$$

Examination of the various P–h charts in Appendix A again confirms that at high volumes in the gas region the h lines coincide with T lines, indicating ideal gas behavior.

Specific heats for an ideal gas

Since u and h depend only upon temperature, so must the specific heats, hence

$$c_v = c_v(T) \qquad \text{for an ideal gas,} \tag{3.22a}$$

and

$$c_P = c_P(T) \qquad \text{for an ideal gas.} \tag{3.22b}$$

Note that, since $h = u + Pv$,

$$c_P = c_v + R \qquad \text{for an ideal gas.} \tag{3.23}$$

For monatomic gases the internal energy is due entirely to molecular translation, and as we show in the next section the translational energy is proportional to the temperature. This means that the specific heats for monatomic ideal gases are constants. For more complex molecules the amount of energy associated with rotation and vibration increases with increasing temperature in a nonlinear way, and hence the specific heats increase somewhat with temperature. However, over limited temperature ranges we often make the approximation that the specific heats are constant, which allows neat analytical solutions to many interesting problems.

Air as an ideal gas

Air, a mixture of gases, can usually be treated as an ideal gas. Table A.2 gives the properties of air according to the ideal gas approximation, with temperature-dependent specific heats.

3.9 A Microscopic Model for the Ideal Gas

Here we will use some simple microscopic analyses to provide a basis for the ideal gas equation and to develop a model for the internal energy of a monatomic ideal gas. This model should help you understand why a low-density gas behaves as it does. It also provides a foundation for discussion of the properties of more complex gases.

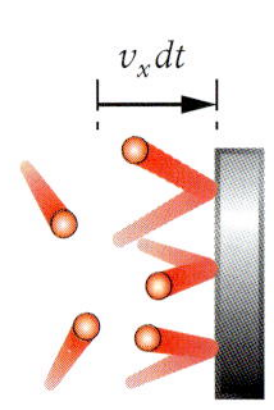

Figure 3.12 Determining the collision rate.

Pressure in an ideal gas

The pressure in an ideal gas is determined by the rate of momentum transfer per unit area as the molecules collide with the container boundary. The molecules move at different speeds and in all possible directions. In order to compute the momentum transfer, let's first consider those particles moving such that their velocity normal to the boundary is $\pm\mathcal{V}_x$ (Figure 3.12). In a small time interval dt, the number of these particles that strike a unit area of the boundary will be half of those contained within striking distance of the boundary, or

$$\frac{1}{2} \times n' \times \mathcal{V}_x \, dt,$$

where n' is the number of molecules per unit volume that have this particular x-speed. The factor 1/2 is needed because only half of the particles are moving towards the boundary. Assume that these particles collide and rebound with the same x-speed in the opposite direction. Therefore, the x-component momentum that they transfer to the wall is

$$\frac{1}{2} n' \mathcal{V}_x \, dt \times 2m\mathcal{V}_x,$$

where the factor of 2 accounts for the change in the x-velocity vector of $2\mathcal{V}_x$, and m is the mass of a molecule. Now we add up the momentum transfers from molecules of all speeds, and divide by dt to get the rate of momentum transfer per unit area, i.e. the pressure

$$P = m \sum_{\text{speeds}} n' \mathcal{V}_x^2 = m\,n\,\overline{\mathcal{V}_x^2}, \qquad (3.24)$$

where n is the total number of particles per unit volume and $\overline{\mathcal{V}_x^2} = \sum(n'/n)\mathcal{V}_x^2$ is the mean-square velocity in the x direction, averaged over all molecules. Now, the total mean-square speed is

$$\overline{\mathcal{V}^2} = \overline{\mathcal{V}_x^2} + \overline{\mathcal{V}_y^2} + \overline{\mathcal{V}_z^2},$$

where $\overline{\mathcal{V}_y^2}$ and $\overline{\mathcal{V}_z^2}$ are the mean-square velocities in the y and z directions, respectively. Since the molecular motions are isotropic,

$$\overline{\mathcal{V}_x^2} = \overline{\mathcal{V}_y^2} = \overline{\mathcal{V}_z^2} = \frac{1}{3}\overline{\mathcal{V}^2},$$

and so (3.24) gives

$$P = \frac{1}{3} mn\overline{\mathcal{V}^2}. \qquad (3.25)$$

Temperature of an ideal gas

If we compare (3.25) with (3.19a), noting that $\rho = mn$, we find

$$RT = \frac{1}{3}\overline{\mathcal{V}^2}. \qquad (3.26)$$

Thus we see that the temperature of an ideal gas provides a measure of its mean-square molecular translational velocity. This same expression holds for the translational energy of a polyatomic ideal gas.

Internal energy of a monatomic ideal gas

If the gas is monatomic, its internal energy is entirely due to the translational kinetic energy of its atoms. The kinetic energy per unit mass is

$$u = \frac{1}{2}\overline{\mathcal{V}^2}.$$

Using (3.26), the internal energy per unit mass (u) and per mole ($\hat{u}$) are found to be

$$u = \frac{3}{2} RT \qquad (3.27a)$$

and

$$\hat{u} = \frac{3}{2} \hat{R}T. \qquad (3.27b)$$

Note that the internal energy of the gas is only a function of its temperature, as discussed above, and that the internal energy per mole is the same for all monatomic ideal gases.

Enthalpy of a monatomic ideal gas

Since $h = u + Pv$, the enthalpy and molar enthalpy for a monatomic ideal gas are

$$h = \frac{5}{2}RT \tag{3.28a}$$

and

$$\hat{h} = \frac{5}{2}\hat{R}T. \tag{3.28b}$$

Note that the enthalpy of the gas is only a function of its temperature, as discussed above, and that the enthalpy per mole is the same for all monatomic ideal gases.

Specific heats of a monatomic ideal gas

Using (3.16) and (3.17), the specific heats, molar specific heats, and specific heat ratio for an ideal monatomic gas are

$$c_v = \frac{3}{2}R, \quad \hat{c}_v = \frac{3}{2}\hat{R}, \quad c_P = \frac{5}{2}R,$$
$$\hat{c}_P = \frac{5}{2}\hat{R}, \quad \gamma \equiv \frac{c_P}{c_v} = \frac{5}{3}.$$

3.10 Extensions to Polyatomic Ideal Gases

Equipartition model

A simple but useful model for polyatomic gases is based on the assumption that each available mode of motion freedom contributes the same amount to the energy of the system. A monatomic molecule has three modes of freedom (translation in the x, y, and z directions) and hence from (3.27b) the energy per degree of freedom is

$$\hat{u} = \frac{1}{2}\hat{R}T \qquad \text{per mode.}$$

This simple rule is useful in estimating the internal energy, enthalpy, and specific heats of more complex molecules, as shown in the following.

Diatomic molecules

A rigid diatomic molecule has these three translational modes of freedom and two additional degrees of rotational freedom, for a total of five modes. If we assume the molecule is rigid (no vibrational energy), the equipartition model says that the internal energy should be

$$\hat{u} = \frac{5}{2}\hat{R}T,$$

for which

$$\hat{h} = \frac{7}{2}\hat{R}T, \qquad \hat{c}_v = \frac{5}{2}\hat{R}, \qquad \hat{c}_P = \frac{7}{2}\hat{R}, \qquad \gamma = 7/5.$$

The table for low-density air in Appendix A.2 shows that $\gamma = 1.4 = 7/5$ at low temperatures, indicating that the equipartition model is valid. However, γ drops below 1.4 as the temperature increases. The equipartition model explains why. As the temperature is increased, the vibrational mode becomes excited, and the molecules can no longer be treated as rigid. When the primary vibrational mode (the "accordion" mode) is fully excited, two more energy modes have been added (one for the vibrational kinetic energy and another for the oscillating potential energy in the intermolecular force field). The equipartition model says that the seven modes should give

$$\hat{u} = \frac{7}{2}\hat{R}T, \qquad \hat{h} = \frac{9}{2}\hat{R}T, \qquad \hat{c}_v = \frac{7}{2}\hat{R},$$
$$\hat{c}_P = \frac{9}{2}\hat{R}, \qquad \gamma = 9/7.$$

The air tables in Appendix A.2 show that γ drops from 1.4 at low temperatures to $\gamma = 1.33 = 4/3 > 9/7$ at about 1000 K, indicating that the vibrational mode is not fully excited at 1000 K.

Complex molecules

Complex molecules have many vibrational modes, each of which can contribute up to $\frac{1}{2}\hat{R}T$ to the molar internal energy. Thus, gases of more complex molecules have higher specific heats, and specific heat ratios closer to unity, than simpler molecules.

EXERCISES

3.1 Calculate the values of specific volume, specific internal energy, and specific enthalpy for water in vapor–liquid equilibrium at 180 °C if its quality is 0.2. Use the tables in Appendix A.3 for data.

3.2 Liquid oxygen (commonly referred to as LOX) in a rocket-propellant tank is kept at a pressure of approximately 1 bar. A certain ullage volume is occupied by saturated vapor for safety. Use the graph in Appendix A.8 to estimate the temperature of the vapor in equilibrium with LOX.

3.3 Using the data of Appendix A.3, estimate β, κ, h, c_v, and c_p for H_2O at 800 K and 1 bar.

3.4 Sketch the u–T and v–T diagrams for H_2O, showing lines of constant pressure. Indicate the saturated liquid and vapor lines.

3.5 Estimate the temperature to which water at the bottom of a 150 m deep lake would have to be raised before it would begin to boil.

3.6 Calculate the density of gaseous H_2O at each of the following states:

a. 0.07 bar, 360 K,

b. 1 atm, 700 K,

c. 200 bar, 700 K,

using the perfect gas model. Compare the values obtained with the values given in the tables of Appendix A.3. What can you observe and explain?

3.7 Calculate the density of gaseous CO_2 at each of the following states:

a. 10 bar, 290 K,

b. 30 bar, 333 K,

c. 100 bar, 410 K,

using the perfect gas model. Compare the values obtained with the values listed in the tables in Appendix A.9. What can you infer?

3.8 A constant volume vessel of 100 liters holds 1 kg of water at an initial pressure of 30 bar. The container is then cooled to 305 K. Determine the initial temperature of the system as well as the pressure after cooling and the energy transfer as heat.

3.9 Determine the energy transfer as heat (J) required to change the state of 2 kg of methane from a saturated liquid at 1 atm to a gas at 300 K and 1 atm, assuming that the methane is confined at constant pressure during this process. Check your read of the tables using the P–h chart, and make a P–h process sketch showing the initial and final states, the vapor dome, and the line $P = 1$ atm. Use the tables in Appendix A.5 for data.

3.10 2000 kg of oxygen is in a 1.8 m^3 container maintained at 90 K. Calculate the liquid and vapor masses (kg) and the liquid and vapor volumes (m^3). Use the tables in Appendix A.8 for data.

3.11 Propane is in a 5 m^3 container as a saturated vapor at 300 K. The sealed tank is then cooled to 260 K by fresh snow. What are the pressure (Pa) and liquid volume in this 260 K state, and how much energy was transferred as heat from the propane? Use the tables in Appendix A.6 for data.

3.12 Refrigerant 134a in a heat pump is compressed in an *adiabatic* piston–cylinder system from 0.14 MPa, 280 K to 1 MPa, 400 K. Calculate the work done on the fluid by the piston. Use the tables in Appendix A.4 for data.

3.13 5 kg of refrigerant 134a at a pressure of 7.5 bar and 0.7 quality is to be cooled inside a sealed rigid container until the pressure reaches 2.1 bar. Referring to the tables in Appendix A.4, calculate the required amount of energy transfer as heat in kJ.

3.14 In order to illustrate the nature of the critical point, one can place CO_2 in a quartz vial and seal the device. Assuming that $v_c = 0.002$ m^3/kg, and that the device is filled at room temperature (290 K), use the T–s chart in Appendix A.9 to find:

a. the pressure in the vial at room temperature,
b. the percent liquid mass and volume ensuring passage through the critical point upon heating (*Hint: Calculate* v^{L} *from* x, v^{LV}, *and* v^{V}).

3.15 3 kg of ammonia at its critical state is placed in a constant-volume container, which is then cooled until the temperature is 300 K. How much energy was transferred as heat from the ammonia in this process, and what tank volume is occupied by liquid at the end of the process? Use the tables in Appendix A.7 for data.

3.16 Initially 10 kg of saturated liquid water at 400 K and 2 kg of saturated water vapor at 300 K are together in a sealed, constant-volume adiabatic container. After a time the water adjusts to a uniform state within the tank. What are the pressure and temperature in this final state, and what are the final volumes of the liquid and vapor? The graphs in Appendix A.3 may be used along with the tables in solving this problem.

3.17 A boiler is fed with 40 kg/s of water at a pressure of 100 bar and a temperature of 250 °C. The velocity of the water at the inlet is negligible. The boiler delivers superheated steam with a pressure of 90 bar and a temperature of 800 K through a 20 cm diameter pipe. Calculate the average velocity of the steam at the exit of the boiler and the energy transfer as heat to the water required to obtain the specified steam conditions. Use the tables in Appendix A.3 for data.

3.18 Two adjacent tanks are connected by a valve. The tank on the left contains 0.5 kg of water at a pressure of 11 bar and a temperature of 600 K. The tank on the right contains 0.25 kg of saturated water at a temperature of 300 K and its quality is 0.4. If the valve between the tanks is opened and the tanks are allowed to reach thermal equilibrium with the environment, what is the amount of energy transfer as heat out of the system assuming the environment temperature is 22 °C? Use the tables in Appendix A.3 for data.

3.19 Carbon dioxide at P_1 and T_1 is contained inside a piston–cylinder. It undergoes polytropic compression according to the process equation $PV^n = \text{constant}$.

a. Derive the formula for the energy transfer as work done by the piston to compress CO_2 to a state with $P_2 > P_1$.

The CO_2 in the initial state is at a pressure of 10 bar, at a temperature of 280 K, and it occupies a volume of 0.001 m^3. The polytropic exponent of this compression process has been determined and is $n = 1.4$. The final pressure reached by CO_2 is 20 bar. Using the tables in Appendix A.9 for data, calculate:

b. the mass of CO_2 inside the cylinder,
c. the specific volume of the fluid in the initial and final states,
d. the distance traveled by the piston,
e. the temperature of the fluid at the final state,
f. the work done by the piston,
g. the energy transfer as heat and its direction.

3.20 A 2 m^3 container for heating water is initially filled with H_2O liquid and vapor at 400 K, with 20% of the *volume* being liquid. Use the tables in Appendix A.3 to work this exercise. Check your numbers using the v–u chart, and show the process representation on a v–u sketch.

a. What is the pressure in the tank?
b. What is the mass of the H_2O in the tank?
c. Suppose that the inflow and outflow pipes are closed and the heating occurs at a rate of 170 kW. What will be the temperature in the tank when the pressure has risen to 10 MPa?
d. How much energy input as heat will be required to reach this condition?
e. How long will it take to reach this condition?

4 Control Volume Energy Analysis

CONTENTS

In any analysis it is very important that the system being analyzed is defined precisely. Careful definition helps the analyst do the work correctly and enables the reader to understand it. In thermodynamic analysis we work with two different types of system definitions: the control *mass* (CM) and the control *volume* (CV). When we apply a basic principle, such as conservation of energy, we always apply it to one of these two types of systems. Each of the energy balances we made in previous chapters was made on a control mass. The primary purpose of this chapter is to help you learn how to make a correct energy balance on a control volume, because in many cases it is natural and simplifies the solution of problems.

4.1 Control Mass and Control Volume

Control mass

A control *mass* is a defined *piece of matter*. The matter may move around, and the boundaries of the control mass follow the matter wherever it goes. By definition, no matter can enter or leave a control mass, and therefore (neglecting relativistic effects) the mass of a control mass is always the same. In a control mass energy analysis, the only forms of energy transfer across the boundary are work and heat.

Control volume

A control *volume* is a defined *region in space*. Matter may enter or leave a control volume, and the matter in a control volume is whatever happens to be in it at any moment. For example, a control volume can be defined as the space around an aircraft jet engine, or the space inside a balloon. The control volume boundary may move through space (as with the jet engine), and may change shape and volume (as with a balloon). The important thing is that the boundary must be well defined at all times. In a control volume energy balance, the possible energy transfers across the boundary include heat, work, and energy transfer associated with mass flow across the boundary. The primary point of this chapter is to derive the correct expression for this mass-associated energy transfer.

4.2 Example of Flow System Analysis: Tank Charging

Flow systems can be analyzed using either a control volume or control mass approach. In either case one must identify the energy inputs, energy outputs, and energy accumulation in the control volume or control mass, and combine these properly in the energy balance. A control mass analysis can always be used, but a control volume analysis is simpler and is the best way to analyze flow systems.

The system

To illustrate the two different approaches, consider the charging of a tank with high-pressure air (Figure 4.1a). We will assume that the air is supplied at a fixed state (T_1, P_1) by an upstream air compressor. The tank pressure will be initially lower than P_1, and will increase over time as mass is pumped into the tank. We will also assume that state 1 exists everywhere in the inlet pipe. The process starts at time $t = t_i$ and ends at time $t = t_f$. We assume that the state of the air inside the tank is uniform throughout the tank at both the initial and final times.

Control volume and control mass

We define the control *volume* (CV) as the space inside the tank (Figure 4.1b), and the control *mass* (CM) as the air that occupies the control volume at the end of the process ($t = t_f$). Figure 4.1b also shows the control mass at the intial and final times. We will begin with a control *mass* analysis that will focus on what happens to this mass.

Mass balance on the control mass

Neglecting relativistic effects, which would be important only if the fluid were moving at close to the speed

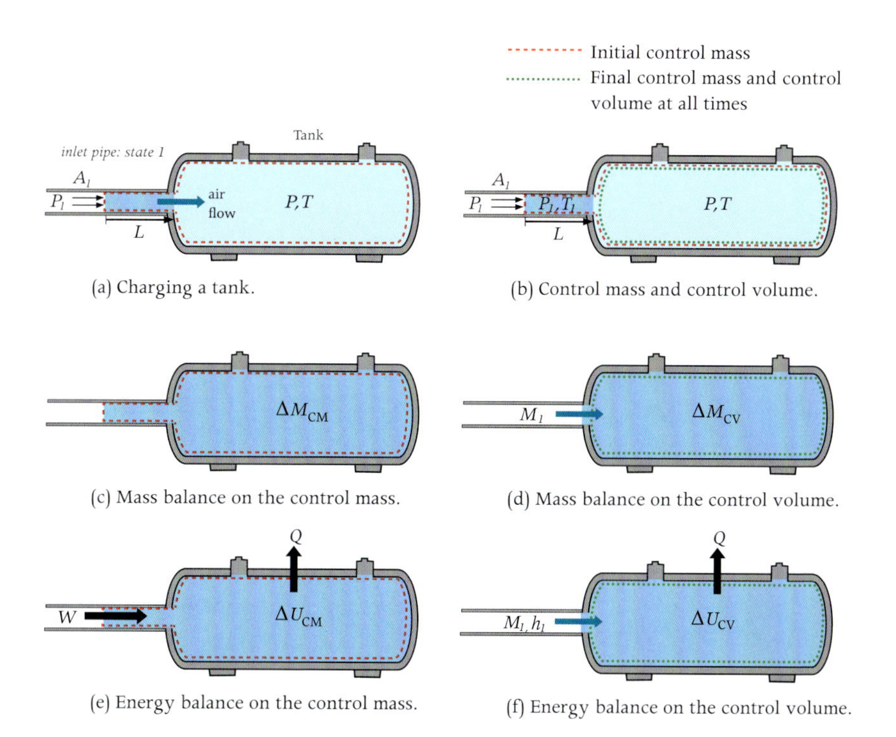

(a) Charging a tank.

(b) Control mass and control volume.

(c) Mass balance on the control mass.

(d) Mass balance on the control volume.

(e) Energy balance on the control mass.

(f) Energy balance on the control volume.

Figure 4.1 Control volume and control mass energy analysis of a tank.

of light (see Chapter 1), the mass of the control mass will be conserved. Denoting the initial and final states by subscripts i and f, respectively, the mass balance on the control mass is simply (see Figure 4.1c)

$$\underbrace{0}_{\text{mass input}} = \underbrace{0}_{\text{mass output}} + \underbrace{M_{\text{CM,f}} - M_{\text{CM,i}}}_{\text{mass accumulation}}. \quad (4.1)$$

Mass balance on the control volume

We will next express (4.1) in terms of quantities pertaining to the control *volume*. That will give us an equation which we can interpret as a mass balance for the control volume. Since the control mass and control volume are the same material at time t_f, we can express the final mass of the control *mass* in terms of the final mass in the control *volume*,

$$M_{\text{CM,f}} = M_{\text{CV,f}}.$$

The inital mass of the control mass is the mass in the tank plus the mass of the air in the inlet pipe that will enter during the process,

$$M_{\text{CM,i}} = M_{\text{CV,i}} + M_1,$$

where M_1 is the mass that entered the tank at state 1 during the process. Thus, (4.1) can be written as

$$0 = M_{\text{CV,f}} - [M_{\text{CV,i}} + M_1],$$

which can be rearranged and reinterpreted as a mass balance on the control *volume*,

$$\underbrace{M_1}_{\text{mass input}} = \underbrace{\Delta M_{\text{CV}}}_{\text{mass accumulation}},$$

where $\Delta M_{\text{CV}} = M_{\text{CV,f}} - M_{\text{CV,i}}$ is the increase in mass inside the control *volume*. The associated mass flow diagram is shown in Figure 4.1d. This *mass* balance for the control volume is obviously correct, and you probably could have written it down directly. We developed it here in detail to help you better understand what follows.

Energy balance on the control mass

The *energy* balance for the control volume is less obvious, so we will again start with a control mass analysis. We allow for energy transfer as heat from the gas in the tank to the cold tank walls; let Q denote the total heat transfer over this period. There is also work done on the control mass where its boundary moves, namely in the pipe. At the left end of the air mass in the pipe there is a pressure force $P_1 A_1$ that pushes the air down the pipe. The work done on the air just inside this boundary by the air just outside is

$$W = P_1 A_1 L,$$

where L is the total distance that this interface moves during the process (see Figure 4.1b). Therefore, the energy balance on the control *mass* is (see Figure 4.1e)

$$\underbrace{W}_{\text{energy input}} = \underbrace{Q}_{\text{energy output}} + \underbrace{\Delta U_{\text{CM}}}_{\text{energy accumulation}}. \quad (4.2)$$

Energy balance on the control volume

We will next express the control mass energy balance (4.2) in terms of quantities pertaining to the control *volume*. This will give us an equation that we can interpret as an energy balance on the control volume. Since the control volume and control mass are the same material at t_f,

$$U_{\text{CM,f}} = U_{\text{CV,f}}.$$

The initial control mass energy is the sum of the initial control volume energy plus the energy of the air mass in the inlet pipe,

$$U_{\text{CM,i}} = U_{\text{CV,i}} + M_1 u_1,$$

so that

$$\Delta U_{\text{CM}} = U_{\text{CV,f}} - \left(U_{\text{CV,i}} + M_1 u_1\right) = \Delta U_{\text{CV}} - M_1 u_1.$$

Note that

$$W = P_1 A_1 L = P_1 V_1 = P_1 v_1 M_1.$$

Finally, since the heat transfer takes place across the portion of the control mass boundary that is the same as the control volume boundary, Q is also the heat transfer from the control volume. Therefore, (4.2) gives

$$M_1 P_1 v_1 = Q + \Delta U_{\text{CV}} - M_1 u_1,$$

which can be rearranged and reinterpreted as the energy balance on the control *volume* (see Figure 4.1f)

$$\underbrace{M_1(u_1 + P_1 v_1)}_{\text{mass-associated energy input}} = \underbrace{Q}_{\text{energy output}} + \underbrace{\Delta U_{\text{CV}}}_{\text{energy accumulation}} .$$

Enthalpy and mass-associated energy transfer

We see that the energy that enters the control volume with the mass M_1 has two parts. The first part, $M_1 u_1$, represents the internal energy carried into the control volume with the mass M_1. The second part, $M_1 P_1 v_1$, represents the work that must be done by fluid outside the control volume to push M_1 across the control volume boundary; this is called *flow work*. These two forms of mass-associated energy transfer occur whenever mass crosses the boundary of a control volume; once you know and understand this fact, you know how to do control volume energy balances. Note that combination $h \equiv u + Pv$, which we defined previously as a property called the *specific enthalpy*, is the energy transfer per unit of mass that crosses the control volume boundary. Thus, the enthalpy always appears in flow system energy analyses, which is the major reason why h is a very important thermodynamic property.

Energy accumulation in the control volume

At any instant when the control volume mass is M and the specific internal energy of the air inside the tank is u, the total internal energy in the control volume is

$$U = Mu.$$

Therefore, the accumulation of internal energy in the control volume is

$$\Delta U_{\text{CV}} = M_{\text{f}} u_{\text{f}} - M_{\text{i}} u_{\text{i}}.$$

Note that the *internal energy* u is the thermodynamic property associated with energy *storage* within the control volume. The *enthalpy* h is the energy *transfer* across the control volume boundary associated with the transfer of mass across the boundary, and has nothing to do with energy storage inside the control volume.

4.3 Generalized Control Volume Energy Analysis

The previous section was an example of the way one can derive the correct energy balance equation for a control *volume* by starting with the energy balance on a control *mass*. Here we consider a more general case using the same methodology. The point of this development is to show you how to make energy balances on control volumes *directly*, without going through the control mass analysis each time.

The device

The device we wish to analyze is shown in Figure 4.2a. Fluid enters at section 1 and leaves at section 2; we denote the fluid states at these two points by subscripts 1 and 2, respectively, and the flow cross-sectional areas by A_1 and A_2. The device receives energy as heat from an electric heater, and energy comes out of the device as work through the rotating shaft.

The control volume

The control volume (CV) is defined as the space inside the device bounded by sections 1 and 2, as indicated by the heavy dashed line in Figure 4.2b. We want to develop the correct mass and energy balances for this control volume, properly accounting for the mass and energy associated with the flow across the CV boundary at 1 and 2. If you understand the previous example very well, you probably can write down these CV balances directly, as we will do for all subsequent CV analyses (try it and compare with our result below). However, in order to help you solidify your understanding so that you can do direct CV balances comfortably, we will go through the full details of deriving the CV balances from balances on a control mass (CM), with which you should by now be comfortable.

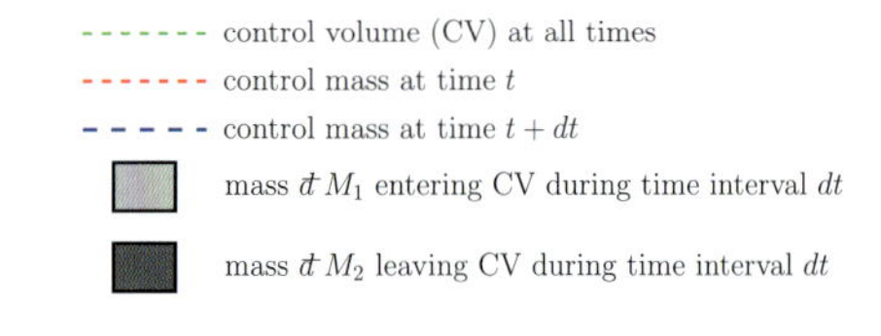

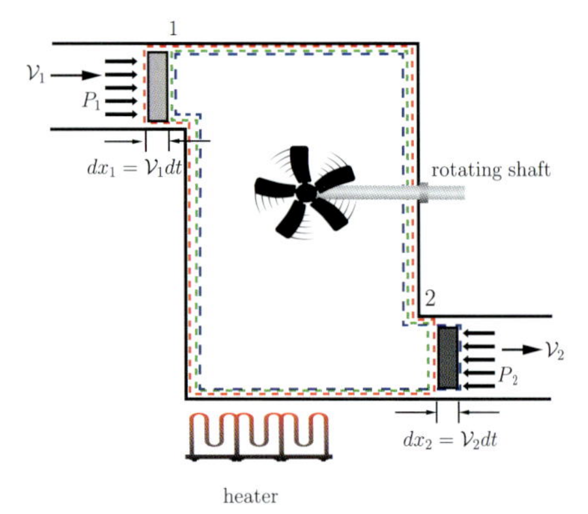

(a) Device to be analyzed.

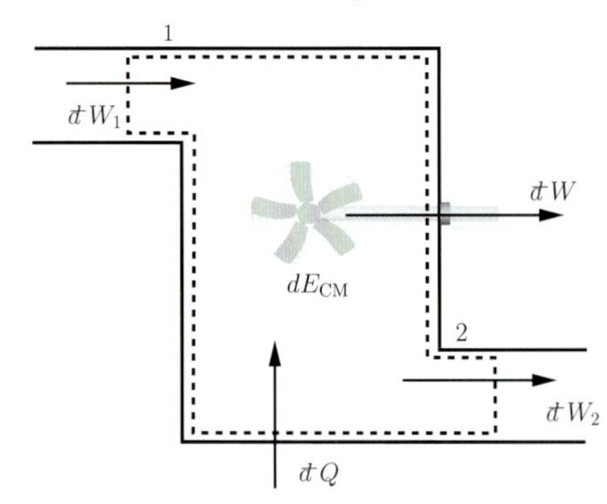

(b) Energy balance on the control mass.

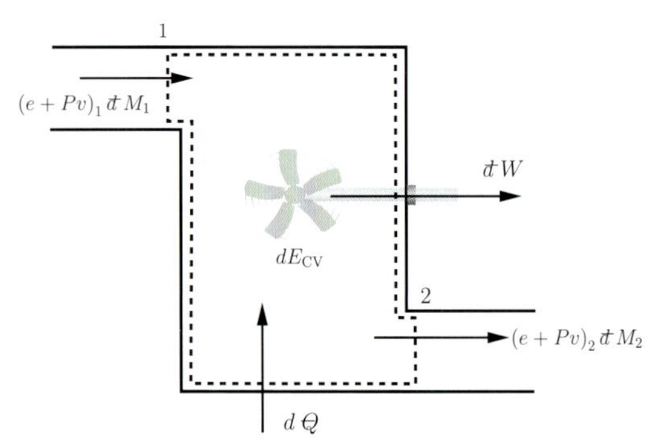

(c) Energy balance on the control volume.

Figure 4.2 Generalized control volume energy analysis.

The control mass

This time we will examine the energy transfers that take place over a small time interval dt. We will analyze a control mass consisting of all the mass that is in the CV at time t plus the mass $\text{đ}M_1$ upstream of section 1 that will enter the CV over the time interval dt, as shown in Figure 4.2b by the CM boundary at the initial time. The boundary of this CM moves as fluid flows through the device, and at time $t + dt$ it fills the CV and extends downstream past section 2 to include $\text{đ}M_2$, as shown in Figure 4.2c.

Mass balance on the control mass

By definition, there are no inputs or outputs of mass to or from the CM, so the mass M_{CM} is constant,

$$\underbrace{dM_{\text{CM}}}_{\substack{\text{mass accumulation}\\ \text{inside CM}}} = M_{\text{CM}}(t + dt) - M_{\text{CM}}(t) = 0. \quad (4.3)$$

In order to express (4.3) in terms of quantities pertaining to the CV, we note that initially the CM consists of the CV plus the little piece of mass $\text{đ}M_1$ that has not yet entered the CV,

$$M_{\text{CM}}(t) = M_{\text{CV}}(t) + \text{đ}M_1, \quad (4.4)$$

where $M_{\text{CV}}(t)$ denotes the mass in the CV at time t. At time $t + dt$ the CM consists of the CV plus the little piece $\text{đ}M_2$ that left the CV during time interval dt,

$$M_{\text{CM}}(t + dt) = M_{\text{CV}}(t + dt) + \text{đ}M_2. \quad (4.5)$$

Substituting (4.4) and (4.5) in (4.3), we obtain

$$M_{\text{CV}}(t + dt) + \text{đ}M_2 - \left[M_{\text{CV}}(t) + \text{đ}M_1\right] = 0,$$

which we rearrange, setting $dM_{\text{CV}} = M_{\text{CV}}(t + dt) - M_{\text{CV}}(t)$, and interpret as the mass balance on the CV:

$$\underbrace{\text{đ}M_1}_{\substack{\text{mass}\\ \text{input}\\ \text{to CV}}} = \underbrace{\text{đ}M_2}_{\substack{\text{mass}\\ \text{output}\\ \text{from CV}}} + \underbrace{dM_{\text{CV}}}_{\substack{\text{mass}\\ \text{accumulation}\\ \text{inside CV}}}. \quad (4.6)$$

Mass balance, rate basis

We divide (4.6) by dt, and define the mass flow rates at the inlet and outlet as follows (remember, in this book an overdot always denotes a *rate of transfer*);

$$\dot{M}_1 \equiv \frac{đ M_1}{dt}, \qquad \dot{M}_2 \equiv \frac{đ M_2}{dt}.$$

Thus we obtain the CV *mass balance on a rate basis,*

$$\underbrace{\dot{M}_1}_{\substack{\text{rate of}\\ \text{mass}\\ \text{input}}} = \underbrace{\dot{M}_2}_{\substack{\text{rate of}\\ \text{mass}\\ \text{output}}} + \underbrace{\frac{dM_{CV}}{dt}}_{\substack{\text{rate of}\\ \text{mass}\\ \text{accumulation}}}. \tag{4.7}$$

This equation should be intuitively obvious if you understand the concept of conservation of mass, and in the future you should not have to do all this work to write down the mass balance on a control volume.

Energy transfers across the control mass boundary

Energy is transferred across the CM boundary in four places. We denote the energy input from the heater during interval dt by $đ Q$, and the energy output as work in the shaft during this period by $đ W$. In addition, there is energy transfer to the CM as work from the fluid outside the CM that pushes it down the inlet pipe. The small amount of work done on the CM when $đ M_1$ is pushed across section 1 is

$$đ W_1 = (P_1 A_1)\, dx_1,$$

which may be rewritten in terms of the pressure P_1, the specific volume $v_1 = 1/\rho_1$, and the mass $đ M_1 = \rho_1 A_1 dx_1$ as

$$đ W_1 = (P_1 A_1)\, dx_1 = P_1\, v_1 \rho_1\, A_1\, dx_1 = P_1 v_1\, đ M_1. \tag{4.8a}$$

Likewise, the CM does work on the fluid downstream as it pushes this fluid down the discharge pipe, and over time interval dt this work is

$$đ W_2 = (P_2 A_2)\, dx_2 = P_2 v_2 đ M_2. \tag{4.8b}$$

These four energy transfers across the CM boundary are shown in Figure 4.2b.

Energy balance on the control mass

Over the time interval dt, the energy balance on the CM is then

$$\underbrace{đ W_1 + đ Q}_{\substack{\text{energy}\\ \text{input}\\ \text{to CM}}} = \underbrace{đ W_2 + đ W}_{\substack{\text{energy}\\ \text{output}\\ \text{from CM}}} + \underbrace{E_{CM}(t + dt) - E_{CM}(t)}_{\substack{\text{energy}\\ \text{accumulation}\\ \text{inside CM}}}, \tag{4.9}$$

where $E_{CM}(t)$ denotes the energy of everything inside the control mass at time t.

Energy balance on the control volume

In order to express (4.9) in terms of quantities applying to the CV, we note that $đ Q$ and $đ W$ cross the common boundary of the CM and CV and so can be regarded as a transfer to the CV. Then, we note that the CM energy at time t consists of the energy in the CV at time t plus the energy in the little piece $đ M_1$ that has not yet entered the CV, that is

$$E_{CM}(t) = E_{CV}(t) + e_1(t) đ M_1. \tag{4.10}$$

Here $e = u + \frac{1}{2}\mathcal{V}^2 + gz$ is the total energy per unit mass of fluid (internal energy plus kinetic energy of bulk motion plus potential energy in the gravitational field) at point 1, evaluated at time t. The energy of the CM at time $t + dt$ is the sum of the CV energy at that time plus the energy of the little mass $đ M_2$,

$$E_{CM}(t + dt) = E_{CV}(t + dt) + e_2(t + dt) đ M_2, \tag{4.11}$$

where $E_{CV}(t)$ denotes the energy of everything inside the CV at time t, and $e_2(t + dt)$ is the energy per unit mass of the fluid at section 2, evaluated at time $t + dt$. Now, $e_2(t + dt) = e_2(t)$ plus a possible infinitesimal change over the time interval dt, and hence to first order in the infinitesimals

$$E_{CM}(t + dt) = E_{CV}(t + dt) + e_2(t) đ M_2. \tag{4.12}$$

Substituting (4.10) and (4.11) into (4.9), we obtain

$$\begin{aligned}(Pv)_1\, đ M_1 + đ Q = (Pv)_2\, đ M_2 + đ W \\ + \left[E_{CV}(t + dt) + e_2 đ M_2\right] \\ - \left[E_{CV}(t) + e_1 đ M_1\right],\end{aligned}$$

which we rearrange and interpret as an energy balance on the CV,

$$\begin{aligned}\underbrace{(e + Pv)_1 đ M_1}_{\substack{\text{mass-associated}\\ \text{energy input}}} + \underbrace{đ Q}_{\substack{\text{other}\\ \text{energy}\\ \text{input}}} = \underbrace{(e + Pv)_2 đ M_2}_{\substack{\text{mass-associated}\\ \text{energy output}}} + \underbrace{đ W}_{\substack{\text{other}\\ \text{energy}\\ \text{output}}} \\ + \underbrace{E_{CV}(t + dt) - E_{CV}(t)}_{\substack{\text{energy accumulation}\\ \text{inside CV}}}.\end{aligned} \tag{4.13}$$

Flow work

We see that the energy associated with mass transfer across the boundary is the sum of the energy e *convected* across the boundary by the mass, plus Pv (per unit mass), which represents the work done on (at 1) or by (at 2) the CV in order to push the mass across the CV boundary. The Pv term in a CV energy balance is called the *flow work*, and it will arise wherever mass crosses the CV boundary.

Enthalpy and mass-associated energy transfer

The total mass-associated energy transfer can be expressed as the sum of the enthalpy and the bulk kinetic and potential energies per unit mass,

$$e + Pv = u + Pv + \frac{\mathcal{V}^2}{2} + gz = h + \frac{\mathcal{V}^2}{2} + gz. \quad (4.14)$$

The enthalpy $h \equiv u + Pv$ always arises in CV energy balances wherever mass crosses the CV boundary, and that is why it is such an important thermodynamic property.

Enthalpy does not accumulate!

The enthalpy is *not* involved in the energy accumulation term, because the energy stored in the CV at any instant is the sum of the e (not h) of all the material inside. The previous example illustrates this point very clearly. Enthalpy is mass-associated energy *transfer*, but it is *not* mass-associated energy *storage*. Associating h (instead of e or u) with energy storage is an error often made by beginning students, which is why this point is emphasized.

Energy balance, rate basis

Dividing (4.13) by dt, the time interval of the analysis, we obtain the control volume energy balance *on a rate basis* (the *power* balance)

$$\underbrace{\dot{M}_1(e + Pv)_1 + \dot{Q}}_{\text{rate of energy input}} = \underbrace{\dot{M}_2(e + Pv)_2 + \dot{W}}_{\text{rate of energy output}} + \underbrace{\frac{dE_{CV}}{dt}}_{\text{rate of energy accumulation}}. \quad (4.15)$$

Here we define the rates of energy transfer by

$$\dot{Q} \equiv \frac{đQ}{dt}, \qquad \dot{W} \equiv \frac{đW}{dt}.$$

They are also often termed *thermal power* and *mechanical power*.

Steady-state assumption

Often in engineering analysis one models a system as being in *steady-state*, which means that the state everywhere inside the system is unchanging with time. Of course the state of the fluid passing through the system may change, but the state at each fixed point does not. That means that the mass and energy of the CV do not change in time, and so these terms can be dropped out from the mass and energy balances (4.7) and (4.15). The rate-basis balances on the control volume then become

$$\underbrace{\dot{M}_1}_{\text{rate of mass input}} = \underbrace{\dot{M}_2}_{\text{rate of mass output}}, \quad (4.16)$$

$$\underbrace{\dot{M}_1(e + Pv)_1 + \dot{Q}}_{\text{rate of energy input}} = \underbrace{\dot{M}_2(e + Pv)_2 + \dot{W}}_{\text{rate of energy output}}. \quad (4.17)$$

Note that these balances involve only terms associated with transfers across the CV boundary, and *no information about the detailed state inside the CV is required*. The steady-state assumption can usually be used for a system that has been warmed up and brought up to steady operating speed. It cannot be used if the changes in mass or energy inside the CV are important, as in the tank example of Section 4.2.

Multiple inputs and outputs

If a control volume has more than the single inputs and outputs considered above, the mass and energy balances will involve all these flows, with each mass flow having a mass-associated energy transfer rate of the form $\dot{M}(e + Pv)$. If you understand why $e + Pv$ is the mass-associated energy transfer, you are ready to do control volume analysis directly, without going through the underlying control mass analysis.

4.4 General Methodology for Energy Analysis

In Section 2.7 we gave a general methodology for energy balance analysis. This is repeated here as it applies in control volume analysis.

1. Draw a diagram identifying the control volume to be analyzed by enclosing it within dotted lines; if the boundary moves, explain what happens to the boundary over time so that both you and the reader are very clear about it.
2. If the control volume moves, show the forces acting on the control volume that you will use in the evaluation of energy transfer as work. Do not show the flow work terms that are included in the enthalpy, or you might include them twice!
3. If the control volume moves, identify the inertial reference frame for evaluation of macroscopic kinetic and potential energy.
4. List all simplifying idealizations.
5. Show all significant energy transfers on your sketch to define the symbols, using arrows to indicate the directions chosen for *positive* energy transfer, and show the energy accumulation if the analysis is not for steady-state.
6. State whether the energy balance is over some definite *time interval*, on a *rate basis*, or on a *per unit mass flow* basis.
7. Write the energy balance; there should be a one-to-one correspondence between the terms in your equation and those on your diagram.
8. Bring in process, state, or other information as necessary to complete the analysis.

Importance of system boundaries

Since energy transfers that occur within the control volume do not cross the control volume boundary, they are not involved in the energy balance. That is why it is so important to have a clear definition of the boundary. Choose the boundary wisely:

- Put the boundary where you know the transfer or want to know the transfer.
- Never put the boundary where you don't know the transfer and don't need to know the transfer!

Unsteady-state balances

If the steady-state idealization is not valid, it is usually best to develop the energy and mass balances on a *rate basis*. You can then integrate these over time to get the balances over a finite time period.

4.5 Example: Supersonic Nozzle

Consider the acceleration of gas from a high pressure and low speed to a lower pressure and very high speed, as for example occurs in the nozzle of a supersonic wind tunnel or in the exhaust nozzle of a rocket (Figure 4.3). The flow enters at section 1 at low (subsonic) speed and leaves at section 2 at very high (supersonic) speed. We define the control volume as the flow region between sections 1 and 2.

Idealizations

We assume that the control volume can be treated as *adiabatic* (negligible heat transfer between the gas and nozzle wall), and that the system is operating at *steady-state* so that no changes occur in the mass

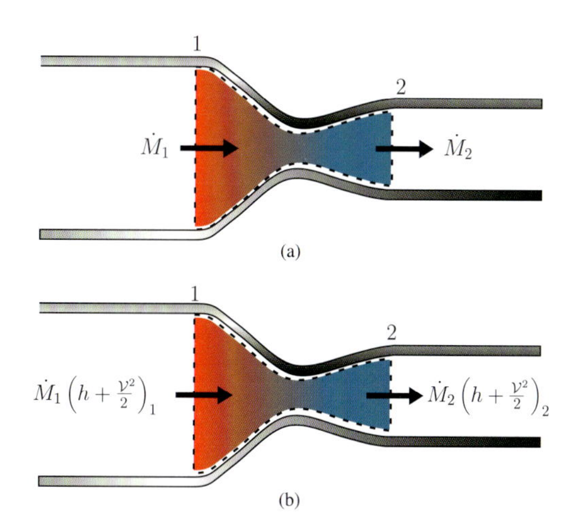

Figure 4.3 Mass balance terms (a) and energy balance terms (b) for a nozzle.

or energy stored within the control volume. The kinetic energy of the moving fluid will be important and must be considered, but we will neglect changes in the gravitational potential energy of the fluid. Friction effects would be confined to a very thin boundary layer on the nozzle surface, which we will neglect by assuming that the velocity and thermodynamic properties do not vary across sections 1 and 2, i.e., that the flow is *one-dimensional* at points 1 and 2.

Mass balance

We use overdots to indicate mass flow rates (Figure 4.3a). Under the steady-state assumption, the mass in the control volume does not change with time, and so the mass balance on a rate basis reduces to a balance of the inflow rate by the outflow rate,

$$\underbrace{\dot{M}_1}_{\text{rate of mass input}} = \underbrace{\dot{M}_2}_{\text{rate of mass output}} + \underbrace{0}_{\text{rate of mass accumulation}}.$$

The mass balance tells us that the mass flow rates in and out of the control volume are identical, as you must have expected, so we can drop the subscripts on the mass flow rates and denote the mass flow rate through the nozzle just by $\dot{M}$.

Energy balance

The rate-basis energy balance on the control *volume* (Figure 4.3) is

$$\underbrace{\dot{M}(e_1 + P_1 v_1)}_{\text{rate of energy input}} = \underbrace{\dot{M}(e_2 + P_2 v_2)}_{\text{rate of energy output}} + \underbrace{0}_{\text{rate of energy accumulation}}, \tag{4.18}$$

Energy balance per unit mass

Dividing (4.18) by $\dot{M}$, we obtain the *energy balance per unit mass flow*,

$$(e + Pv)_1 = (e + Pv)_2\,. \tag{4.19}$$

Since here $e = u + \mathcal{V}^2/2$, and by definition $h \equiv u + Pv$, (4.19) is simply

$$\left(h + \frac{\mathcal{V}^2}{2}\right)_1 = \left(h + \frac{\mathcal{V}^2}{2}\right)_2.$$

Further simplification

If we assume that the kinetic energy of the inlet flow is much less than that of the supersonic discharge ($\mathcal{V}_1^2 << \mathcal{V}_2^2$), we can express the discharge velocity in terms of the fluid enthalpy change,

$$\mathcal{V}_2 = \sqrt{2\,(h_1 - h_2)}.$$

This shows that the nozzle is a device in which some of the random molecular energy is converted to organized kinetic energy of the flow. If we now add the ideal gas model, with $h_2 - h_1 = c_P\,(T_2 - T_1)$, we see that the lower we can get the temperature at 2 the higher will be the flow velocity $\mathcal{V}_2$. The lowest we could possibly get T_2 is absolute zero, so the maximum velocity we can possibly get is

$$\mathcal{V}_{2,\max} = \sqrt{2 \cdot c_P \cdot T_1}.$$

This shows that if we want very high velocity we must have very hot gas upstream. Notice how this simple energy balance, together with information on the fluid properties, gives an important and simple theoretical result that is of great practical importance in engineering design, a situation that illustrates the power and utility of thermodynamics.

4.6 Example: Hydraulic Turbine

Consider the hydraulic turbine of Figure 4.4. Water enters the turbine inlet pipe from a reservoir high in the mountains and flows down to the turbine. In the turbine the water is first accelerated in nozzles to high speed. The water then impinges on the buckets of the turbine rotor, causing the rotor to turn. The rotor in essence converts most of the kinetic energy of the water into shaft work. The water leaves the turbine at relatively low velocity through the outlet pipe. Energy is transferred as work through the rotating turbine shaft to an electrical generator, where

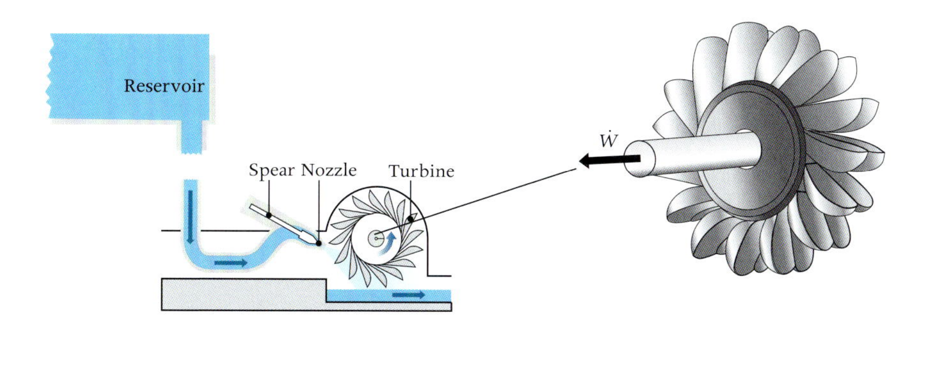

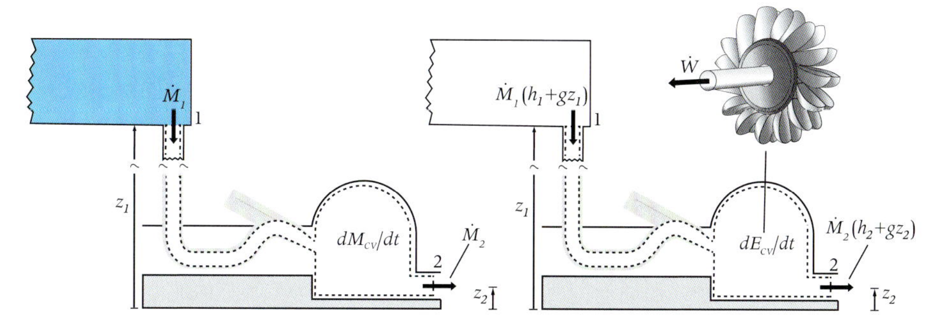

Figure 4.4 Mass balance terms (a) and energy balance terms (b) for a hydraulic turbine.

it is converted to electrical energy for transmission. We define the control volume as the region between sections 1 and 2.

Idealizations

We will make the following idealizations:

1. One-dimensional flow at sections 1 and 2.
2. No significant energy transfer as heat.
3. Flow kinetic energy at 1 and 2 negligible.

If the fluid is idealized as *incompressible* (constant density) and the volume of the control volume is fixed, then the mass in the control volume will be constant. However, if the turbine speed or the flow rate are changing, then the energy stored in the control volume will not be constant. We will allow for changes in both the mass and energy in the control volume in writing the mass and energy balances. We will (as always) carefully distinguish between rates of flow, which we denote by overdots, and rates of change, which we express as derivatives.

Mass balance

The rate-basis mass balance on the control volume (Figure 4.4a) is given by

$$\underbrace{\dot{M}_1}_{\substack{\text{rate of}\\ \text{mass input}}} = \underbrace{\dot{M}_2}_{\substack{\text{rate of}\\ \text{mass output}}} + \underbrace{\frac{dM_{CV}}{dt}}_{\substack{\text{rate of mass}\\ \text{accumulation}}} .$$

As discussed above, the rate of mass accumulation in the control volume will be zero if the fluid is incompressible and the volume is fixed.

Energy balance

The rate-basis energy balance on the control volume is (Figure 4.4b)

$$\underbrace{\dot{M}_1\left(h_1+gz_1\right)}_{\substack{\text{rate of}\\ \text{energy input}}} = \underbrace{\dot{M}_2\left(h_2+gz_2\right)+\dot{W}}_{\substack{\text{rate of}\\ \text{energy output}}} + \underbrace{\frac{dE_{CV}}{dt}}_{\substack{\text{rate of energy}\\ \text{accumulation}}} ,$$

where $\dot{W}$ is the *rate* of energy transfer as work out the shaft.

Control volume energy change

The control volume energy E_{CV} will be the sum of the internal, potential, and kinetic energy of everything inside the control volume. If the turbine is changing speed or warming up, the change in E_{CV} may be important. If the system is at steady-state, $dE_{CV}/dt = 0$, and we do not have to evaluate E_{CV} in order to use the control volume energy balance.

Friction effects

Does the analysis neglect friction? To answer this question, consider the work done by the flow friction force on the long downcomer pipe. This force might be quite large if the pipe is long, but since the pipe does not move the frictional work is zero. The effects of friction would show up as an increase in the internal energy of the water ($u_2 > u_1$), and we have not assumed that this is zero. So we have indeed allowed for friction.

Power output and design calculations

Under the steady-state assumption, $\dot{M}_1 = \dot{M}_2$, and the energy balance reduces to

$$\dot{W} = \dot{M}[g(z_1 - z_2) + (h_1 - h_2)].$$

In a design situation one would know the two elevations z_1 and z_2, and so if the enthalpy change $h_2 - h_1$ were known then the mass flow rate required for a given power output could be determined from this equation. From the Second Law of Thermodynamics we will learn that this enthalpy change is zero for an ideal (100% efficient) hydraulic turbine, and so if we have an idea about what sort of efficiency real turbines have we can easily determine the required flow for a given power. This is the way that thermodynamic analysis is used in engineering design.

4.7 Example of System Analysis: Heat Pump

We will illustrate the methodology and provide an interesting set of examples by considering the thermodynamic analysis of a realistic heat pump. This example will illustrate the analysis of three types of components and give you an early glimpse of a thermodynamic system analysis of the sort that you will be able to do in more detail with just a little more knowledge.

The system

Figure 4.5a shows the system (or process) flow diagram, and Figure 4.5b shows the process representation. The working fluid is refrigerant 134a.

> Refrigerant 134a (1,1,1,2-tetrafluoroethane ($C_2H_2F_4$)) is one of several synthetic chemicals developed in the twentieth century that was highly welcomed because it was very safe compared with other refrigerants used in home refrigerators in earlier days. In the past, refrigerants containing chlorine, like R22 ($CHClF_2$), were used also because they allowed high conversion efficiency. Unfortunately, in the late twentieth century it was discovered that these chlorine-containing fluids were damaging the Earth's ozone layer, thus they were replaced by refrigerants like R134a. However, since the beginning of the twenty-first century it has been known that most fluids containing fluorine atoms contribute to the global warming of the Earth (greenhouse effect). As a consequence, refrigerants like R22 are now only used in special applications under very controlled conditions, while fluids like R134a are to be phased out in the near future because of their global warming effect. Currently, research on suitable replacements is therefore very actively pursued.

The fluid enters the compressor as a saturated vapor (state 1), and is compressed to a higher pressure and temperature. Next it flows through a heat exchanger (condenser), where it is condensed to a saturated liquid (state 3) by transferring energy as heat to the household air. The fluid then flows through a flow restriction (valve), which causes a controlled pressure drop. Since the liquid is saturated at state 3, the drop in pressure causes some of the liquid to evaporate; a kilogram of vapor at a given pressure has more energy than a kilogram of liquid, and the extra

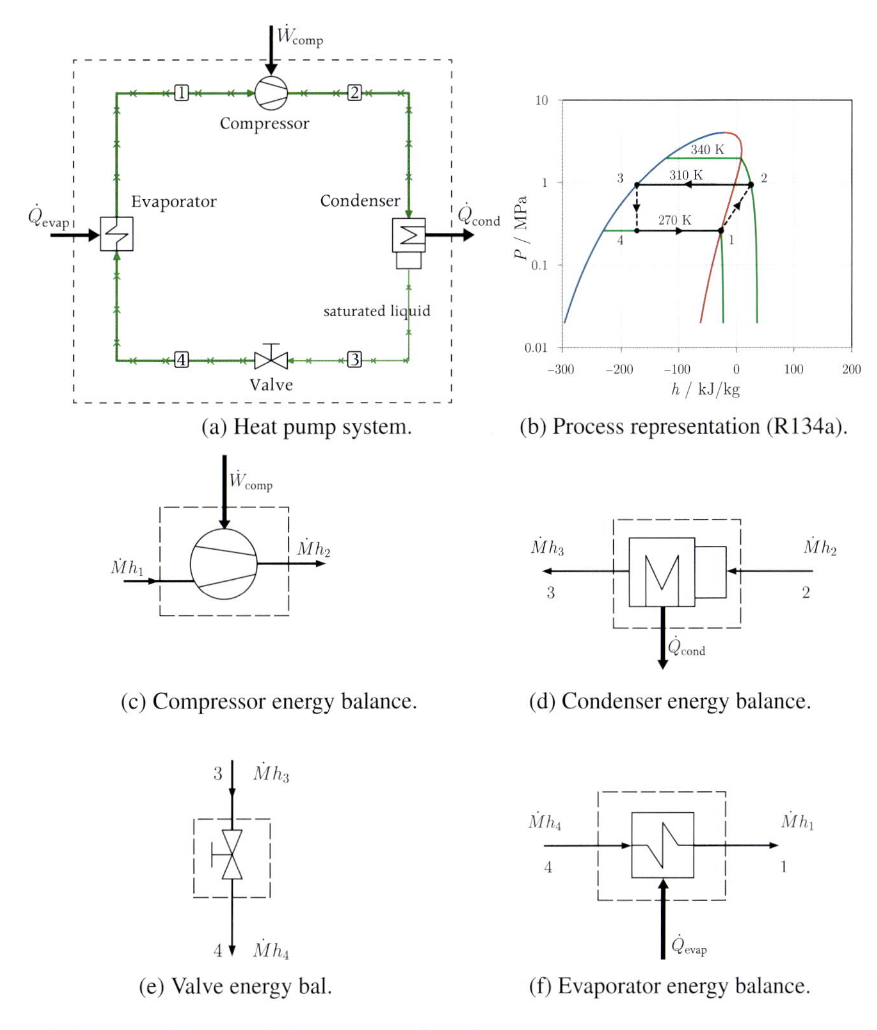

Figure 4.5 Mass balance and energy balance terms for a heat pump system.

energy is taken from the remaining liquid, with the result that the temperature of the liquid–vapor mixture drops significantly. The cold mixture (state 4) then enters another heat exchanger (the evaporator), where it is heated back to the initial saturated vapor state (1) by heat transfer from the ambient air. This completes the fluid cycle, which is repeated endlessly in steady-state operation.

Operating conditions

We specify that the evaporation will occur at 270 K. The energy transfer as heat comes from the outdoor air, which must be at a somewhat higher temperature (perhaps 280 K, or 7 °C). We also specify that condensation occurs at 310 K, which will be slightly above the temperature of the house air being heated (perhaps 300 K, or 27 °C). This is almost all we need to specify in order to determine all four states; we are missing information that will fix state 2, for which we need to know more about entropy. So for the purposes of this example we will specify that state 2 is at 340 K.

Common idealizations

Throughout the analysis we will employ three idealizations:

1. Steady-state.
2. One-dimensional at state points.
3. Negligible kinetic and potential energy at state points.

The *steady-state* idealization means that there are no changes in the mass or energy stored within the control volume under study. This idealization eliminates the accumulation terms from the mass and energy balances, and greatly simplifies the analysis because nothing need be known about the details inside the control volume. The *one-dimensional* idealization means that the only dimension in which properties vary is in the direction of flow, and hence that every element of fluid that passes that point has the same state; the flow certainly is not one-dimensional within any of the components, but idealization is made only at the intercomponent states. We assume *negligible kinetic and potential energy* at the intercomponent states because these energy forms will not be very important (compared to the dominant energy forms). The kinetic energy certainly is important within the valve, and possibly within the compressor, but not typically at the intercomponent states. In a later stage of the design analysis, when the flow rates and pipe sizes are known, we could determine the kinetic and potential energies at the intercomponent states and include them in the energy balance analysis; they would have very little effect in a system like this.

Heat exchanger process model

We shall idealize that the heat exchanger processes occur at *constant pressure*. The assumption of constant pressure is reasonable for a heat exchanger because the fluid in the exchanger is free to move about and equalize any pressure differences that may develop (friction causes a slight pressure drop that we could consider in a more refined analysis).

Compressor analysis

For the compressor analysis we make the following idealizations:

1. Steady-state.
2. One-dimensional at 1 and 2.
3. Negligible kinetic and potential energy at 1 and 2.
4. Negligible heat transfer.

Heat transfer is neglected because the compressor is designed to raise the pressure and will be insulated to guard against heat transfer. Figure 4.5c shows the control volume and energy terms that are important in this problem. The energy balance, on a rate basis, is

$$\underbrace{\dot{M}h_1 + \dot{W}_{\text{comp}}}_{\substack{\text{rate of} \\ \text{energy input}}} = \underbrace{\dot{M}h_2}_{\substack{\text{rate of} \\ \text{energy output}}},$$

where $\dot{W}_{\text{comp}}$ is the compressor shaft power input (provided by an electric motor), and $\dot{M}$ is the R134a mass flow rate throughout the system. Note that we have used the results of a mass balance on the compressor in saying that $\dot{M} \equiv \dot{M}_1 = \dot{M}_2$. Thus, the compressor power is given by

$$\dot{W}_{\text{comp}} = \dot{M}(h_2 - h_1).$$

Dividing by $\dot{M}$, we find the *compressor work per unit of mass flow* (or *specific work*),

$$w_{\text{comp}} \equiv \frac{\dot{W}_{\text{comp}}}{\dot{M}} = h_2 - h_1.$$

The operating condition specifications fix states 1 and 2. From the R134a saturation table, h_1 is read as the enthalpy of saturated vapor at 270 K, or $h_1 = -26.99$ kJ/kg. Condensation occurs at 310 K, for which the pressure is read from the saturation tables as 0.9343 MPa. State 2 is fixed by the intersection of this pressure line and the 340 K isotherm. From the R134a chart we read $h_2 = 25$ kJ/kg. Therefore, the compressor work per unit mass is

$$w_{\text{comp}} = (25 - -27)\ \text{kJ/kg} = 52\ \text{kJ/kg}.$$

Condenser analysis

For the condenser analysis we make the following idealizations:

1. Steady-state.
2. One-dimensional at 2 and 3.
3. Negligible kinetic and potential energy at 2 and 3.
4. Constant pressure process.

Figure 4.5d shows the control volume and energy flows for this analysis. The energy balance, on a rate basis, is

$$\underbrace{\dot{M}h_2}_{\substack{\text{rate of}\\ \text{energy input}}} = \underbrace{\dot{M}h_3 + \dot{Q}_{\text{cond}}}_{\substack{\text{rate of}\\ \text{energy output}}},$$

so

$$\dot{Q}_{\text{cond}} = \dot{M}(h_2 - h_3).$$

Dividing by $\dot{M}$, the *condenser heat transfer per unit of mass flow* is

$$q_{\text{cond}} \equiv \frac{\dot{Q}_{\text{cond}}}{\dot{M}} = h_2 - h_3.$$

State 3 is saturated liquid at 310 K, for which the saturation tables give $h_3 = -173.55$ kJ/kg. Therefore,

$$q_{\text{cond}} = (25 - -174) = 199 \text{ kJ/kg}.$$

Coefficient of performance

We now know enough to determine the coefficient of performance,

$$\text{COP} \equiv \frac{\dot{Q}_{\text{cond}}}{\dot{W}_{\text{comp}}} = \frac{q_{\text{cond}}}{w_{\text{comp}}} = 199/52 = 3.8.$$

Again we see the advantage of heating your house with a heat pump; the energy we get into the house will be more than five times the energy that we have to buy from the utility company to run the compressor!

Sizing the system

If we now specify the house heating rate $\dot{Q}_{\text{cond}}$ we can determine the mass flow rate. For example, if $\dot{Q}_{\text{cond}} =$ 40 kW, then

$$\dot{M} = \frac{\dot{Q}_{\text{cond}}}{q_{\text{cond}}} = \frac{40 \text{ kJ/s}}{199 \text{ kJ/kg}} = 0.20 \text{ kg/s}.$$

The compressor power requirement is then

$$\dot{W}_{\text{comp}} = \dot{M}w_{\text{comp}} = 0.20 \text{ kg/s} \times 52 \text{ kJ/kg} = 10.4 \text{ kW}.$$

If this is all we wanted to know in the preliminary analysis, we are done. In the interests of completeness we will analyze the two remaining components and then check our work with an overall energy balance.

Valve analysis

For the valve analysis we make the following idealizations:

1. Steady-state.
2. One-dimensional at 3 and 4.
3. Negligible kinetic and potential energy at 3 and 4.
4. Negligible heat transfer.

The control volume and energy diagram for the valve is shown in Figure 4.5e. The energy balance, on a rate basis, is

$$\underbrace{\dot{M}h_3}_{\substack{\text{rate of}\\ \text{energy input}}} = \underbrace{\dot{M}h_4}_{\substack{\text{rate of}\\ \text{energy output}}},$$

which gives

$$h_3 = h_4.$$

Note that we anticipated this result in drawing the process representation, Figure 4.5b.

Evaporator analysis

For the evaporator analysis we make the following idealizations:

1. Steady-state.
2. One-dimensional at 4 and 1.
3. Negligible kinetic and potential energy at 4 and 1.
4. Constant pressure process.

Figure 4.5f shows the control volume and energy flows for this analysis. The energy balance, on a rate basis, is

$$\underbrace{\dot{M}h_4 + \dot{Q}_{\text{evap}}}_{\substack{\text{rate of}\\ \text{energy input}}} = \underbrace{\dot{M}h_1}_{\substack{\text{rate of}\\ \text{energy output}}},$$

so

$$\dot{Q}_{\text{evap}} = \dot{M}(h_1 - h_4).$$

Dividing by $\dot{M}$, the *evaporator heat transfer per unit of mass flow* is

$$q_{\text{evap}} \equiv \frac{\dot{Q}_{\text{evap}}}{\dot{M}} = h_1 - h_4.$$

Since $h_4 = h_3 = -174$ kJ/kg, and $h_1 = -27$ kJ/kg,

$$q_{\text{evap}} = (-27 - -174) \text{ kJ/kg} = 147 \text{ kJ/kg}.$$

Overall energy balance check

We can check our work with an overall energy balance on the control volume shown in Figure 4.5a, for which we make the following idealizations:

1. Steady-state.
2. Compressor and valve are adiabatic.

The energy balance, on a rate basis, is then

$$\underbrace{\dot{Q}_{\text{evap}} + \dot{W}_{\text{comp}}}_{\substack{\text{rate of} \\ \text{energy input}}} = \underbrace{\dot{Q}_{\text{cond}}}_{\substack{\text{rate of} \\ \text{energy output}}} .$$

Dividing by the mass flow rate, we get the overall energy balance on a *per unit mass basis*,

$$\underbrace{q_{\text{evap}} + w_{\text{comp}}}_{\text{energy input}} = \underbrace{q_{\text{cond}}}_{\text{energy output}} .$$

Using the numbers computed above,

$$(147 + 52)\ \text{kJ/kg} \equiv 199\ \text{kJ/kg},$$

which checks, indicating that we did our arithmetic correctly. In doing a complex system analysis, it is always a good idea to make checks like this to establish confidence in your work.

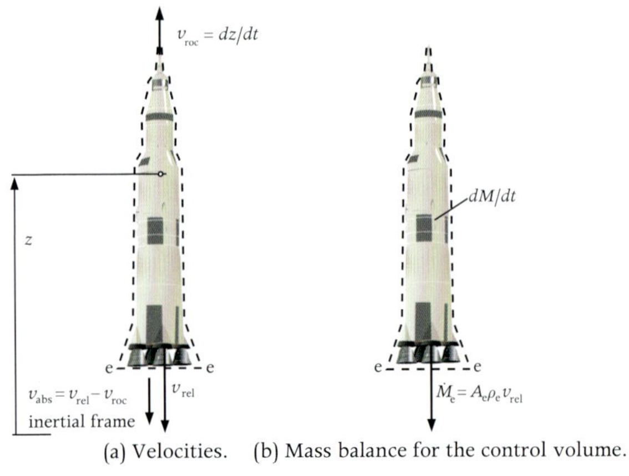

(a) Velocities. (b) Mass balance for the control volume.

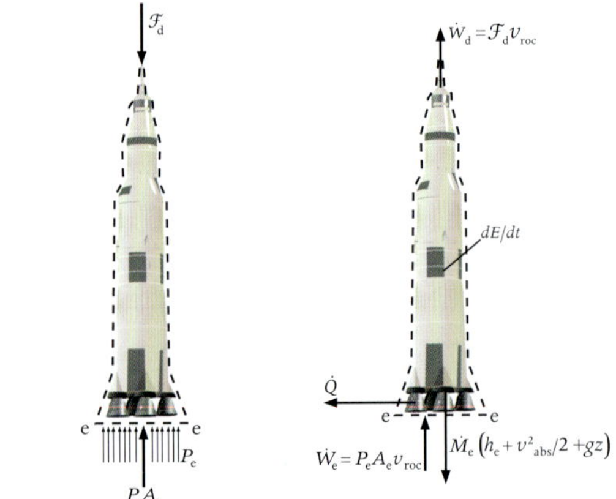

(c) Forces on the control volume. (d) Energy balance on the control volume.

Figure 4.6 Rocket analysis: an unsteady, moving control volume.

4.8 Example with Unsteady and Moving Control Volume: Rocket

If the control volume moves with respect to the inertial reference frame, one must be very careful to use the correct flow velocities at various points in the analysis. The energy analysis of an ascending rocket provides an excellent example. Figure 4.6a shows the control volume, defined by the outer surface of the rocket and the nozzle exit plane e–e. We make the following simplifying idealizations:

1. The Earth is a suitable inertial reference frame.
2. The motion is entirely vertical.
3. Changes in the gravitational acceleration g are negligible.
4. The flow in the nozzle exit plane is one-dimensional.
5. The amount of mass inside the control volume moving at a speed other than $\mathcal{V}_{\text{roc}}$ is negligible.
6. The variations in gravitational potential energy per unit mass within the control volume are negligible.

Note that we do *not* assume steady-state because the energy and mass contained within the control volume both vary with time.

Relevant velocities

Assuming that the Earth is a suitable inertial plane (it might be for the launch phase, but certainly is not for interplanetary flight), the upward velocity of the control volume relative to the inertial frame is

$\mathcal{V}_{\text{roc}} = dz/dt$. The downward velocity of the ejected fluid, relative to the nozzle exit plane, is $\mathcal{V}_{\text{rel}}$. For vertical flight, the absolute downward velocity of the ejected fluid in the inertial frame is $\mathcal{V}_{\text{abs}} = \mathcal{V}_{\text{rel}} - \mathcal{V}_{\text{roc}}$. We must be careful to use the proper velocity in different terms in the analysis, as described below.

Mass balance

Figure 4.6b shows the terms in the mass balance on the control volume. Note that the mass flow rate at the exit is evaluated in terms of the fluid velocity *relative* to the boundary, ($\dot{M}_e = A_e \rho_e \mathcal{V}_{\text{rel}}$). The control volume mass balance, on a rate basis, is

$$\underbrace{0}_{\substack{\text{rate of}\\ \text{mass inflow}}} = \underbrace{\dot{M}_e}_{\substack{\text{rate of}\\ \text{mass outflow}}} + \underbrace{\frac{dM}{dt}}_{\substack{\text{rate of mass}\\ \text{accumulation}}},$$

where M is the instantaneous mass inside the control volume; note that $dM/dt < 0$.

Mass-associated energy transfer

As demonstrated in the tank charging example (Section 4.2), the enthalpy h_e represents the sum of internal energy carried out of the control volume per unit mass of fluid, plus the work done by the fluid inside the control volume in expelling the mass against the exit pressure P_e. To h_e we must add the kinetic and potential energy of the fluid, *as measured in the inertial reference frame* (kinetic energy is only $M\mathcal{V}^2/2$ in an *inertial* frame). Therefore, the total mass-associated energy transfer out of the control volume (per unit mass) is $h_e + \mathcal{V}_{\text{abs}}^2/2 + gz_e$.

Forces and energy transfers as work

Figure 4.6c shows the forces acting on the control volume. The energy transfers associated with these forces must be carefully considered. The drag force $\mathcal{F}_d$ is the net pressure and shear force acting on the outer skin of the rocket, opposing the rocket motion. As the rocket moves against this force, it does work on the air at the rate $\dot{W}_d = \mathcal{F}_d \mathcal{V}_{\text{roc}}$, as shown in Figure 4.6d. Since the boundary moves with respect to the reference frame, there will be energy transfer as work *to* the control volume as a result of the exit plane pressure force $P_e A_e$ acting through the control volume boundary displacement. The energy transfer due to the pressure pushing fluid across the exit plane is included in the enthalpy term as discussed above. If you have difficulty seeing this, do the analysis by the control mass method used in the first three examples, and verify that you obtain the control volume equations written here. The energy transfer rate associated with the pressure acting on the displaced boundary is thus $\dot{W}_e = P_e A_e \mathcal{V}_{\text{roc}}$ (see Figure 4.6d). The energy transfer associated with the gravitational force is considered by the inclusion of potential energy in the energy transfer and energy accumulation terms.

Energy balance

The energy balance on the control volume, on a rate basis, is (see Figure 4.6d)

$$\underbrace{P_e A_e \mathcal{V}_{\text{roc}}}_{\substack{\text{rate of}\\ \text{energy input}}} = \underbrace{\mathcal{F}_d \mathcal{V}_{\text{roc}} + \dot{Q} + \dot{M}_e \left(h_e + \mathcal{V}_{\text{abs}}^2/2 + gz_e\right)}_{\substack{\text{rate of}\\ \text{energy output}}} + \underbrace{\frac{dE}{dt}}_{\substack{\text{rate of energy}\\ \text{accumulation}}}.$$

Here E represents the total instantaneous energy inside the control volume. We can express this as

$$E = E_{\text{fuel}} + E_{\text{structure}} + E_{\text{payload}}.$$

Each of these energy terms can be expressed as the sum of internal energy, kinetic energy, and potential energy for the component; for example,

$$E_{\text{fuel}} = \left[U + \frac{1}{2}M\mathcal{V}_{\text{roc}}^2 + Mgz\right]_{\text{fuel}}.$$

Note that the absolute velocity of all the material inside the control volume is assumed to be $\mathcal{V}_{\text{roc}}$ (idealization 5), and the z variations inside are neglected (idealization 6). This energy balance would be part of the analysis involved in designing the rocket system.

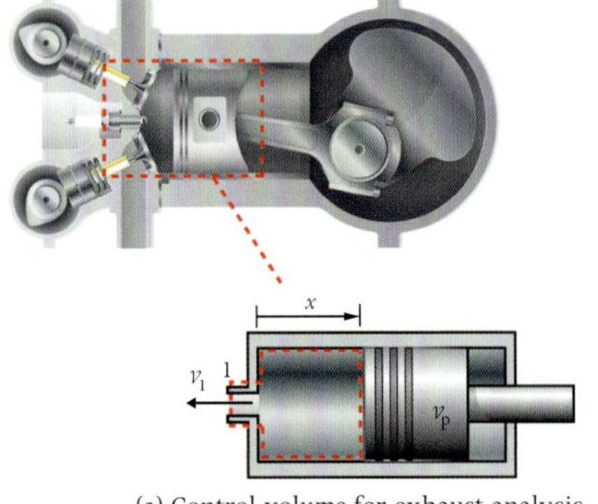

(a) Control volume for exhaust analysis.

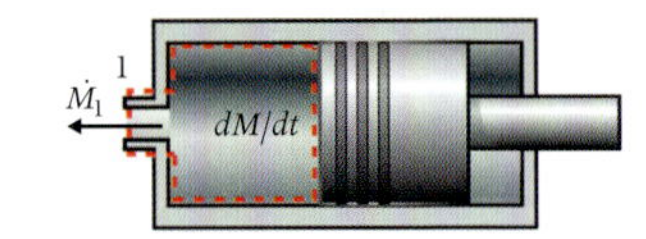

(b) Mass balance for the control volume.

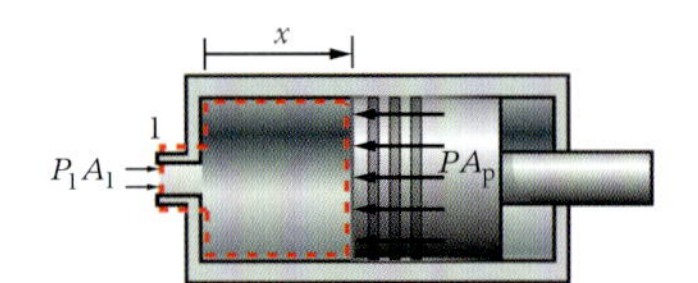

(c) Forces on the control volume.

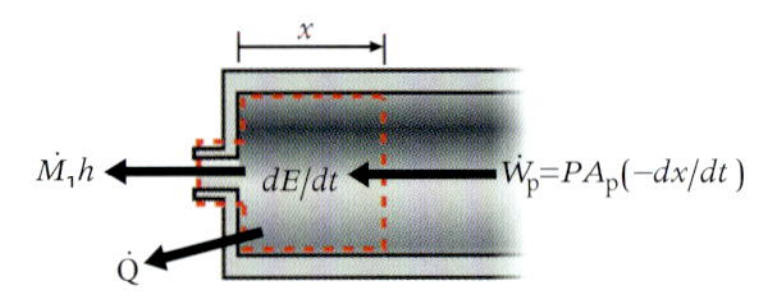

(d) Energy balance on the control volume.

Figure 4.7 Analysis of an exhaust stroke of an IC engine, a distorting control volume.

4.9 Example with Distorting Control Volume: IC Engine

In the previous examples the control volume always had the same size and shape, but this need not be the case. The internal combustion engine provides an excellent example where the control volume size changes in time. Figure 4.7 shows the control volume used in the analysis of the exhaust stroke. The right-hand boundary is attached to the piston face, which moves to the left during the exhaust stroke. The hot products of combustion are pushed out of the exhaust port by the piston, which moves to the left with the instantaneous velocity $\mathcal{V}_p$. The instantaneous velocity at the exit plane 1 determines the instantaneous mass flow rate $\dot{M}_1$ through the exhaust port. We simplify the problem with the following idealizations:

1. Negligible kinetic and potential energy everywhere.
2. Uniform state inside the cylinder at any instant.
3. One-dimensional flow at section 1.
4. Instantaneous state at section 1 is the same as in the cylinder.

Mass balance

The mass balance, on a rate basis, is (see Figure 4.7b)

$$\underbrace{0}_{\substack{\text{rate of}\\ \text{mass inflow}}} = \underbrace{\dot{M}_1}_{\substack{\text{rate of}\\ \text{mass outflow}}} + \underbrace{\frac{dM}{dt}}_{\substack{\text{rate of mass}\\ \text{accumulation}}}, \tag{4.20}$$

where M is the instantaneous gas mass in the cylinder. This ordinary differential equation would be one of several that would comprise a simple model for the engine.

Mass-associated energy transfer

Under the simplifying idealizations, the rate of energy transfer across the exit plane is $\dot{M}h$ since $h_1 = h$ by idealization 4. Remember that the enthalpy h_1 includes the work done by the gas inside in pushing the exhaust gas across the exit plane against the pressure P_1.

Forces and energy transfers as work

Figure 4.7c shows the pressure forces on the control volume boundary. The work associated with P_1 pushing gas across the exit plane is included in the enthalpy (see above). We must also include the energy transfer as work from the piston to the gas in the cylinder as a result of the motion of the control volume boundary at the piston face. This energy transfer rate is

$$\dot{W}_p = PA_p\left(-\frac{dx}{dt}\right).$$

Note that $dx/dt < 0$ during the exhaust process, so $\dot{W}_p > 0$.

Energy balance

Denoting the instantaneous heat transfer rate from the gas to the colder cylinder walls as $\dot{Q}$, the energy balance on the control volume, on a rate basis, is

$$\underbrace{-PA_P\frac{dx}{dt}}_{\substack{\text{rate of}\\ \text{energy input}}} = \underbrace{\dot{M}_1 h + \dot{Q}}_{\substack{\text{rate of}\\ \text{energy output}}} + \underbrace{\frac{dE}{dt}}_{\substack{\text{rate of energy}\\ \text{accumulation}}}, \tag{4.21}$$

where E is the instantaneous energy in the control volume. This is another ordinary differential equation that would be involved in the engine analysis. Under idealization 2, we can express E in terms of M and the instantaneous specific internal energy of the gas in the cylinder,

$$E = Mu,$$

then we have

$$\frac{dE}{dt} = M\frac{du}{dt} + u\frac{dM}{dt}. \tag{4.22}$$

If we make use of the ideal gas property relations to relate u and h to T, the energy balance and mass balance become a coupled pair of ordinary differential equations in T and M, which would be solved numerically as part of the analysis. Models like this are very useful in IC engine design.

Differential equations for an adiabatic system

To do the analysis with heat transfer would require a model for the heat transfer rate, which is beyond the scope of thermodynamics. However, we can take the analysis to completion if we neglect the heat transfer, which might be acceptable in some circumstances. We use the ideal gas equations of state (see Section 3.8) with constant specific heats,

$$Pv = RT, \qquad u = c_v T, \qquad h = c_P T, \qquad c_P - c_v = R. \tag{4.23}$$

Substituting for $\dot{M}$ from (4.20), and using (4.22), (4.21) becomes

$$-\frac{MRT}{V}\frac{dV}{dt} = -c_P T\frac{dM}{dt} + Mc_v\frac{dT}{dt} + c_v T\frac{dM}{dt}, \tag{4.24}$$

where V is the instantaneous cylinder volume (we neglect the small volume in the discharge pipe). For a given engine running at a known speed, $V(t)$ is known, and so (4.24) is an ordinary differential equation in the unknowns T and M. We need two equations to solve for two unknowns, and the second would be the mass balance (4.20), recast as

$$\frac{dM}{dt} = \dot{M}(P, T, P_{\text{manifold}}), \tag{4.25}$$

where $\dot{M}(P, T, P_{\text{manifold}})$ would be a model for the mass flow rate through the discharge valve as a function of the pressure and temperature of the inlet flow and the discharge pressure P_{manifold}. This would come from a fluid mechanics analysis, and is beyond the scope of thermodynamics. Given the flow model, we would integrate (4.24) and (4.25) numerically using a standard computer program for solving systems of ordinary differential equations like these.

Partial solution for the adiabatic system

Although we do not have a specific mass flow rate model, we can work out an expression for the temperature in the cylinder as a function of the mass and volume during the discharge process. We multiply (4.24) by $dt/(MRT)$, which gives

$$-\frac{dV}{V} = -\frac{c_P}{R}\frac{dM}{M} + \frac{c_v}{R}\frac{dT}{T} + \frac{c_v}{R}\frac{dM}{M},$$

which we rearrange using (4.23) as

$$\frac{dM}{M} = \frac{c_v}{R}\frac{dT}{T} + \frac{dV}{V}.$$

The variables are now separated, so this equation can be integrated from an initial state *in* to some final state *fin*,

$$\int_{\text{in}}^{\text{fin}} \frac{dM}{M} = \int_{\text{in}}^{\text{fin}} \frac{c_v}{R}\frac{dT}{T} + \int_{\text{in}}^{\text{fin}} \frac{dV}{V}.$$

Noting that $c_v/R = 1/(\gamma - 1)$, this gives

$$\ln(M_{\text{fin}}/M_{\text{in}}) = \frac{1}{\gamma - 1}\ln(T_{\text{fin}}/T_{\text{in}}) + \ln(V_{\text{fin}}/V_{\text{in}}),$$

or

$$\frac{M_{\text{fin}}}{M_{\text{in}}} = \frac{V_{\text{fin}}}{V_{\text{in}}}\left(\frac{T_{\text{fin}}}{T_{\text{in}}}\right)^{1/(\gamma-1)},$$

so

$$\frac{T_{\text{fin}}}{T_{\text{in}}} = \left(\frac{M_{\text{fin}} V_{\text{in}}}{M_{\text{in}} V_{\text{fin}}}\right)^{\gamma-1}.$$

Neat analytical solutions like this are very useful in providing insight into the behavior of many systems. The energy and mass balance methodology illustrated here forms the basis for such analysis. Thermodynamic properties information, and ancillary information from heat transfer or fluid mechanical analysis, are then brought in to complete the analysis. Put to good use in designing or improving important devices, this is the sort of work done by a competent engineering thermoscientist.

EXERCISES

4.1 One kilogram of R134a is heated in a constant-volume container from the critical point to 450 K. How much energy is transferred as heat to the fluid? Use the diagrams or data in Appendix A.4.

4.2 One kilogram of carbon dioxide is expanded adiabatically in a piston–cylinder system from 10 bar and 500 K to the saturated-vapor state at 5 bar. How much work is done by carbon dioxide and on what? Use the data in Appendix A.9.

4.3 In a coal-fired power station, steam (H_2O) enters the nozzle in a turbine at $T_1 = 800$ K, $P_1 = 2$ MPa at negligible velocity, and emerges at high velocity as a liquid–vapor mixture having quality $x_2 = 0.97$ at $P_2 = 0.07$ MPa. Assuming steady-state and negligible heat transfer, calculate the nozzle exit velocity (m/s).

4.4 Steam (H_2O) enters the steam turbine in a nuclear power station at $T_1 = 800$ K, $P_1 = 2$ MPa at negligible velocity, and emerges at $T_2 = 370$ K, $P_2 = 0.07$ MPa. Assuming steady-state and neglecting heat transfer, and neglecting the potential and kinetic energy of the fluid at the inlet and outlet of the turbine, determine the turbine shaft work per kg of H_2O processed (J/kg) and the mass flow rate $\dot{M}$ (kg/s) required to deliver 20 MW of shaft power.

4.5 20 kg/s of water/steam enters the condenser in a central power station at $T_1 = 342$ K with a quality of $x_1 = 0.98$, and emerges as a saturated liquid at the same pressure ($P_2 = P_1$). Assuming steady-state, and neglecting the potential and kinetic energy of the fluid at the inlet and outlet of the condenser, calculate the rate of energy transfer as heat from the steam (to cooling water) in W.

4.6 Calculate the mass flow rate of the steam that enters an adiabatic duct with a velocity of 10 m/s, at a pressure of 1.5 bar and a temperature of 400 °C. The duct has a diameter of 0.16 m at the inlet and the steam outlet temperature is 650 K. Plot the exit velocity as well as the exit diameter versus the outlet pressure from 0.1 to 1 bar. Use Appendix A.3 for data.

4.7 Your company asks you, the process engineer, to estimate the rate at which energy is transferred as heat through a steam turbine casing. This turbine operates with an inlet pressure of 16 bar and an inlet temperature of 377 °C, while the steam exits the turbine as saturated vapor at 30 °C. Its shaft power is 9 MW. The mass flow rate of steam is 15 kg/s. Use Appendix A.3 for data.

4.8 An adiabatic piston–cylinder system contains a 1000 W immersion heater and 4 kg of water. The water is initially at 1 atm and 96% quality. The heater is operated for 7 minutes. Calculate the final volume of the system using Appendix A.3 for data.

4.9 Potassium, an alkali metal, enters a condenser in the vapor phase with a mass flow rate of 0.02 kg/s and at a temperature and pressure of 1370 K and 0.4 MPa. It exits as a saturated liquid at 0.2 MPa. Calculate the rate of energy transfer as heat from the condenser in kW using Appendix A.11 for data.

4.10 Oxygen flows through an adiabatic steady-flow compressor as saturated vapor at a rate of 1000 kg/hr. The saturated vapor enters at 2.5 bar and exits at 17.5 bar and 175 K. Calculate the shaft work per unit mass of oxygen and the required power to drive the compressor. Use Appendix A.8 for data.

4.11 At a tee junction in the piping of a chemical plant, 5 kg/s of methane enters on one side at $P_1 = 1$ MPa, $T_1 = 400$ K, and 1 kg/s of methane enters on the other as a saturated liquid at $P_2 = 1$ MPa. The two streams are mixed and exit from the junction at $P_3 = 1$ MPa. Assuming steady-state, neglecting the kinetic and potential energies of the flow streams, and treating the mixer as adiabatic, determine the temperature of the discharge stream (T_3). This will require both an energy balance and a mass balance.

4.12 In a two-fluid heat exchanger of a methane-cooling system, 4 kg/s of methane at $T_1 = 350$ K, $P_1 = 0.5$ MPa enters the tubes, emerging at $T_2 = 225$ K with negligible loss in pressure ($P_1 = P_2$). Refrigerant R134a flows on the outside of the tubes, entering as a liquid–vapor mixture having quality $x_3 = 0.4$ at $T_3 = 200$ K and emerging as a saturated vapor at the same pressure ($P_4 = P_3$). Assuming steady-state, neglecting the kinetic and potential energies of the flow streams, and assuming that the heat exchanger is adiabatic (no heat transfer to the environment, but of course much heat transfer taking place internally), determine the required mass flow rate $\dot{M}_{R134a}$ of the refrigerant (kg/s).

4.13 The throttling calorimeter is a device for determining the thermodynamic state of a fluid in vapor–liquid equilibrium. The procedure to estimate the vapor fraction of the fluid consists in bleeding off a small amount of the two-phase fluid, throttle it through a valve, and take the temperature and pressure measurements as shown in the figure.

Explain why P_1 and T_1 are not sufficient to fix the thermodynamic state of the *wet* fluid, and how P_2 and T_2 measurements allow state 1 to be determined. How much throttling (pressure drop caused by the valve) is necessary for this quality measurement technique to work? Do you see limitations in terms of measurable states?

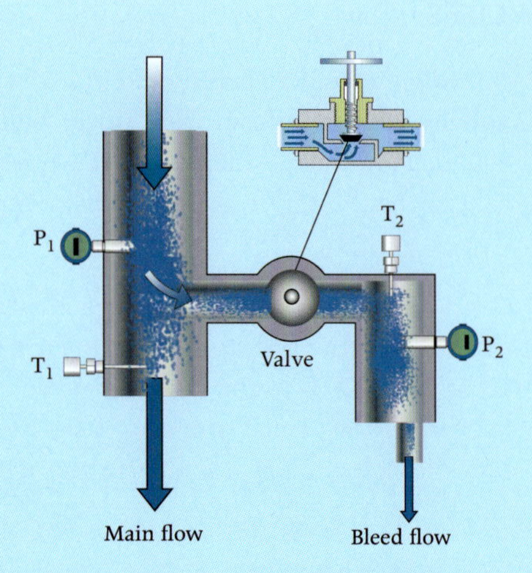

In the case that the fluid is water, calculate the quality if the readings of the calorimeter are

$P_1 = 10$ bar, $T_1 = 179.9$ °C,

$P_2 = 1.2$ bar, $T_2 = 120$ °C.

Refer to the tables in Appendix A.3 for data.

4.14 The cooling system for a small private jet airplane consists of the hardware shown here.

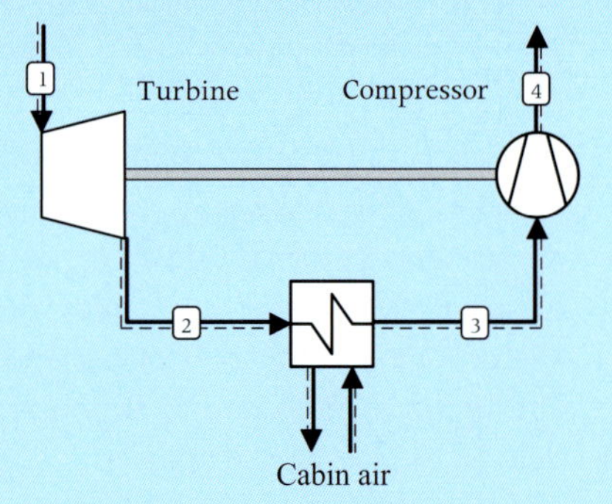

High-pressure air from the aircraft engine compressor is fed to a small turbine at state 1 (temperature T_1). The turbine extracts energy from the air, lowering the air temperature to state 2. This cold air is then passed through a heat exchanger, where it picks up energy transferred as heat from the warmer cabin air, thereby cooling the cabin air. The air at state 3 is then fed to a compressor, which is driven by the turbine. The air emerges from the compressor at state 4, and is fed back into the jet engine at temperature T_4. The mass flow rate of engine air through the cooling system is $\dot{M}$. Using the perfect gas equation of state with constant specific heats, derive an expression for the heat transfer rate $\dot{Q}$ in terms of the mass flow rate $\dot{M}$, the inlet and discharge temperatures T_1 and T_4, and constants for the gas. Treat the turbine and compressor as adiabatic, and neglect kinetic and potential energy of 3 and 4. *Hint: Choose your control volume(s) wisely!*

4.15 A small solar engine is used for water pumping in a desert area. Water is the working fluid.

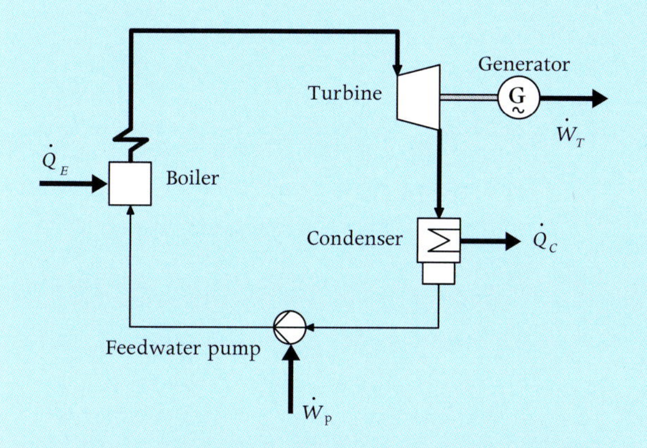

Water enters the pump as saturated liquid at 50 °C and is pumped up to 2 bar by a small centrifugal pump. The boiler evaporates the water at 2 bar, and saturated vapor at this pressure enters a small expander. The steam leaves the expander at 50 °C and with a liquid fraction of 6%, and is subsequently condensed. The mass flow rate is 140 kg/hr and the pump is driven by a 0.4 kW motor. Determine the power output of this solar engine, the energy conversion efficiency taken as the net shaft work output divided by the energy transfer as heat to the fluid in the boiler, and estimate the area of the solar collectors, assuming the collectors can absorb 800 W/m^2. Use the tables in Appendix A.3 for data.

4.16 An alkali metal, potassium, is the working fluid in a nuclear power system that was proposed in the 1960s for space application, either on board of a spacecraft or in power stations on the Moon or Mars. Given recent developments, it might be considered again. It employs similar hardware to that in Exercise 4.15, except that a nuclear reactor replaces the Sun as the energy source. The potassium enters the pump as a saturated liquid at 1000 K, is compressed, is then evaporated at 2 bar and superheated to 1350 K in the reactor-boiler. The turbine discharge is at 1000 K with 97% quality. Neglect the pressure drop through the reactor and condenser for this simplified thermodynamic analysis. The power required to operate the pump is 2% of the turbine-shaft power. Determine

a. the energy conversion efficiency of this system, i.e., the net shaft power divided by the energy transfer rate as heat to the working fluid from the reactor,
b. the potassium mass flow rate,
c. the reactor power required for 500 kW$_e$ of net power, assuming that a 97% efficient electric generator is employed.

Use Appendix A.11 for data.

4.17 R134a flows in a pipe of a process plant at 3 m/s. Its pressure is 2 MPa and its temperature is 182 °C. A plant component needs the working fluid at specific conditions. Design a steady-flow device to deliver the fluid at these conditions, namely a device capable of accelerating the fluid up to 150 m/s and at a pressure of 1.4 MPa. What hardware is required and what is the exit temperature?

4.18 The inlet of the jet engine of a supersonic aircraft is a carefully shaped flow passage that takes in flow at very high speed $\mathcal{V}_1$ (relative

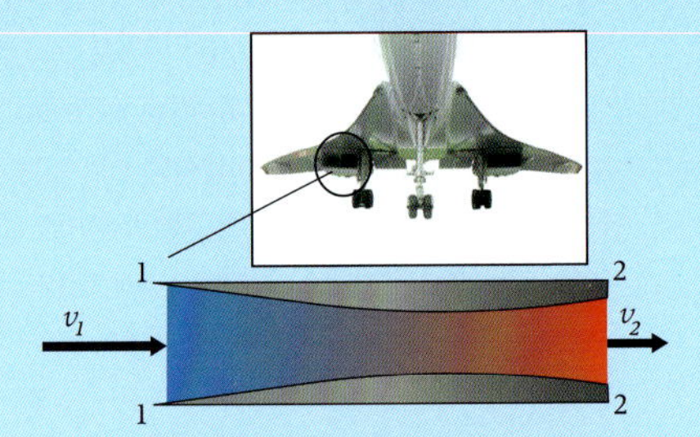

to the engine) (section 1) and discharges the air at much lower velocity $\mathcal{V}_2 \ll \mathcal{V}_1$ and higher pressure (section 2) to the inlet of the compressor.

Treating the air as a perfect gas with constant specific heats, derive an expression for the temperature T_2 at section 2 in terms of the temperature T_1 at section 1, the inlet velocity $\mathcal{V}_1$, and constants for the gas. Heat transfer between the air and the walls may be neglected. Use a reference frame attached to the engine.

4.19 An impact press consists of a piston that is accelerated rapidly by pumping high-pressure gas into the cylinder. The high-speed piston slams into the material to be flattened.

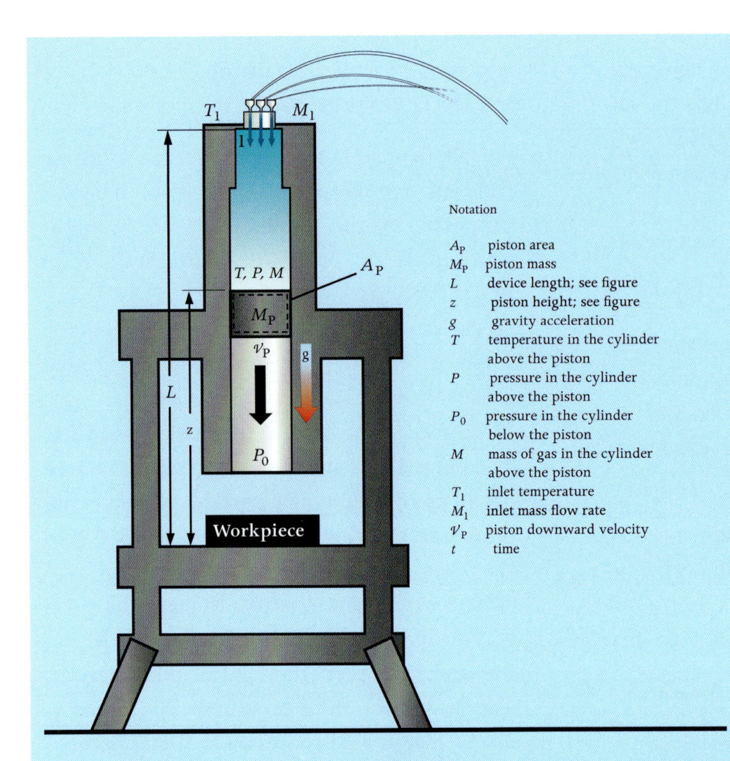

Derive the equations needed to perform the design and analysis of this system according to the following suggested procedure and assumptions.

Assume that the gas enters at a constant and known flow rate $\dot{M}_1$ at constant and known state 1 (T_1, P_1), and that the inlet gas is mixed instantaneously with the fluid inside, which at any instant is at T, P, which vary with time. We will neglect the kinetic and potential energy of the gas, but must consider the kinetic and potential energies of the piston. We will treat the piston as frictionless, and assume that the pressure on the workpiece side of the piston is constant at P_0, the atmospheric pressure. We will assume that heat transfer is negligible, and that the internal energy of the piston does not change. Denote the piston mass by M_P, the piston area by A_P, and the downward piston velocity by $\mathcal{V}_P$ (be sure not to confuse $\mathcal{V}_P$ with the gas volume $V = A_P(L - z)$). Write the *rate-basis* energy and mass balances at time t for the control volume above the piston; this will give you two differential equations involving the unknowns.

Then, write the rate-basis energy balance on the piston, which will give you another differential equation (the same as would be obtained from Newton's law). Using the ideal gas equations, these three balances and the definition $\mathcal{V}_p = -dz/dt$ give you a coupled set of four ordinary differential equations for the variables z, $\mathcal{V}_p$, T, and M. Express all of the terms in these equations as a function of these four variables and known quantities, such as $\dot{M}_1$, T_1, A_p, L, M_p, P_0, the acceleration of gravity g, and constants for the gas. These are the equations that would be solved (analytically or numerically) to design the press.

4.20 A hydraulic turbine is fed with water from a reservoir whose free surface is at $z_1 = 22$ m a.s.l. (above sea level). Water enters the turbine inlet pipe, flows through the turbine and its diffuser, and is discharged downstream of the dam at $z_2 = 14$ m a.s.l.

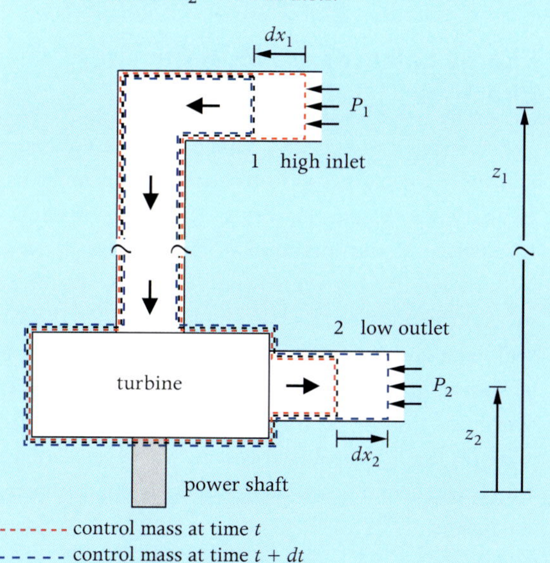

Calculate the average power that the hydraulic turbine can generate over a year. The mean annual water mass flow rate is 200 m^3/s. Choose carefully between control mass or control volume for your analysis.

Assume that the following idealizations hold:

(a) One-dimensional flow at 1 and 2.
(b) No energy transfer as heat occurs anywhere.
(c) The flow kinetic energy at 1 and 2 is negligible.
(d) Flow friction forces along the inlet pipe and the diffuser are negligible, except for those acting on the blades of the turbine, which reduce the amount of kinetic energy of the water that is converted into mechanical power at the shaft. Take into account this thermodynamic loss by assuming that the hydraulic efficiency of the turbine is 94%.

Is the control mass concept useful in this case? Justify your answer.

5 Entropy and the Second Law

CONTENTS

The Second Law of Thermodynamics is one of the great principles of modern science. It deals with a very special property of matter, *entropy*, which is as important as energy in thermodynamic analysis. There are almost as many ways to think about entropy as there are different textbooks on thermodynamics. Most textbooks take a very abstract (and purely macroscopic) mathematical approach, which certainly can be rigorous and elegant but does not provide much physical insight. Therefore, we will take a very physical approach to entropy, developing the concept of entropy as a measure of the degree of *microscopic* randomization, disorder, and unpredictability. These ideas provide the basis for a *macroscopic* hypothesis (the *entropy hypothesis*) about entropy as a property of matter. Simple mathematical developments based on this postulate will then show how we can obtain values for the entropy of substances. Then, with quantitative information about entropy as a thermodynamic property, we can do entropy balances to determine what can happen in various complex situations. Thus, the analysis that we will do with entropy will be macroscopic and very similar to what one finds in other textbooks on engineering thermodynamics. The microscopic underpinnings should help you understand what this analysis means and why it must be true.

5.1 The Concept of Entropy

Energy balances are insensitive to direction

When applied to a control mass or control volume, an energy balance gives us one equation that can be used to analyze a system. However, the energy balance is insensitive to the direction of the process; it would be satisfied if we reversed the directions of time and all energy transfers. For example, consider the control mass consisting of a flywheel immersed in a gas as shown in Figure 5.1. In the initial state (1) the flywheel is spinning rapidly in cool gas, and in the final state (2) the flywheel is at rest in warm gas. Treating the flywheel and gas as an isolated system, the energy balance is

$$\underbrace{0}_{\substack{\text{energy}\\\text{input}}} = \underbrace{0}_{\substack{\text{energy}\\\text{output}}} + \underbrace{E_2 - E_1}_{\substack{\text{energy}\\\text{accumulation}}},$$

where $E = U + KE$ is the sum of the internal energy plus the kinetic energy of all the material inside. Thus, the energy balance says

$$U_1 + KE_1 = U_2.$$

In the initial state (1), significant energy resides in the organized motion of the flywheel molecules rotating

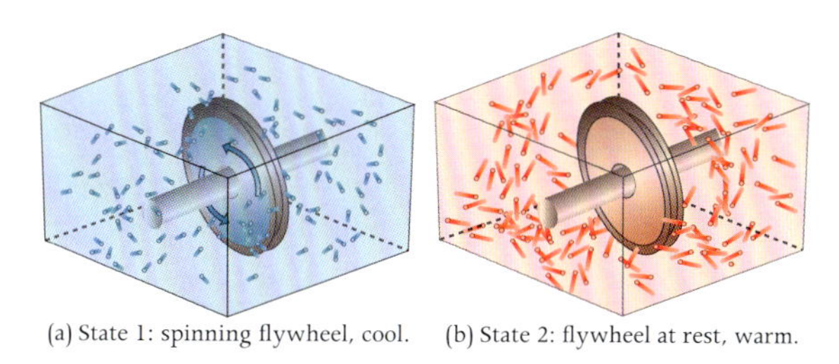

Figure 5.1 Flywheel immersed in a gas.

in tandem around the axis. In the final state (2), all the energy resides in randomly oriented molecular motions of the gas and flywheel molecules (U_2). The randomization is a result of the friction that occurs between the flywheel and the gas and within the gas itself.

But now suppose that somehow the initial and final states could be interchanged, so that we started initially at state 2 (warm, stationary flywheel) and ended up finally at state 1 (cool, spinning flywheel). The energy balance would again require that $E_1 = E_2$, as above, and this would be exactly satisfied. But we know very well that the process $2 \rightarrow 1$ never will happen. Once randomization has set in at the microscopic level, and the system has reached state 2, the system will not become microscopically organized on its own. But the energy balance does not tell us that the process $2 \rightarrow 1$ is impossible, because the energy balance is insensitive to direction.

Another principle is needed

It is clear from this example that conservation of energy by itself does not provide a complete scientific framework for analyzing thermodynamic processes. We also need a basic principle that will allow us to distinguish between processes that are possible ($1 \rightarrow 2$) and processes that are not ($2 \rightarrow 1$). *Entropy* (a property of matter denoted by S) and the *Second Law of Thermodynamics* (which rules out impossible processes) fill this need.

Entropy and microscopic randomness

Entropy can be viewed as a quantitative measure of microscopic randomness. In the example above, state 1 is a state of low entropy, because a large amount of the energy is due to the organized rotation of the flywheel molecules. In state 2 this energy has been converted into molecular motions, which occur in every possible direction, with the instantaneous direction of any given molecule being very random. State 2 is therefore a state of higher entropy, and so $S_2 > S_1$. Randomness is sometimes measured by counting the number (Ω) of microscopic states in which a macroscopic system of given energy might be found. You may have seen a statistical approach to entropy that started with a definition of entropy in terms of this number (e.g., $S = k \ln \Omega$). This is a very special statistical definition that only works if all available states are equally probable. We will take a less restrictive view.

Entropy and microscopic uncertainty

At the microscopic level, the motions of molecules and the sub-atomic particles they comprise can be described by the dynamics of systems of many particles. Such systems are known to exhibit *chaos*, which means that the precise position and location of each particle can be tracked accurately only for a relatively short time. Beyond that time, the instantaneous states of the particles are very strongly dependent on the initial conditions, and infinitesimal differences in initial conditions give rise to vastly different particle locations and velocities. This means that as time progresses one becomes more and more uncertain about the detailed microscopic state of the system. Entropy can be viewed as a measure of this uncertainty. In the flywheel example, we know more about the microscopic details in state 1 than we do in state 2, and so $S_2 > S_1$. You may have seen a statistical treatment of entropy in which the entropy was defined in terms of the probabilities of the possible microscopic states (i.e., $S = -k \sum p_i \ln p_i$). This definition arises from very fundamental ideas in statistics and information theory. In Section 5.8 we will introduce this statistical definition of entropy and show how it relates to the macroscopic formulation that we will develop in this chapter.

Entropy is extensive

Essential to the concept of entropy is the idea that it is *extensive*, i.e., the entropy of a system is the sum of the entropy of its parts. This seems reasonable, because there must be more microscopic disorder if there is more material to be disordered. But we cannot theoretically or experimentally prove the extensivity of entropy (any more than we could prove the extensivity of energy) because it is deeply embedded in the entropy concept, and we use the extensivity to determine values for entropy.

Entropy change in an isolated system

In an isolated system, such as that of Figure 5.1, the natural tendency is for the entropy of the system to increase. If the entropy were to decrease, this would mean that somehow the molecules had organized themselves. We do not expect to see this happen, and so we expect the entropy of the isolated system to increase. The notion that the entropy of an isolated system will never decrease ($\Delta S_{\text{isolated system}} \geq 0$) is a special form of the *Second Law of Thermodynamics*.

Production

As noted in Section 1.2, the general form of any balance equation involves three types of terms; (1) transfers (in or out), (2) accumulation, and (3) production:

$$\text{input} + \text{production} = \text{output} + \text{accumulation}.$$

Production ($\mathcal{P}$) is the creation that occurs inside the system on which the balance is made. In the daily automobile-balance on an automobile factory, production is the number of vehicles built inside the plant during the day; it should not be confused with the accumulation of automobiles within the plant (the increase in inventory) because some automobiles are shipped out and some may be shipped in each day. Energy is (by concept) conserved; the conservation of energy can be expressed simply by saying that energy production must always be zero, or

$$\mathcal{P}_E = 0.$$

Likewise, the principle of conservation of mass can be expressed as

$$\mathcal{P}_M = 0.$$

Entropy production

In an entropy balance, $\mathcal{P}_S$ is the amount of entropy produced inside the system. Entropy production arises from microscopic randomization, which is what always happens naturally. The *Second Law of Thermodynamics* is the statement that entropy production is never negative,

$$\mathcal{P}_S \geq 0. \tag{5.1}$$

Negative $\mathcal{P}_S$ would mean that the microscopic behavior was somehow self-organizing, a situation never found in an inanimate system.

> Living systems do organize themselves at the molecular level, but at the expense of increased disorganization in the surrounding inanimate media; students often raise the interesting question whether living systems exhibit positive or negative entropy production. Most scientists would say that the second law applies to both living and inanimate systems; but are they right? Engineers do not need to worry about this question... until they start to design systems that use living matter!

Heat and entropy transfer

Consider a system containing a liquid (Figure 5.2). Imagine transferring energy as heat from the liquid, causing it to solidify. In the liquid state the molecules roam randomly, but in the solid state the molecules are neatly arranged in a crystal structure. The solid state is clearly more organized microscopically, and hence must have a lower entropy than the liquid. What happened to the entropy that was in the liquid before it solidified? Either it disappeared (negative production), or it was transferred somewhere else. Since entropy production cannot be negative, the entropy must have been transferred out of the fluid. Where did it go? Imagine that the solidification was accomplished by transferring energy as heat from the liquid A being solidified to a different material B, which melted in this process (Figure 5.2). The entropy of the the melted material B increased while the entropy of the solidified material A decreased,

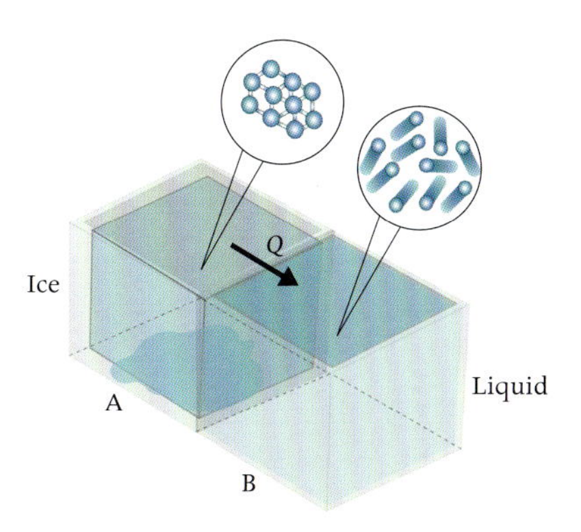

Figure 5.2 Solidification reduces entropy.

and so the decrease in entropy of A was a result of entropy transfer from A to B. Entropy is always transferred whenever energy is transferred as heat. Since heat is a microscopically disorganized energy transfer process, it seems quite reasonable that some disorganization (entropy) should accompany the energy flow. We will soon see exactly how much.

Work and entropy transfer

Work is a microscopically organized energy transfer. When work is done on a piston by a gas, the piston molecules move together in a fully organized response to the push from the gas. Because energy transfer as work is microscopically organized, *there is no entropy transfer associated with energy transfer as work*. This idea is an essential part of the concept of entropy. The idea is used in developing the equations that allow quantitative evaluations of entropy, and so it cannot be tested by entropy measurements; the idea is deeply embedded in the entropy concept (a fact not often clearly recognized in many presentations of the second law). To the sophisticated thermodynamicist, entropy allows what is probably the best distinction between heat and work, with work being the *entropy-free* energy transfer.

Reversible process

A process is called *reversible* if it does not produce any entropy. Entropy is *conserved* in a reversible process. The reversible process is an idealized limit of real processes, and requires such fictions as wires with no electrical resistance, frictionless pulleys, and compression or expansion processes that are sufficiently slow that no variations in pressure or temperature occur within the substance undergoing the process. These are all very reasonable and useful idealizations of limiting behavior, and so the reversible process is a very useful limit.

5.2 The Entropy Hypothesis

The discussion above was intended to provide insight and justification for what we call the *entropy hypothesis*:

- Matter has a thermodynamic property, called entropy (S), that measures the amount of microscopic randomness or uncertainty about the microscopic state.
- Entropy is an extensive property; the entropy of a system is the sum of the entropy of its parts.
- Entropy can be produced, or in the limit of a reversible process be conserved, but entropy can never be destroyed; $\mathcal{P}_S \geq 0$.
- Entropy is transferred with heat, but there is no entropy transfer associated with energy transfer as work; *work is entropy free*.

In the sections that follow, we will see how these four ideas, together with the concept of temperature as the indicator of thermal equilibrium, allow one to develop equations relating entropy to measurable quantities like pressure, temperature, density, and internal energy. In addition, we will find a simple expression for the entropy transfer with heat. With this information you will have all that you need to make entropy balances and invoke the Second Law of Thermodynamics (5.1) to find out what is possible, impossible, or the best one can do in a given situation.

5.3 Entropy Change in a Reversible Adiabatic Process

Consider a control mass that undergoes an adiabatic (no heat transfer) process (Figure 5.3). The only energy exchange it can have with its surroundings is through energy transfer as work. According to the fourth statement of the entropy hypothesis, the work does not transport any entropy, and since there is no energy transfer as heat there is no entropy transfer to or from the control mass. Hence the entropy balance is

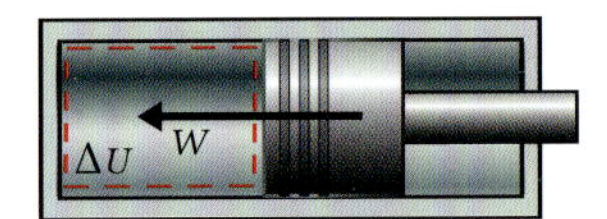

Figure 5.3 Entropy does not change in a reversible adiabatic process.

$$\underbrace{0}_{\text{entropy input}} + \underbrace{\mathcal{P}_S}_{\text{entropy production}} = \underbrace{0}_{\text{entropy output}} + \underbrace{\Delta S}_{\text{entropy accumulation}} .$$

Since by (3) of the entropy hypothesis the entropy production must be positive, *for an adiabatic control mass*

$$\Delta S = \mathcal{P}_S \geq 0.$$

If the process is reversible, no entropy is produced, and so *for a reversible adiabatic process in a control mass*

$$\Delta S = 0.$$

Thus, *the entropy of a substance will not change if it is put through a reversible adiabatic process*. This result has many important applications in thermodynamic theory, and will be very useful in the following section.

5.4 Entropy of a Simple Compressible Substance

Entropy and state

The State Principle tells us that, for a given amount M of a simple compressible substance in an equilibrium state, all thermodynamic properties are fixed when two independently variable properties (e.g. U and V) are fixed. Since by (1) of the entropy hypothesis S is a thermodynamic property, for a given mass M of a given substance in equilibrium,

$$S = \mathcal{S}(U, V). \tag{5.2}$$

Since S is extensive, we can express the *specific entropy* or *entropy per unit mass*, $s \equiv S/M$, for a simple compressible substance as

$$s = \mathcal{s}(u, v). \tag{5.3}$$

In this section we use the ideas about entropy and temperature outlined above to develop a differential equation that reveals how s must vary with u and v. This equation enables one to calculate entropy differences between different equilibrium states, and is the basis for the entropy values that you have seen in thermodynamic property tables and charts.

Entropy derivatives

The derivatives of the function $\mathcal{S}(U, V)$ in (5.2) are thermodynamic properties. For the moment let's denote them as follows:

$$X \equiv \left(\frac{\partial S}{\partial U}\right)_V, \tag{5.4a}$$

$$Y \equiv \left(\frac{\partial S}{\partial V}\right)_U. \tag{5.4b}$$

Using the chain rule of calculus, the change dS arising due to changes dU and dV is then

$$dS = \left(\frac{\partial S}{\partial U}\right)_V dU + \left(\frac{\partial S}{\partial V}\right)_U dV = XdU + YdV. \tag{5.5}$$

Since S, U, and V are all extensive properties (proportional to the mass M), the coefficients X and Y must be intensive properties (independent of the mass). Our immediate goal is to relate these properties to other intensive properties that we can measure.

Entropy of a two-part system

Consider a system C consisting of two separate control masses A and B of simple compressible substances in thermal contact (Figure 5.4). We assume that A and B are each in internal equilibrium, but that they are not necessarily in equilibrium with one another. Because each is in equilibrium, (5.2) applies for each, but with different functions for each if the substances or masses are different,

$$S_A = \mathcal{S}_A(U_A, V_A), \qquad S_B = \mathcal{S}_B(U_B, V_B).$$

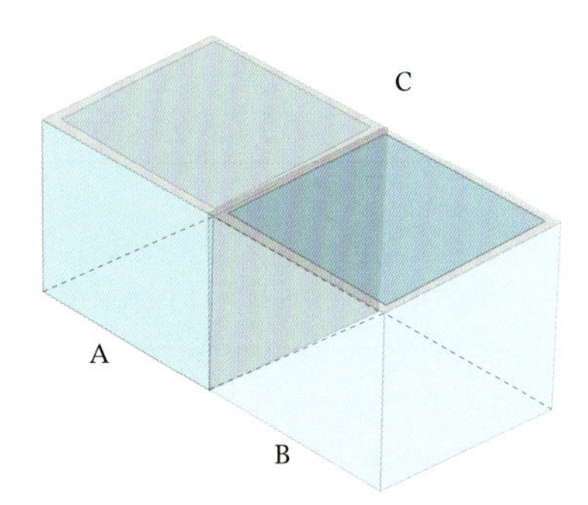

Figure 5.4 The combined system is isolated.

By (2) of the entropy hypothesis, the total entropy of the combined system C is

$$S_C = S_A(U_A, V_A) + S_B(U_B, V_B).$$

The masses M_A and M_B are fixed, but changes in S_C may result from changes in any of U_A, V_A, U_B, or V_B. Using (5.5), the entropy change dS_C for infinitesimal changes of the energies and volumes is

$$dS_C = dS_A + dS_B = (X_A\,dU_A + Y_A\,dV_A) + (X_B\,dU_B + Y_B\,dV_B)\,. \qquad (5.6)$$

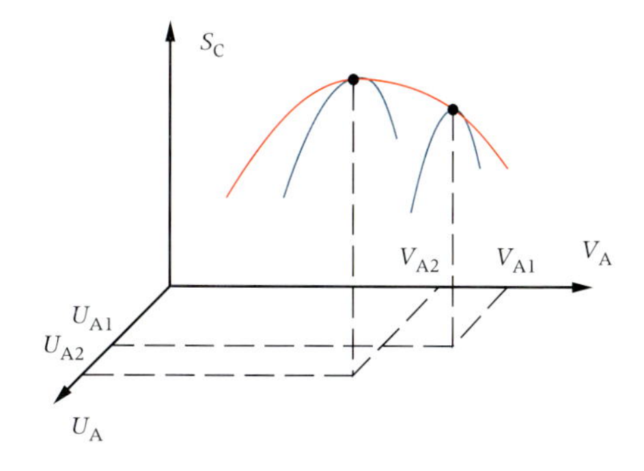

Figure 5.5 Entropy of the combined system.

Entropy change under isolation

Now imagine isolating the combined system C of Figure 5.4. The isolation will fix the values for the total energy and volume, so

$$U_C = U_A + U_B = \text{constant},$$
$$V_C = V_A + V_B = \text{constant}.$$

Hence, for the isolated system,

$$dU_B = -dU_A,$$
$$dV_B = -dV_A.$$

Therefore, (5.6) can be written as

$$dS_C = (X_A - X_B)\,dU_A + (Y_A - Y_B)dV_A. \qquad (5.7)$$

The entropy of the isolated system will change only if U_A or V_A changes. The second law requires that $dS_C \geq 0$ if any changes occur within the isolated system. Thus, changes that do occur will increase S_C, and the maximum possible S_C will be obtained when A and B finally reach equilibrium with one another. We can think of S_C, the entropy of the isolated system C, as being a function of the free variables U_A and V_A, represented by a surface (Figure 5.5). The equilibrium state is at the top of this entropy hill, which can be approached from different directions by starting with different initial distributions of the energy and volume between A and B.

Condition for thermal equilibrium

Now suppose that we further constrain the system so that the volumes of A and B are fixed, for example at V_{A1} on Figure 5.5. Now the entropy S_C will vary only with the energy distribution, which is determined by U_A. Any energy exchange between A and B must be as heat, because the constant volumes prevent energy exchange as work. We know that heat transfer will occur if the temperatures of A and B are different, and will stop when thermal equilibrium has been attained and the two temperatures are equal. Equilibrium will be attained when the entropy has risen as high as it can, given the volume constraints, and this occurs on Figure 5.5 at U_{A1}. At the point of maximum entropy, $dS_C/dU_A = 0$, and (5.7) shows that this requires (remember $dV_A = 0$ here)

$$X_A = X_B.$$

Thus, when A and B have reached thermal equilibrium with one another, the X values for the two will be the same. This is the conceptual attribute of temperature, and so X must somehow be related to temperature. When two systems have the same temperature they have the same X. Therefore, X must be a function of T, and it must be the same function for all substances, $X = X(T)$. With a little more analysis we will identify that function.

Approach to thermal equilibrium

As thermal equilibrium is approached, the entropy of the isolated system C must increase. Therefore, for the case where V_A and V_B are both fixed

$$dS_C = (X_A - X_B)\,dU_A > 0.$$

Hence,

$$\text{if} \quad X_A > X_B \quad \text{then} \quad dU_A > 0,$$
$$\text{if} \quad X_A < X_B \quad \text{then} \quad dU_A < 0,$$

so that energy will be transferred as heat *from* the system with the *smaller* X *to* the one with the *larger* X. This means that $X(T)$ must *decrease* with increasing temperature, because temperature has been defined such that energy transfer as heat flows from high temperature to low temperature. You might correctly guess that $X = 1/T$, where T is the absolute temperature defined by the constant volume gas thermometer (Section 3.3). We will demonstrate this very shortly.

Condition for mechanical equilibrium

We now relax the constraint that the individual volumes of A and B are fixed, retaining the constraint of fixed volume for the combined system C. A and B may now exchange energy as work, with one expanding and the other contracting a corresponding amount. They can also exchange energy as heat. Eventually equilibrium will be attained at the state of maximum S_C, at U_{A2} and V_{A2} in Figure 5.5. At the maximum,

$$\left(\frac{\partial S_C}{\partial U_A}\right)_{V_A} = 0 \quad \text{and} \quad \left(\frac{\partial S_C}{\partial V_A}\right)_{U_A} = 0.$$

From (5.7), this requires

$$X_A = X_B, \qquad (5.8a)$$
$$Y_A = Y_B. \qquad (5.8b)$$

Equation (5.8a) is the condition of thermal equilibrium, and will be satisfied if $T_A = T_B$. Equation (5.8b) is the condition of mechanical equilibrium. A force balance on a piece of the boundary separating A and B shows that the pressures must be equal in A and B for the interface to remain stationary in the final equilibrium state. Therefore, Y must somehow involve the pressure. If you guessed correctly that $X = 1/T$, then you have deduced that the dimensions of entropy are those of energy/temperature (J/K), and hence the dimensions of Y are those of

$$\text{energy}/\left(\text{temperature} \cdot \text{length}^3\right)$$
$$= \text{force}/\left(\text{temperature} \cdot \text{length}^2\right)$$
$$= \text{pressure/temperature;}$$

this means that $Y = \text{constant} \times P/T$, where the constant might be positive or negative. With a little more analysis we will know this constant.

Approach to mechanical equilibrium

As mechanical equilibrium is approached, the entropy of the isolated system C must increase. Therefore, if the system is in thermal equilibrium so that $T_A = T_B$ and $X_A = X_B$,

$$dS_C = (Y_A - Y_B)\, dV_A > 0.$$

Hence,

$$\text{if} \quad Y_A > Y_B \quad \text{then} \quad dV_A > 0,$$
$$\text{if} \quad Y_A < Y_B \quad \text{then} \quad dV_A < 0.$$

If the pressure of A is bigger than the pressure of B, then A will expand and $dV_A > 0$. This suggests that Y should be a monotonically increasing function of pressure. If you guessed correctly that $X = 1/T$, and thus deduced that the dimensions of Y are those of pressure/temperature, you might have guessed correctly that $Y = P/T$. We will demonstrate this very shortly.

Reversible adiabatic expansion

We can gain additional information about Y by considering the reversible adiabatic expansion of a simple compressible substance, which could be either A or B (Figure 5.6). In Section 5.3 we showed that, as a consequence of the entropy hypothesis, the entropy of the substance will remain constant during any reversible adiabatic process. Thus, using (5.5), during this process

$$dS = 0 = XdU + YdV. \qquad (5.9)$$

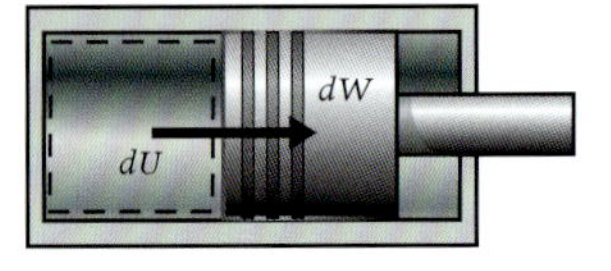

Figure 5.6 Reversible adiabatic expansion.

The energy transfer as work from the substance for a small step in this process will be $đW = PdV$, so the energy balance (see Figure 5.6) gives

$$\underbrace{0}_{\substack{\text{energy}\\\text{input}}} = \underbrace{PdV}_{\substack{\text{energy}\\\text{output}}} + \underbrace{dU}_{\substack{\text{energy}\\\text{accumulation}}}.$$

Substituting $dU = -PdV$ in (5.9), for this process,

$$0 = (-XP + Y)dV,$$

which tells us that

$$Y = X(T)P. \tag{5.10}$$

Since we can start a reversible adiabatic expansion from any state, (5.10) must hold at every state. As expected, Y is proportional to pressure, but it also depends upon temperature.

Identification of X using the ideal gas

The function $X(T)$ must be the same for all substances. We can evaluate it for the special case of an ideal gas if we use the fact that the internal energy of an ideal gas depends only on temperature (see Section 3.8). Recalling the definitions of X and Y as partial derivatives of $S(U, V)$ (see 5.4), we use the fact (from calculus) that differentiation order is unimportant for mixed derivatives,

$$\frac{\partial}{\partial V}\left(\frac{\partial S}{\partial U}\right) = \frac{\partial}{\partial U}\left(\frac{\partial S}{\partial V}\right).$$

Hence,

$$\left(\frac{\partial X}{\partial V}\right)_U = \left(\frac{\partial Y}{\partial U}\right)_V.$$

Now, for an ideal gas $U = U(T)$, and so if U is held constant, T cannot change, and hence X cannot change, so

$$\left(\frac{\partial X}{\partial V}\right)_U = 0 = \left(\frac{\partial Y}{\partial U}\right)_V. \tag{5.11}$$

Equation (5.10) established that $Y = XP$. For an ideal gas, $P = MRT/V$, where T is the absolute temperature as measured by a constant volume gas thermometer (see Section 3.8). So, for an ideal gas,

$$Y = \frac{MRTX}{V}.$$

Substituting in (5.11),

$$\left(\frac{\partial Y}{\partial U}\right)_V = \frac{MR}{V}\left(\frac{\partial (TX)}{\partial U}\right)_V = 0. \tag{5.12}$$

TX depends only upon T, and (5.12) tells us that TX does not change when we change U. But, for an ideal gas, the only way to change U is to change T; therefore, (5.12) says that TX does not change when we change T. That means that TX must be a constant, or

$$X = \text{constant} \times \frac{1}{T}.$$

We are free to choose the constant, which sets the scale for entropy just as the arbitrary constant in Newton's law sets the scale for force. For simplicity (and consistency with everyone else!) we take

$$X = \frac{1}{T}, \tag{5.13a}$$

where T is the temperature (K) as measured by an ideal gas thermometer. Since X must be the same function for all substances, this is the proper choice for X in general. Then, from (5.10)

$$Y = \frac{P}{T}. \tag{5.13b}$$

The Gibbs equation

We now know both of the derivatives involved in the differential of the entropy. Substituting (5.13) in (5.5), we obtain the *Gibbs equation*,

$$dS = \frac{1}{T}dU + \frac{P}{T}dV. \tag{5.14}$$

This is a differential equation that we can integrate to determine the difference in S between any two states. Expressed in intensive form, the Gibbs equation is

$$ds = \frac{1}{T}du + \frac{P}{T}dv. \tag{5.15}$$

The intensive Gibbs equation provides the basis for evaluation of the entropy of simple compressible substances. Similar equations can be developed in the same general way for dielectrics, ferrofluids, and other substances.

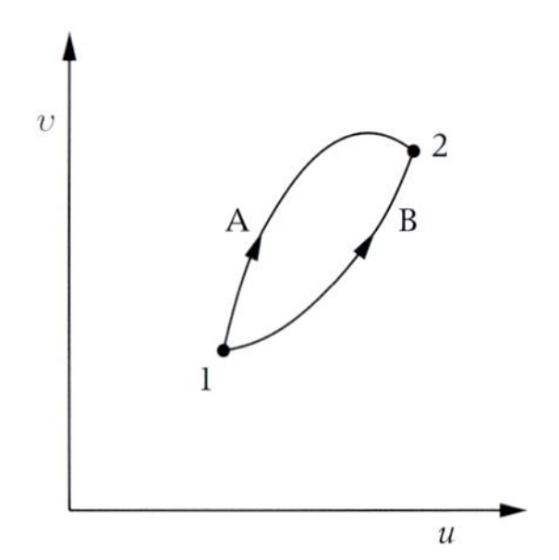

Figure 5.7 The entropy difference $s_2 - s_1$ is independent of the path of integration from state 1 to state 2.

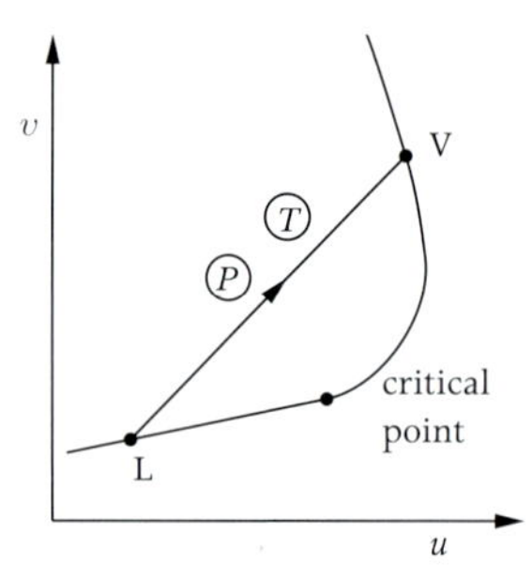

Figure 5.8 The pressure and temperature are both constant along the path of integration from state L to state V. The circle-T and circle-P symbols are used to indicate lines of constant T and P.

Units of entropy

Equation (5.14) shows that the dimensions of entropy S are those of energy/temperature, and so in the SI system the units of entropy are J/K. The SI units for specific entropy s are therefore $\mathrm{J}/\left(\mathrm{kg}\cdot\mathrm{K}\right)$, a combination that sometimes is given the alias *Gibbs*.

Integration of the Gibbs equation

Differences in entropy between two states can be determined by integrating (5.15) from one state to the other. The integration must be carried out along some path in the v–u plane, as indicated in Figure 5.7. If we denote the states by 1 and 2,

$$\int_1^2 ds = s_2 - s_1 = \int_1^2 \frac{1}{T}du + \int_1^2 \frac{P}{T}dv. \tag{5.16}$$

The values of the two integrals on the right will depend upon the path of integration in the v–u plane; but the sum of the two integrals will be independent of the path because the entropy is a property and therefore $s_1 = s_1(u_1, v_1)$ and $s_2 = s_2(u_2, v_2)$. The relationships between u, v, T, and P, must be consistent with this path-independent behavior, and this fact is extremely useful in generating thermodynamic property information from experimental data or theoretical models.

Example: entropy change for evaporation. Consider the evaporation of a liquid at a constant pressure. As discussed in Chapter 3, the temperature is also constant for this process. Therefore, the Gibbs equation is easily integrated from the saturated liquid state L to the saturated vapor state V along a line of constant T and P (Figure 5.8),

$$\int_L^V ds = \int_L^V \frac{1}{T}du + \int_L^V \frac{P}{T}dv,$$

which gives

$$s^V - s^L = \frac{1}{T}\left(u^V - u^L\right) + \frac{P}{T}\left(v^V - v^L\right).$$

Since $P^L = P^V$, this is equivalent to

$$s^{LV} = \frac{1}{T}\left(h^V - h^L\right) = \frac{h^{LV}}{T}, \tag{5.17}$$

where $s^{LV} \equiv s^V - s^L$ and $h^{LV} \equiv h^V - h^L$. You can verify this equation using tabulated thermodynamic property data. Equation (5.17) was used in the computer codes making FLUIDPROP, therefore the property tables in the appendix also comply with it; in many other tables the equation was not used and so you may detect tiny errors due to slight numerical inconsistencies in data fitting.

Example: entropy differences for an ideal gas. For an ideal gas where $u = u(T)$,

$$du = c_v(T)dT \qquad \text{and} \qquad Pv = RT,$$

which, when substituted in (5.16), give

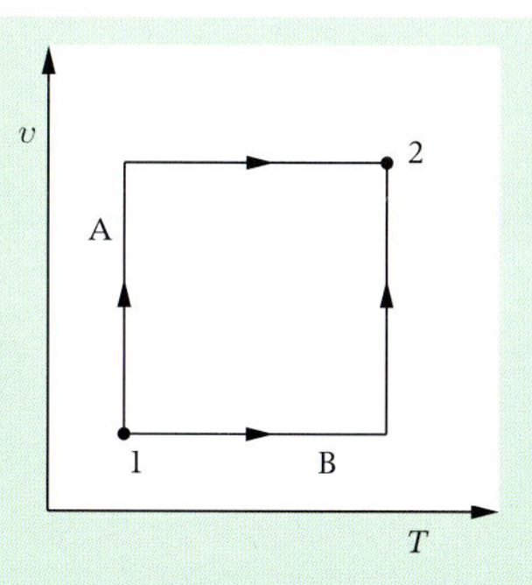

Figure 5.9 The T and v integrals are both the same for paths A and B from state 1 to state 2.

$$s_2 - s_1 = \int_1^2 \frac{c_v(T)}{T} dT + R \int_1^2 \frac{1}{v} dv. \quad (5.18)$$

Here it is easy to see that the integrals along paths A and B in Figure 5.9 are the same because the T integration is independent of v and the v integration is independent of T. If we can treat c_v as constant over the temperature range involved, then (5.18) gives

$$s_2 - s_1 = c_v \ln (T_2/T_1) + R \ln (v_2/v_1). \quad (5.19)$$

This equation is one of the *equations of state* of an ideal gas with constant specific heats, and is very useful in doing analyses of the performance of devices involving gases that can be modeled in this manner.

Entropy datum states

In situations without chemical reactions, the entropy balances only involve differences in entropies of the substance. The charts and tables use arbitrarily chosen datum states, where $s = 0$ at the datum state (0). When chemical reactions are considered, it is important to know the entropy s_0 at the datum state, and you need to remember that the values of s in the charts and tables really are $s - s_0$.

Thermodynamic definitions of temperature and pressure

The identifications of X and Y give (see (5.4))

$$\frac{1}{T} = \left(\frac{\partial s}{\partial u}\right)_v, \quad (5.20)$$

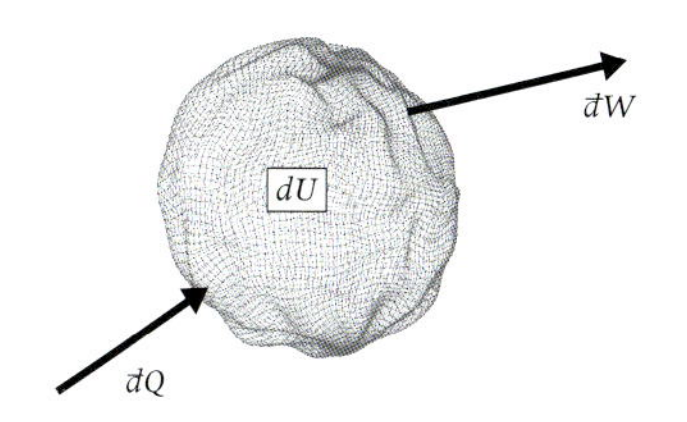

Figure 5.10 The control mass is in equilibrium at a uniform temperature T and pressure P, and undergoes a reversible process.

and

$$P = T\left(\frac{\partial s}{\partial v}\right)_u. \quad (5.21)$$

In some presentations, one begins with these as *the thermodynamic definitions of temperature and pressure*. Using the reverse of the arguments above that identified X and Y, one can then show that the thermodynamic temperature is the same as the ideal gas temperature, and that the thermodynamic pressure is the same as the pressure measured with a mechanical pressure gauge, provided that the substance is in an equilibrium state (assumed above when we considered the adiabatic expansion as being reversible).

5.5 Entropy Transfer with Heat

Entropy transfer for a reversible process

Consider a control mass consisting of a simple compressible substance in internal equilibrium, having uniform temperature T and a uniform pressure P (Figure 5.10). Suppose it receives heat $đQ$ and does work $đW = PdV$. An energy balance on the mass gives

$$\underbrace{đQ}_{\text{energy input}} = \underbrace{PdV}_{\text{energy output}} + \underbrace{dU}_{\text{energy accumulation}}. \quad (5.22)$$

The change in the entropy of the system is given by the Gibbs equation (5.15). Substituting for dU in the Gibbs equation using (5.22), the pressure terms cancel and one is left with

$$dS = \frac{1}{T} đQ. \quad (5.23)$$

If the process inside the control mass is reversible, then no entropy is produced inside and any entropy changes must be due to entropy transfer across the boundary. By concept, work does not transport any entropy, and therefore the only possible entropy transport is that associated with the heat transfer $đQ$. Therefore, for a reversible process we can rearrange (5.23) and interpret the result as the entropy balance on the control mass:

$$\underbrace{\frac{đQ}{T}}_{\substack{\text{entropy}\\\text{input}}} + \underbrace{0}_{\substack{\text{entropy}\\\text{production}}} = \underbrace{0}_{\substack{\text{entropy}\\\text{output}}} + \underbrace{dS}_{\substack{\text{entropy}\\\text{accumulation}}}.$$

Thus, for this reversible process, *the amount of entropy transferred with the heat* is $\dfrac{đQ}{T}$.

Where are we headed?

The purpose of this section is to show that in *all* situations the entropy transfer with heat is $đQ/T$, where T is the temperature on the boundary where the transfer takes place. The entropy transfer can be viewed as a transfer of microscopic disorder brought about by microscopically disorganized energy transfer (energy transfer as heat). We begin these developments by introducing the two simple conceptual systems shown in Figures 5.11 and 5.12.

Thermal energy reservoir

A *thermal energy reservoir* (TER) is imagined as a large mass of some simple compressible substance that is maintained at constant volume and is in internal equilibrium at a uniform temperature T. A TER acts as a source or sink for energy transfer as heat (disorganized energy transfer). We can compute the entropy change of a TER when we transfer a small amount of energy $đQ$ to the TER as heat using the Gibbs equation and energy balance (see Figure 5.11),

$$dS_{\text{TER}} = \frac{dU_{\text{TER}}}{T_{\text{TER}}} = \frac{đQ_{\text{in}}}{T_{\text{TER}}}. \tag{5.24}$$

Note that this is the same as (5.23) for a simple compressible substance undergoing a general reversible process, and hence there is no entropy produced in a TER.

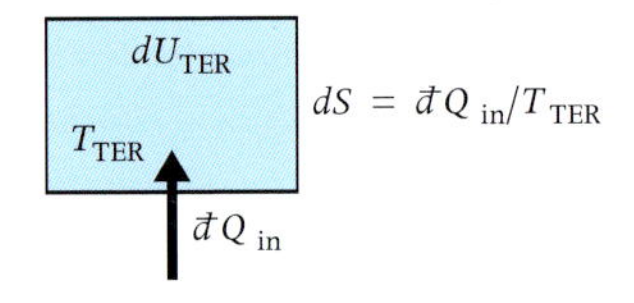

Figure 5.11 Thermal energy reservoir.

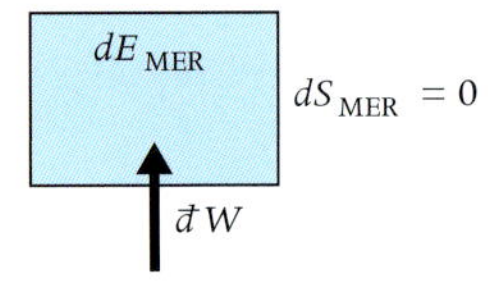

Figure 5.12 Mechanical energy reservoir.

Mechanical energy reservoir

A *mechanical energy reservoir* (MER) is imagined as a frictionless flywheel that can store energy it receives as work in a microscopically fully organized manner (molecules rotating around the axis in tandem), and return that energy without any loss through the output of energy as work (see Figure 5.12). There is no change in the entropy (microscopic disorder) in a MER when it receives or delivers energy, so by concept

$$dS_{\text{MER}} = 0.$$

Since its entropy does not change, and there is no entropy transferred in or out, the entropy production in a MER is zero.

Entropy production by heat transfer

Now let's imagine the transfer of energy as heat from a TER at a higher temperature to another TER at a lower temperature (Figure 5.13a), separated by a conducting medium. The TERs are assumed to be so large that their temperatures change negligibly during this transfer, and the conducting medium operates in steady-state, with no change in its entropy. The combined system is isolated, so no entropy flows in or out. The entropy balance on the combined system is therefore

$$\underbrace{0}_{\substack{\text{entropy}\\\text{input}}} + \underbrace{đ\mathcal{P}_S}_{\substack{\text{entropy}\\\text{production}}} = \underbrace{0}_{\substack{\text{entropy}\\\text{output}}} + \underbrace{dS_{\text{TER1}} + dS_{\text{TER2}}}_{\substack{\text{entropy}\\\text{accumulation}}}.$$

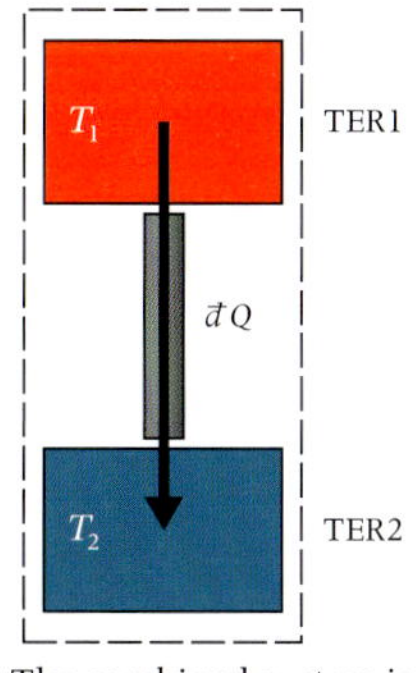

(a) The combined system is isolated.

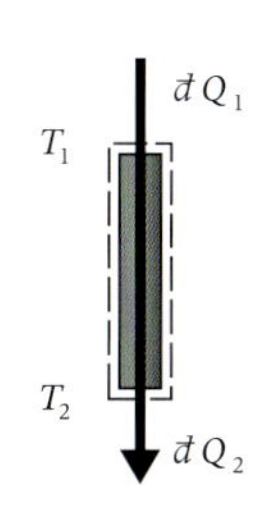

(b) Balance on the conducting medium.

Figure 5.13 Transfer of energy as heat from a TER at a higher temperature to a TER at a lower temperature.

Note that we use $đ$ to denote the infinitesimality of the entropy production, rather than d, since $đ\mathcal{P}_S$ is a small *amount* and not a small *change*. Since no entropy is produced in the TERs, all of the entropy production arises in the conducting medium. We evaluate the entropy change of each TER using (5.24), noting that $đQ_{\text{in,TER1}} = -đQ$ and $đQ_{\text{in,TER2}} = +đQ$,

$$dS_{\text{TER1}} = -\frac{đQ}{T_1} \qquad dS_{\text{TER2}} = \frac{đQ}{T_2}.$$

Therefore, the entropy produced in the conducting medium is

$$đ\mathcal{P}_S = \left(\frac{1}{T_2} - \frac{1}{T_1}\right) đQ \geq 0, \tag{5.25}$$

where $đQ$ is the energy transfer as heat from the warmer TER1 to the colder TER2. Note that $đ\mathcal{P}_S \geq 0$, as required by the Second Law of Thermodynamics, if $đQ > 0$ and $T_1 > T_2$. We can rearrange (5.25) and interpret the result as an *entropy balance on the conducting medium* (Figure 5.13b),

$$\underbrace{\frac{đQ}{T_1}}_{\substack{\text{entropy}\\\text{input}}} + \underbrace{đ\mathcal{P}_S}_{\substack{\text{entropy}\\\text{production}}} = \underbrace{\frac{đQ}{T_2}}_{\substack{\text{entropy}\\\text{output}}} + \underbrace{0}_{\substack{\text{entropy}\\\text{accumulation}}}. \tag{5.26}$$

Note that again $đQ/T$ emerges as the entropy transfer with heat, with T being the temperature at the point where the transfer is evaluated.

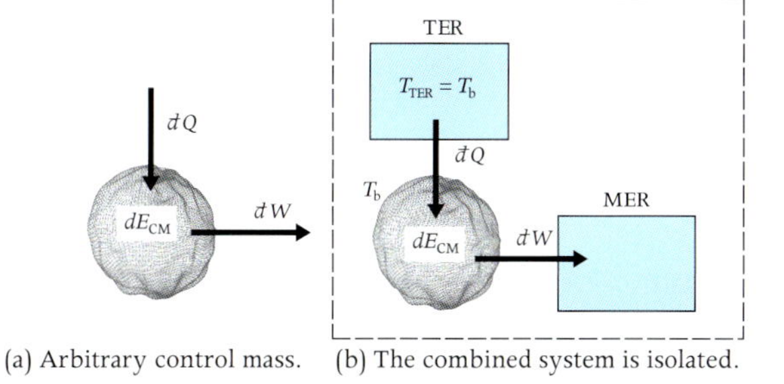

(a) Arbitrary control mass. (b) The combined system is isolated.

Figure 5.14 Entropy transfer with heat supplied by a TER and the work delivered by a MER.

Reversible heat transfer

From (5.26), the ratio of the entropy production to the entropy transfer from TER1 is

$$\frac{đ\mathcal{P}_S}{đQ/T_1} = \left(\frac{T_1}{T_2} - 1\right).$$

In the limiting case when T_1 and T_2 become equal, the entropy production per unit of entropy transfer vanishes; this means that, in the limit of vanishing temperature difference across the conductor, the *heat transfer becomes reversible*. In real problems there is always a finite temperature difference driving the heat transfer, and this is usually an important source of entropy production; but, in the limit of zero driving temperature difference, the heat transfer becomes reversible and does not produce a significant amount of entropy (compared to the entropy transfer).

Entropy transfer as heat

Consider the control mass of Figure 5.14a, for which the energy balance is

$$\underbrace{đQ}_{\substack{\text{energy}\\\text{input}}} = \underbrace{đW}_{\substack{\text{energy}\\\text{output}}} + \underbrace{dE_{CM}}_{\substack{\text{energy}\\\text{accumulation}}}.$$

We would like to write the entropy balance for this control mass, but as yet we do not know what to write for the entropy input with the heat. So, instead let's write the entropy balance on an isolated system in which we imagine the heat $đQ$ as being supplied by a TER and the work $đW$ delivered to a MER (Figure 5.14b) and see what it tells us. The entropy balance on the combined system is

$$\underbrace{0}_{\text{entropy input}} + \underbrace{đ\mathcal{P}_S}_{\text{entropy production}} = \underbrace{0}_{\text{entropy output}} + \underbrace{dS_{\text{CM}} + dS_{\text{TER}} + dS_{\text{MER}}}_{\text{entropy accumulation}}. \tag{5.27}$$

We will configure the TER and MER so that all of the entropy production is due to processes within the control mass. To achieve this we must make the heat transfer reversible, and we can imagine doing this by making the temperature difference infinitesimal between the TER and the point on the control mass boundary where the heat is transferred. Therefore, we take the TER temperature as the temperature T_b on the control mass boundary where the energy comes in as heat. Using the results above for the entropy changes of the TER and MER, taking care to note that $đQ_{\text{in,TER}} = -đQ$, (5.27) becomes

$$đ\mathcal{P}_{S,\text{CM}} = dS_{\text{CM}} - \frac{đQ}{T_B}. \tag{5.28}$$

We rearrange (5.28) and interpret the result as the *entropy balance on the control mass* (Figure 5.14a):

$$\underbrace{\frac{đQ}{T_B}}_{\text{entropy input}} + \underbrace{đ\mathcal{P}_{S,\text{CM}}}_{\text{entropy production}} = \underbrace{0}_{\text{entropy output}} + \underbrace{dS_{\text{CM}}}_{\text{entropy accumulation}}.$$

Again we find that $đQ/T$ is the entropy transfer associated with heat transfer, with T being the temperature on the boundary where the transfer occurs.

In summary, the amount of entropy transfer with heat is determined by the amount of heat transfer $đQ$ and the temperature T at the point on the boundary where the heat transfer occurs,

$$\text{amount of entropy transfer with heat} = \frac{đQ}{T}. \tag{5.29}$$

The entropy transfer is in the same direction as the energy transfer.

Rate of entropy transfer with heat

Dividing (5.29) by dt, the time interval for transfer of the energy $đQ$ as heat, the *rate* of entropy transfer as heat is the heat transfer rate $\dot{Q}$ divided by the boundary temperature T,

$$\text{rate of entropy transfer with heat} = \frac{\dot{Q}}{T}. \tag{5.30}$$

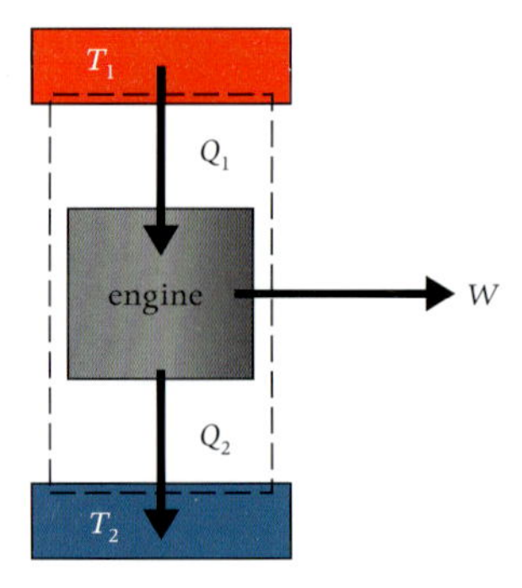

Figure 5.15 A simple engine.

It is important to remember that the T in the entropy transfer is the temperature at the point *on the boundary* where the transfer is evaluated.

5.6 Example Uses of Control Mass Entropy Balances

Carnot efficiency for energy conversion

An engine receives energy as heat Q_1 from a high-temperature source (e.g., flames) at temperature T_1, rejects energy as heat Q_2 to a sink (e.g., river water) at lower temperature T_2, and puts out shaft work W (Figure 5.15). These energy transfers are either for a definite period of steady-state operation, or if the engine executes cyclic internal processes are for an integer number of identical cycles, so that the accumulation of energy and entropy within the engine are zero over the time period being examined. What is the maximum possible *energy conversion efficiency* ($\eta \equiv W/Q_1$) for such an engine?

This sort of problem is analyzed using a combination of energy and entropy balances. Since there is no energy or entropy accumulation, the energy and entropy balances on the indicated control mass are

$$\underbrace{Q_1}_{\text{energy input}} = \underbrace{Q_2 + W}_{\text{energy output}} \tag{5.31}$$

and

$$\underbrace{\frac{Q_1}{T_1}}_{\text{entropy input}} + \underbrace{\mathcal{P}_S}_{\text{entropy production}} = \underbrace{\frac{Q_2}{T_2}}_{\text{entropy output}}. \tag{5.32}$$

Since the question involves W and Q_1, we use (5.31) to eliminate Q_2, and substitute in (5.32), obtaining

$$\frac{Q_1}{T_1} + \mathcal{P}_S = \frac{Q_1 - W}{T_2},$$

from which we solve for the efficiency as

$$\eta = \frac{W}{Q_1} = \left(1 - \frac{T_2}{T_1}\right) - \frac{T_2 \mathcal{P}_S}{Q_1}.$$

The second law requires $\mathcal{P}_S \geq 0$, and $T_2 > 0$, so the last term is positive and hence reduces the efficiency. The maximum efficiency is therefore obtained for a reversible device, for which $\mathcal{P}_S = 0$. Thus

$$\eta_{\max} = 1 - \frac{T_2}{T_1}$$

is called the *Carnot efficiency*, honoring an early scientist who pioneered the development of second law concepts. It sets a limit on what can be achieved with real devices, *without saying anything about how they might be made!* The power and importance of this sort of general thermodynamic analysis should be evident.

Some engines must operate with much lower source temperatures, like the Ocean Thermal Energy Conversion (OTEC) power plants currently under investigation. These power plants use warm ocean surface water as the energy source and cold deep water as the sink. In this case typical temperatures are T_1 = 315 K and T_2 = 290 K, for which the Carnot efficiency would be 1 – 290 / 315 = 0.08 (8%). Although the energy (here from the Sun) would be "free," that does not necessarily make these power plants economical. With such an efficiency, possibly as low as 4%, an enormous amount of warm water would have to be processed to produce the power for a typical city, and this would require huge heat exchangers and massive capital investment for equipment that would have a relatively short lifetime in the harsh ocean environment. OTEC systems are a wonderful topic for design studies, and someday they may become economically attractive *if* low-cost, high-performance durable materials can be developed through research.

For example, suppose that $T_1 = 600$ K and $T_2 = 300$ K. The Carnot efficiency is then $1 - 300/600 = 0.5$ (50%). A real device operating at these temperatures would produce entropy and hence would have a lower efficiency, perhaps 20%.

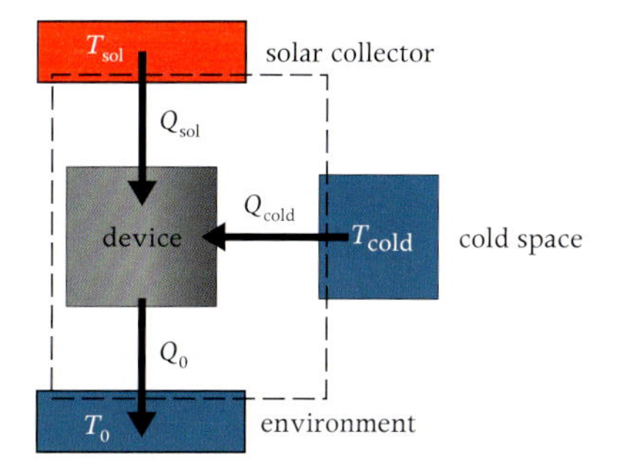

Figure 5.16 Solar powered heat pump.

Example: Solar powered refrigerator. An inventor seeking venture capital claims to have a refrigeration system whose simplified sketch is shown in Figure 5.16. It is powered by solar energy transferred as heat Q_{sol} from a solar collector at temperature T_{sol}, pulls energy as heat Q_{cold} from the cold space at temperature T_{cold}, dumps energy as heat Q_0 to the environment at temperature T_0, with $T_{sol} > T_0 > T_{cold}$, and operates steadily (or in a cycle). The inventor claims that, under typical conditions where $T_{sol} = 400$ K, $T_0 = 300$ K, and $T_{cold} = 250$ K, the *coefficient of performance*, COP $\equiv Q_{cold}/Q_{sol}$, of the device is more than 2, indicating that the cooling rate is twice the energy collection rate. Is this possible?

We start our analysis by applying basic energy and entropy balances. Because of the steady (or cyclic) operation, we assume that there is no energy or entropy accumulation, and so the energy and entropy balances on the indicated control mass are

$$\underbrace{Q_{sol} + Q_{cold}}_{\text{energy input}} = \underbrace{Q_0}_{\text{energy output}} \tag{5.33}$$

and

$$\underbrace{\frac{Q_{sol}}{T_{sol}} + \frac{Q_{cold}}{T_{cold}}}_{\text{entropy input}} + \underbrace{\mathcal{P}_S}_{\text{entropy production}} = \underbrace{\frac{Q_0}{T_0}}_{\text{entropy output}}. \tag{5.34}$$

Substituting in (5.34) for Q_0 from (5.33),

$$\frac{Q_{sol}}{T_{sol}} + \frac{Q_{cold}}{T_{cold}} + \mathcal{P}_S = \frac{Q_{sol} + Q_{cold}}{T_0}.$$

Solving for the COP,

$$\text{COP} = \frac{Q_{cold}}{Q_{sol}} = \left(\frac{1 - T_0/T_{sol}}{T_0/T_{cold} - 1}\right) - \frac{T_0 \mathcal{P}_S/Q_{sol}}{T_0/T_{cold} - 1}.$$

The second law requires $P_S \geq 0$, so the last term contains factors that are all positive and hence the entropy production term will reduce the COP below that given by the first term. Therefore, the maximum COP would be obtained for a reversible device that did not produce entropy,

$$\text{COP}_{max} = \frac{1 - T_0/T_{sol}}{T_0/T_{cold} - 1}.$$

For the inventor's reported temperatures,

$$\text{COP}_{max} = \frac{1 - 300/400}{300/250 - 1} = 1.25.$$

Therefore, without knowing anything in detail about the supposed device, we can be sure that the inventor cannot possibly have such a device with a COP = 2. The inventor is making claims that simply cannot be true. You can now protect your friends from making foolish investments in can't-work systems by doing a simple analysis like this. However, it might be possible for a real device to achieve a COP of (say) 0.5 with these temperatures. The engineering of such a system would require very clever design and very careful analysis. Analysis of real systems requires knowledge of what is inside them and is very case-specific. After just one more chapter you will be ready to tackle such analysis.

Adiabatic compression

A gas is to be compressed adiabatically in a piston–cylinder system from a given state 1 to some state 2 at a given smaller volume. What is the minimum work required for the compression?

The energy and entropy balances on the control mass (Figure 5.17) are

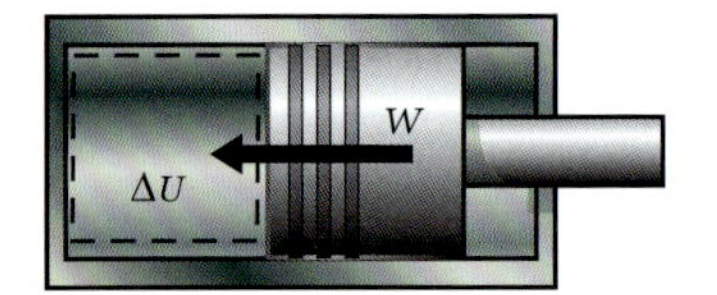

Figure 5.17 Adiabatic compression analysis.

$$\underbrace{W}_{\text{energy input}} = \underbrace{M(u_2 - u_1)}_{\text{energy accumulation}}$$

and

$$\underbrace{\mathcal{P}_S}_{\text{entropy production}} = \underbrace{M(s_2 - s_1)}_{\text{entropy accumulation}}. \tag{5.35}$$

Using the ideal gas property relation $u = c_v T$,

$$W = Mc_v(T_2 - T_1). \tag{5.36}$$

Since T_1 is fixed, the least work will be required for a device having a process that yields the lowest possible T_2. The entropy change of the gas is given by (5.19), which can be rewritten as

$$\frac{T_2}{T_1} = \left(\frac{v_1}{v_2}\right)^{R/c_v} \exp\left(\frac{s_2 - s_1}{c_v}\right). \tag{5.37}$$

Since v_1 and v_2 are fixed, the smallest T_2 will be obtained for the smallest s_2. The second law requires $\mathcal{P}_S \geq 0$, so the entropy balance (5.35) gives $s_2 \geq s_1$, with the equality holding for the limiting case of a reversible compression. Therefore, the minimum work for an (adiabatic) compresson is obtained for an *isentropic* (constant entropy) process. Using (5.37) in (5.36), this work is

$$W_{min} = Mc_v T_1 \left[\left(\frac{v_1}{v_2}\right)^{R/c_v} - 1\right].$$

If we compress the gas rapidly, the gas molecules will pile up near the face of the piston (Figure 5.18a). This will cause a higher pressure on the piston face, and hence more work done by the piston, than if the compression is carried out slowly (Figure 5.18b). The entropy production in rapid compression arises because of the heat transfer and viscous effects required to smooth out the non-uniformities. In order for the process to be modeled as reversible, the piston speed must be slow compared to the speed of sound in the gas, which typically is about 300 m/s, if the gas is air. Piston speeds in typical devices are much slower, and hence it is often reasonable to model a process like this as reversible.

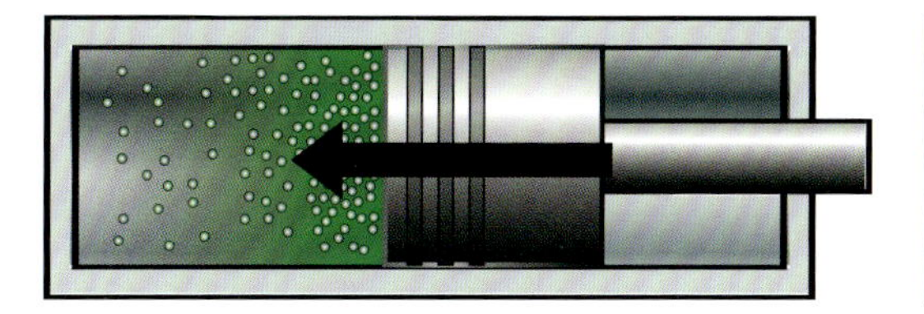

(a) Rapid compression requires more work. (b) Slow compression is ideally reversible.

Figure 5.18 Compression in a piston–cylinder system.

5.7 Example Uses of Control Volume Entropy Balances

Methane liquefaction

A plant is to be designed to turn pipeline gas (methane) into liquid for ocean tanker transport. The methane will be supplied to the plant at 1 atm, 300 K and must be discharged as a saturated liquid at 1 atm. What is the minimum power requirement for a steady-state plant processing 10 kg/s?

Figure 5.19 shows the pertinent energy flows for this analysis. Entropy flows in and out of the control volume with the mass. Since the liquid methane is microscopically less chaotic than gaseous methane, the methane will leave with less entropy than it had when it entered, and so entropy must leave the plant in some other way. The only other way entropy can be transported out is with heat, and so there must be some heat transfer from the plant. $\dot{Q}_0$ denotes this heat transfer rate to the environment at temperature T_0. $\dot{W}$ is the net input power required by the compressors, pumps, and whatever else is inside the plant. The steady-state rate-basis energy and entropy balances are

$$\underbrace{\dot{M}h_1 + \dot{W}}_{\substack{\text{rate of}\\ \text{energy input}}} = \underbrace{\dot{M}h_2 + \dot{Q}_0}_{\substack{\text{rate of}\\ \text{energy output}}} \tag{5.38}$$

and

$$\underbrace{\dot{M}s_1}_{\substack{\text{rate of}\\ \text{entropy input}}} + \underbrace{\dot{\mathcal{P}}_S}_{\substack{\text{rate of}\\ \text{entropy}\\ \text{production}}} = \underbrace{\dot{M}s_2 + \dot{Q}_0/T_0}_{\substack{\text{rate of}\\ \text{entropy output}}} . \tag{5.39}$$

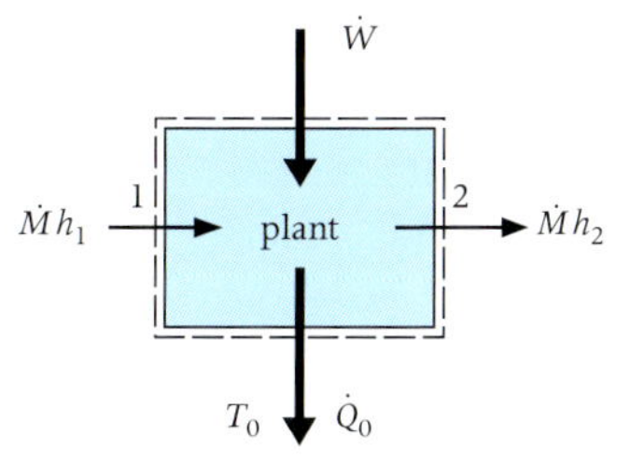

Figure 5.19 Methane liquefaction plant.

The inlet and discharge states are fixed. Different designs will have different $\dot{W}$, $\dot{Q}_0$, and $\dot{\mathcal{P}}_S$. Substituting in (5.39) for $\dot{Q}_0$ from (5.38), and solving for $\dot{W}$, one finds

$$\dot{W} = \dot{M}\left[(h_2 - T_0 s_2) - (h_1 - T_0 s_1)\right] + T_0 \dot{\mathcal{P}}_S .$$

The second law requires $\dot{\mathcal{P}}_S \geq 0$. The entropy production term therefore increases $\dot{W}$ for given inlet and discharge states. Hence, the minimum power requirement is

$$\dot{W}_{\text{min}} = \dot{M}\left[(h_2 - T_0 s_2) - (h_1 - T_0 s_1)\right] .$$

Using thermodynamic property values from TPSI,

$$\begin{aligned}\dot{W}_{\text{min}} &= 10\ \text{kg/s} \times [(70.70 - 300 \times 0.6987) \\ &\quad - (984.69 - 300 \times 7.3878)]\ \text{kJ/kg} \times 10^{-3} \\ &= 10.9\ \text{MW}.\end{aligned}$$

Available energy

The combination $h - T_0 s$ that arose in the example above is not a property of the fluid because it contains T_0, and hence it is not in the property tables. It always arises in steady-flow energy-entropy analysis, which is often called *exergy analysis*.

We will discuss exergy analysis in more detail in Chapter 9.

Turbine analysis: turbine isentropic efficiency

A turbine is a device that extracts mechanical power from a fluid. In a typical gas turbine, hot high-pressure gas is first accelerated in the turbine nozzle to high speed, and then impinged on the blades of the turbine rotor, causing it to spin. Often there are several nozzle-rotor stages in series. Power is transmitted by the rotating turbine shaft. Heat transfer from the turbine casing is usually not very large compared to the power generated, and so a turbine is usually idealized as being adiabatic in preliminary design analysis.

Figure 5.20a shows the pertinent energy flows. The rate-basis energy and entropy balances are

$$\underbrace{\dot{M}h_1}_{\text{rate of energy input}} = \underbrace{\dot{M}h_2 + \dot{W}}_{\text{rate of energy output}}$$

and

$$\underbrace{\dot{M}s_1}_{\text{rate of entropy input}} + \underbrace{\dot{\mathcal{P}}_S}_{\text{rate of entropy production}} = \underbrace{\dot{M}s_2}_{\text{rate of entropy output}}.$$

The energy balance gives the work output per unit mass (specific work) as

$$w \equiv \frac{\dot{W}}{\dot{M}} = h_1 - h_2.$$

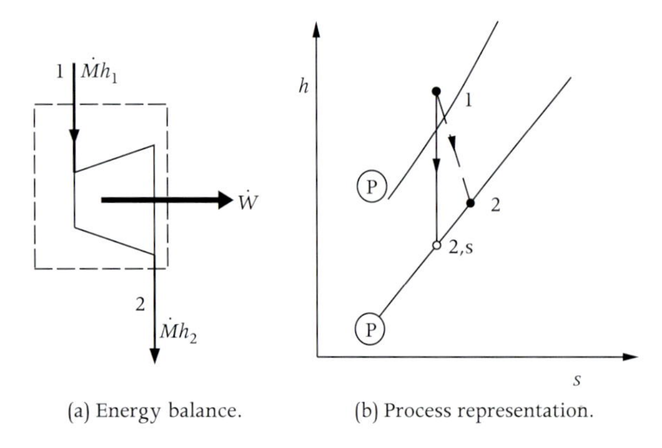

(a) Energy balance. (b) Process representation.

Figure 5.20 Turbine analysis.

Since the second law requires $\mathcal{P}_S \geq 0$, the entropy balance gives

$$s_2 \geq s_1.$$

The *hs* thermodynamic plane (called the *Mollier diagram*) is extremely useful in turbine analysis. Figure 5.20b shows the turbine *process representation* on the *hs* plane. The inlet state 1 is fixed, and the turbine discharge pressure is fixed, so state 2 must lie somewhere on the line $P_2 = const$. The energy balances showed that the work output per unit mass is $h_1 - h_2$, so the best turbine would have the least possible h_2. The second law analysis showed that $s_2 \geq s_1$, so state 2 cannot be to the left of the reference state (2, *s*), which has the same entropy as state 1. Hence, h_2 cannot be less than $h_{2,s}$. Therefore, the best adiabatic turbine is *isentropic*, with discharge at state (2, *s*).

The *isentropic efficiency* of a turbine is defined as the ratio of the work output from the actual turbine to the work output of an ideal (isentropic) turbine *having the same inlet state and discharge pressure*. The work output (per unit mass) from the ideal turbine would be $w_s = h_1 - h_{2,s}$, while the work from the actual turbine is $h_1 - h_2$, so the isentropic efficiency of a turbine is

$$\eta_s \equiv \frac{w}{w_s} = \frac{h_1 - h_2}{h_1 - h_{2,s}} \qquad \text{(turbine or expander).}$$

For example, suppose that a steam turbine has an isentropic efficiency of 0.82, with an inlet state of 600 K and 1 MPa and a discharge pressure of 0.1 MPa. Using IF97, the calculation goes as follows:

$$w_s = h_1 - h_{2,s} = (3108 - 2623)\text{ kJ/kg} = 485\text{ kJ/kg},$$
$$w = \eta w_s = 0.82 \times 485\text{ kJ/kg} = 398\text{ kJ/kg},$$
$$h_2 = h_1 - w = (3108 - 398)\text{ kJ/kg} = 2710\text{ kJ/kg}.$$

Finally, h_2 and P_2 fix state 2. One finds that state 2 is in the superheat (gas) region at about 390 K.

Compressor analysis: compressor isentropic efficiency

A compressor is a device designed to raise the pressure of a fluid. This takes mechanical power input. Small compressors use pistons in cylinders to do the

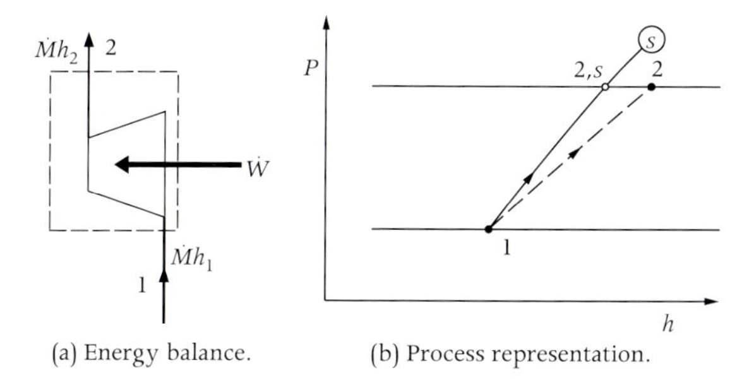

(a) Energy balance. (b) Process representation.

Figure 5.21 Compressor analysis.

compression. Large compressors, such as those in aircraft jet engines, use a series of rotating blades that accelerate the flow and stationary blades that slow the flow down, causing a rise in pressure. Heat transfer is usually small in comparison to the mechanical energy input, and so compressors are often analyzed as adiabatic.

Figure 5.21a shows the significant energy terms for an adiabatic compressor. The rate-basis energy and entropy balances are

$$\underbrace{\dot{M}h_1 + \dot{W}}_{\substack{\text{rate of} \\ \text{energy input}}} = \underbrace{\dot{M}h_2}_{\substack{\text{rate of} \\ \text{energy output}}}$$

and

$$\underbrace{\dot{M}s_1}_{\substack{\text{rate of} \\ \text{entropy} \\ \text{input}}} + \underbrace{\dot{\mathcal{P}}_S}_{\substack{\text{rate of} \\ \text{entropy} \\ \text{production}}} = \underbrace{\dot{M}s_2}_{\substack{\text{rate of} \\ \text{entropy} \\ \text{output}}} .$$

The energy balance gives the work per unit mass as

$$w \equiv \frac{\dot{W}}{\dot{M}} = h_2 - h_1. \tag{5.40}$$

Since the second law requires $\mathcal{P}_S \geq 0$, the entropy balance gives

$$s_2 \geq s_1.$$

Thus, the discharge state 2 cannot lie to the left of the reference state $(2, s)$ in the P–h process representation ($s_{2,s} = s_1$), cf. Figure 5.21b. From (5.40) we see that the best compressor would have the least possible h_2. Therefore, the best adiabatic compressor is isentropic, and discharges the fluid at state $(2, s)$.

The *isentropic efficiency* of a compressor is defined as the ratio of the work required by an ideal isentropic compressor to the work required by the actual compressor, where the ideal compressor has the *same inlet state and discharge pressure* as the actual compressor,

$$\eta_s \equiv \frac{w_s}{w} = \frac{h_{2,s} - h_1}{h_2 - h_1} \qquad \text{(compressor or pump)}.$$

Note that this is the inverse of the definition of the isentropic efficiency of a turbine. In both cases the isentropic efficiency is defined so that it will be less than unity for a real device, provided of course that the device is truly adiabatic.

5.8 Entropy in Non-equilibrium States

Entropy

The First and Second Laws of Thermodynamics are the fundamental tools for engineering analysis. The first involves conservation of energy and the second production of entropy. The *energy* of a set of molecules is the sum of their energies; if the system is isolated, on its approach to equilibrium the energy may shift around, but the total energy will *not change*. The *entropy* of this set of molecules depends upon the randomness of the energy distribution among the molecules; if the system is isolated, the total entropy will *increase* during the relaxation to equilibrium, ultimately reaching its maximum for the given constraints (e.g., total energy and volume for a simple compressible substance). The differential equation for entropy change, i.e., the Gibbs equation (5.14) developed previously, applies for changes *between* thermodynamic equilibrium states. It does *not* describe entropy change for material not yet in thermodynamic equilibrium. The purpose of this section is to recast the formula for entropy so that it can be evaluated in *non-equilibrium* states. It will help the reader to understand the entropy concept, to be aware of the entropy change associated with relaxation to equilibrium, and to know that useful *microscopic* features of equilibrium thermodynamic states not available from classical thermodynamic analysis can be obtained from this general definition of entropy. This is presented here to alert the reader to other ideas about entropy that may be encountered in other reading. It will not be used in our analysis, but it might help a reader of other texts.

A helpful analogy

Suppose we have a large tray containing a large number of jumping beans, half red and half white, initially organized with the red all on one side and the white on the other. Observers viewing the tray from a great distance could not see individual beans and would characterize the object as red on one side and white on the other. As the beans began to jump, the observer would see reddish coloring appear on the side that was white and the redness of the other side diminish, and eventually both sides would become pink and the color change would cease. The beans would still go on jumping, but from the great distance no significant change would be noted. This analogy has many points of contact with the notions of entropy and the Second Law of Thermodynamics. We can replace the beans by atoms of argon and helium, the tray by an isolating wall, and the observer by ourselves. The quantitative measure of the microscopic randomness, of our uncertainty as to the exact microscopic state when we know only the macroscopic state, is the *entropy*. We can explain all processes observed in nature with the postulate that the microscopic randomness, as measured by the entropy, can be produced but is never destroyed.

Quantum states

A key idea from quantum mechanics is that the states that atoms, molecules, and entire systems have are *discretely quantized*. For example, a photon providing radiation of a particular frequency ν can have but one energy, given by the Einstein–Planck equation,

$$\varepsilon = \mathsf{h}\nu.$$

Planck's constant h is 6.626×10^{-34} J · s. The energy traveling with radiation is therefore said to be *quantized*. Other evidence for quantization is provided by measurements of the energy of electrons orbiting atoms, where it is found that the electrons can possess only very particular energies.

The quantum theory of matter and energy has met every test that has been devised. By postulating that any microscopic particle or system of particles constrained in some manner (such as being required to orbit a nucleus, to be in a magnetic field, or to be in a box of specified size) can exist only in certain *allowed quantum states*, we can obtain amazingly complete and precise descriptions of all natural phenomena. The allowed quantum states are determined by the nature of the particles and the circumstances imposed on them, and are calculated from the *Schrödinger equation*. For the present we need only the idea that the allowed states are quantized, but do not need to evaluate them.

It is important to realize that quantization refers to very detailed *microscopic* descriptions of state. At any instant a piece of matter must be in one of its allowed quantum states. Macroscopic instruments reflect averages over the sequence of quantum states and are much less detailed. For example, a *microscopic* description of state might require values for the energy and position of *each particle*, while *macroscopic* descriptions might involve only the total energy and volume, the average temperature and pressure, and the entropy.

Quantum states and entropy

An isolated system of many interacting molecules will undergo continuous change in its quantum state. Macroscopic instruments are unable to follow these rapid changes, hence tend to average over a relatively long sequence of quantum states. Repeated experiments starting out of equilibrium can be used to study the approach to equilibrium, but each experiment will involve its own sequence of microscopic states. A large set of such repetitions (real or hypothetical) is called an *ensemble* of experiments. The quantum state probability $p_i(t)$ is defined as the fraction of the ensemble of experiments in which quantum state i would be realized at time t. The sum of the probabilities of all of the allowed quantum states must be unity,

$$\sum p_i = 1.$$

If at time $t = 0$ the same quantum state $i = 1$ is always obtained, then $p_1(0) = 1$ and $p_j(0) = 0$ for $j > 1$; there is no microscopic randomness in this ensemble, and its entropy is zero. As the molecules move the state will change; more states will appear with non-zero probabilities, and the entropy will increase. The entropy of this isolated system will grow until its microscopic state is as random as possible. In Section 5.4 we developed the Gibbs equation to calculate changes in the equilibrium entropy, and the microscopic entropy model developed here must be consistent with these equilibrium states. Our goal now is to extend the mathematics for entropy introduced earlier in this chapter to microscopic non-equilibrium situations. This will also provide important information for equilibrium situations not derivable from thermodynamics.

Entropy and microstate probabilities

A list of all the quantum states and their probabilities p_i would convey a very complete picture of the randomness and uncertainty as to the instantaneous microscopic state. But the list would be far too long to be useful; it would be much better if we could use a *single number* to measure the amount of randomness and uncertainty reflected by the entire list of p_i. Therefore, we will seek a definition of entropy in terms of the probabilities of all of the microscopic states. The product of the state probabilities $S = p_1 \times p_2 \times \cdots \times p_n$ won't do because it is zero when any of the state probabilities is zero. The sum of the p_i will not do either because it is always unity. There are many other ideas that you can show do not work; but there is one idea that will. The randomness measure must be constructed so as to give greater values for S when the microscopic state is more random. We look for a definition of entropy in the form

$$S = \sum_i p_i f(p_i)$$

where the function $f(p_i)$ still has to be found. Since entropy by concept is an extensive property, the entropy of system C, composed of two separate parts A and B, must be given by the linear sum of the entropies of A and B,

$$S_C = S_A + S_B,$$

where

$$S_A = \sum_{i=1}^{n} p_i f(p_i), \qquad S_B = \sum_{j=1}^{m} p_j f(p_j).$$

The entropy of the combined system, expressed in terms of the probabilities of its two components, is

$$S_C = \sum_{i=1}^{n} \sum_{j=1}^{m} p_{ij} f(p_{ij}).$$

Since the i and j states are completely independent, the joint probability is $p_{ij} = p_i p_j$, hence

$$S_C = \sum_{i=1}^{n} \sum_{j=1}^{m} p_i p_j f(p_i p_j) = \sum_{i=1}^{n} p_i f(p_i) + \sum_{j=1}^{m} p_j f(p_j).$$

Note that f must be such that this is true regardless of the values of p_i and p_j. A function that will work is $f(\ ) = \ln(\ )$. With this choice the left-hand side becomes

$$\sum_{i=1}^{n} \sum_{j=1}^{m} p_i p_j \ln p_i + \sum_{i=1}^{n} \sum_{j=1}^{m} p_i p_j \ln p_j.$$

The j sum in the first term is unity as is the i sum in the second, so this reduces to

$$\sum_i p_i \ln p_i + \sum_j p_j \ln p_j.$$

Since $p_i \leq 1$ and $\ln p_i \leq 0$, we put a minus sign to get a positive S and a constant to adjust the scale, and define the entropy as

$$S = -k \sum_i p_i \ln p_i.$$

This is the definition of the entropy in terms of the microstate probabilities. It is adjusted to match the macroscopic entropy changes involved in equilibrium situations, and it is useful for the evaluation of entropy in non-equilibrium situations. Its use is rather complex for beginning thermodynamics, and so we avoid its use when not needed. It is mentioned here to help readers who encounter it in other reading, from which more can be learned about it.

EXERCISES

5.1 Consider a control mass undergoing a cyclic process, thus returning to its initial state after a time interval. Prove that, if there is a net positive input of energy as heat $đQ$ into the control mass,

$$\oint \frac{đQ}{T} \leq 0.$$

(This is the so-called *inequality of Clausius*, and is an intermediate theorem in many treatments of the second law).

5.2 Calculate the thermodynamic pressure of saturated water vapor at 100 °C and compare the result with what can be obtained from tabulated values of vapor–liquid equilibrium pressures in Appendix A.3.

5.3 It has been proposed that energy be taken as heat from the atmosphere around Madagascar and used to run a power plant. Energy would be rejected as heat to the Indian Ocean. Estimate the maximum conversion efficiency of thermal energy into electric power that might be obtained with such a power plant.

5.4 An inventor claims to have built a device that receives 1 kW of energy as heat and converts it into 750 W of electric power. The rest of the energy is put out as mechanical work and dissipated outside the device by friction. Discuss the validity of these claims.

5.5 Thermal power systems for use in space normally reject energy as heat by radiating it into space. Since fluid-filled radiators are generally quite heavy, it has been suggested that this energy be converted into electricity and the current run through lighter resistors to dissipate the energy as heat. What do you think of this idea? Energy transfer as heat from a space vehicle must take place by radiation. The weight of a space radiator is proportional to its area, which is determined by the rate at which the energy must be radiated as heat. The rate of energy radiation is proportional to the product of the area and the fourth power of the radiator temperature. Consider a reversible engine providing a fixed amount of power output and operating with a fixed source temperature. Show that the least radiator weight is obtained when the radiator temperature is 0.75 times the source temperature.

5.6 The average amount of solar energy received at the earth's surface in a certain location is approximately 1350 W/m^2. Not all this energy can be used in a solar power plant, because of the reradiation of energy by the collector surface. Assume that the reradiation is described by

$$Q = A\sigma\varepsilon T^4,$$

where A is the collector area, ε is the emissivity, and σ is the Stefan–Boltzmann constant (equal to 5.67×10^{-8} W m^{-2} K^{-4}).

a. Calculate the maximum collector surface temperature at which all the incident solar energy is reradiated and the emissivity is equal to 0.5.

b. Suppose that the efficiency of the power plant equals the Carnot efficiency. If T_{amb} is the environmental temperature to which energy is rejected as heat and T_c is the collector temperature, determine the collector temperature at which the system should operate in order to achieve the most power output for a collector of a fixed size. Assume the environmental temperature is 25 °C.

5.7 Steam enters the nozzle of a turbine with a velocity of 3 m/s at a pressure of 35 bar and a temperature of 800 K. At the exit of the nozzle the pressure and temperature are found to be 1 bar and 400 K. Calculate the velocity at the exit of the nozzle and the entropy-production rate for the control volume enclosing the nozzle if the mass flow rate is 4500 kg/hr. Use Appendix A.3 for data.

5.8 Steam enters an adiabatic diffuser at 800 m/s as a saturated vapor at 0.4 MPa. What is the maximum possible discharge pressure? Use Appendix A.3 for data.

5.9 Compute the maximum quality of oxygen that can be achieved by expanding it adiabatically in a piston–cylinder system from the saturated vapor state at 10 bar to twice the volume. Use Appendix A.8 for data.

5.10 Referring to Appendix A.4, compute the minimum amount of power required by an adiabatic compressor that handles 5 kg/min of R134a, compressing it from a temperature and pressure of 320 K and 0.6 bar to a pressure of 4 bar.

5.11 Liquid methane is to be converted into gas for insertion into the natural gas pipeline by a steady-state gasification plant. The methane will enter the plant as a saturated liquid at 1 atmosphere (state 1: *very* cold) and will leave at the environment temperature T_0 and pipeline pressure P_0 (state 2). The device will use a heat exchanger to take energy from the environment. Starting from basic principles, derive an expression for the maximum amount of useful shaft power $\dot{W}$ that could be generated by such a plant. Express your result in terms of thermodynamic properties at states 1 and 2. Assume steady-state, negligible kinetic and potential energies, and one-dimensional flow at states 1 and 2.

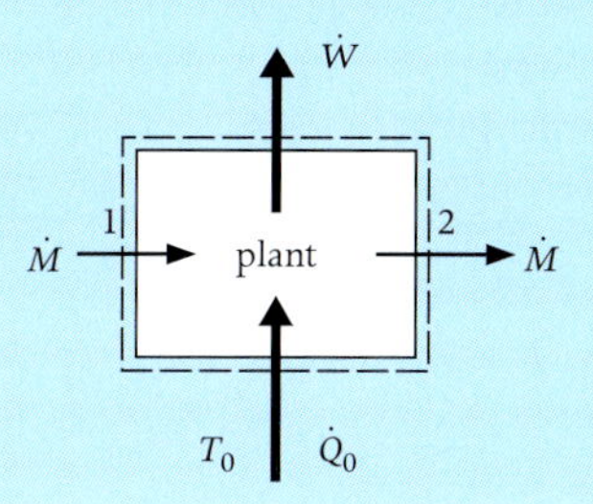

5.12 A control mass consists of a mixture of fluids in a piston–cylinder system.

The weight of the piston maintains a *constant pressure* P on the fluid mixture, which is also maintained at a *constant temperature* T by heat transfer to the cylinder walls. Irreversible chemical reactions within the fluid cause changes in the composition, resulting in changes in the volume, internal energy, enthalpy, and entropy of the mixture that eventually lead to an equilibrium state. Denote these changes by ΔV, ΔU, ΔH, and ΔS (Δ = final − initial). Kinetic and potential energy changes in the mixture are negligible. The enthalpy is defined as $H \equiv U + PV$ so that (at constant P) $\Delta H = \Delta U + P\Delta V$. The Gibbs function is defined as $G \equiv H - TS$, so that (at constant T) $\Delta G = \Delta H - T\Delta S$. Starting from basic principles, develop an expression for the amount of entropy produced ($\mathcal{P}_S$) in terms of the changes defined above, T and P. Then, use this to show $\Delta G \leq 0$ for any changes that occur in this system at constant T and P. This result forms the basis for analysis of equilibria in vapor–liquid mixtures and in mixtures of reacting fluids.

5.13 An *adiabatic* device is to be designed that will take in two steady flows of gas at given states 1 and 2, with $T_2 > T_1$, and discharge the gas at a given pressure P_3 and undetermined temperature T_3. The pressure of all three states will be the same ($P_1 = P_2 = P_3$). The purpose of the device is to generate useful (shaft) power.

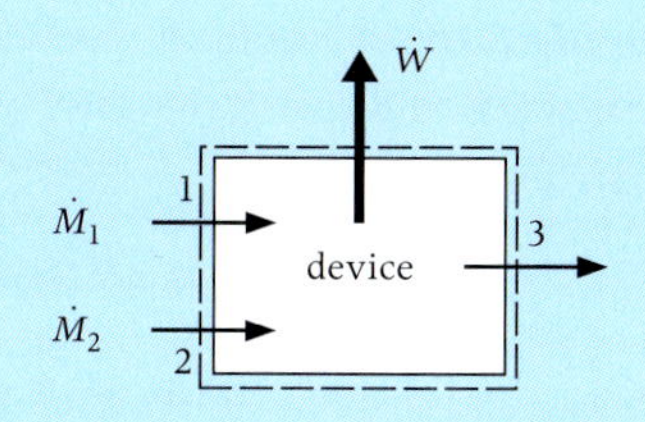

Kinetic and potential energy of the flow at states 1–3 may be neglected, one-dimensional flow may be assumed at states 1–3, steady-state may be assumed, and the gas may be assumed to be an ideal gas with constant specific heats, for which the entropy is given by

$$s_2 - s_1 = c_P \ln(T_2/T_1) - R\ln(P_2/P_1).$$

Using an energy balance, show that the shaft power output $\dot{W}$ for given mass flow rates, states 1 and 2, and P_3 decreases *linearly* with increasing T_3, and hence that the maximum power output will be obtained for a design that gives the lowest possible T_3. Then, using an entropy balance, develop the expression for the lowest possible T_3 in terms of T_1, T_2, and the mass flow rate ratio $f \equiv \dot{M}_1/\dot{M}_3$. List any idealizations that you make *other* than those given above.

5.14 The inlet of a supersonic aircraft engine is a supersonic diffuser. Air enters at high speed $\mathcal{V}_1$ at state 1 (T_1, P_1), and exits at high pressure P_2 at much lower speed $\mathcal{V}_2$. The diffuser is adiabatic and has no moving parts.

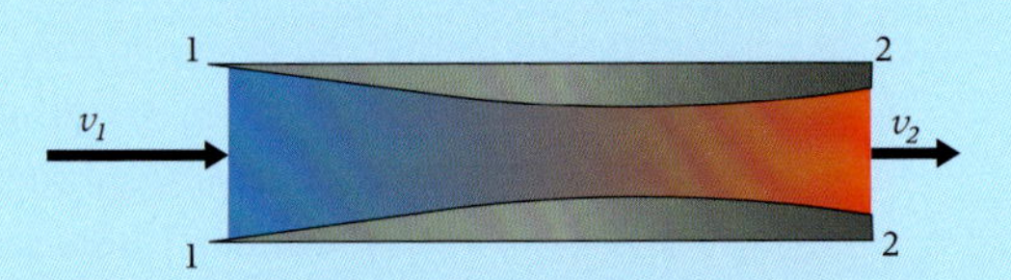

The Mach number M is defined as the ratio of the flow velocity to the speed of sound c, which for an ideal gas is $c = \sqrt{\gamma RT}$, so $M_1 = \mathcal{V}_1/a_1 = \mathcal{V}_1/\sqrt{\gamma RT_1}$, where M_1 is the flight Mach number. Starting from basic principles, and using the ideal gas equations of state, show that

$$\frac{T_2}{T_1} = 1 + \frac{\gamma - 1}{2} M_1^2.$$

Then, derive an expression for the discharge pressure P_2/P_1 in terms of M_1, $\mathcal{P}_S$ (the entropy production per unit mass), and constants for the gas (neglect the kinetic energy at 2 compared to that at 1). From this determine the expression for the maximum possible P_2/P_1 that could be obtained, in terms of M_1 and γ. Sketch the process representation for an ideal and an actual diffuser with the same inlet conditions on a Ts diagram, showing the pertinent lines of constant P for both cases.

5.15 A heat pump consists of the hardware shown, with refrigerant 134a as the working fluid. The valve and compressor are well insulated and may be treated as adiabatic. Measurements during operation are as follows:

- $T_1 = 265$ K, $P_1 = P_4 = 0.2$ MPa,
- $T_2 = 350$ K, $P_2 = P_3 = 1.5$ MPa,
- compressor power input $\dot{W}_{comp} = 10$ kW,
- saturated liquid at state 3.

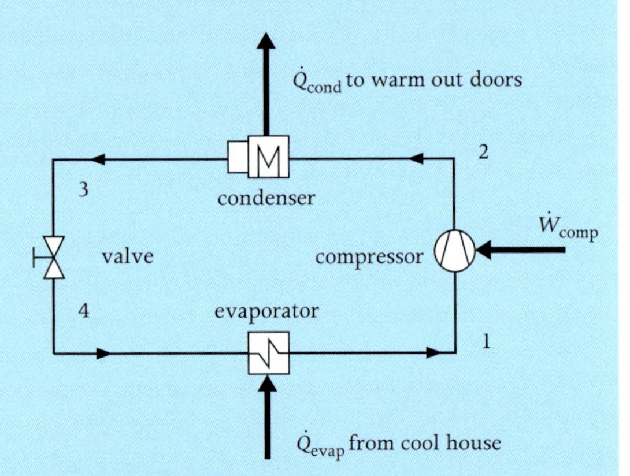

Applying basic principles and suitable idealizations, determine the evaporator heat transfer rate $\dot{Q}_{evap}$ (kW) and the compressor isentropic efficiency.

5.16 A steady flow of air at the rate $\dot{M}$ will enter a steady-state device at T_1, P_1 and emerge at T_2, P_2. The device requires shaft power $\dot{W}$ and is cooled by heat transfer at the rate $\dot{Q}_0$ to the environment at T_0.

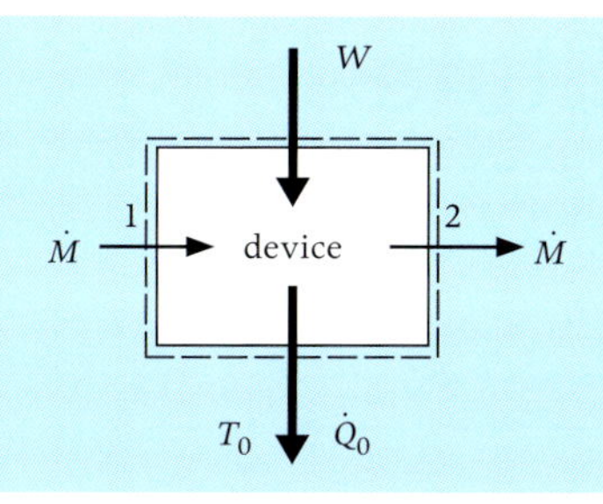

The specifications are that $T_1 = T_2 = T_0$ and $P_2 > P_1$.

Starting from basic principles, and using the ideal gas equations of state, derive an expression for the minimum $\dot{W}$ required by this device, in terms of $\dot{M}$, T_0, the pressure ratio $\beta_P = P_2/P_1$, and constants for the gas.

6 Thermodynamics of State

CONTENTS

The First and Second Laws of Thermodynamics provide the basic tools for the analysis of energy systems. Examples in the previous chapter show the power of this fundamental analysis. The goal of this chapter is to illustrate how the first and second laws can be used to obtain general thermodynamic models of fluids, and to introduce some equations of state that are useful in engineering.

6.1 Equation of State for the Ideal Gas

Definition

We review and extend here some concepts first introduced in Section 3.8. An ideal gas is defined as any fluid whose P–v–T relationship is

$$Pv = RT. \tag{6.1}$$

The gas constant R is related to the universal gas constant $\hat{R}$ and the molar mass $\hat{M}$ by

$$R = \frac{\hat{R}}{\hat{M}}.$$

$\hat{R}$ has the experimental value

$$\hat{R} = 8.3143 \text{ J}/\left(\text{mol} \cdot \text{K}\right).$$

The Boltzmann constant k, often called the gas constant per molecule, is defined in terms of $\hat{R}$ and Avogadro's number N_{A} by

$$k = \hat{R}/N_A. \tag{6.2}$$

Values of $\hat{M}$ and R are given for several gases in Table A.1.

The molar mass is the mass per mole of material. For carbon 12 the molar mass is 12 kg/kmol = 12 g/mol. The mole is defined such that 1 kmol of a substance contains the same number of molecules as 12 kg of carbon 12; likewise, one mole contains the same number of molecules as 12 g of carbon 12; The Avogadro's number N_A is the number of molecules in each case.

Other forms

The defining equation (6.1) can be put in other forms. Multiplying (6.1) by mass yields

$$PV = MRT. \tag{6.3}$$

Denoting the number of moles of gas by N and observing that the mass M is related to the number of moles and the molar mass $\hat{M}$ by

$$M = N\hat{M},$$

we can write (6.3) as

$$PV = N\hat{M}RT = N\hat{R}T.$$

Various forms of the ideal gas equation of state are given in Table 6.1.

Temperature of an ideal gas

The Gibbs equation is

$$ds = \frac{1}{T}du + \frac{P}{T}dv.$$

Since ds is an exact differential, the equality of the cross-derivative gives

$$\left(\frac{\partial(1/T)}{\partial v}\right)_u = \left(\frac{\partial(P/T)}{\partial u}\right)_v = \left(\frac{\partial(R/v)}{\partial u}\right)_v = 0.$$

T is therefore independent of v on any line of constant u, and consequently, for a perfect gas, $T = T(u)$, or $u = u(T)$.

Table 6.1 Forms of the ideal gas equation of state.

$Pv = RT$	R = gas constant
$Pv = \frac{\hat{R}}{\hat{M}}T$	$\hat{R}$ = universal gas constant, and $\hat{M}$ = molar mass, mass/mole
$P\hat{v} = \hat{R}T$	$\hat{v} = \hat{M}v$, volume/mole
$PV = NRT$	V = volume of N gas moles
$PV = MRT$	V = volume of a gas mass M

Internal energy and temperature

Recalling the definition of specific heat at constant volume (3.16), we have that

$$du = c_v dT. \tag{6.4}$$

Since u depends only on T, c_v must be a function only of temperature, i.e.,

$$c_v = c_v(T). \tag{6.5}$$

Integrating (6.4) gives the specific internal energy difference between two states 1 and 2 as

$$u_2 - u_1 = \int_{T_1}^{T_2} c_v(T)\,dT.$$

Enthalpy and temperature

The enthalpy of an ideal gas is a function only of temperature,

$$h(T) = u + Pv = u(T) + RT.$$

Hence, from the definition of specific heat at constant pressure,

$$dh = c_P(T)dT. \tag{6.6}$$

Integrating (6.6) gives the specific enthalpy difference between two states as

$$h_2 - h_1 = \int_{T_1}^{T_2} c_P(T)\,dT. \tag{6.7}$$

Entropy

The differential equation for the entropy is obtained from the Gibbs equation as

$$ds = \frac{du}{T} + \frac{P}{T}dv = \frac{c_v}{T}dT + \frac{R}{v}dv. \quad (6.8)$$

Alternatively, using the differential $dh = du + Pdv + vdP$, the entropy differential can be written as

$$ds = \frac{dh}{T} - \frac{v}{T}dP = \frac{c_P}{T}dT - \frac{R}{P}dP. \quad (6.9)$$

Integrating (6.8) gives the specific entropy difference between two states as

$$s_2 - s_1 = \int_{T_1}^{T_2} \frac{c_v\{T\}}{T}dT + R\ln\frac{v_2}{v_1}, \quad (6.10)$$

while integrating (6.9) gives the specific entropy difference between two states as

$$s_2 - s_1 = \int_{T_1}^{T_2} \frac{c_P\{T\}}{T}dT - R\ln\frac{P_2}{P_1}. \quad (6.11)$$

Temperature dependence

For this ideal substance the enthalpy equation depends only on temperature, therefore

$$c_P\,dT = c_v\,dT + R\,dT,$$

which requires that at any temperature

$$c_P = c_v + R. \quad (6.12)$$

Since R is constant, there is only one free variable in (6.12), which we can take as c_P. For simple calculations involving the ideal gas model, it is useful to define a property ϕ by

$$\phi\{T\} \equiv \int_{T_0}^{T} \frac{c_P\{T'\}}{T'}dT'.$$

Equation (6.11) is then

$$s_2 - s_1 = \phi\{T_2\} - \phi\{T_1\} - R\ln\frac{P_2}{P_1}. \quad (6.13)$$

We can also take c_v as a free variable and define a property ψ by

$$\psi\{T\} \equiv \int_{T_0}^{T} \frac{c_v\{T'\}}{T'}dT'.$$

Equation (6.10) can thus be written as

$$s_2 - s_1 = \psi\{T_2\} - \psi\{T_1\} + R\ln\frac{v_2}{v_1}. \quad (6.14)$$

Values of enthalpy and the property $\phi\{T\}$ and $\psi\{T\}$ have been obtained for many gases; see Table A.2 for low-density air.

The GasMix library implements the ideal gas model with temperature-dependent specific heats. Tables are useful for hand calculations and you should practice with them and get a feel for it. Quick calculations by hand to check ideas are very useful, and being able to do it is a great skill for an engineer. Use a computer and property software for more complex or repetitive problems.

Reduced pressure and volume

The reduced pressure P_r and reduced volume v_r are defined in air tables as

$$\ln P_r = \frac{\phi\{T\}}{R}, \quad (6.15)$$

and

$$\ln v_r = -\frac{\psi\{T\}}{R}. \quad (6.16)$$

Note that both are dimensionless quantities depending on T.

Isentropic process analysis

By setting the entropy difference to zero in (6.13), (6.11) gives

$$\ln\frac{P_2}{P_1} = \frac{\phi_2 - \phi_1}{R}. \quad (6.17)$$

By using (6.15), the right-hand side of (6.17) can be written as

$$\ln P_{r_2} - \ln P_{r_1} = \ln\frac{P_{r_2}}{P_{r_1}},$$

so that (6.17) becomes

$$\frac{P_2}{P_1} = \frac{P_{r_2}}{P_{r_1}} \quad \textit{for an isentropic process.}$$

Similarly, it may be shown that

$$\frac{v_2}{v_1} = \frac{v_{r_2}}{v_{r_1}} \quad \textit{for an isentropic process.}$$

Example: use of the ideal gas reduced pressure. In order to illustrate the use of the ideal gas reduced pressure, let us work out this simple problem. Suppose we want to compress air from 0.1 MPa to 0.5 MPa isentropically, and need to know the final temperature. From Table A.2 we have that at 290 K, $P_r = 65.19$. We can thus calculate P_{r_2} as

$$P_{r_2} = \frac{P_2}{P_1} P_{r_1} = \frac{0.5}{0.1} 65.19 = 325.9.$$

Table A.2 can give us the temperature corresponding to $P_r = 325.9$, and we see that its value is about 458.2 K. The values we have obtained are for air considered as an ideal gas, but with temperature-dependent specific heat, which might be important depending on the thermodynamic states, and on the desired accuracy.

Constant specific heat (or polytropic) model

Simplified forms of the perfect-gas equations of state are obtained if c_P, and hence c_v, are also considered as constants, which leads to

$$Pv = RT, \tag{6.18}$$

$$u_2 - u_1 = c_v(T_2 - T_1), \tag{6.19}$$

$$h_2 - h_1 = c_p(T_2 - T_1), \tag{6.20}$$

$$s_2 - s_1 = c_v \ln \frac{T_2}{T_1} + R \ln \frac{v_2}{v_1}, \tag{6.21}$$

$$s_2 - s_1 = c_P \ln \frac{T_2}{T_1} - R \ln \frac{P_2}{P_1}, \tag{6.22}$$

$$s_2 - s_1 = c_P \ln \frac{v_2}{v_1} + c_v \ln \frac{P_2}{P_1}, \tag{6.23}$$

$$c_P - c_v = R. \tag{6.24}$$

These equations are especially useful in gas dynamics, where neat closed-form algebraic expressions for the properties of a one-dimensional flow field can be derived.

Example: isentropic compression of air. In order to illustrate the use of these simplified equations, let us compute the temperature rise for the isentropic compression of air from 0.1 MPa and 290 K to 0.5 MPa, and compare that with the previous result. Using (6.22), we set $s_2 = s_1$ and have

$$\frac{T_2}{T_1} = \left(\frac{P_2}{P_1}\right)^{R/c_P} = \left(\frac{P_2}{P_1}\right)^{(\gamma-1)/\gamma}, \tag{6.25}$$

where $\gamma = c_P/c_v$ is the ratio of the specific heats. For air $\gamma = 1.4$, hence

$$\frac{T_2}{T_1} = \left(\frac{0.5}{0.1}\right)^{0.286} = 1.584.$$

Then, with $T_1 = 290$ K, we have that $T_2 = 1.584 \times 290$ K $= 459.3$ K. The table-based calculation of the previous example predicts a $T_2 = 458.2$ K. The slight difference is due to the error associated with the assumption of constant specific heats.

We can also compute the work required to carry out the compression process. If it takes place in an adiabatic steady-flow compressor (Figure 6.1), then the energy balance gives

$$\dot{W} = \dot{M}(h_2 - h_1).$$

Using (6.20) yields

$$\begin{aligned}\frac{\dot{W}}{\dot{M}} &= h_2 - h_1 = c_P(T_2 - T_1) \\ &= 1.01\,\text{kJ/(kg}\cdot\text{K)} \times (459.3 - 290)\,\text{K} = 171\,\text{kJ/kg}.\end{aligned}$$

The value of c_P is set to 1.01 kJ/(kg · K), which is the arithmetic average of the values of c_P at 290 and 460 K obtained from Table A.2.

Note that the 1.1 K difference in the calculated T_2 obtained by assuming constant c_P causes a difference in the calculated compressor work of 1 percent. Whether this difference is tolerable for an engineering analysis depends on the objective of the analysis.

Polytropic process

A useful relation between the pressures and volumes at two states can be derived from (6.23). Dividing by c_P, we have

$$\frac{s_2 - s_1}{c_v} = \ln\left[\left(\frac{v_2}{v_1}\right)^{\gamma} \frac{P_2}{P_1}\right].$$

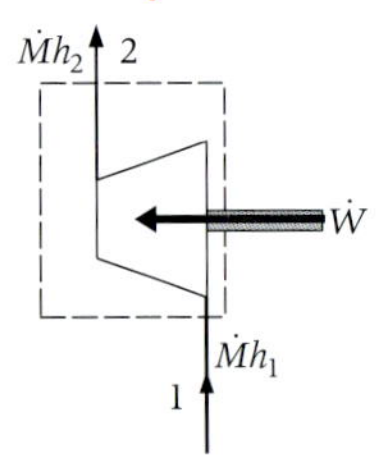

Figure 6.1 Representation of an adiabatic compressor for control volume energy analysis.

Taking the antilog, results in

$$P_2 v_2^{\gamma} = P_1 v_1^{\gamma} \exp\left(\frac{s_2 - s_1}{c_v}\right). \tag{6.26}$$

From the equation above we see that two states having the same entropy will have the same values of their associated property Pv^{γ}. For an *ideal gas with constant specific heats undergoing a reversible adiabatic process, i.e., isentropic,*

$$Pv^{\gamma} = \text{constant}. \tag{6.27}$$

The mentioned restrictions must be carefully verified whenever this relation is employed.

We have now seen two examples of $Pv^n = \text{constant}$ processes. $Pv^{\gamma} = \text{constant}$ represents the isentropic process for the perfect gas with constant c_P, while $Pv = \text{constant}$ represents the isothermal (constant-temperature) process. The general process represented by $Pv^n = \text{constant}$ is called the *polytropic process*. The polytropic process equation is a useful generalization. In summary, the various polytropic processes are characterized by the value of the exponent n, as shown in Table 6.2.

Other processes can often be approximated with an appropriate value for the *polytropic exponent n*. If $Pv^n = \text{constant}$ during some process, it follows that

$$P_1 v_1^n = P_2 v_2^n$$

or

$$\frac{P_2}{P_1} = \left(\frac{v_1}{v_2}\right)^n. \tag{6.28}$$

Using the perfect-gas equation we can also show that

$$\frac{T_2}{T_1} = \left(\frac{P_2}{P_1}\right)^{(n-1)/n} = \left(\frac{v_1}{v_2}\right)^{n-1}. \tag{6.29}$$

Table 6.2 Polytropic exponent for several processes.

Process	n
Isobaric process	$n = 0$
Isothermal process	$n = 1$
Isentropic process	$n = \gamma$
Constant-volume process	$n \to \infty$

This expression is particularly important for the isentropic case when $n = \gamma$.

Reference (or datum) state

Equations (6.1) and (6.5) are sufficient to calculate all thermodynamic properties, and in particular allow us to calculate the differences in internal energy, enthalpy, and entropy between any two states. Sometimes it is convenient to set the internal energy and entropy to zero at some arbitrarily chosen reference or datum state. In thermodynamic analyses not involving chemical reaction, we work exclusively with differences in u, h, and s, so the reference state is completely arbitrary. Since the values of h and u differ by Pv at the datum temperature, it is not permissible to select reference states for both enthalpy and internal energy; only one of these may be set to zero at the datum state. Sometimes it is more convenient to select a datum state for enthalpy rather than for internal energy. In case chemical reactions are involved in the analysis, the choice of the reference state is not arbitrary, as we shall see in Chapter 10.

6.2 Thermodynamic Functions and Property Relations

The equation of state for the ideal gas ($Pv = RT$) is the simplest *volumetric* equation of state and relates quantities that can be directly measured: pressure, temperature, mass and volume. For our thermodynamic analyses, we also need properties like h, s, u and quantities like h^{LV}, s^{LV}, etc.

Exact thermodynamic relations among volumetric (P, v, T) and calorimetric properties (u, h, s, etc.) can be obtained by combining convenient definitions and appropriate mathematical transformations. We have already seen how we can define a thermodynamic property in order to easily evaluate energy transfer and calculate it from other primary properties. For example, enthalpy was introduced when deriving the generalized energy balance for an open system (4.14). Entropy can be obtained in a similar manner from volumetric and calorimetric properties, see (5.16). Other thermodynamic properties become particularly useful in the development of equations of state, and in the treatment of thermodynamic equilibrium. They are more easily introduced here, limited to the case of pure and simple compressible substances.

Helmholtz energy

Consider a closed isothermal, isochoric (constant volume) system. The system evolves from state 1 to state 2. The energy balance for the CV (same as for the CM) is

$$U_2 = U_1 + Q + W,$$

while the entropy balance is

$$S_2 = S_1 + \frac{Q}{T} + \mathcal{P}_S.$$

Eliminating Q from these equations, and noting that $T = T_1$ and $T = T_2$, we can write the energy transfer as work to the system as

$$W = (U_2 - T_2 S_2) - (U_1 - T_1 S_1) + T\mathcal{P}_S.$$

It is possible thus to define a new thermodynamic property

$$A \equiv U - TS, \tag{6.30}$$

such that its difference gives the energy transfer as work for an isochoric, isothermal process for a closed system. This property is called *Helmholtz energy*. The work involved in the reversible process with $T = \text{const.}$ and $V = \text{const.}$ to bring the system from state 1 to state 2 is therefore

$$W = A_2 - A_1.$$

For the same process, if dissipation is taken into account, the required work becomes

$$W = A_2 - A_1 + T\mathcal{P}_S.$$

Similarly to the other extensive properties described so far, the specific Helmholtz energy for a simple compressible substance $a = A/M$ is defined as

$$a \equiv u - Ts. \tag{6.31}$$

Gibbs energy

In analogy with the treatment leading to the definition of Helmholtz energy, it is possible to define a thermodynamic function related to the reversible energy transfer as work in an isobaric and isothermal process for a closed system. It is now apparent how we can obtain other thermodynamic state functions by changing the constraints on the considered system. In this case the energy and entropy balances give

$$U_2 = U_1 + Q + W - (P_2 V_2 - P_1 V_1),$$

and

$$S_2 = S_1 + \frac{Q}{T} + \mathcal{P}_S.$$

Eliminating Q from these equations gives

$$W = (U_2 + P_2 V_2 - TS_2) - (U_1 + P_1 V_1 - TS_1) + T\mathcal{P}_S.$$

If the process is carried out reversibly, it is possible to define another state function, named Gibbs energy, as

$$G \equiv U + PV - TS = H - TS, \tag{6.32}$$

so that

$$W = G_2 - G_1.$$

The specific Gibbs energy for a simple compressible substance is therefore

$$g \equiv u + Pv - Ts = h - Ts. \tag{6.33}$$

The energy transfer as work for a reversible process occurring at $T = \text{const.}$ and $P = \text{const.}$ in a closed system is given by the difference in Gibbs energy between initial and final state. The Gibbs energy is particularly useful in the study of phase equilibria and of the equilibrium of reacting systems (Section 8.5 and Chapter 10).

Maxwell relations

As will become clear, relations between thermodynamic properties are repeatedly used when defining and deriving other thermodynamic properties, when analyzing an energy conversion system or process, or even when correlating experimental data aimed at the development of thermodynamic models. Several useful rules from calculus applying to partial derivatives are reported in Appendix B as an aid to the understanding of how the following relations are obtained.

The Gibbs equation (5.15) may be written in another useful form as

$$du = Tds - Pdv. \tag{6.34}$$

As du is an exact differential,

$$du = \left(\frac{\partial u}{\partial s}\right)_v ds + \left(\frac{\partial u}{\partial v}\right)_s dv.$$

Since u is a regular function of its arguments, the order in which a second partial derivative is taken is unimportant, i.e.,

$$\left(\frac{\partial^2 u}{\partial s \partial v}\right)_{v,s} = \left(\frac{\partial^2 u}{\partial v \partial s}\right)_{s,v},$$

it follows that

$$\left(\frac{\partial T}{\partial v}\right)_s = -\left(\frac{\partial P}{\partial s}\right)_v. \tag{6.35}$$

A differential equation for the specific enthalpy can be obtained by differentiating its definition (3.9), i.e.,

$$dh = du + Pdv + vdP,$$

and combining it with (6.34), which yields

$$dh = Tds + vdP. \tag{6.36}$$

Similarly, another important relation among partial derivatives of primary thermodynamic properties can be obtained by equating the second mixed derivatives of the enthalpy,

$$\left(\frac{\partial^2 h}{\partial s \partial P}\right)_{P,s} = \left(\frac{\partial^2 h}{\partial P \partial s}\right)_{s,P},$$

which gives

$$\left(\frac{\partial T}{\partial P}\right)_s = \left(\frac{\partial v}{\partial s}\right)_P. \tag{6.37}$$

The procedure used to obtain (6.35) and (6.37), can be applied to the differentials of the specific Helmholtz and Gibbs functions, namely,

$$da = du - Tds - sdT$$

and

$$dg = dh - Tds - sdT,$$

to obtain two other relevant property relations. These derivations are left to the student for practice.

In summary, four relations linking the partial derivatives of primary thermodynamic properties can be obtained from the four differential equations of state for du, dh, da and dg,

$$\left(\frac{\partial T}{\partial v}\right)_s = -\left(\frac{\partial P}{\partial s}\right)_v, \tag{6.38a}$$

$$\left(\frac{\partial T}{\partial P}\right)_s = \left(\frac{\partial v}{\partial s}\right)_P, \tag{6.38b}$$

$$\left(\frac{\partial P}{\partial T}\right)_v = \left(\frac{\partial s}{\partial v}\right)_T, \tag{6.38c}$$

$$\left(\frac{\partial v}{\partial T}\right)_P = -\left(\frac{\partial s}{\partial P}\right)_T. \tag{6.38d}$$

These are known as the *Maxwell relations* and are valid for simple compressible substances.

6.3 Properties from $P = P(v, T)$ and $c_{P\ (\text{or}\ v)}^{IG}$

We have seen in Section 3.8 that the ideal gas equation of state can only be applied in a limited region of the thermodynamic space of a substance. Figure 6.2 shows in color the ideal gas, the dense vapor/supercritical and the liquid thermodynamic regions in the P–v and T–s state diagrams for two exemplary fluids, water and toluene. The ideal gas equation does not hold for liquid states, nor can it be employed to calculate the properties of vapor–liquid equilibrium states. The deviation of volumetric properties from ideal gas behavior is often expressed in terms of the so-called *compressibility factor*

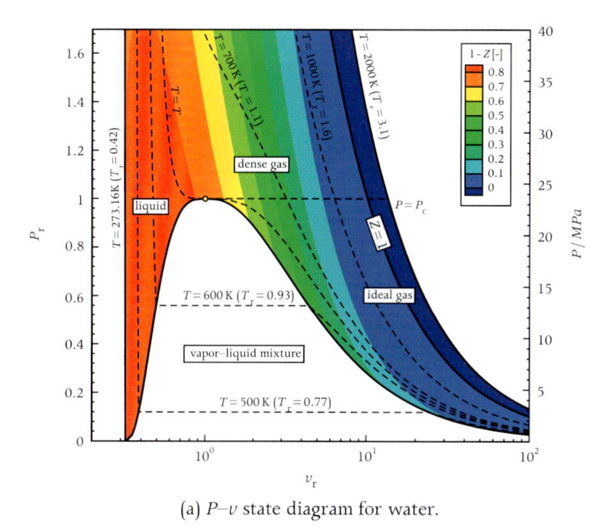

(a) P–v state diagram for water.

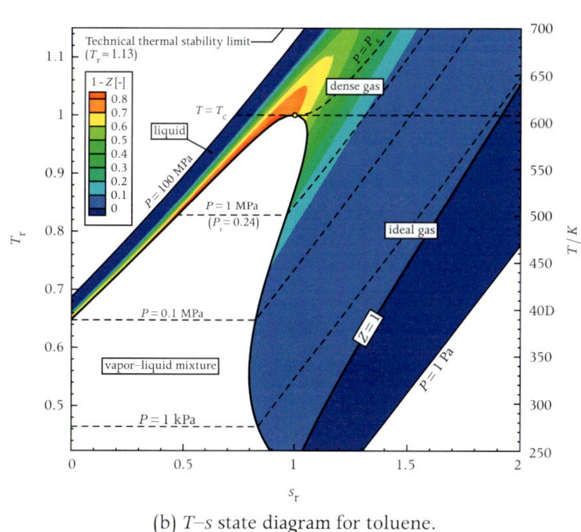

(b) T–s state diagram for toluene.

Figure 6.2 The charts show the thermodynamic regions where the ideal gas law applies ($Z - 1 \approx 0$) and where more complex property models must be employed. The two exemplary diagrams are obtained with FLUIDPROP (Model = REFPROP, Fluid = water, toluene).

$$Z \equiv \frac{Pv}{RT}, \tag{6.39}$$

which is equal to 1 for the ideal gas.

In order to overcome the limitations of the equation of state for a perfect gas, many equations of state have been developed over time and are still being developed. A great number of these equations are expressed analytically in the form $P = P(\rho, T)$, or $P = P(v, T)$, see Section 6.9, where the $P(\rho, T)$ function has a varying degree of complexity, depending on the sought accuracy. P, T and ρ (or v) are chosen because they can be accurately and directly measured and because experimental data can be directly fitted with these relations. By complementing the $P = P(\rho, T)$ equation with a temperature-dependent relation for the specific heat in the ideal gas state, either the isochoric or the isobaric ideal gas heat capacity as they are related by (6.12), a complete and consistent thermodynamic model for a simple compressible substance can be obtained. This means that all thermodynamic properties, like the ones listed in Table 6.3, can be calculated from $P = P(\rho, T)$ and $c_v^{\mathrm{IG}} = c_v^{\mathrm{IG}}(T)$, or $c_P^{\mathrm{IG}} = c_P^{\mathrm{IG}}(T)$, by using exact thermodynamic relations.

We shall show in the following how we can derive the necessary equations in order to obtain all the primary and secondary thermodynamic properties from $P = P(\rho, T)$ and $c_v^{\mathrm{IG}} = c_v^{\mathrm{IG}}(T)$.

Internal energy, entropy, and enthalpy

Let us start, for example, by writing the entropy s as a function of T and v,

$$s = s(T, v).$$

The exact differential for s is therefore

$$ds = \left(\frac{\partial s}{\partial T}\right)_v dT + \left(\frac{\partial s}{\partial v}\right)_T dv.$$

By applying the chain rule (B.8) to $s = s(T, v)$ we obtain

$$\left(\frac{\partial s}{\partial T}\right)_v = \left(\frac{\partial s}{\partial u}\right)_v \left(\frac{\partial u}{\partial T}\right)_v. \tag{6.40}$$

From the Gibbs equation (5.15), we have that

$$\left(\frac{\partial s}{\partial u}\right)_v = \frac{1}{T},$$

and from the definition of isochoric specific heat

$$\left(\frac{\partial u}{\partial T}\right)_v = c_v.$$

Equation (6.40) therefore becomes

$$\left(\frac{\partial s}{\partial T}\right)_v = \frac{c_v}{T}.$$

Table 6.3 Intensive thermodynamic properties of a simple fluid.

Property	Definition
Conceptual	
Internal energy u, J/kg	energy associated with molecular and atomic motions and forces
Entropy s, J/(kg · K)	a measure of the average disorder of matter on the macroscopic scale
Defined	
Temperature T, K	$T \equiv \left(\frac{\partial u}{\partial s}\right)_v$, equal to the empirical temperature
Pressure P, Pa	$P \equiv T\left(\frac{\partial s}{\partial v}\right)_u$, equal to the mechanical pressure
Specific volume v, m^3/kg	volume per unit mass
Density ρ, kg/m^3	mass per unit volume
Enthalpy h, J/kg	$h \equiv u + Pv$
Helmholtz free energy a, J/kg	$a \equiv u - Ts$
Gibbs free energy g, J/kg	$g \equiv h - Ts$
Specific heat at constant volume c_v, J/(kg · K)	$c_v \equiv \left(\frac{\partial u}{\partial T}\right)_v$
Specific heat at constant pressure c_P, J/(kg · K)	$c_P \equiv \left(\frac{\partial h}{\partial T}\right)_P$
Sound speed c, m/s	$c \equiv \left(\frac{\partial P}{\partial \rho}\right)_s$
Isothermal compressibility κ, 1/Pa	$\kappa \equiv -\frac{1}{v}\left(\frac{\partial v}{\partial P}\right)_T$
Isobaric compressibility β, 1/K	$\beta \equiv -\frac{1}{v}\left(\frac{\partial v}{\partial T}\right)_P$
Isentropic compressibility α, 1/Pa	$\alpha \equiv -\frac{1}{v}\left(\frac{\partial v}{\partial P}\right)_s$

From the third Maxwell relation (6.38c) we have that

$$\left(\frac{\partial s}{\partial v}\right)_T = \left(\frac{\partial P}{\partial T}\right)_v.$$

We can therefore write the differential of the specific entropy as

$$ds = \frac{c_v}{T}dT + \left(\frac{\partial P}{\partial T}\right)_v dv. \tag{6.41}$$

By recalling that, from the Gibbs equation $du = Tds - Pdv$, it follows that the differential of the specific internal energy is

$$du = T\left[\frac{c_v}{T}dT + \left(\frac{\partial P}{\partial T}\right)_v dv\right] - Pdv,$$

which can be simplified as

$$du = c_v dT + \left[T\left(\frac{\partial P}{\partial T}\right)_v - P\right]dv. \tag{6.42}$$

Since the specific internal energy is a thermodynamic property, thus a function of the state of the substance, (6.42) can be integrated along an arbitrary path. If the path is the one indicated in Figure 6.3, and using $dv = -d\rho/\rho^2$, this results in

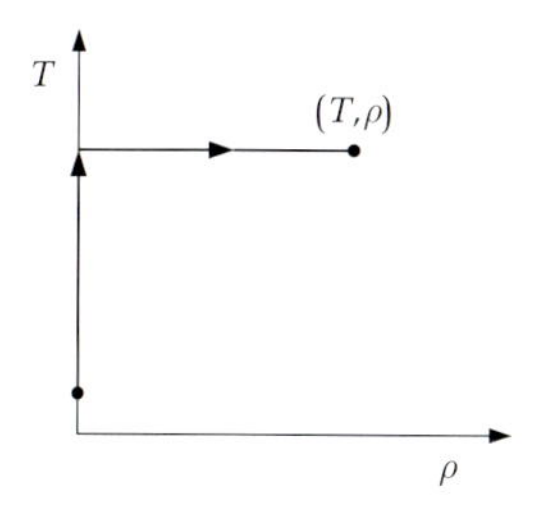

Figure 6.3 Integration path for Equations (6.41) **and** (6.42).

$$u - u_0 = \int_{T_0}^{T} c_v^{\text{IG}}(T)dT + \int_{\rho\to 0}^{\rho} \frac{1}{\rho^2}\left[P - T\left(\frac{\partial P}{\partial T}\right)_\rho\right]d\rho. \tag{6.43}$$

Note that the first integration is carried out with $\rho \to 0$, allowing c_v to be replaced by c_v^{IG} and that the second integration is carried out at constant temperature. T_0 is chosen as the reference or datum temperature. The constant u_0 is simply a term used to set the datum for u as desired. The equation for the internal energy difference (6.43) can be evaluated if $P = P(\rho, T)$ and $c_v^{\text{IG}} = c_v^{\text{IG}}(T)$ are given analytical functions, our stated objective. Alternatively, the specific internal energy in terms of specific volume in place of density reads

$$u - u_0 = \int_{T_0}^{T} c_v^{\text{IG}}(T)dT + \int_{\infty}^{v}\left[T\left(\frac{\partial P}{\partial T}\right)_v - P\right]dv. \tag{6.44}$$

The differential of the specific entropy (6.41) can be integrated along the same path shown in Figure 6.3. To this purpose it is useful to add and subtract to (6.41) a term $-Rd\rho/\rho$. The result of such integration is therefore

$$s - s_0 = \int_{T_0}^{T} \frac{c_v^{\text{IG}}(T)}{T}dT - R\ln\rho + \int_{\rho\to 0}^{\rho}\frac{1}{\rho^2}\left[R\rho - \left(\frac{\partial P}{\partial T}\right)_\rho\right]d\rho. \tag{6.45}$$

Here s_0 is a constant that can be chosen to set arbitrarily the datum for s. Note that the second integral in (6.45) vanishes for $\rho \to 0$, therefore the equation for the specific entropy difference valid for the perfect gas (6.11) can be recovered.

Finally, from the definition of enthalpy, we have that

$$h = u + \frac{P}{\rho}. \tag{6.46}$$

By looking at the equations giving thermodynamic properties u, s, h, we notice that all of them can be written as a summation in which the first term gives a relation valid for the ideal gas. The summation of the remaining terms is called a *departure* or *residual function* because it helps in quantifying the departure from ideal gas behavior. For example, the specific internal energy can be written as

$$u - u_0 = u^{\text{IG}} + u^{\text{R}},$$

where the ideal gas contribution is

$$u^{\text{IG}} = \int_{T_0}^{T} c_v^{\text{IG}}(T)dT,$$

and the residual function is given by

$$u^{\text{R}} = \int_{0}^{\rho}\frac{1}{\rho^2}\left[P - T\left(\frac{\partial P}{\partial T}\right)_\rho\right]d\rho. \tag{6.47}$$

The concept of deviation from an idealized behavior is often employed in the realm of thermodynamic properties, and more generally in engineering.

Isochoric and isobaric specific heat and their ratio

Let us continue with working out the equations for other secondary (or derived, or derivative) thermodynamic properties, and obtain those for the specific heat at constant volume and pressure. From (6.42) we have that

$$\left(\frac{\partial u}{\partial T}\right)_v = c_v,$$

and

$$\left(\frac{\partial u}{\partial v}\right)_T = T\left(\frac{\partial P}{\partial T}\right)_v - P.$$

Therefore the second mixed derivative of $u = u\{T, v\}$ is given by

$$\left(\frac{\partial^2 u}{\partial v \partial T}\right)_{T,v} = \frac{\partial}{\partial T}\left[T\left(\frac{\partial P}{\partial T}\right)_v - P\right]_v = \left(\frac{\partial P}{\partial T}\right)_v + T\left(\frac{\partial^2 P}{\partial T^2}\right)_v - \left(\frac{\partial P}{\partial T}\right)_v = T\left(\frac{\partial^2 P}{\partial T^2}\right)_v .$$

We notice, though, that

$$\left(\frac{\partial^2 u}{\partial v \partial T}\right)_{T,v} = \left(\frac{\partial c_v}{\partial v}\right)_T ,$$

therefore

$$\left(\frac{\partial c_v}{\partial v}\right)_T = T\left(\frac{\partial^2 P}{\partial T^2}\right)_v ,$$

which allows integrating at constant temperature between two volumes v_1 and v_2, thus obtaining

$$c_v\{v_2, T) - c_v\{v_1, T) = T \int_{v_1,T}^{v_2,T} \left(\frac{\partial^2 P}{\partial T^2}\right)_v dv. \quad (6.48)$$

The specific heat at constant volume can also be written as the summation of an ideal gas contribution and a departure function, if in (6.48) v_1 is taken as ∞. This results in

$$c_v = c_v^{\mathrm{IG}}\{T) + T \int_{\infty}^{v} \left(\frac{\partial^2 P}{\partial T^2}\right)_v dv. \quad (6.49)$$

The isobaric heat capacity can be conveniently expressed as a function of c_v, as shown in the following. From (6.36), we have that

$$T = \left(\frac{\partial h}{\partial s}\right)_P . \quad (6.50)$$

The specific entropy can be written as a function of T and P, therefore

$$ds = \left(\frac{\partial s}{\partial T}\right)_P dT + \left(\frac{\partial s}{\partial P}\right)_T dP. \quad (6.51)$$

From (6.50) and the definition of isobaric heat capacity, the first partial derivative in (6.51) can be written as

$$\left(\frac{\partial s}{\partial T}\right)_P = \left(\frac{\partial s}{\partial h}\right)_P \left(\frac{\partial h}{\partial T}\right)_P , \quad (6.52)$$

which leads to

$$\left(\frac{\partial s}{\partial T}\right)_P = \frac{c_P}{T}. \quad (6.53)$$

Equation (6.41) can be differentiated with respect to T at constant P and this gives

$$\left(\frac{\partial s}{\partial T}\right)_P = \frac{c_v}{T}\left(\frac{\partial T}{\partial T}\right)_P + \left(\frac{\partial P}{\partial T}\right)_v \left(\frac{\partial v}{\partial T}\right)_P = \frac{c_v}{T} + \left(\frac{\partial P}{\partial T}\right)_v \left(\frac{\partial v}{\partial T}\right)_P . \quad (6.54)$$

Eliminating $\left(\frac{\partial s}{\partial T}\right)_P$ from (6.53) and (6.54) yields

$$c_P = c_v + T\left(\frac{\partial P}{\partial T}\right)_v \left(\frac{\partial v}{\partial T}\right)_P .$$

Now, using the triple product rule,

$$\left(\frac{\partial P}{\partial T}\right)_v \left(\frac{\partial v}{\partial P}\right)_T \left(\frac{\partial T}{\partial v}\right)_P = -1,$$

produces

$$\left(\frac{\partial v}{\partial T}\right)_P = -\left(\frac{\partial P}{\partial T}\right)_v \left(\frac{\partial v}{\partial P}\right)_T ,$$

therefore

$$c_P = c_v + T\left(\frac{\partial P}{\partial T}\right)_v \left[-\left(\frac{\partial P}{\partial T}\right)_v \left(\frac{\partial v}{\partial P}\right)_T\right].$$

Finally, as an expression that can be easily evaluated analytically from c_v and an equation of state in the form $P = P\{T, v)$, we can write

$$c_P = c_v - T\frac{\left(\frac{\partial P}{\partial T}\right)_v^2}{\left(\frac{\partial P}{\partial v}\right)_T}. \quad (6.55)$$

A derived thermodynamic property often encountered in thermodynamic analyses is the ratio between the constant-volume and the constant-pressure specific heats

$$\gamma \equiv \frac{c_P}{c_v}. \quad (6.56)$$

We can therefore get an expression for γ from

$$\gamma = 1 - \frac{T}{c_v}\frac{\left(\frac{\partial P}{\partial T}\right)_v^2}{\left(\frac{\partial P}{\partial v}\right)_T}. \quad (6.57)$$

Speed of sound

An important derived thermodynamic property is the speed of sound, defined as

$$c^2 \equiv \left(\frac{\partial P}{\partial \rho}\right)_s = -v^2 \left(\frac{\partial P}{\partial v}\right)_s. \tag{6.58}$$

The speed of sound in a fluid is the velocity at which a weak pressure wave propagates isentropically in the medium. It has great importance in fluid dynamics. In addition, the speed of sound can be measured very accurately, and this kind of measurement is used to develop equations of state for fluids.

Example: the derivation of the speed of sound. Also, the speed of sound can be obtained in a form that can be evaluated from a volumetric equation of state and the specific heats. Since $P = f(v, T)$, and considering T as a function of v and s, if we use the composite derivative rule for $s =$ const., we get

$$\left(\frac{\partial P}{\partial v}\right)_s = \left(\frac{\partial P}{\partial T}\right)_v \left(\frac{\partial T}{\partial v}\right)_s + \left(\frac{\partial P}{\partial v}\right)_T. \tag{6.59}$$

The triple product rule yields

$$\left(\frac{\partial T}{\partial v}\right)_s \left(\frac{\partial v}{\partial s}\right)_T \left(\frac{\partial s}{\partial T}\right)_v = -1,$$

therefore

$$\left(\frac{\partial T}{\partial v}\right)_s = -\left(\frac{\partial s}{\partial v}\right)_T \left(\frac{\partial T}{\partial s}\right)_v. \tag{6.60}$$

From the Gibbs equation (5.15)

$$\left(\frac{\partial u}{\partial s}\right)_v = T.$$

In analogy with (6.52), we can write

$$\left(\frac{\partial s}{\partial T}\right)_v = \left(\frac{\partial s}{\partial u}\right)_v \left(\frac{\partial u}{\partial T}\right)_v = \frac{c_v}{T},$$

therefore

$$\left(\frac{\partial T}{\partial s}\right)_v = \frac{T}{c_v}. \tag{6.61}$$

From the Maxwell relation (6.38c)

$$\left(\frac{\partial s}{\partial v}\right)_T = \left(\frac{\partial P}{\partial T}\right)_v. \tag{6.62}$$

Equations (6.62) and (6.61) can be substituted in (6.60), which gives

$$\left(\frac{\partial T}{\partial v}\right)_s = -\frac{T}{c_v}\left(\frac{\partial P}{\partial T}\right)_v. \tag{6.63}$$

Substituting (6.63) in (6.59), and recalling the expression for γ (6.57) leads to

$$\left(\frac{\partial P}{\partial v}\right)_s = -\frac{T}{c_v}\left(\frac{\partial P}{\partial T}\right)_v^2 + \left(\frac{\partial P}{\partial v}\right)_T = \gamma \left(\frac{\partial P}{\partial v}\right)_T.$$

Finally, the sound speed can be obtained analytically from a volumetric equation of state and from the specific heats as

$$c = \sqrt{-v^2 \gamma \left(\frac{\partial P}{\partial v}\right)_T}. \tag{6.64}$$

Note that in the ideal-gas case this expression reduces to the well known

$$c = \sqrt{\gamma R T}. \tag{6.65}$$

Isentropic, isobaric, and isothermal compressibility

The isentropic compressibility, defined as

$$\alpha \equiv -\frac{1}{v}\left(\frac{\partial v}{\partial P}\right)_s, \tag{6.66}$$

can also be expressed as $\alpha = \dfrac{v}{c^2}$, therefore

$$\alpha = \frac{-1}{\gamma v}\left(\frac{\partial v}{\partial P}\right)_T. \tag{6.67}$$

The two other coefficients, already introduced in Section 3.7, the isothermal compressibility $\kappa \equiv -\dfrac{1}{v}\left(\dfrac{\partial v}{\partial P}\right)_T$ and the isobaric compressibility $\beta \equiv -\dfrac{1}{v}\left(\dfrac{\partial v}{\partial T}\right)_P$, can also be easily evaluated analytically from a complete thermodynamic model expressed in terms of $P = P(T, v)$ and $c_v^{\text{IG}} = c_v^{\text{IG}}(T)$.

Additional partial derivatives and integral

A complete and consistent thermodynamic model, and therefore all primary and secondary thermodynamic properties, can thus be derived from the volumetric equation of state and the ideal gas specific heat capacity by combining the few partial derivatives of the pressure

$$\left(\frac{\partial P}{\partial v}\right)_T, \quad \left(\frac{\partial P}{\partial T}\right)_v, \quad \left(\frac{\partial^2 P}{\partial T^2}\right)_v, \qquad (6.68)$$

and the integral

$$\int_{\infty}^{v} \left(\frac{\partial^2 P}{\partial T^2}\right)_v dv. \qquad (6.69)$$

This observation is particularly relevant when coding a complete thermodynamic model into a computer program, like, for example, the models implemented in FLUIDPROP .

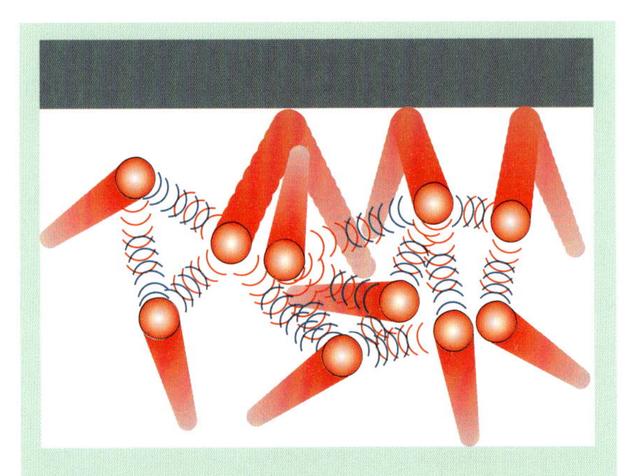

Figure 6.4 Intermolecular forces tend to reduce the pressure produced by a fluid onto a containing wall.

Example: the van der Waals equation of state. On the basis of molecular arguments, Johannes Diderik van der Waals (1837–1923) devised a modification to the perfect-gas equation of state that accounts for weakly attractive intermolecular forces and is able to qualitatively describe both the vapor and the liquid phase of a fluid. Attractive forces are present if the molecules are close enough, as opposed to the ideal gas hypothesis, which entails that they are so far apart as not to experience any mutual interaction.

A molecule about to strike the wall will experience a net attraction by molecules within the fluid, and this will lessen the impulse it delivers to the wall (Figure 6.4). The attractive force is proportional to the number of molecules per unit volume hitting the wall in a unit of time. Therefore the effect of molecular attraction should be to lessen the pressure by an amount roughly proportional to the square of the gas density. This suggests that we should replace P in the ideal gas expression by $P + \left(\mathsf{a}/v^2\right)$, where a is a constant. In addition, the molecules of a dense gas occupy some volume, which suggests that v should be replaced by $v - \mathsf{b}$, where b is a constant roughly indicative of the volume occupied by a unit of mass in a dense (liquid) state. These qualitative observations led van der Waals to propose an equation of state that bears his name, which he wrote in the form

$$\left(P + \frac{\mathsf{a}}{v^2}\right)(v - \mathsf{b}) = RT. \qquad (6.70)$$

The constants a and b have been determined for many fluids by fitting experimental data. It must be noted though that the accuracy of the predictions made with the van der Waals equation of state is quite low, and more complex functional forms are often adopted to calculate fluid properties in engineering problems. The van der Waals equation is *cubic* in v. As a consequence, corresponding to any given P and T there are either one or three real values of v. Figure 6.5 shows several isotherms calculated with the van der Waals equation of state and plotted on the P-v plane. We know that $Pv = RT$ must hold for an ideal gas, hence isotherms must have a negative slope. However, there is one particular isotherm having an inflection point, where $(\partial P/\partial v)_T = 0$ (see Section 3.5); we can interpret this point as the critical point.

We now see that the parameters a and b of the van der Waals equation of state are determined solely by the critical pressure and temperature of a fluid. Let us write Eq. (6.70) as

$$P = \frac{RT}{v - \mathsf{b}} - \frac{\mathsf{a}}{v^2}. \qquad (6.71)$$

Both the slope and the curvature must vanish at the critical point, therefore

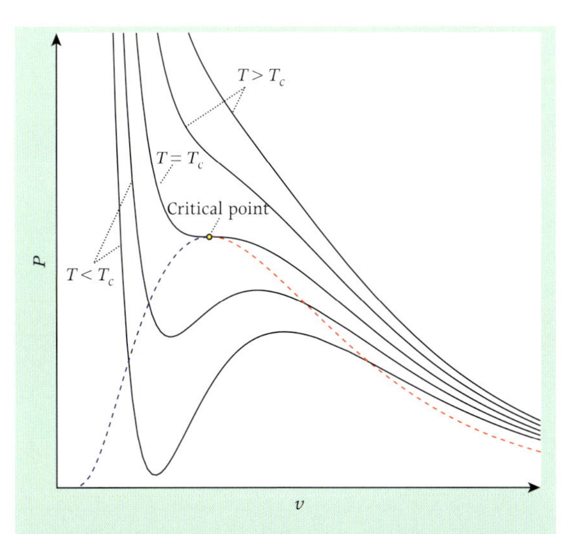

Figure 6.5 Several isotherms calculated with the van der Waals equation of state in the P–v plane.

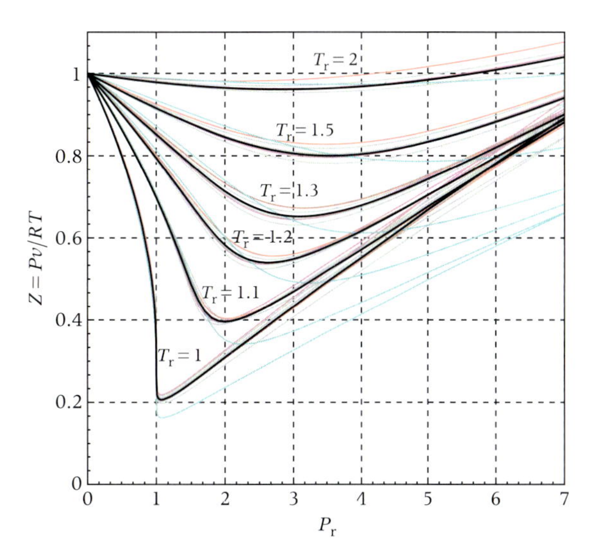

Figure 6.6 Generalized compressibility chart: the black lines are calculated with $Z = Z(P_r, T_r)$, obtained by fitting properties of alkanes only. The colored lines are computed with highly accurate reference equations of state implemented in REFPROP . The black lines interpolate quite well the lines referring to fluids of the same class and having a similar acentric factor, in this case hydrocarbons (methane, ethylene, ethane, propane, butane, isopentane). The $Z = Z(P_r, T_r)$ lines of more complex hydrocarbons like heptane (grey), or for carbon dioxide (red), nitrogen (magenta) and water (cyan) are increasingly far apart, indicating that the application of the principle of corresponding states cannot be used to accurately predict properties of all fluids over the entire thermodynamic space.

$$\left(\frac{\partial P}{\partial v}\right)_T = \frac{-RT_c}{(v_c - b)^2} + \frac{2a}{v_c^3} = 0,$$
$$\left(\frac{\partial^2 P}{\partial v^2}\right)_T = \frac{2RT_c}{(v_c - b)^3} - \frac{6a}{v_c^4} = 0.$$

Solving this system of equations for a, b, and using $Z_c \equiv \dfrac{P_c v_c}{RT_c}$ and (6.70) at the critical point gives

$$a = a(T_c) = \frac{27}{64}\frac{R^2 T_c^2}{P_c}, \tag{6.72a}$$

$$b = \frac{RT_c}{8P_c}, \tag{6.72b}$$

$$Z_c = \frac{P_c v_c}{RT_c} = \frac{3}{8} = 0.375. \tag{6.72c}$$

We also see that the van der Waals equation predicts the same value of $\dfrac{P_c v_c}{RT_c}$ for all substances. While it is not verified experimentally that the *critical compressibility factor* Z_c has the same value of 0.375 for any fluid, it is true that its value is in the range 0.25 to 0.4 for almost all fluids (Figure 6.6).

We can now apply what we have learned in this chapter so far to derive expressions for all thermodynamic properties starting from the volumetric van der Waals equation of state and an expression for the isochoric ideal gas specific heat $c_v^{\text{IG}}(T)$. We thus work out the necessary building blocks given by (6.68) and (6.69), therefore

$$\left(\frac{\partial P}{\partial v}\right)_T = \frac{-RT}{(v - b)^2} + \frac{2a}{v^3},$$
$$\left(\frac{\partial P}{\partial T}\right)_v = \frac{R}{v - b},$$
$$\left(\frac{\partial^2 P}{\partial T^2}\right)_v = 0,$$

and

$$\int_v^{\infty} \left(\frac{\partial^2 P}{\partial T^2}\right)_v dv = 0.$$

Let us start from the ideal gas part of the internal energy (6.44), and entropy (6.45): if the isochoric ideal gas specific heat is given by the commonly used polynomial function $c_v^{IG}(T) = k_1 + k_2 T + k_3 T^2 + k_4 T^3$, its integral with respect to temperature is

$$\int_{T_0}^{T} c_v^{IG}(T) dT = k_1(T - T_0) + \frac{k_2}{2}(T^2 - T_0^2) + \frac{k_3}{3}(T^3 - T_0^3) + \frac{k_4}{4}(T^4 - T_0^4),$$

and its integral over T with respect to temperature is

$$\int_{T_0}^{T} \frac{c_v^{IG}(T)}{T} dT = k_1 \ln \frac{T}{T_0} + \frac{k_2}{2}(T - T_0) + \frac{k_3}{2}(T^2 - T_0^2) + \frac{k_4}{3}(T^3 - T_0^3).$$

The volume integral in (6.44) can be evaluated as

$$\int_{v}^{\infty} \left[T \left(\frac{\partial P}{\partial T} \right)_v - P \right] dv = \int_{v}^{\infty} \left[\underbrace{T \frac{R}{v - \mathsf{b}} - \frac{RT}{v - \mathsf{b}}}_{=0} - \frac{\mathsf{a}}{v^2} \right] dv = -\frac{\mathsf{a}}{v}.$$

The internal energy is therefore given by

$$u = u_0 + k_1(T - T_0) + \frac{k_2}{2}(T^2 - T_0^2) + \frac{k_3}{3}(T^3 - T_0^3) + \frac{k_4}{4}(T^4 - T_0^4) - \frac{\mathsf{a}}{v}.$$

The volume integral in the equation for the specific entropy (6.45) can be evaluated as

$$\int_{v}^{\infty} \left[\left(\frac{\partial P}{\partial T} \right)_v - \frac{R}{v} \right] dv = \int_{v}^{\infty} \left[\frac{R}{v - \mathsf{b}} - \frac{R}{v} \right] dv = R \ln(1 - \mathsf{b}/v).$$

The entropy is therefore

$$s = s_0 + k_1 \ln \frac{T}{T_0} + \frac{k_2}{2}(T - T_0) + \frac{k_3}{2}(T^2 - T_0^2) + \frac{k_4}{3}(T^3 - T_0^3) + R \ln(v - \mathsf{b}).$$

The specific enthalpy is, according to its definition

$$h \equiv u + Pv = u_0 + k_1(T - T_0) + \frac{k_2}{2}(T^2 - T_0^2) + \frac{k_3}{3}(T^3 - T_0^3) + \frac{k_4}{4}(T^4 - T_0^4) + v \left(\frac{RT}{v - \mathsf{b}} - \frac{2\mathsf{a}}{v^2} \right).$$

The isochoric specific heat capacity is

$$c_v = c_v^{IG}(T) + T \int_{v}^{\infty} \left(\frac{\partial^2 P}{\partial T^2} \right)_v dv = c_v^{IG}(T),$$

> The fact that the isochoric specific heat capacity is equal to its perfect-gas value at a given temperature is a peculiarity of the van der Waals model. For all other more realistic thermodynamic models the residual part is not zero, and the isochoric specfic heat capacity depends both on temperature and specific volume.

while the isobaric specific heat capacity is

$$c_p = c_v - T \frac{\left(\frac{\partial P}{\partial T} \right)_v^2}{\left(\frac{\partial P}{\partial v} \right)_T} = c_v^{IG}(T) - T \frac{\left(\frac{R}{v - \mathsf{b}} \right)^2}{\frac{-RT}{(v - \mathsf{b})^2} + \frac{2\mathsf{a}}{v^3}} = c_v^{IG}(T) + \frac{R^2 T v^3}{RTv^3 - 2\mathsf{a}(v - \mathsf{b})^2},$$

and their ratio is

$$\gamma \equiv \frac{c_p}{c_v} = 1 + \frac{R^2 T v^3}{c_v^{IG}(T) \left(RTv^3 - 2\mathsf{a}(v - \mathsf{b})^2 \right)}.$$

The derivation of the speed of sound, isentropic, isobaric, and isothermal compressibility is for students to train themselves.

6.4 The Principle of Corresponding States

The van der Waals equation of state, as any other volumetric equation of state, may be written in terms of the critical compressibility factor Z_c. Equation (6.70) therefore becomes

$$\left(\frac{Pv}{RT}+\frac{27}{64}\frac{R^2T_c^2}{P_cRTv}\right)\left(1-\frac{RT_c}{8P_cv}\right)=1.$$

By using the nondimensional pressure and temperature, reduced with their respective values at the critical point, $P_r = \frac{P}{P_c}$ and $T_r = \frac{T}{T_c}$, the van der Waals equation becomes

$$\left(Z+\frac{1}{Z}\frac{27}{64}\frac{P_r}{T_r^2}\right)\left(1-\frac{P_r}{8ZT_r}\right)=1, \tag{6.73}$$

which suggests that a plot in the form

$$Z = Z\{P_r, T_r)$$

might correlate non-ideal gas data for all fluids. This leads us to the principle of corresponding states: as we see in Figure 6.6, though not perfect, the $Z = Z\{P_r, T_r)$ correlation of experimental data for many fluids is quite good. This fact is known as the *principle of corresponding states*; it is not a fundamental principle in the sense of the first and the second laws of thermodynamics, but merely a convenient approximation. The $Z = Z\{P_r, T_r)$ chart is called the *generalized compressibility chart* and it can be used to predict the properties of fluids for which more accurate equations of state have not yet been obtained. The chart can be made more accurate than the one obtained with (6.73) by fitting a large number of fluid data with an optimized functional form for $Z = Z\{P_r, T_r)$, like, for example, the equation reported in W. C. Reynolds, *Thermodynamic Properties in SI*, Dept. of Mechanical Engineering, Stanford University, 1979, p. 158.

6.5 Some Other Useful Relations

Engineering analyses dealing with single-component simple compressible substances are commonly expressed by means of a subset of the eight variables $P, T, v\,(\text{or } 1/\rho), s, u, h, a, g$. We have already seen that, depending on the problem, the choice of certain combinations of variables simplifies the setting up of the balance equations (see for instance Sections 2.8, 4.5–4.9, 5.6, 5.7). Moreover, in order to change from one thermodynamic variable to another, expressions for some partial derivatives must be derived. The number of (nontrivial) partial derivatives which relate the mentioned variables is quite large,[1] but experimental data and thermodynamic models are usually available as values for P, v, T and $c_P^{IG} = c_P^{IG}\{T)$, therefore balances are also usually expressed by utilizing these variables as independent. For this reason, just the Maxwell relations (6.38a)–(6.38d), together with the following relations

$$\left(\frac{\partial h}{\partial s}\right)_P = \left(\frac{\partial u}{\partial s}\right)_v = T, \tag{6.74a}$$

$$\left(\frac{\partial g}{\partial P}\right)_T = \left(\frac{\partial h}{\partial P}\right)_s = v, \tag{6.74b}$$

$$\left(\frac{\partial u}{\partial v}\right)_s = \left(\frac{\partial a}{\partial v}\right)_T = -P, \tag{6.74c}$$

$$\left(\frac{\partial a}{\partial T}\right)_v = \left(\frac{\partial g}{\partial T}\right)_P = -s, \tag{6.74d}$$

are useful for a great number of thermodynamic variable transformations.

Note that, if needed, it is possible to relate the change of any thermodynamic variable among P, T, v, s, u, h, a, g to any change in two other variables, by working out expressions for partial derivatives in the form $\left(\frac{\partial x}{\partial y}\right)_z$. The problem of working out all the relations among derivatives of primary thermodynamic variables up to the second order has resulted in the so-called Bridgman table.[1]

Example: derivation of thermodynamic relations. We can gain some experience in the manipulation of thermodynamic relations by

[1] See P. W. Bridgman, "A collection of thermodynamic formulas", *Phys. Rev.*, 3, (4), pp. 273–281, 1914.

obtaining (6.74a) and (6.74b) from basic definitions. From the Gibbs equation (6.34) we immediately have that

$$\left(\frac{\partial u}{\partial s}\right)_v = T.$$

By combining the Gibbs equation and the definition of enthalpy we see that

$$dh = du + Pdv + vdP = Tds + vdP,$$

consequently,

$$\left(\frac{\partial h}{\partial s}\right)_P = T,$$

and

$$\left(\frac{\partial h}{\partial P}\right)_s = v.$$

Now, substituting $dh = Tds + vdP$ into the differential of the Gibbs energy, $dg = dh - Tds - sdT$, yields

$$dg = vdP - sdT,$$

which allows us to conclude that

$$\left(\frac{\partial g}{\partial P}\right)_T = v.$$

6.6 Properties from Fundamental Equations

Primary thermodynamic functions like h, u, s, g, and a are defined in order to easily evaluate energy transfer in thermodynamic systems formed by simple compressible substances, depending on how the system is constrained. For example, the specific Gibbs function g is useful to describe a thermodynamic process occurring at fixed T and P, the specific Helmholtz function a is related to an isothermal–isochoric transformation, and so forth.

These functions, if expressed in terms of their associated variables, are also called *fundamental equations*. They are called fundamental because any other thermodynamic property can be derived from such an equation by partial differentiation. This cannot be done for an equation of state in the form $P = P\{v, T\}$ because, as we have shown, this can be done only if the volumetric equation of state is complemented by a function for the ideal-gas specific heat.

The variables associated with a fundamental equation of state are usually called *natural* or *canonical* variables. T and P, for example, are natural variables for the specific Gibbs function. Fundamental equations are commonly written as

$$u = u\{s, v\}, \tag{6.75a}$$

$$a = a\{T, v\}, \tag{6.75b}$$

$$g = g\{T, P\}, \tag{6.75c}$$

$$h = h\{s, P\}. \tag{6.75d}$$

We demonstrate here that all thermodynamic properties can be obtained by partial differentiation from an equation of state in the Helmholtz energy form

$$a = a\{T, v\}.$$

First of all, as is the case with a volumetric equation of state, the Helmholtz function can also be solved for its canonical variables T and v. Moreover from (6.74c) we get

$$P = -\left(\frac{\partial a}{\partial v}\right)_T, \tag{6.76}$$

and from (6.74d)

$$s = -\left(\frac{\partial a}{\partial T}\right)_v. \tag{6.77}$$

From the definition of Helmholtz energy

$$u = a + Ts = a - T\left(\frac{\partial a}{\partial T}\right)_v, \tag{6.78}$$

and from the definition of enthalpy

$$h = u + Pv = a - T\left(\frac{\partial a}{\partial T}\right)_v - v\left(\frac{\partial a}{\partial v}\right)_T. \tag{6.79}$$

Let's now derive the secondary properties. The isochoric specific heat is simply

$$c_v = \left(\frac{\partial u}{\partial T}\right)_v = -T\left(\frac{\partial^2 a}{\partial T^2}\right)_v, \tag{6.80}$$

and the isobaric specific heat is given by, see (6.55),

$$c_P = c_v - T\frac{\left(\dfrac{\partial P}{\partial T}\right)_v^2}{\left(\dfrac{\partial P}{\partial v}\right)_T}$$

$$= -T\left(\frac{\partial^2 a}{\partial T^2}\right)_v - T\frac{-\left(\dfrac{\partial^2 a}{\partial T \partial v}\right)_{T,v}^2}{-\left(\dfrac{\partial^2 a}{\partial v^2}\right)_T}$$

$$= -T\left[\left(\frac{\partial^2 a}{\partial T^2}\right)_v + \frac{\left(\dfrac{\partial^2 a}{\partial T \partial v}\right)_{T,v}^2}{\left(\dfrac{\partial^2 a}{\partial v^2}\right)_T}\right]. \qquad (6.81)$$

The speed of sound can be derived from the Helmholtz function as

$$c = \sqrt{-v\gamma^2\left(\frac{\partial P}{\partial v}\right)_T} = \sqrt{v\gamma^2\left(\frac{\partial^2 a}{\partial v^2}\right)_T}. \qquad (6.82)$$

Now that you know the tricks, try and derive other common properties like α, β, and κ.

The practical usefulness of fundamental equations derives from the fact that, with modern computers, it is possible to fit experimental data of various types (P, v, T, heat of vaporization, speed of sound, etc.) to such functions by performing a multivariable constraint optimization procedure. Thermodynamic models expressed in terms of fundamental equations of state are easily and efficiently implemented in computer programs. In particular, the Helmholtz function is often chosen to express modern multiparameter equations of state (Section 6.9), which are valid for a large thermodynamic region encompassing subcooled liquid, vapor–liquid, superheated and supercritical states.

> The REFPROP library of FLUIDPROP implements a number of such Helmholtz-based equations of state for several fluids.

Note that the Helmholtz function is the best choice among (6.75a)–(6.75d) if one wants to develop a fluid model that is valid over the entire thermodynamic space because $u = u(s, v)$ and $h = h(s, P)$ depend on one natural variable that cannot be directly measured, and $g = g(T, P)$ depends on natural variables that are not independent if the state is saturated.

6.7 Virial Equation of State for Gases

So far, apart from the ideal gas equation of state, we have dealt only with the volumetric equation of state of van der Waals. The famous scientist derived this equation from physical understanding at microscopic level and related its parameters to the attractive-repulsive interaction between molecules. In principle, a complete thermodynamic model for fluids could be obtained from a small set of parameters characterizing the molecules and their interaction at microscopic level. This is the direction of a very lively field of research: molecular thermodynamics or statistical mechanics. Fluid properties can also be obtained by following the statistical approach, i.e., by numerically solving equations describing the motion and interaction of molecules on a statistical basis. This method is not currently suitable for technical applications, because in these applications fluid properties must be calculated for a large number of times, and the computational time to obtain one single thermodynamic state with statistical mechanics methods is excessive. However, even if the understanding of molecular interaction is still incomplete, hence a generalized theory is not available, macroscopic models (equations of state) derived from molecular interaction information have been developed and are successfully used, as described in Section 6.9.

The direct relation between equation of state parameters and molecular interaction characteristics is a very desirable feature. In principle, one can obtain experimental information about a certain molecule and its interaction with other molecules, and use that information to develop an equation of state describing the thermodynamic properties of that fluid in all possible equilibrium states. This approach is actively pursued in current research. The information about all possible macroscopic equilibrium states is contained in the microscopic interaction between molecules, but such interaction is extremely complex, thus a complete theory leading to general equation-of-state models does not exist, though much progress has recently been achieved.

However, there is one *physically based* equation which is relatively simple and can be used to illustrate the concept: the virial equation of state.[2] This thermodynamic model is valid for the gas phase up to moderate reduced density. The virial equation relates the compressibility factor to several temperature-dependent coefficients and to a power series of the inverse specific volume and is

$$Z = 1 + \frac{B\{T)}{v} + \frac{C\{T)}{v^2} + \cdots . \tag{6.83}$$

In this equation $B\{T)$ and $C\{T)$ are the second and the third virial coefficients. Availability of fluid data is often limited to the first coefficient in the expansion. Explicit expressions for B and C can be obtained in terms of the potential function describing molecular interaction, as shown in textbooks of statistical mechanics. These treatments demonstrate that the first virial coefficient is representative of the interaction between two molecules, the second between three and so on.

The virial equation of state is important also because it can be extended to mixtures of fluids. The second virial coefficient for a gaseous mixture is

$$B = \sum_{i=1}^{n}\sum_{j=1}^{n} y_i y_j B_{ij} , \tag{6.84}$$

where i, j are the running indexes associated with the n molar fractions of the components of the fluid. The theoretically-based relation between the second virial coefficient of the mixture and the coefficients of the components allows the development of equation of state models for mixtures. These models are therefore founded on a microscopic description of the interaction between molecules of different substances.

[2] Virial comes from the Latin word "vis", meaning "force", and it translates into "of the force". It refers therefore to the interaction between molecules.

Despite its theoretical basis, the virial equation of state is not widely used, firstly because it can model only the gas phase, and only for mildly non-ideal states, secondly because only the second virial coefficient has been studied extensively enough, and is tabulated for some simple fluids. In addition, at high pressure the convergence of the series is very slow, if it converges at all. For these reasons the virial equation is mostly useful for the theoretical verification of other models, or if precise property estimates are needed for a rather small range of pressures and temperatures in the gas phase.

Noting that

$$\begin{aligned} B\{T) &= \lim_{v\to\infty} (Z-1)v = RT \lim_{P\to 0}\left(\frac{Z-1}{P}\right) \\ &= RT \lim_{P\to 0}\left(\frac{\partial Z}{\partial P}\right)_T , \end{aligned} \tag{6.85}$$

the second virial coefficient can be obtained from experimental PvT data by plotting values of Z_{exp} vs P at constant T: for a certain $\bar{T}$, $B\{\bar{T})/R\bar{T}$ is equal to the slope of the $Z = Z\{P, \bar{T})$ curve at $P = 0$. Many experimental data for various fluids have been used to obtain second virial coefficients and $B\{T)$ correlations have also been developed in numerical or graphical form.

Example: computation of B from experimental data. Let us see here how we can compute temperature-dependent second virial coefficients for a gas from experimental data. The virial equation of state for that gas can then be obtained by fitting several values of $B\{T)$.

Values of Z_{exp} that were obtained from PvT data along several isotherms are reported as a function of pressure in Figure 6.7. These values align very well along a line $Z = mP$ starting at $Z = 1$, where m is the slope. The slope of the $Z = Z(P)$ line at zero pressure is related to the second virial coefficient by (6.85), therefore we can calculate $B = \{T)$ by evaluating the slope of each isotherm in the $Z - P$ plane.

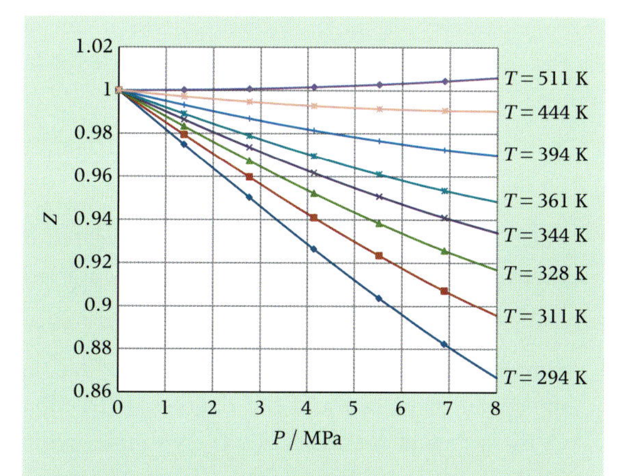

Figure 6.7 Plot of the compressibility factor Z as a function of the pressure P for methane. The isothermal lines are obtained by joining values taken from the experimental P–v–T data in R. H. Olds, H. H. Reamer, B. H. Sage, *et al.*, "Volumetric behavior of methane," *Ind. Eng. Chem.*, **35**, (*8*), 922–924, 1943.

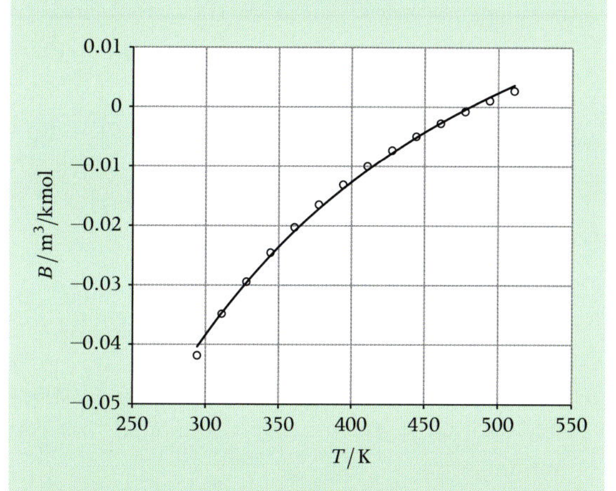

Figure 6.8 The second virial coefficients B as a function of the temperature T for methane. The points represent the data from Table 6.4 and the continuous line the values of the fitted function (6.86).

The slope can be obtained by performing a simple linear regression on the experimental data of the isotherms with a least square approach, and evaluating $m = B\{\bar{T}\}$ for each $\bar{T}$. The resulting values of $B\{\bar{T}\}$ are reported in Table 6.4 and plotted in Figure 6.8 (circles).

Table 6.4 Second virial coefficients B as function of the temperature, calculated from experimental P–v–T data in R. H. Olds, H. H. Reamer, B. H. Sage, *et al.*, "Volumetric behavior of methane," *Ind.Eng.Chem.*, **35**, (*8*), 922–924, 1943, and from Eq. (6.86).

	$B\{T\}$ / m^3/kmol	
T [K]	From exp. data	Eq. (6.86)
294	−0.04187	−0.04039
311	−0.03486	−0.03468
328	−0.02944	−0.02958
344	−0.02455	−0.02501
361	−0.02028	−0.02089
378	−0.01647	−0.01715
394	−0.01309	−0.01374
411	−0.01003	−0.01062
428	−0.00732	−0.00776
444	−0.00495	−0.00512
461	−0.00278	−0.00268
478	−0.00080	−0.00042
494	0.00106	0.00169
511	0.00270	0.00365

A functional form that is often adopted to fit $B\{T\}$ values obtained from experimental data is

$$B\{T\} = a - b\exp\left(\frac{c}{T}\right). \tag{6.86}$$

If the B values for methane of Table 6.4 are used in a least-square fit to obtain the parameters in (6.86), the results are

$$a = 0.7261 \text{ m}^3/\text{kmol},$$
$$b = 0.6666 \text{ m}^3/\text{kmol},$$
$$c = 41.06 \text{ K}.$$

These are therefore the parameters of the virial equation of state for methane

$$Z = 1 + \frac{1}{v}\left[a - b\exp\left(\frac{c}{T}\right)\right],$$

which allows us to calculate quite accurately its volumetric properties for $0.95 \lesssim Z \lesssim 1.05$.

For example, we can calculate the specific volume of a state even if it is slightly outside the range of pressure and temperature of the experimental data of Figure 6.7 used to obtain the equation of state. For $P = 1.5$ MPa and $T = 260$ K we obtain $\hat{v} = 13.84$ m^3/kmol, which is approximately 4.0 % different from the value calculated with the ideal gas equation of state (14.41 m^3/kmol).

6.8 Process Fluids and their Characteristics

The conversion of energy for useful purposes and other industrial processes are often based on one or more fluids undergoing several thermodynamic transformations. The founding block of the analysis of these processes is the calculation of the thermodynamic properties of fluids. Any energy conversion process is characterized by one or more process fluids. The fluid whose transformations produce the useful effect in a thermodynamic system is often called *working fluid*. The working fluid of a Rankine power plant is water (or an organic fluid). The working fluid in a vapor-compression refrigeration system can be R134a or any other refrigerant. The working fluids of a gas turbine are air, fuel, and combustion products, all of them mixtures.

In order to simplify the treatment of the various processes, working fluids are often classified. This can be done in many different ways. Fluid classification can be based on chemical formulas. In this case the fundamental subdivision is between organic fluids (fluids formed by molecules containing carbon atoms) and inorganic fluids. Fluids can also be classified more specifically as hydrocarbons (HC – fluids formed by molecules containing only carbon and hydrogen atoms), hydrofluorocarbons (HFC – fluids composed of molecules containing hydrogen, fluorine and carbon atoms), etc. The classification of fluids can be helpful when selecting a fluid for a certain application, because fluids of the same class share many characteristics.

An exemplary list of fluid features which can be used to subdivide process fluids in categories is as follows:

- Natural/synthetic (e.g. water vs refrigerant R134a).
- Polar/non-polar (polarity is a property related to intermolecular forces and molecule orientation. H_2O is a polar molecule, while CH_4 is not).
- Simple/complex (the complexity depends on the molecular structure and the degrees of freedom of the atoms in the molecule. O_2 is simple, while MDM, a siliconic fluid whose chemical formula is $C_8H_{24}Si_3O_2$, is a complex fluid, or, better, a fluid formed by complex molecules. The higher the molecular complexity, the higher the specific heat of the fluid.
- Heavy/light (depending on the molecular weight). Lighter molecules can be more complex than heavier molecules, whenever they are formed by a larger number of atoms. Consider, e.g., methane (CH_4, $\hat{M} = 16.04$ kg/kmol) and molecular oxygen (O_2, $\hat{M} = 32.00$ kg/kmol). The number of atoms is higher for methane, though its molecular weight is lower, therefore methane is more complex than oxygen and has a higher heat capacity.
- Linear/cyclic (this is a feature of polymers and gives a description of the molecular structure).
- Organic (hydrocarbons, fluorocarbons, silicones, etc.)/inorganic (water, air, ammonia, oxygen, helium, etc.).
- Metallic (e.g. sodium, potassium, etc.). They have been studied and applied in high-temperature topping Rankine cycles (Section 7.9) as these metals liquefy and vaporize at suitable temperatures.
- Mixture (a fluid made by combining two or more substances).
- Solution (a mixture in which it is possible to distinguish a solute which is dissolved in a solvent).

Examples of applications and related working fluids are reported in Table 6.5. Characteristics which are considered when selecting a fluid

Table 6.5 Examples of energy conversion applications and typical working fluids.

Application	Working fluids
Steam power plant	water
Gas turbine (open cycle)	air, hydrocarbons and combustion products
Gas turbine (closed cycle)	air, helium, carbon dioxide
Stirling engine	helium, nitrogen, air and light gases in general
Fuel cell	air, hydrogen, natural gas (a methane-rich mixture)
Organic Rankine cycle turbogenerator	organic fluid (hydrocarbons, fluorocarbons, siloxanes)
Refrigeration plant (compression cycle)	hydrocarbons (e.g., propane, butane), fluorocarbons (e.g., R134a, R410a), NH_3, CO_2
Refrigeration plant (absorption cycle)	lithium bromide/water mixture or ammonia/water mixture
Internal combustion engine	air/hydrocarbons and combustion products
Heat pump	light hydrocarbons, fluorocarbons, ammonia, CO_2

for an engineering application are: effect on the environment, achievable thermodynamic performance, achievable heat transfer performance, flammability, toxicity, thermal stability, cost, etc.

6.9 Complex Equations of State

In technical applications involving thermodynamic processes, it is often necessary to compute fluid properties across all possible phases (liquid, vapor–liquid, vapor) and over a wide range of pressures and temperatures. The thermodynamic analysis of energy systems, like the analysis of refrigeration or power cycles, requires mainly solving balance equations applied to fluid control volumes. For this reason the modeling of fluids is our main focus, while solid or other states or phases of matter are not included in this treatment.

Nowadays fluid models are implemented in computer programs and the use of tables or other manual procedures is obsolete. Thermodynamic charts, which are usually available as a graphical output of these computer programs – have a look at the *Sample Thermodynamic Diagrams* subfolder in the FluidProp installation folder – are often useful in preliminary studies and for general thermodynamic assessment because they can provide visual information which, with some experience, can be related to the thermodynamic performance of energy conversion cycles and other technical processes. The correct choice of the fluid model is an important task when dealing with system modeling and analysis, therefore relevant models are illustrated in some detail here.

Often, an important requirement for a fluid model is that it should consistently describe the liquid and vapor phases, the so-called supercritical region, and enthalpy differences between states, also across phases. This implies the use of one single thermodynamic model for all the thermodynamic regions of interest, that is, for subcooled, saturated, superheated, and supercritical states. Figure 6.9 shows the P–v state diagram for water, indicating the subcooled liquid, vapor–liquid, superheated, and supercritical regions. The points forming the diagram have been calculated with the so-called *Industrial Formulation 97* (IF97) model issued by the International Association for the Properties of Water and Steam (IAPWS).

IAPWS is an international not-for-profit association of national organizations concerned with the properties of water and steam. More specifically, thermophysical properties and other aspects of high-temperature

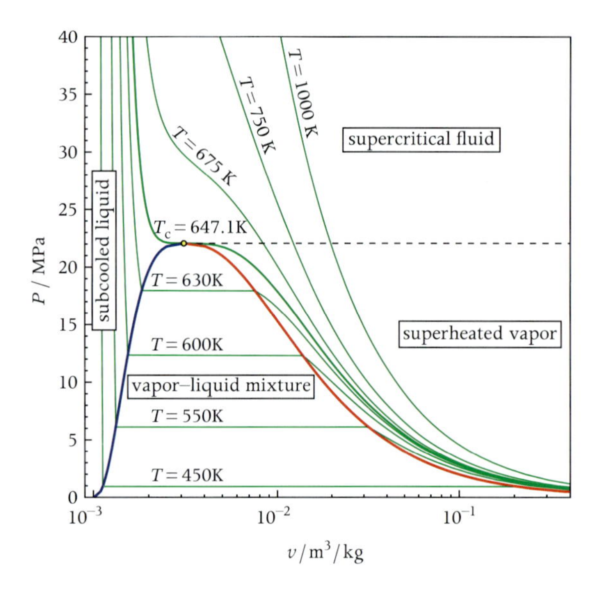

Figure 6.9 *P–v* **state diagram of water indicating the saturated liquid and vapor lines, several isotherms and the critical point. All the lines are calculated with the IF97 model. Note that the triple point temperature of water is** 276.16 K.

steam, water and aqueous mixtures, which are relevant to thermal power systems and other industrial applications.

Equations of state are expressed in terms of intensive properties. As explained in Section 3.4, for a simple compressible fluid in the vapor or liquid phase, two independent properties are necessary and sufficient to determine its state. Moreover, as we have seen in Sections 6.3 and 6.6, the complete thermodynamic model of a simple fluid can be formulated either as

$$P = P\{v, T\}, \text{ or } P = P\{\rho, T\}$$

together with

$$c_P^{\mathrm{IG}} = c_P^{\mathrm{IG}}\{T\}, \text{ or } c_v^{\mathrm{IG}} = c_v^{\mathrm{IG}}\{T\},$$

or as

$$a = a\{T, v\}.$$

The compressibility factor Z is often used as the explicit thermodynamic variable in place of P in volumetric equations of state, as it is well suited for reducing round-off errors in computer calculations, given that it is a small number close to unity.

Let us now have a closer look at the form of these functions and its implications. A characteristic shared among all equations of state is that these analytical functions contain parameters that depend on the fluid. Observe that also the very simple ideal-gas equation of state,

$$P = P\{v, T\} = \frac{RT}{v},$$

contains the fluid parameter

$$R = \frac{\hat{R}}{\hat{M}}.$$

The determination of fluid-specific parameters requires, in the majority of cases, experimental information. This is often a set of macro-properties determined by performing measurements on a certain amount of that fluid (e.g., the critical pressure and temperature), or micro-properties obtained by measuring or estimating molecular interactions (e.g., parameters which can be used to describe the attractive-repulsive molecular interaction).

One of the most common fluid parameters used in equations of state is the so-called acentric factor, defined as

$$\omega \equiv -\log \frac{P\{T_r = 0.7\}}{P_c} - 1, \qquad (6.87)$$

where $T_r \equiv \dfrac{T}{T_c}$ is the *reduced temperature*. Though it has a simple form, it is a very useful empirical parameter because it is related to the shape and physical behavior of the molecule. The acentric factor is an index related to the slope of the saturation pressure curve in the PT plane (Figure 6.10). The saturation pressure curve is extremely sensitive to molecular characteristics, it is relatively easy to measure and databases reporting the value of the acentric factors for a great number of fluids are common. Note also that its name refers to a molecular property: the acentric factor is an index of how "non-centric" or "acentric" the molecule is.

For a simple fluid (Ar, Kr, Xe), i.e., for almost spherical molecules, the logarithm of the reduced vapor pressure plotted versus the inverse of the reduced temperature is linear, i.e.,

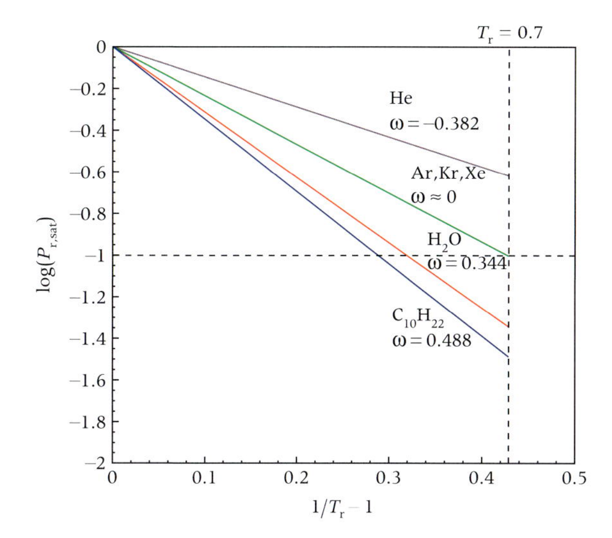

Figure 6.10 Evaluation of the acentric factor $\omega \equiv -\log \frac{P\{T_r = 0.7)}{P_c} - 1$ from vapor pressure data.

$$\frac{d\left[\log\left(P_r^{sat}\right)\right]}{d\left(1/T_r\right)} = \alpha,$$

and $\log\left(P_r^{sat}\right) = -1$ for $T_r = 0.7$, that is $\omega = 0$. For the majority of the fluids the dependence is also almost linear, but the slope is different: the acentric factor is a measure of the deviation of $\log\left(P_r^{sat}\right)$ from -1 at $T_r = 0.7$.

Cubic equations of state

We have treated the simplest of the cubic equations of state, that is the equation of van der Waals. A large family of equations of state are called cubic because they can be expressed as a cubic function of the specific volume.

The most general formulation of a cubic equation of state is

$$P = \frac{RT}{v - b} - P_{att}\{T, v), \qquad (6.88a)$$

with

$$P_{att}\{T, v) = \frac{a\{T)}{v\,(v + d) + c\,(v - d)}, \qquad (6.88b)$$

the term giving the contribution to the pressure due to attractive intermolecular forces.

The cubic dependence on the specific volume can be made explicit by solving (6.88a) for v, thus obtaining

$$v^3 + \alpha\{T, P)v^2 + \beta\{T, P)v + \gamma\{T, P) = 0, \quad (6.89a)$$

where

$$\alpha = d + c - b - \frac{RT}{P}, \qquad (6.89b)$$

$$\beta = -cd - bd - bc - \frac{RT}{P}\left(c + d - \frac{a\{T)}{RT}\right), \qquad (6.89c)$$

$$\gamma = bcd + \frac{RT}{P}\left(cd - \frac{a\{T)b}{RT}\right). \qquad (6.89d)$$

The fluid-dependent parameters a, b, c and d in (6.88a) can be constants, a function of the temperature, and/or of other fluid properties. For example the van der Waals equation is obtained from (6.88a) for a = const and $c = d = 0$. Cubic equations of state are often called *van der Waals-type equations* because they share the same physical basis: the first term in (6.88a), the so-called *covolume* term, is related to the volume occupied by the molecule, i.e., to its dimension (b is called the covolume parameter). The second term is related to the intermolecular attraction forces. Note also that the ideal gas behavior is recovered for value of the parameters for which $P \to 0$ and $v \to \infty$ (for example, a=0 and b=0 in the van der Waals equation of state).

There exists a great number of cubic equations of state, possibly greater than four hundred. They were developed in an effort to make them more accurate over a particular thermodynamic region, or to cover the whole thermodynamic range, or for a particular class of fluids. Widely used equations of state are of the two-parameters type. In these equations, the attractive parameter a is temperature-dependent to correctly capture the physical behavior of molecules.

The Soave–Redlich–Kwong equation is given by

$$P = \frac{RT}{v - b} - \frac{a\{T)}{v\,(v + b)}, \qquad (6.90)$$

with

$$a\{T) = a_c \cdot \alpha\{T), \qquad (6.91)$$

where

$$\alpha\{T) = \left[1 + \left(0.48 + 1.574\omega - 0.176\omega^2\right)\left(1 - \sqrt{T_r}\right)\right]^2,$$

and is obtained by correlating many fluid data.

The fluid-specific a_c constant in (6.91) is obtained by imposing the physical constraint that the critical point is an inflection point of the critical isotherm, see Figure 6.9, i.e.,

$$\left(\frac{\partial P}{\partial v}\right)_{T_c} = 0,$$

and

$$\left(\frac{\partial^2 P}{\partial v^2}\right)_{T_c} = 0.$$

This results in

$$a_c = a\left(T_c\right) = 0.42748\frac{R^2 T_c^2}{P_c},$$

and

$$b = 0.08664\frac{RT_c}{P_c}.$$

The condition that the critical point is an inflection point of the critical isotherm is valid for any equation of state, see Section 6.12. Furthermore the first non-zero derivative must be odd and have a negative value.

The Soave–Redlich–Kwong equation can be written in terms of compressibility factor as

$$Z^3 - Z^2 + \left(A - B^2 - B\right) Z - AB = 0,$$

with

$$B = \frac{Pb}{RT}, \quad A = \frac{aP}{R^2T^2}. \tag{6.92}$$

Note also that the Soave–Redlich–Kwong equation (6.90) is obtained from (6.88a) by setting

$$\begin{aligned} c &= 0 \\ d &= b. \end{aligned} \tag{6.93}$$

The Soave–Redlich–Kwong equation of state was originally developed for light hydrocarbons.

Another widely adopted thermodynamic model is given by the Peng–Robinson equation of state. It has a different specific volume dependence which allows for more accurate liquid density estimations. It is given by

$$P = \frac{RT}{(v-b)} - \frac{a\left(T\right)}{v(v+b)+b(v-b)}. \tag{6.94}$$

In this case the α function is

$$\alpha\left(T\right) = \left[1 + \left(0.37464 + 1.5422\omega - 0.26992\omega^2\right)\left(1-\sqrt{T_r}\right)\right]^2.$$

The a_c and b constants, obtained by imposing the critical constraints, become

$$a_c = a\left(T_c\right) = 0.45724\frac{R^2T_c^2}{P_c},$$

$$b = 0.07780\frac{RT_c}{P_c}.$$

In terms of compressibility factor (6.94) can be written as

$$Z^3 - (1-B)\,Z^2 + \left(A - 3B^2 - 2B\right) Z - \left(AB - B^2 - B^3\right) = 0,$$

with A and B as in (6.92).

The Peng–Robinson equation of state (6.94) is obtained from the general-form cubic equation of state (6.88a) by setting

$$c = b$$

$$d = b.$$

One major limitation of two-parameter cubic equations of state is that they predict a compressibility factor at the critical point Z_c which is not dependent on the fluid. This does not comply with theoretical and experimental evidence. The critical compressibility factor for the Soave–Redlich–Kwong equation of state is

$$Z_c = 0.333,$$

while for the Peng–Robinson equation of state it is

$$Z_c = 0.307.$$

A lower value of Z_c is closer to the majority of the actual values: for example $Z_c = 0.12$ for hydrogen fluoride, 0.229 for water, $Z_c \approx 0.27$ for most hydrocarbons and 0.286 – 0.311 for noble gases. Therefore the Peng–Robinson equation of state is often preferable, also because it is somehow less inaccurate than the Soave–Redlich–Kwong close to the critical point. Appendix A.12 shows the derivation of a complete thermodynamic model starting from a variant of the Peng–Robinson equation of state. This model is implemented in the StanMix library.

Variants of these two popular equations of state are obtained by: i) modifying the functional form of the attractive parameter a; ii) introducing a temperature dependence in the co-volume b; and iii) inserting a *translation term* in the covolume to obtain better liquid density estimates.

In general, if the modifications are well conceived and respect thermodynamic constraints, it can be argued that the more complex the equation of state, the more accurate it is for a broader range of thermodynamic properties and different classes of fluids. Complexity has its counterpart in the difficulty of the computer implementation. Moreover, it might require input parameters which are not available for a certain fluid or class of fluids, thus leading to a less general applicability.

The advantage of simpler equations of state, like the Peng–Robinson and the Soave–Redlich–Kwong, is that they need fewer input data (just two) with respect to more complex functional forms, and that they can be easily and consistently extended to multicomponent fluids, see Chapter 8. For many engineering applications, like, for example, for preliminary thermodynamic cycle calculations, they are usually well suited. For the design of sophisticated equipment (e.g., heat exchangers, turbines, compressors, etc.) a greater accuracy might be demanded. When justified by the popularity of a certain application or by the required accuracy, a more complex equation of state is usually available or its development is undertaken. This is the case for instance for widespread technological fluids like, water, ammonia, methane, or refrigerants, to name a few, for which accurate equations of state are available and continually improved.

Multiparameter volumetric equations of state

A number of accurate multiparameter equations of state have been developed from the 1970s onwards for fluids widely used in technical and scientific applications, like water, ammonia, hydrocarbons, refrigerants, and light gases. They are empirical in the sense that they are obtained by selecting a functional form that is suitable to fit experimental data of a fluid or a class of fluids and the parameters are than fitted on selected measured data. The functional form of these volumetric equations comes usually from a modification of a virial series expansion (6.83). At that time, the selection of the optimal functional form was based mainly on the ability of the developer of the equation of state, while recently mathematical methods to optimize the functional form have been formulated.

One of the most used volumetric multiparameter equations of state is that for water developed by Keenan, Keyes, Hills, and Moore, given by

$$P = \rho R T \left[1 + \rho Q + \rho^2 \left(\frac{\partial Q}{\partial \rho} \right)_T \right], \tag{6.95}$$

where

$$Q = (\tau - \tau_c) \sum_{j=1}^{7} (\tau - \tau_{aj})^{j-2} \times \left[\sum_{i=1}^{8} A_{ij} \left(\rho - \rho_{aj} \right)^{i-1} + e^{-E\rho} \sum_{i=9}^{10} A_{ij} \rho^{i-9} \right],$$

and

$$\tau = \frac{T_a}{T},$$
$$\tau_{a1} = \tau_c,$$
$$\tau_{aj} = 2.5, \; j > 1.$$

T_a is a reference temperature and it is equal to 1000 K, if the properties are in SI units.

> In the past, thermodynamic properties were made available as tables because computers were not common, therefore the collections of values computed with the thermodynamic model of Keenan, Keyes, Hills, and Moore are also known as *steam tables*.

We report here the entire $P = P(\rho, T)$ equation to provide a sense of the complication of the analytical function. A complex functional form is needed to achieve high accuracy for all the points in the thermodynamic region of interest, i.e., subcooled liquid, vapor–liquid equilibrium, superheated vapor and supercritical states. The number of parameters (A_{ij}, E, T_a, ρ_{aj}), all of them valid only for water, is quite large. A similar but different equation was developed, for example, to describe the properties of ammonia.

Volumetric multiparameter equations of state can also cover a *class of fluids*. The Starling equation of state has been developed for light hydrocarbons (butane, propane, pentane, hexane, etc.), and it is given by

$$P = \rho RT + \left(B_0 RT - A_0 - \frac{C_0}{T^2} + \frac{D_0}{T^3} + \frac{E_0}{T^4}\right)\rho^2 + \left(\mathsf{b}RT - \mathsf{a} - \frac{\mathsf{d}}{T}\right)\rho^3 + \alpha\left(\mathsf{a} + \frac{\mathsf{d}}{T}\right)\rho^6 + \mathsf{c}\frac{\rho^3}{T^2}\left(1+\theta\rho^2\right)e^{-\theta\rho^2}. \tag{6.96}$$

Other equations of state can be applied even more broadly, like the Martin–Hou equation of state

$$P = \frac{RT}{v-\mathsf{b}} + \sum_{i=2}^{5}\frac{1}{(v-\mathsf{b})^i}\left(A_i + B_i T + C_i e^{-kT/T_c}\right) + \frac{A_6 + B_6 T + C_6 e^{-kT/T_c}}{e^{\alpha v}(1+\mathsf{c}\cdot e^{\alpha v})}, \tag{6.97}$$

which is often adopted for refrigerants.

All these equations of state are implemented, for several fluids, in the TPSI library. Figure 6.11 shows P–v diagrams for ammonia and pentane calculated with TPSI .

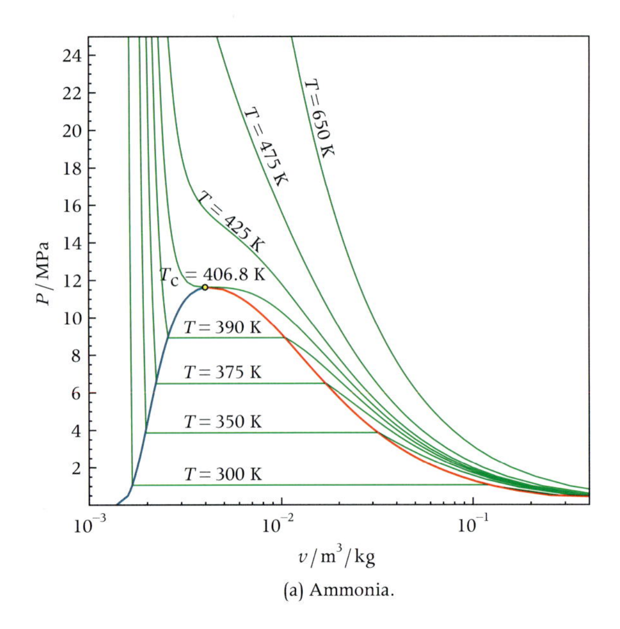

(a) Ammonia.

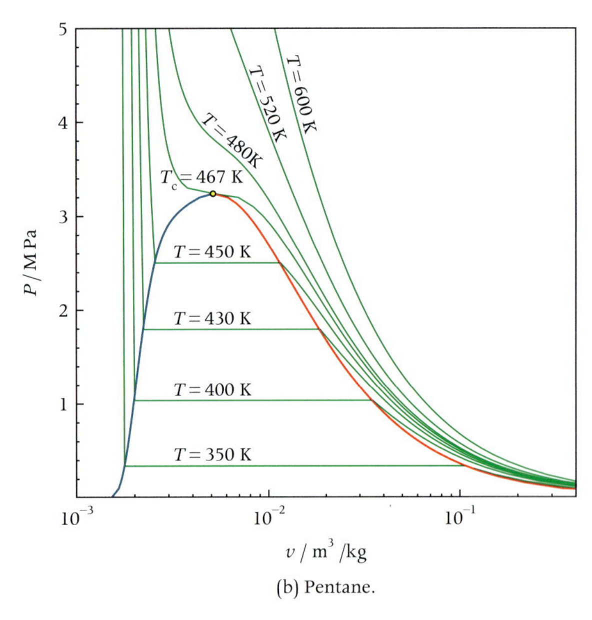

(b) Pentane.

Figure 6.11 ***P*–*v* diagrams for ammonia and pentane, reporting the vapor–liquid coexistence curve (blue-red), several isotherms (green) and the critical point (yellow circle). All the lines are calculated with the multiparameter volumetric equations of state implemented in the TPSI library. Note that the triple point temperature of ammonia is** 195.5 K, **and the triple point temperature of pentane is** 143.47 K.

Helmholtz equations of state

The most recent and accurate multiparameter equations of state are formulated in terms of the specific reduced Helmholtz energy, therefore they are *fundamental equations*, as explained in Section 6.6. The specific reduced Helmholtz energy is given by

$$\frac{a(T,\rho)}{RT} = \frac{a^{\mathrm{IG}}(T,\rho) + a^{\mathrm{R}}(T,\rho)}{RT} = \alpha^{\mathrm{IG}}(\tau,\delta) + \alpha^{\mathrm{R}}(\tau,\delta),$$

where superscripts IG and R stand for ideal gas and residual, respectively, and δ and τ represent the dimensionless density $\delta \equiv \rho/\rho_{\mathrm{ref}}$ and the inverse dimensionless temperature $\tau \equiv T_{\mathrm{ref}}/T$. The reference parameters are usually the critical values.

The ideal gas Helmholtz energy is obtained from the isobaric ideal gas heat capacity correlation (Section 6.10) by considering that

$$a^{\mathrm{IG}} = a_0 + \int_{T_0}^{T} (c_P^{\mathrm{IG}} - R) dT - T \int_{T_0}^{T} \frac{c_P^{\mathrm{IG}} - R}{T} dT + RT \ln\left(\frac{\rho}{\rho_0}\right). \tag{6.98}$$

Subscript 0 refers to an arbitrarily chosen reference state.

The residual part is given by the general empirical function

$$\alpha^{\mathrm{R}}(\tau, \delta) = \sum_{i=1}^{I_{\mathrm{Pol}}} n_i \tau^{t_i} \delta^{d_i} + \sum_{i=I_{\mathrm{Pol}}+1}^{I_{\mathrm{Pol}}+I_{\mathrm{Exp}}} n_i \tau^{t_i} \delta^{d_i} \exp(-\delta^{p_i}). \tag{6.99}$$

In this equation, the first terms of the summation are polynomials, while the second are exponentials. The coefficients n_i are substance-specific and are determined once the functional form is defined by a complex fitting procedure, which is driven by a multivariable constraint nonlinear optimization algorithm.

Modern numerical techniques allow for the optimization of the functional form by testing all possible combinations and reducing the number of terms in (6.99). This is one of the main advancements compared to the volumetric equations of state of the previous generation. The functional form can be specialized to one fluid for which a large set of accurate measurements are available, or to a class of fluids, but this entails a decrease in the overall accuracy. Two families of Helmholtz-based equations of state can therefore be defined, the *reference* equations, valid for a specific substance, and the *technical* equations, applicable to a class of fluids.

An equation of state can be called a *reference equation* if it can represent all the most accurate experimental data of any kind (volumetric and caloric) for a certain fluid over the entire thermodynamic range of existence for that substance within the stated experimental uncertainty. Reference equations have currently been developed only for methane, ethane, propane, n-butane, isobutane, carbon dioxide, nitrogen, oxygen, argon, ethylene, propylene, sulfur hexafluoride, and water, because only for these fluids is a sufficient number of very accurate experimental data available. Reference equations of state completely substitute the use of measured values, even for the most accuracy-demanding applications, like, for example, the calibration of special instruments. These equations are capable of correctly predicting the very peculiar behavior of substances extremely close to the critical point, something that is impossible for any other general equation of state. In this case some additional terms with respect to the ones in (6.99) are necessary.

Technical equations have been developed for a greater number of substances. In this case the functional form is optimized with an algorithm which considers data sets for various fluids simultaneously. An example of these equations is the Span–Wagner functional form for non-polar fluids, whose residual term is given by

$$\begin{aligned} \alpha^{\mathrm{R}}(\tau, \delta) &= n_1 \delta \tau^{0.250} + n_2 \delta \tau^{1.125} + n_3 \delta \tau^{1.500} + n_4 \delta^2 \tau^{1.375} \\ &+ n_5 \delta^3 \tau^{0.250} + n_6 \delta^7 \tau^{0.875} + n_7 \delta^2 \tau^{0.625} e^{-\delta} \\ &+ n_8 \delta^5 \tau^{1.750} e^{-\delta} + n_9 \delta \tau^{3.625} e^{-\delta^2} + n_{10} \delta^4 \tau^{3.625} e^{-\delta^2} \\ &+ n_{11} \delta^3 \tau^{14.5} e^{-\delta^3} + n_{12} \delta^4 \tau^{12.0} e^{-\delta^3}. \end{aligned} \tag{6.100}$$

This equation can be used when a smaller number of measurements is available. Remarkably, its extrapolation behavior is correct to the point that these equations are accurate enough for most technical applications. Moreover, the number of fluid parameters is so small (12) that also computational intensive applications like CFD simulations can nowadays rely on these models.

Helmholtz-based equations of state for many fluids are implemented in the REFPROP program developed and maintained by the National Institute of Standards and Technology (NIST). REFPROP is also available as a library within FLUIDPROP .

Figure 6.12 presents two exemplary thermodynamic charts for toluene (obtained with a technical equation), and carbon dioxide (obtained with a reference equation).

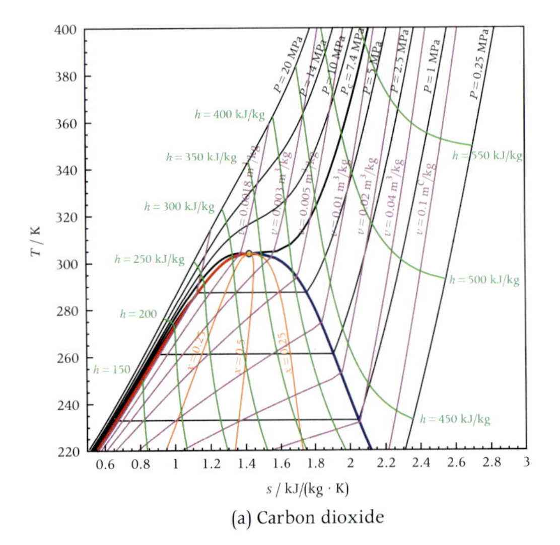

(a) Carbon dioxide

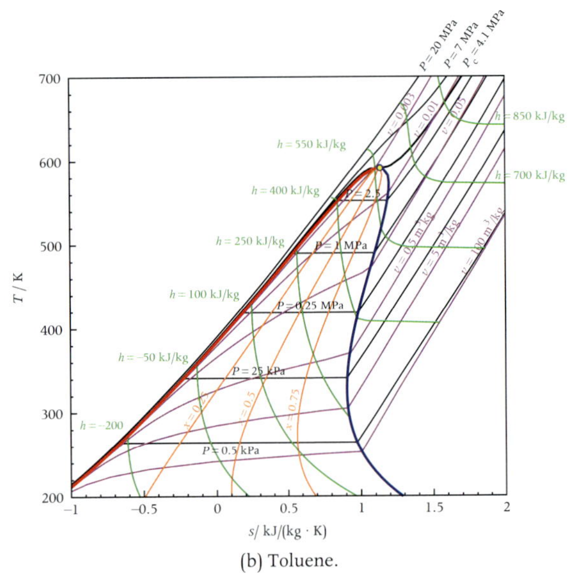

(b) Toluene.

Figure 6.12 ***T–s* state diagrams for carbon dioxide and toluene. All the lines are calculated with multiparameter Helmholtz equations of state implemented in the RefProp library. The triple point temperature of carbon dioxide is 216.592 K, while that of toluene is 178.0 K.**

Equations of state from statistical thermodynamics, or "molecular" equations of state

Many fluids cannot be modeled with the empirical equations of state we have just described, because of the lack of experimental information (critical temperature and pressure, vapor pressures, P–v–T data, etc.). The development of new technologies or modifications to existing processes often relies on adopting less conventional fluids, and their properties must be calculated nevertheless.

The problem of estimating the thermodynamic properties of these fluids can be tackled in two ways. Experimental information about the fluid can be obtained from measurements or, and this is a general trend in engineering, sophisticated physically-based models can be used as a partial substitute to costly and lengthy experimental campaigns.

Empirical equations of state are available mainly for fluids in which the most relevant molecular interactions are indeed repulsion and dispersion forces (van der Waals forces), together with weak electrostatic forces due to dipoles, quadrupoles, etc. This is the case, for example, for simple inorganic substances, many hydrocarbons, refrigerants, etc. Many fluids and mixtures belonging to the class of electrolytes, polar solvents, hydrogen-bonded fluids, polymers, and plasmas, display a range of other peculiar molecular interactions: strong polar and complexing forces, induction forces, Coulombic forces, and forces related to the flexibility of molecular chains. These interactions are stronger than those due to dispersion or weak electrostatic forces, but still weaker than those characterizing chemical bonds. Figure 6.13 gives an idea of the difference in strength among possible molecular interactions. Multiparameter equations of state are available for ammonia and water, falling in the category of strongly associating fluids, but this is not the case for other, less widespread compounds belonging to the same category. The application of cubic equations of state to associating fluids can lead to large inaccuracy.

Information about molecular interaction can be used to develop equations of state which cover the entire range of interest. Molecular interaction can be described by an energy potential. Figure 6.14 shows a simple molecular potential function, the Lennard–Jones potential, which can describe the intermolecular forces occurring between simple molecules, namely short-range repulsion and long-range attraction. The radial interaction force

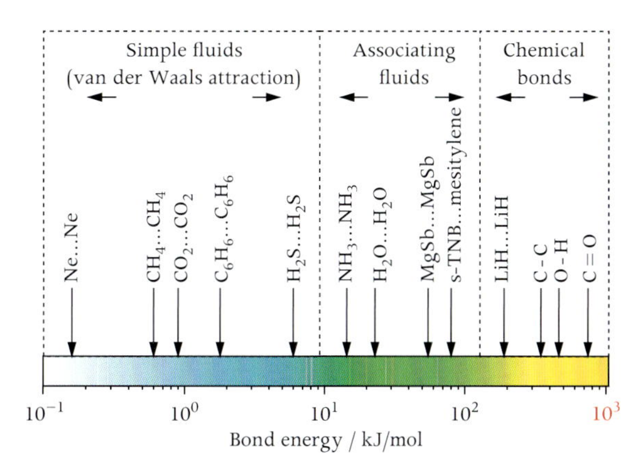

Figure 6.13 Bond energies. Data are taken from I. Klotz and R. M. Rosenberg, *Chemical Thermodynamics*, 5th ed. New York: John Wiley & Sons, 1994, p. 55, and from J. E. Huheey, E. A. Keiter, and R. L. Keiter, *Inorganic Chemistry*, 4th ed. New York: Harper Collins, 1993, Table E.1.

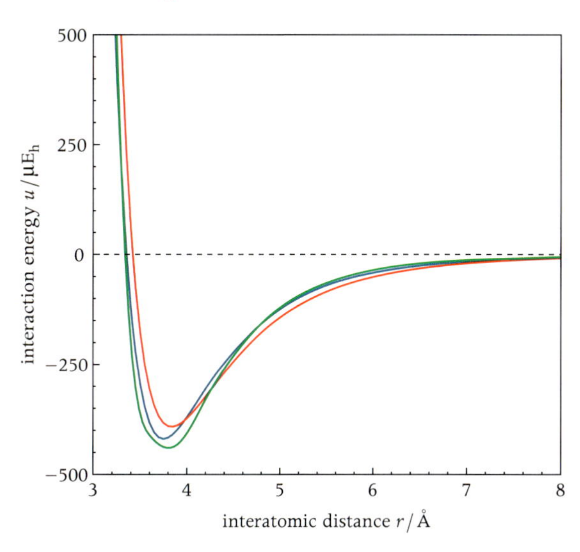

Figure 6.14 The Lennard–Jones potential for the argon molecule (green line). Given the almost ideal behavior of argon, the Lennard–Jones potential can accurately represent the interaction between argon molecules. The potential obtained with ab initio calculations, blue line, from K. Leonhard and U. K. Deiters, "Monte Carlo simulations of neon and argon using *ab initio* potentials," *Mol. Phys.*, 98, (*20*), 1603–1616, 2000, and accurate experimental data, red line, from R. A. Aziz, "A highly accurate interatomic potential for argon," *J. Chem. Phys.*, 99, (6), 4518–4525, 1993, are also reported for comparison.

$F\{r\}$ is obtained from the potential $W\{r\}$ as $F\{r\} = -\nabla W \cdot \{r\}$.

A large enough number of molecules interacting according to the laws of mechanics can be treated statistically to obtain the value of thermodynamic properties at equilibrium. The branch of science which deals with this problem is called statistical thermodynamics. It lays the bridge between molecular interaction and thermodynamic properties. For example, the pressure of a system at a certain temperature and occupying a certain volume can be calculated as the statistical result of the collision of the molecules, which is determined by their intermolecular potential. We have already invoked this concept in Section 3.9.

A fluid made of hard spheres interacting without even attractive forces, allows one to obtain, by means of statistical mechanics manipulation and simulation, the Carnhan–Starling equation of state

$$Z = \frac{Pv}{RT} = Z\{\eta\} = \frac{1+\eta+\eta^2-\eta^3}{(1-\eta)^3}, \qquad (6.101)$$

where $\eta = \dfrac{\pi\rho\sigma^3}{3}$, σ is the hard sphere diameter and ρ the density. This equation shows the link between thermodynamic properties and molecular information.

Statistical mechanics establishes a link between the Helmholtz free energy and molecular physics. Let's consider a system formed by N molecules, at temperature T and occupying a volume V. It can be demonstrated that all the thermodynamic information is contained in the canonical partition function

$$Q\{N, V, T\} = \sum_{\text{states } i} e^{-E_i\{N,V\}/kT}, \qquad (6.102)$$

where E_i is the energy of the ith quantum state, k is the Boltzman constant and the summation is carried out over all the states of the molecules consistent with N and V. It can also be demonstrated that the canonical partition function is related to the Helmholtz free energy by

$$A\{N, V, T\} = -kT \ln Q\{N, V, T\}. \qquad (6.103)$$

These two equations connect molecular information (E_i) and thermodynamic properties (A).

Among a variety of approaches to the development of thermodynamic models based on molecular information, only the statistical associating fluid theory (SAFT) is briefly addressed here, as we think that preliminary knowledge about this quickly developing field should enter the luggage of the modern mechanical engineer.

The idea underlying the SAFT approach is the so-called thermodynamic perturbation theory. According to this theory, a model fluid (e.g., a homomorphic, nonassociating fluid) is first defined, and the properties of real fluids are obtained through a perturbation expansion. Real fluids are called "associating fluids" because all the other interactions promote association among molecules. In addition, a molecule has a repulsive core, one or more short-range attractive sites, and all other interactions (e.g., dispersion, long-range dipolar forces) are treated as approximate mean-field terms. Figure 6.15 presents this idea in a graphical fashion.

By looking at (6.102) and (6.103) we can infer that the Helmholtz free energy can be expressed in terms of contributions of molecular forces. This idea leads to the SAFT theory. According to SAFT, the Helmholtz free energy is written as the sum of contributions due to hard-sphere repulsive interactions, chain formation through bonding of a number of hard spheres, and association, i.e.,

$$a(T, \rho) = a^{\text{segments}} + a^{\text{chain}} + a^{\text{association}}.$$

All the contributions can be expressed in terms of molecular parameters, which can be obtained from a description of the molecular potential. Some of these molecular properties, like multipole moments (dipole and quadrupole moments), and polarizabilities are directly related to an observable property of the molecule. Other parameters, e.g., shape, energy, and size parameters, cannot be linked directly to an observable molecular property, although they have a clear physical meaning.

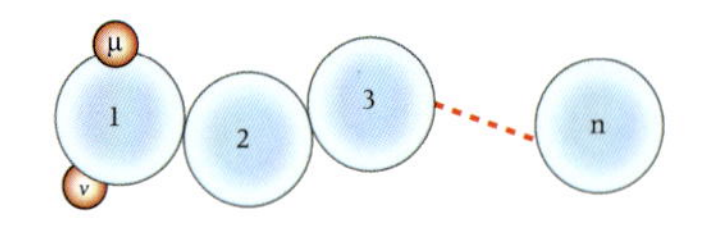

Figure 6.15 Simplified representation of the molecule of an alkane according to the SAFT theory. The molecule is formed by n spherical segments representing the monomers, and two associating sites ν and μ.

> A variant of the SAFT model, the PCP-SAFT model, is implemented into the PCP-SAFT library of the professional version of FLUIDPROP .

The problem is that molecular interaction is very complex and many parameters must be used in order to obtain an accurate description. Experimental data can simplify the problem of estimating these parameters, but in this way the difference from empirical equations of state is less marked. On the other hand, these parameters have a well-defined physical meaning, therefore they can be obtained from direct measurements (e.g., Fourier transform infrared spectra) or from so-called *ab initio* calculations. These are computational-chemistry methods based on first principles and allow, in the case of simple molecules, computation of molecular properties starting from the Schrödinger equation.

As a last remark, it is important to know that methods derived from statistical mechanics and molecular dynamics can be implemented in computer programs. There are two approaches to molecular simulations: one consists in describing a population of molecules over a period of time by means of mechanics laws, and averaging the quantities over time to obtain thermodynamic information. The other involves the averaging over a large number of randomly generated systems of molecules with the same macroscopic constraints, and it usually goes by the name of Monte Carlo simulation. Molecular simulations can be used to compute properties associated with a given state and their ever increasing accuracy allows us nowadays to consider their results as "virtual experiments," which can provide data (e.g. P–v–T data) for the development of empirical equations of state.

6.10 Ideal Gas Heat Capacity

We have seen that in order to obtain the complete thermodynamic model of a fluid, a relation for the ideal gas heat capacity as a function of temperature is necessary. Often empirical calorimetric data are correlated with a simple polynomial function, like

$$c_P^{\text{IG}}(T) = \alpha + \beta T + \gamma T^2 + \eta T^3, \tag{6.104}$$

where α, β, γ and η are fluid-specific parameters obtained by fitting experimental data. This power expansion series can be derived by applying rigorous quantum statistical mechanics theory and appropriate simplifications. Currently, very accurate speed-of-sound measurements can be used to obtain values of c_P as a function of temperature in the limit of very low pressure. Parameters for the correlation can also be obtained from so-called group contribution methods. With these methods the value of a certain molecular property can be obtained by appropriately summing the tabulated contributions of the groups of atoms (e.g. the methyl group CH_3, etc.) forming that molecule. The application of group-contribution methods leads to less accurate values, though it is applicable to a large variety of fluids.

Statistical mechanics can be employed to obtain more complex and potentially more accurate functional forms for the temperature dependence of the ideal gas heat capacity. This is the case, for example, for the equation of Aly and Lee

$$c_P^{\mathrm{IG}}\{T\} = A + B\left[\frac{C/T}{\sinh\left(C/T\right)}\right]^2 + D\left[\frac{E/T}{\cosh\left(E/T\right)}\right]^2. \tag{6.105}$$

A fundamental approach, like the one described in the previous paragraph, can also be adopted to estimate the ideal gas heat capacity. In this case no molecular interaction has to be accounted for, because, in the ideal gas approximation, molecules are distant enough not to interact.

Example: ideal gas heat capacity from speed of sound measurements. The parameters for the polynomial correlation of the ideal gas heat capacity can be calculated from speed of sound data measured in the ideal gas region of a fluid, i.e., in the limit of very low pressure. By using (6.56) and (6.12), (6.65) (all valid for ideal gases) can be rewritten so that

$$\hat{c}_P^{\mathrm{IG}}\{T\} = \frac{1}{\dfrac{1}{\hat{R}} - \dfrac{T}{\hat{M}c^2}}. \tag{6.106}$$

By measuring the speed of sound of a fluid in the ideal gas state at several temperatures, $c_P^{\mathrm{IG}}\{T\}$ can be calculated from (6.106). Next, the values of the coefficients of the polynomial correlation (6.104) can be computed so that differences (residuals) between the c_P data derived from experimental sound speeds and the values calculated with (6.104) are minimal. To this purpose, we can use a multivariable optimization algorithm available for example in EXCEL or MATLAB to calculate the coefficients such that the sum of the squares of the residuals is minimum.

In Table 6.6 measured speed of sound data for siloxane D_4 (octamethylcyclotetrasiloxane) are summarized together with the c_Ps computed with (6.106) and the polynomial correlation (6.104). Other values that are needed are $\hat{R} = 8.3143$ J/(mol · K) and $\hat{M} = 296.62$ kg/kmol for D_4.

Table 6.6 Determination of the coefficients of (6.104) for siloxane D_4, using experimental speed of sound measurements from N. R. Nannan, P. Colonna, C. M. Tracy, *et al.*, "Ideal-gas heat capacities of dimethylsiloxanes from speed-of-sound measurements and *ab initio* calculations," *Fluid Phase Equilib.*, *257*, (1), 102–113, 2007.

T/K	Measured c / m/s	$\hat{c}_P$ / J/(mol · K) from (6.106)	Coefficients of polynomial function (6.104)	$\hat{c}_P$ / J/(mol · K) (6.104)	Squares of residuals / [J/(mol · K)]^2
450	113.32	467.62	$\alpha = 85.52$	470.12	6.25
465	115.17	478.25	$\beta = 1.081$	479.50	1.57
480	116.99	489.05	$\gamma = -5.130\times10^{-4}$	488.67	0.14
495	118.78	500.70	$\eta = 2.231\times10^{-8}$	497.62	9.47
				sum of squares:	17.44

6.11 Vapor–Liquid Equilibrium

Section 3.5 presents the concepts of saturation, phases, and vapor–liquid equilibrium. In Section 5.4 we have seen that the condition for thermodynamic equilibrium of a closed and isolated two-part system is that its entropy is maximum. The conditions for thermal and mechanical equilibrium are therefore that the temperature and the pressure of the two parts are the same. This observation can be exploited to obtain the equation which allows one to rigorously calculate the vapor–liquid equilibrium states from an equation of state.

If the two parts of a closed and isolated system as in Figure 6.16 are volumes filled with liquid and vapor of the same simple compressible substance, we can write

$$T^{\mathrm{L}} = T^{\mathrm{V}}$$
$$P^{\mathrm{L}} = P^{\mathrm{V}}.$$

From the integration of the Gibbs equation $ds = \frac{1}{T}du + \frac{P}{T}dv$ at fixed T and P, see Section 5.4, we have that

$$s^{\mathrm{V}} - s^{\mathrm{L}} = \frac{h^{\mathrm{V}} - h^{\mathrm{L}}}{T},$$

which can be rearranged and gives

$$h^{\mathrm{L}} - Ts^{\mathrm{L}} = h^{\mathrm{V}} - Ts^{\mathrm{V}}.$$

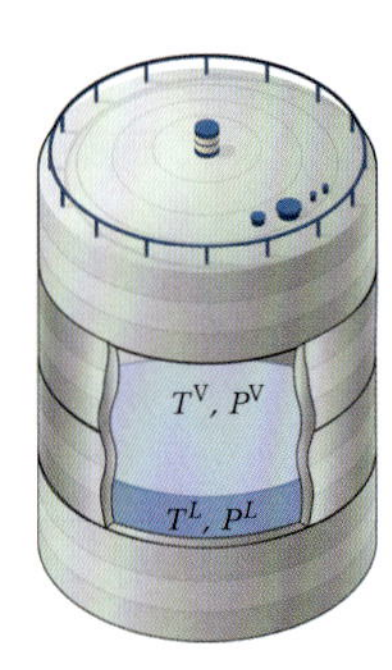

Figure 6.16 A closed and isolated control volume filled with vapor and liquid in equilibrium.

By recalling the definition of specific Gibbs energy, $g \equiv h - Ts$, we notice that a condition for vapor–liquid equilibrium is given by

$$g^{\mathrm{L}} = g^{\mathrm{V}}. \tag{6.107}$$

It is outstanding that the conditions for vapor–liquid equilibrium of a pure compressible fluid

$$T^{\mathrm{L}} = T^{\mathrm{V}},$$
$$P^{\mathrm{L}} = P^{\mathrm{V}},$$
$$g^{\mathrm{L}} = g^{\mathrm{V}},$$

are obtained by applying the first and second principle of thermodynamics. This is the case for all the most important equations illustrated in this book.

There is another way to show how these vapor–liquid equilibrium conditions can be obtained from the two laws of thermodynamics. If the volume occupied by the vapor–liquid mixture is in equilibrium, the second law prescribes that $dS = 0$, that is, that the entropy of the vapor and liquid inside the control volume is maximal. From the definition of Gibbs energy, $G \equiv H - TS$, and the Gibbs equations (5.15) we have that

$$dG = dH - TdS - SdT$$

and

$$TdS = dH - VdP.$$

The resulting relation

$$dG = VdP - SdT \tag{6.108}$$

assures us that at fixed T and P also dG must be equal to 0. Note also that $G = G^{\mathrm{L}} + G^{\mathrm{V}}$ and that $dG = dG^{\mathrm{L}} + dG^{\mathrm{V}}$. From the definition of specific Gibbs energy $G^{\mathrm{L}} = g^{\mathrm{L}}M^{\mathrm{L}}$, therefore, at fixed T and P,

$$dG^{\mathrm{L}} = g^{\mathrm{L}}dM^{\mathrm{L}},$$

and the same expression holds for the vapor phase. The differential of the Gibbs energy for a closed and isolated system where the vapor and liquid of a certain substance are in equilibrium is given by

$$dG = g^{\mathrm{L}}dM^{\mathrm{L}} + g^{\mathrm{V}}dM^{\mathrm{V}}. \tag{6.109}$$

The mass conservation prescribes

$$M = M^{\mathrm{L}} + M^{\mathrm{V}},$$

and, for the closed system,

$$dM = dM^{\mathrm{L}} + dM^{\mathrm{V}} = 0,$$

therefore

$$dM^{\mathrm{L}} = -dM^{\mathrm{V}},$$

which says that in a state infinitely close to vapor–liquid equilibrium, a small portion of fluid that evaporates is counterbalanced by an equal small portion of fluid that condenses.

By substituting $dM^{\mathrm{L}} = -dM^{\mathrm{V}}$ into (6.109) we obtain

$$dG = \left(g^{\mathrm{L}} - g^{\mathrm{V}}\right) dM^{\mathrm{L}} = 0,$$

leading to the condition of equality of the specific Gibbs energy for the liquid and vapor phase that we have just seen.

The saturation line and fugacity

The conditions for vapor–liquid equilibrium can be exploited to calculate saturation states from an equation of state (or to optimize the equation of state parameters in order to fit a measured saturation curve).

The specific Gibbs energy for the liquid and the vapor phase can be calculated from a thermodynamic model (Sections 6.3 and 6.6), and the equation

$$g^{\mathrm{L}} - g^{\mathrm{V}} = 0 \tag{6.110}$$

can be solved at fixed T for P or vice versa. The algebraic expressions resulting by working out (6.110) using complex equations of state cannot be made explicit in T or P, hence the calculation of saturated states is performed numerically. The saturation line of a pure fluid, also known as vapor-pressure line, can be obtained by solving (6.110) for all the states from the triple point up to the critical point. This curve is very relevant for technical and scientific applications.

If we consider (6.110) a bit more carefully, we see that using it to calculate the saturation condition is not an optimal choice from a numerical point of view. From (6.108) we notice that

$$dg = vdP - sdT. \tag{6.111}$$

At fixed T, $dg = vdP$, therefore

$$\lim_{\tilde{P}\to 0} \tilde{g} = g - \int_{\tilde{P}}^{P} vdP = g - \infty = -\infty.$$

At low pressure g becomes a large number and this negatively affects the accuracy of numerical calculations.

Numerical phase equilibrium calculations are simplified if the saturation condition is expressed in terms of another thermodynamic function which is more well-behaved. This function is called fugacity and it is defined as

$$f \equiv P \exp\left[\frac{g(T,P) - g^{\mathrm{IG}}(T,P)}{RT}\right]. \tag{6.112}$$

The fugacity coefficient is defined as

$$\phi \equiv \frac{f}{P}, \tag{6.113}$$

such that in the limit of $P \to 0$ the fugacity coefficient goes to 1. We recall that also the specific Gibbs energy can be written as the summation of an ideal gas term and a departure term as

$$g - g_0 = g^{\mathrm{IG}} + g^{\mathrm{R}}.$$

Equating the specific Gibbs energy of the liquid and of the vapor is therefore equivalent to

$$g^{\mathrm{R,L}} = g^{\mathrm{R,V}}.$$

We see therefore that

$$\frac{g^{\mathrm{IG}}(T,P) - g(T,P)}{RT} = \frac{g^{\mathrm{R}}}{RT} = \ln\phi.$$

Hence the vapor–liquid equilibrium condition is also conveniently expressed by

$$\phi^{\mathrm{L}}(T,P) = \phi^{\mathrm{V}}(T,P). \tag{6.114}$$

If the thermodynamic model is formulated as $P = P(v,T)$, an analytical expression for the fugacity coefficient can be obtained as follows. First we notice that

$$\ln\phi = \frac{g^{\mathrm{R}}}{RT} = \frac{g - g^{\mathrm{IG}}(T,P)}{RT} = \frac{h - h^{\mathrm{IG}} - T\left(s - s^{\mathrm{IG}}\right)}{RT} = \frac{h^{\mathrm{R}} - Ts^{\mathrm{R}}}{RT}. \tag{6.115}$$

In analogy with (6.47), but using the specific volume as an independent variable, we obtain

$$h^{R} = RT(Z-1) + \int_{v=\infty}^{v} \left[T\left(\frac{\partial P}{\partial T}\right)_v - P \right] dv,$$

and

$$s^{R} = R\ln Z + \int_{v=\infty}^{v} \left[\left(\frac{\partial P}{\partial T}\right)_v - \frac{R}{v} \right] dv.$$

Substituting these expressions for h^R and s^R in (6.115) gives

$$\ln\phi = \frac{1}{RT}\int \left[\frac{RT}{v} - P\right] dv - \ln Z + (Z-1), \tag{6.116}$$

which can be used to obtain an analytical expression for $\ln\phi$ from a volumetric equation of state $P = P\{v, T\}$.

If the thermodynamic model is expressed in terms of specific Helmholtz energy, the fugacity coefficient is

$$\ln\phi = \alpha^{R} + \delta\alpha^{R}_{\delta} - \ln\left(1 + \delta\alpha^{R}_{\delta}\right), \tag{6.117}$$

in which $\alpha^R \equiv \frac{a}{RT}$ represents the dimensionless, reduced residual Helmholtz energy and $\delta \equiv \frac{\rho}{\rho_{\text{ref}}}$ the reduced or dimensionless density. The reference density is usually the critical density. Its derivation is left to the brave student!

Example: saturation pressure at a given temperature using the van de Waals equation of state. Let us see how we can compute the saturation pressure given a certain temperature using the van de Waals equation of state. First of all we can obtain the logarithm of the fugacity coefficient $\ln\phi$ for the van der Waals equation of state by substituting (6.71) into (6.116) and obtaining

$$\ln\phi = \frac{1}{RT}\int_{v=\infty}^{v} \left[\frac{RT}{v} - \frac{RT}{v-\mathsf{b}} + \frac{\mathsf{a}}{v^2}\right] - \ln Z + (Z-1)$$

$$= \frac{1}{RT}\left[RT\ln v - RT\ln(v-\mathsf{b}) - \frac{\mathsf{a}}{v}\right]_{v=\infty}^{v} - \ln Z + (Z-1)$$

$$= \left[\ln\frac{v}{v-\mathsf{b}} - \frac{\mathsf{a}}{vRT}\right]_{v=\infty}^{v} - \ln Z + (Z-1)$$

$$= \ln\frac{v}{v-\mathsf{b}} - \frac{\mathsf{a}}{vRT} - \ln Z + (Z-1).$$

Given a certain T between the triple point and the critical point values, we can calculate $\ln\phi^L$ and $\ln\phi^V$ with a tentative value of P and iterate on P until

$$\ln\phi^L = \ln\phi^V,$$

thus obtaining the saturation pressure for the given temperature.

Vapor–liquid equilibrium in charts

Let's have a look at these fugacity functions and at a graphical representation of vapor–liquid equilibrium. As we have seen, at constant temperature $dg = vdP$, which can be integrated between the vapor and the liquid equilibrium states and gives

$$g^L - g^V = \int_{P^L}^{P^V} vdP.$$

If we plot an isotherm calculated with a cubic equation of state in the P-v plane (Figure 6.17) we notice that the saturated liquid and vapor states are obtained with two of the three roots of the cubic equation $v\{P, \tilde{T}\} = 0$ satisfying the condition that the area subtended by the isotherm is 0, i.e., $\int_{P^L}^{P^V} vdP = 0$.

These states are also characterized by $\ln\phi^L = \ln\phi^V$. The third specific volume root in between the saturated liquid and vapor state does not correspond to any equilibrium state. The shape of the isotherms of cubic equations is called "van der Waals loop."

Phase rule for pure substances

Consider a vapor–liquid mixture of a simple substance. The condition of equilibrium applied to the liquid and vapor phases is

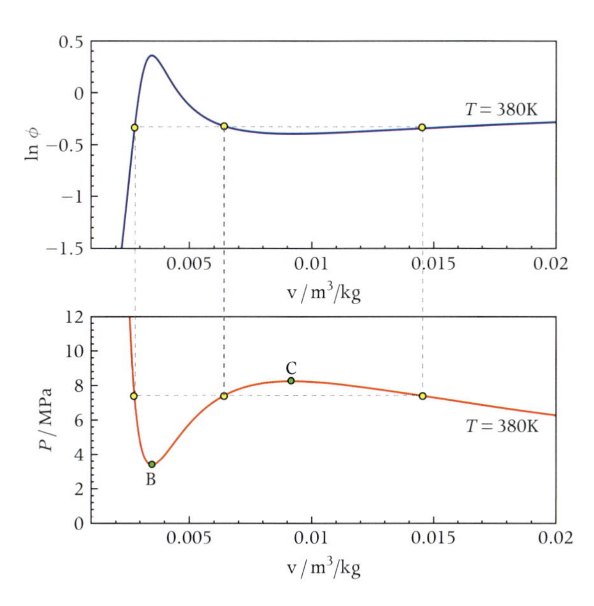

Figure 6.17 Isotherm in the P–v and $\ln\phi - v$ state diagrams calculated with the iPRSV cubic equation of state (StanMix) for ammonia.

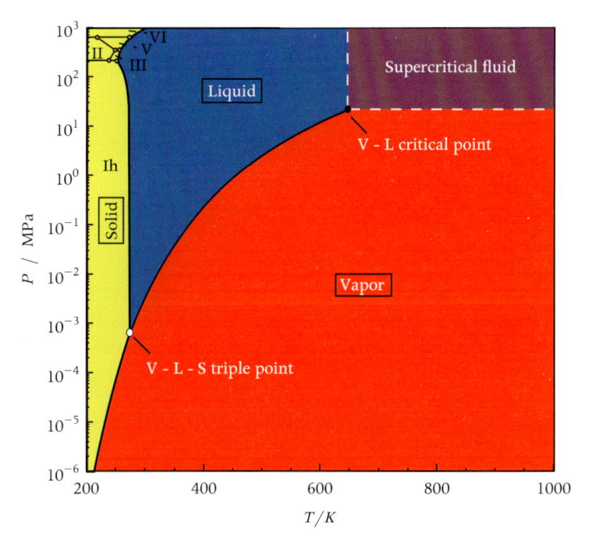

Figure 6.18 Phase diagram of H_2O. Several solid phases (different types of ice) are indicated, together with the corresponding triple points. The points of the vapor–liquid equilibrium line are calculated with RefProp. The points of the melting and sublimation lines are calculated with equations available at www.iapws.org, and published in W. Wagner, T. Riethmann, R. Feistel, *et al.*, "New equations for the sublimation pressure and melting pressure of H_2O ice Ih," *J. Phys. Chem. Ref. Data*, **40**, (*4*), 043103-1–11, 2011. The data of the various ice triple points are taken from M. Choukroun and O. Grasset, "Thermodynamic model for water and high-pressure ices up to 2.2 GPa and down to the metastable domain," *J. Chem. Phys.*, **127**, (*12*), 124506, 2007.

$$g^L(T, P) = g^V(T, P).$$

This condition provides a relation between T and P, which must be satisfied for all the thermodynamic states featuring vapor–liquid equilibrium. This condition can be rewritten as

$$T = T^{sat}(P),$$

showing that the temperature and pressure of a vapor–liquid equilibrium mixture of a single substance cannot vary independently.

Suppose that the substance is in an equilibrium state with three phases. This is possible, for example, if a solid phase is also present. The equilibrium condition becomes

$$g^L(T, P) = g^V(T, P) = g^S(T, P). \qquad (6.118)$$

This amounts to two equations in the variables T and P, which suffice to determine the unique value of T and P at which the three phases can coexist. Thus, while it is possible to have a two-phase mixture of a pure substance over a range of temperatures, a mixture of three specified phases can exist only at a single temperature and pressure, called the *triple point* of the substance. A substance that can exist in several phases may have several triple points; seven triple points are shown in the phase diagram for water of Figure 6.18.

While a single substance can exist in different phases, only three phases can coexist in equilibrium, namely liquid, vapor, and solid. Coexistence of four phases would require that three conditions of equilibrium in two variables (P and T) be satisfied, and no solution would be possible in general. We might restate these observations as follows: given a single substance, the number of properties from the set (T, P) which may be independently fixed is equal to three minus the number of phases. This is a special case of the *Gibbs phase rule*,

$$f = n - p + 2,$$

where f is the number of intensive variables that can be independently varied, n is the number of species in the fluid, p is the number of phases. The general case is presented in Section 8.5 for a mixture.

Clapeyron equation

A useful equation describing the variation of saturation pressure with saturation temperature can be derived with the help of (6.111), by applying it to the condition for thermodynamic equilibrium $g^{\mathrm{L}} = g^{\mathrm{V}}$, giving

$$dg^{\mathrm{V}}\left(T^{\mathrm{V}}, P^{\mathrm{V}}\right) = dg^{\mathrm{L}}\left(T^{\mathrm{L}}, P^{\mathrm{L}}\right).$$

We therefore have that

$$\left(v^{\mathrm{V}} - v^{\mathrm{L}}\right) dP^{\mathrm{sat}} = \left(s^{\mathrm{V}} - s^{\mathrm{L}}\right) dT^{\mathrm{sat}},$$

which can be rewritten as

$$\frac{dP^{\mathrm{sat}}}{dT^{\mathrm{sat}}} = \frac{s^{\mathrm{V}} - s^{\mathrm{L}}}{v^{\mathrm{V}} - v^{\mathrm{L}}}. \tag{6.119}$$

This relation gives the change in the saturation pressure with respect to changes in temperature in terms of saturated liquid and vapor properties. Now, since $g^{\mathrm{L}} = g^{\mathrm{V}}$,

$$h^{\mathrm{L}} - Ts^{\mathrm{L}} = h^{\mathrm{V}} - Ts^{\mathrm{V}}$$

or

$$s^{\mathrm{V}} - s^{\mathrm{L}} = \frac{h^{\mathrm{V}} - h^{\mathrm{L}}}{T}, \tag{6.120}$$

which, combined with (6.119), results in

$$\frac{dP^{\mathrm{sat}}}{dT^{\mathrm{sat}}} = \frac{h^{\mathrm{V}} - h^{\mathrm{L}}}{T\left(v^{\mathrm{V}} - v^{\mathrm{L}}\right)}. \tag{6.121}$$

This equation is known as the *Clapeyron equation*; it relates the variation of pressure with temperature along the saturated-vapor (or liquid) to the enthalpy and volume of vaporization. This equation tells us that $\left(\frac{dP}{dT}\right)^{\mathrm{sat}}$ is always positive, since enthalpies of vaporization are always positive, as are the volume changes undergone during phase change. Consequently the vapor pressure of any substance increases with temperature. You can verify that the thermodynamic models implemented in FLUIDPROP correctly comply with the Clapeyron equation, which is an exact thermodynamic relation obtained from first principles, by calculating numerically the pressure derivative in (6.121) and comparing it with the evaluation of the right-hand side of the equation.

Example: latent heat of vaporization from P–v–T data. As an additional illustration of the use of the Clapeyron equation, let us calculate the enthalpy of vaporization $\Delta h^{\mathrm{LV}} = h^{\mathrm{V}} - h^{\mathrm{L}}$ in a case where only saturated P–v–T data are available. The enthalpy of vaporization is also often called *latent heat of vaporization*.

We take water as the exemplary fluid, and assume that we have measured the saturated volume properties at 450 K obtaining

$$v^{\mathrm{L}} = 0.001\ \mathrm{m^3/kg}$$
$$v^{\mathrm{V}} = 0.208\ \mathrm{m^3/kg}.$$

We have also measured the saturation pressure at 451 K and 449 K, obtaining $P^{\mathrm{sat}}(451\ \mathrm{K}) = 954$ Pa and $P^{\mathrm{sat}}(449\ \mathrm{K}) = 910$ Pa. We can now calculate the derivative of the saturation pressure with respect to temperature at 450 K using finite differences as

$$\left(\frac{dP}{dT}\right)^{\mathrm{sat}} = \frac{P^{\mathrm{sat}}(T + \Delta T) - P^{\mathrm{sat}}(T - \Delta T)}{2\Delta T} = \frac{954 - 910}{2} = 22\ \mathrm{kPa/K}.$$

The enthalpy of vaporization at 450 K according to the Clapeyron equation is therefore

$$\Delta h^{\mathrm{LV}} = T\left(v^{\mathrm{V}} - v^{\mathrm{L}}\right)\left(\frac{dP}{dT}\right)^{\mathrm{sat}} = 450 \times (0.208 - 0.001) \times 22 = 2049\ \mathrm{kJ/kg},$$

which compares well with the value that can be obtained directly from the IF97 model, that is

$$h^{\mathrm{LV}} = 2774.41 - 749.293 = 2025.12\ \mathrm{kJ/kg}. \tag{6.122}$$

The difference between the value calculated with experimental P–v–T data plus the Clapeyron equation and that calculated with the IF97 model is only 1 %.

6.12 Stability for Simple Compressible Fluids

The study of equilibrium is common to many branches of physics and it is often associated with the study of the stability of equilibrium points or states. In general, a system is in equilibrium when it does not change in time. In this book, the concept of thermodynamic equilibrium has been first introduced in Section 3.1, while in Section 5.4 we have seen that a condition for thermodynamic equilibrium is that the entropy is at a maximum. In Section 6.11 the equilibrium condition is used to obtain an equation that allows us to calculate vapor–liquid equilibrium states of a pure fluid from an equation of state. Now we shall investigate how the concept of stability is related to thermodynamic equilibrium and how the study of stability can yield useful information.

The reader might be familiar with the concept of equilibrium applied to the mechanics of rigid bodies. With reference to Figure 6.19, we can conceptually classify four types of equilibrium, namely, a) stable, b) unstable, c) metastable, and d) neutral. Figure 6.19a depicts a stable equilibrium position: no matter how far the ball is pushed to the left or to the right, it will return to its initial position at the bottom of the curved plane. Figure 6.19b shows the ball in an unstable equilibrium state, because it will fall far from its initial position as soon as it receives a small perturbation. The ball in Figure 6.19c is in a metastable equilibrium state: for small enough perturbations the ball falls back towards its initial position, but a larger perturbation permanently displaces the ball. Finally, if the plane is flat, no matter how large the perturbation is, the ball achieves an equilibrium state, which is identical to the initial state (Figure 6.19d). We can argue that in reality almost all states are metastable, if the time of observation is long enough. The analysis of the equilibrium states of a rigid body can be performed in terms of potential energy. These examples of equilibrium states are a good starting point for the analysis of thermodynamic equilibrium states in fluids, which is relevant to several applications.

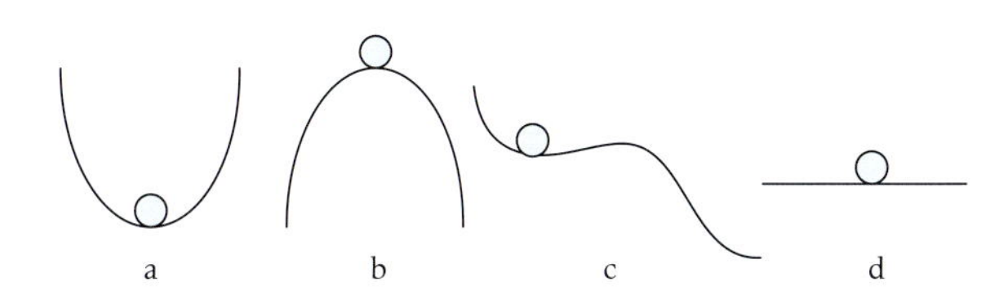

Figure 6.19 The four types of equilibrium: a) stable, b) unstable, c) metastable, and d) neutral

First of all let us have a look at what metastable states are in thermodynamics. Metastable states can be obtained in a simple fluid, both in the liquid and in the vapor phase. If a fluid is highly purified so that it does not contain the microscopic solid impurities needed to initiate phase transitions, it is possible to obtain liquid metastable states at a certain temperature, featuring a pressure which is lower than the saturation pressure at that temperature. Conversely, for the same conditions, it is possible to obtain vapor metastable states for that same temperature characterized by a pressure that is higher than the saturation pressure. These thermodynamic states are metastable because a large-enough perturbation in temperature or pressure brings the system to its equilibrium state, namely the saturated state.

As we are about to investigate, metastable subcooled liquid states cannot be attained for pressures lower than a limiting value and metastable superheated vapor states are also characterized by a maximum pressure. This situation is depicted in Figure 6.20, which reports measurements of stability-limit states for several fluids. Concepts related to stability analysis are employed in the treatment of multiphase equilibria, that is, for example, the equilibrium of fluid mixtures consisting of more than two immiscible liquid phases, and in the development of algorithms for the calculation of equilibrium states.

The stability of thermodynamic states is studied by analyzing small virtual perturbations from an equilibrium state. The Second Law of Thermodynamics applied to an isolated system yields that the system has reached equilibrium when its entropy is at a maximum. The specific entropy s is a continuous function. This implies that the first variation of s must be zero at equilibrium, i.e.,

$$\delta s = 0.$$

The *equilibrium* is *stable*, therefore every small perturbation brings the system to its previous equilibrium state, if the entropy is at a maximum, therefore the second variation of the entropy function is

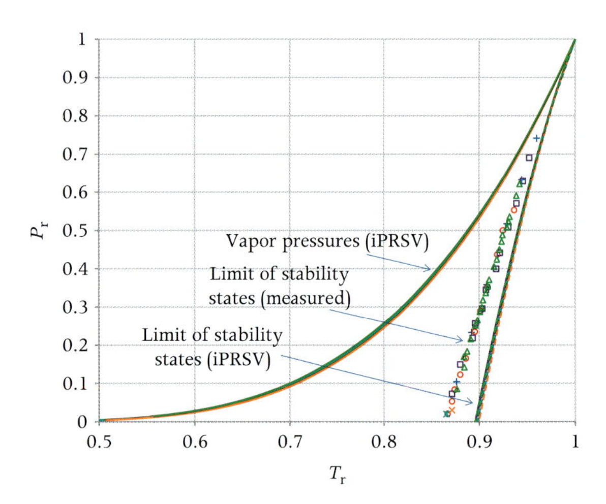

Figure 6.20 Reduced pressure as a function of the reduced temperature for measured limiting metastable states of several fluids: methane (+), oxygen (∗), nitrogen (×), argon (○), krypton (□), xenon (△). The data are taken from E. M. de Sá, E. Meyer, and V. Soares, "Adiabatic nucleation in the liquid–vapor phase transition", *J. Chem. Phys.*, 114, (*19*), 8505–8510, 2001. The spinodal curves calculated with the iPRSV equation of state model available in StanMix (dashed lines) correlate fairly well with the experimental information. The calculated saturation lines for the selected fluids are also shown.

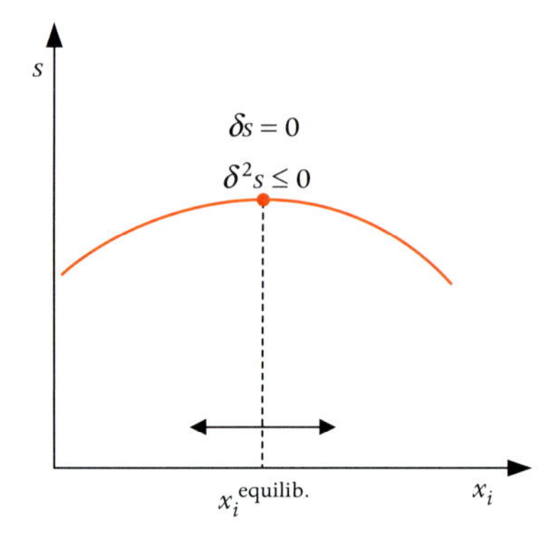

Figure 6.21 Entropy of an isolated system as a function of a thermodynamic variable x_i. The state having maximum entropy is an equilibrium state and any variation of x_i leads to a decrease in entropy. According to the second law of thermodynamics, the perturbed system evolves toward the state of maximum entropy. In the state of maximum entropy $\delta s = 0$ and $\delta^2 s \leqslant 0$.

less than or equal to zero. A small perturbation can be a fluctuation of a fluid property, like pressure and temperature. Figure 6.21 qualitatively shows a small variation of entropy due to a perturbation in a single thermodynamic variable.

Note that, in general, if the second variation of the entropy is also zero, the third variation has to be taken into account, and so forth. The conditions for stable equilibrium can therefore be written as

$$\delta^2 s \leqslant 0,$$
$$\text{if } \delta^2 s = 0 \Rightarrow \delta^3 s \leqslant 0,$$
$$\text{etc.}$$

Legendre transformations, a mathematical operation that can be applied to continuous functions, allow one to obtain equivalent stability conditions in terms of other fundamental thermodynamic functions, and translate from one condition to another using partial derivatives. It can be demonstrated in this way that an equivalent stability condition is given by $\delta^2 u \geqslant 0$, therefore the specific internal energy is at a minimum for a state in stable thermodynamic equilibrium.

Single-phase states of a simple fluid are *intrinsically stable*. By applying the stability criterion $\delta^2 s \leqslant 0$ to two portions of a single-phase substance in equilibrium at constant mass, internal energy, and specific volume, two useful relations can be obtained. The first relation is the *thermal stability condition*

$$c_v > 0, \tag{6.123}$$

and the second is the *mechanical stability condition*

$$\left(\frac{\partial P}{\partial v}\right)_T < 0. \tag{6.124}$$

We omit the lengthy derivation of the stability conditions, which can be found in advanced thermodynamics textbooks, like, for example, J. W. Tester and M. Modell, *Thermodynamics and its Applications*, 3rd ed. Prentice Hall, 1997, Chapter 7.

Any equation of state for a pure substance must satisfy these criteria. It is important to notice that thermodynamics by itself cannot be used to calculate the specific heat of a substance (statistical mechanics and/or experiments are needed), but it does provide consistency relations that thermodynamic properties must satisfy.

Now we consider a subcritical isotherm in the P–v plane (Figure 6.17). We observe that all the states between B and C violate the mechanical stability condition $\left(\frac{\partial P}{\partial v}\right)_T < 0$, therefore they cannot be obtained in reality. If we consider the liquid side, we could expand the liquid isothermally to a pressure that is lower than the saturation pressure, without violating the mechanical stability condition. Note, though, that the variation must be small, because a larger variation would trigger the vaporization of the superheated liquid. The description of this experiment fits the definition of *metastable state*. The same holds for states obtained starting from saturated vapor by carefully lowering the pressure at constant temperature. The line obtained by joining all the states having $\left(\frac{\partial P}{\partial v}\right)_T = 0$, is called a *spinodal curve* and represents the limit of intrinsic stability. Figure 6.22 shows the spinodal curve in the P–v plane. Figure 6.20 shows furthermore that states of intrinsic stability limit can be calculated quite accurately with cubic equations of state. This feature provides an additional semi-empirical justification to the cubic functional form at the basis of a large number of pressure-explicit equations of state. On the contrary, other more complex empirical equations of state predict isotherms with a non-physical shape in the liquid–vapor region.

As a last remark, we can infer from Figure 6.22 that, at low temperature, metastable states can occur at negative pressure (!). This means that liquid water in these metastable states can sustain a tensile force. These states have been observed in capillary tubes.

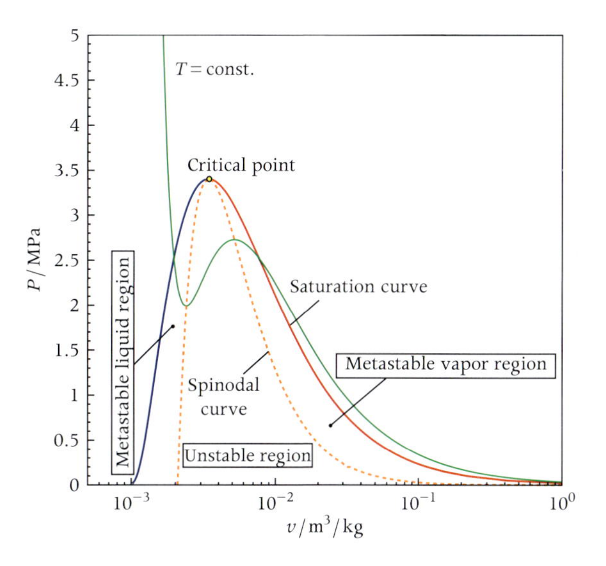

Figure 6.22 P–v state diagram of an exemplary simple fluid (nitrogen, states calculated with StanMix) showing stable, unstable and metastable states in the vapor–liquid region. The liquid metastable region is delimited by the saturated liquid line (blue) and by the left branch of the spinodal curve (dashed, orange). The vapor metastable region is bounded by the saturated vapor line (red) and by the right branch of the spinodal curve (dashed, orange). A subcritical temperature line (black) helps illustrating that the spinodal curve is formed by states featuring $\left(\frac{\partial P}{\partial v}\right)_T = 0$.

6.13 The Choice of a Thermodynamic Model

The calculation of thermodynamic properties of fluids is the basis for the engineering analysis of a wide range of processes and systems. A variety of computer programs for process and system analysis are available and they make use of software libraries implementing many fluid thermodynamic models. It is therefore important for the prospective engineer to know about these models in order to make an appropriate choice. A given problem might require the use of one of these programs, or, more simply, the calculation of a property starting from few fluid data.

FluidProp, the computer program accompanying this book, implements a number of these models, and you can experiment with this tool in order to solve many of the exercises related to this and following chapters. GasMix encodes the ideal gas model, StanMix the iPRSV equation of state, TPSI several fluid-specific equations of state, IF97 the widely adopted model for water used in industrial applications, RefProp state-of-the-art reference and technical equations of state, and PCP-SAFT

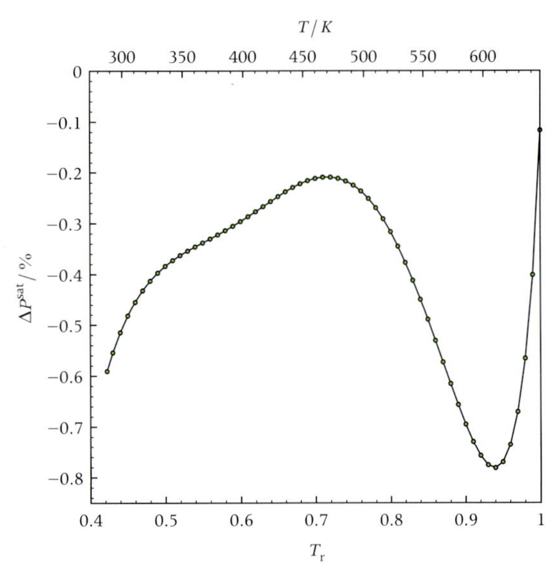

(a) Deviation of saturation pressures of water calculated with the iPRSV equation of state (StanMix) *versus* values obtained with the reference equation of state (RefProp).

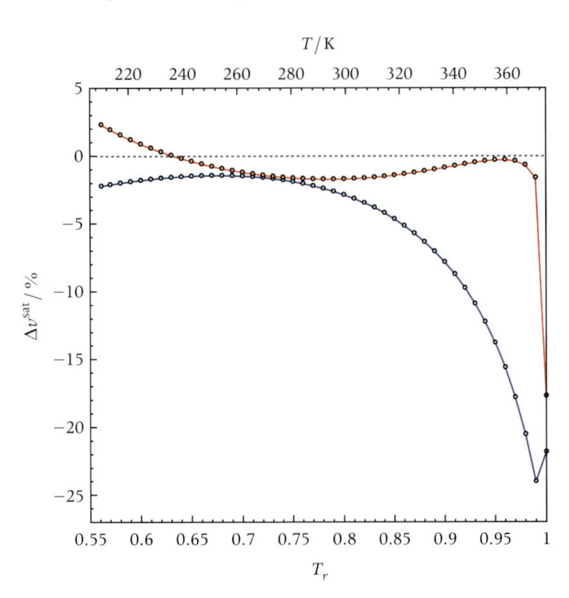

(b) Deviation of saturated specific volume values of R134a calculated with the iPRSV equation of state (StanMix) *versus* values obtained with a technical equation of state (RefProp). Red = saturated vapor; Blue = saturated liquid.

Figure 6.23 Accuracy of equations of state: volumetric cubic versus multiparameter Helmholtz-based equations of state.

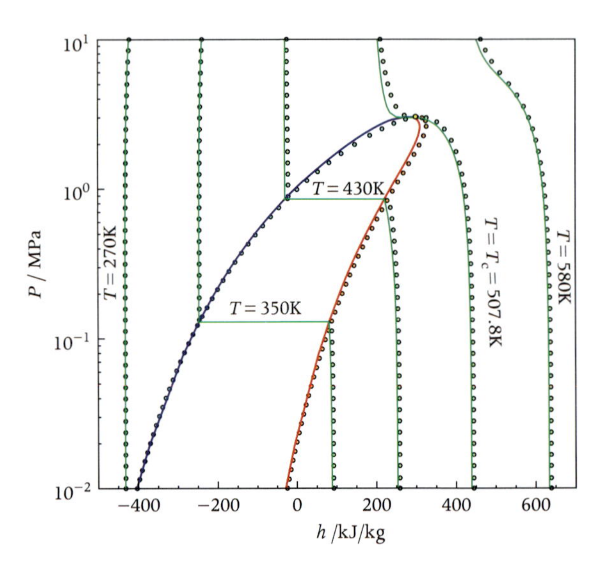

Figure 6.24 *P*–*h* state diagram for hexane. Continuous lines are calculated with the multiparameter Helmholtz-based equation of state implemented in RefProp, while circles are calculated with StanMix.

a molecular equation of state, based on statistical thermodynamics.

The complexity of a thermodynamic model impacts both its accuracy and the computational efficiency. Usually, but not always, higher accuracy can be attained at the expense of computational efficiency. Computational efficiency becomes an issue when simulating complex processes, such as, for example, fluid dynamics through complex geometries, or the dynamics of complex energy systems. To give a quantitative idea of the accuracy of a cubic equation of state, Figure 6.23a reports the deviations between saturation pressures of water calculated with StanMix and those calculated with the reference equation implemented in RefProp. Values of saturated liquid and vapor volumes computed with StanMix are less accurate, especially close to the critical point, and for reduced temperatures greater than 0.8 K (Figure 6.23b). Note, for example, that a less accurate estimation of the saturated liquid density does not appreciably affect the estimation of the latent heat of vaporization, an important quantity in many engineering problems. Finally, Figure 6.24 provides an indication of the comparison of the iPRSV model implemented in StanMix with a model based on a multiparameter technical equation of state implemented in RefProp .

A point of attention is that some thermodynamic models are not consistent: properties are calculated with several ad hoc equations, depending on the thermodynamic region (one equation for superheated vapor states, one for subcooled liquid states, one for supercritical states, etc.). Properties at the boundary of a region can therefore be numerically discontinuous. This is the case, for example, of the IF97 model. Often this has no consequences, as these discontinuities are quite small, but small jumps in property values could generate numerical problems, if, for instance, the model is used for fluid dynamic simulations.

EXERCISES

6.1 In cases where viscous stresses and thermal conduction can be neglected, the Euler equations describe the motion of a fluid. The one dimensional form of the Euler equations (mass, momentum, and energy conservation accounting) is

$$\frac{\partial \rho}{\partial t} = -\frac{\partial \rho \mathcal{V}_x}{\partial x},$$
$$\frac{\partial \rho \mathcal{V}_x}{\partial t} = -\frac{\partial \rho \mathcal{V}_x^2}{\partial x} - \frac{\partial P}{\partial x},$$
$$\frac{\partial \rho e}{\partial t} = -\frac{\partial \rho e \mathcal{V}_x}{\partial x} - \frac{\partial P \mathcal{V}_x}{\partial x},$$

where $e = u + \frac{1}{2}\mathcal{V}_x^2$ is the specific total energy per unit mass and $\mathcal{V}_x$ is the velocity of the fluid. This is a system of equations in the four unknowns $\rho, P, u, \mathcal{V}_x$.

Show that, if the fluid is a polytropic ideal gas, a fourth simple algebraic equation relating u to P and ρ neatly closes the system of equations.

6.2 Assume that for an ideal gas, within a temperature interval, the dependence of the isobaric specific heat on temperature can be approximated by a quadratic equation $c_P\{T\} = K_1 + K_2 T + K_3 T^2$, where $K_{1,2,3}$ are constants.

(a) Obtain the equation for $u\{T\}$ and $h\{T\}$.

(b) The coefficients for $c_P\ \{T\}$ of carbon dioxide in the range 100–1000 K are $K_1 = 23.87$, $K_2 = 52.85 \times 10^{-3}$ and $K_3 = -23.47 \times 10^{-6}$. Calculate the difference in internal energy and enthalpy of carbon dioxide between 340 K and 290 K at 0.1 MPa.

6.3 Calculate the entropy difference for the same substance and conditions of Exercise 6.2(b), and compare it with the values you can obtain with GasMix. Why is there a small difference? (hint: look up the equation for the specific heat in the documentation of GasMix).

6.4 Starting from the Gibbs equation, show that for a polytropic (constant specific heat) gas undergoing an isentropic process $\frac{T_2}{T_1} = \frac{P_2}{P_1}^{(\gamma-1)/\gamma}$, where $\gamma \equiv c_P/c_v$.

6.5 Consider steam flowing through an adiabatic valve for the two different process conditions:

(a) inlet at $P_{\text{in}} = 70$ bar and $T_{\text{in}} = 730$ K, and outlet at $P_{\text{out}} = 20$ bar;

(b) inlet at $P_{\text{in}} = 4$ bar and $T_{\text{in}} = 700$ K, and outlet at $P_{\text{out}} = 1$ bar.

Examine and discuss the validity of using the ideal gas model in analyzing these processes.

6.6 When designing the air-conditioning system of an airplane, the work required to isentropically compress air must be calculated. Consider the process at steady state, the inlet compressor pressure is 0.5 bar and the outlet is at 5 bar. In addition to calculating the isentropic work, determine the air outlet temperature. Should you consider the air specific heat as temperature-dependent (Appendix A.2)?

6.7 A piston–cylinder device contains 1 kg of a gaseous refrigerant. The gas undergoes a compression from $P_1 = 1.25$ bar and $v_1 = 42.77\ \text{m}^3/\text{kg}$ to $P_2 = 10$ bar and $v_1 = 10.67\ \text{m}^3/\text{kg}$. The compression process can be described with the equation $Pv^n = \text{constant}$. Determine the value of the polytropic coefficient n.

6.8 Consider an idealized power system operating according to the Carnot cycle. The working fluid can be modeled as a polytropic ideal gas.

(a) Draw qualitatively the cycle in the P–v diagram of the working fluid (states 1, 2, 3, and 4). The process 1–2 is the isothermal heating, while 3–4 is the isothermal cooling.

(b) Demonstrate that $v_4 v_2 = v_1 v_3$.

(c) Demonstrate that $\frac{T_2}{T_3} = \left(\frac{P_2}{P_3}\right)^{\frac{\gamma-1}{\gamma}}$.

(d) Demonstrate that $\frac{T_2}{T_3} = \left(\frac{v_3}{v_2}\right)^{\gamma-1}$.

6.9 Determine the change in specific entropy in kJ/(kg · K) between the states:

(a) air, $P_1 = 0.1$ MPa, $T_1 = 20$ °C, $P_2 = 0.1$ MPa, $T_2 = 120$ °C;

(b) air, $P_1 = 1$ bar, $T_1 = 27$ °C, $P_2 = 3$ bar, $T_2 = 377$ °C;

(c) carbon dioxide, $P_1 = 150$ kPa, $T_1 = 30$ °C, $P_2 = 300$ kPa, $T_2 = 300$ °C;

(d) carbon monoxide, $T_1 = 300$ K, $v_1 = 1.1$ m^3/kg, $T_2 = 500$ K, $v_2 = 0.75$ m^3/kg;

(e) nitrogen, $P_1 = 2$ MPa, $T_1 = 800$ K, $P_2 = 1$ MPa, $T_2 = 300$ K.

Employ the ideal gas model and the data in Appendix A.1. For air, use the method of Appendix A.2.

6.10 Determine the reference state for the entropy and enthalpy of R134a chosen for the model implemented in STANMIX by inspecting the thermodynamic T–s or P–h diagram in Appendix A. Determine the value of the internal energy of R134a at 410 K and 0.5 MPa if one datum state were the enthalpy of R134a of saturated vapor at $T = 220$ K. What would be the value of the enthalpy at this point?

6.11 Obtain the Maxwell relations (6.38c) and (6.38d) starting from the Gibbs equation.

6.12 In analogy with the way the Gibbs and Helmholtz energy are obtained by considering processes occurring in a closed system, obtain the thermodynamic function that most simply allows for the calculation of the energy transfer as work in an isobaric and isentropic process.

6.13 Work out the differential relation between specific entropy of a simple compressible fluid and its temperature and pressure, such that it can be evaluated with the help of c_P and an equation of state in the form $P = P\{T, v\}$.

6.14 How can the residual part of c_P of a simple compressible fluid be evaluated from an an equation of state in the form $P = P\{T, v\}$?

6.15 What is the relation between the isothermal compressibility and the speed of sound for a simple compressible fluid?

6.16 The Joule–Thomson (or Joule–Kelvin) coefficient for a simple compressible substance is defined as $\mu_{JT} \equiv \left(\frac{\partial T}{\partial P}\right)_h$.

(a) Prove that $\mu_{JT} = \frac{1}{c_P}\left[T\left(\frac{\partial v}{\partial T}\right)_P - v\right]$;

(b) Discuss how this equation might be used;

(c) Show that the Joule–Thomson coefficient for a perfect gas is zero.

6.17 Obtain equations for the speed of sound and isothermal, isobaric, and isentropic compressibility valid for the van der Waals equation of state.

6.18 Evaluate the specific volume of heptane for $T_r = 1.1$ and $P_r = 3.5$ using

(a) the van der Waals equation of state ($T_c = 540.13$ K, $P_c = 27.36$ bar);

(b) the generalized compressibility chart of Figure 6.6;

(c) TPSI.

Which is more accurate among the values obtained with (a) and (b)? Consider the flow of 5 kg/s heptane in these thermodynamic conditions through a circular pipe. The flow velocity must be lower than or equal to 2 m/s in order to limit the friction losses over the long pipe. Calculate the pipe cross-sectional area using the Van der Waals equation of state and TPSI. What is the consequence of using an inaccurate thermodynamic model?

6.19 Derive the thermodynamic relations (6.74c) and (6.74d) from basic definitions.

6.20 Obtain the expression for the isothermal compressibility κ from the Helmholtz energy $a = a\{T, v\}$.

6.21 Show that the isothermal compressibility is greater than or equal to isentropic compressibility

6.22 Assume that the equation of state for a certain fluid is given as $g = g\{T, P)$ (for example, this is the case for the IF97 equation of state, but it has also been applied for mixture equations of state like ammonia/water and LiBr/water). Show that all primary properties can be obtained by appropriate differentiation. Derive also the expressions for the specific heats c_P, c_v, and the speed of sound c.

6.23 Given the truncated virial equation for methane $Z = 1 + \frac{B\{T)}{v}$ with

$$B\{T) = \frac{RT_c}{P_c}\left[0.083 - \frac{0.422}{T_r^{1.6}} + \omega\left(0.139 - \frac{0.172}{T_r^{4.2}}\right)\right],$$

calculate the pressure for $T = 180$ K and $v = 0.03$ m^3/kg and calculate the specific volume for $P = 0.3$ MPa and $T = 320$ K. Compare the values with the values obtained with the ideal gas equation of state. Indicate the points corresponding to this states in the P-h diagram of Appendix A.5 and highlight the thermodynamic region where the ideal gas assumption applies. The critical temperature and pressure of methane are $T_c = 190.6$ K and $P_c = 4.60$ MPa, respectively, the molecular weight $\hat{M} = 16.04$ kg/kmol, the acentric factor $\omega = 0.0114$, and $T_r = T/T_c$.

6.24 The vapor pressure of refrigerant R134a can be approximated with the Antoine equation

$$\ln P^{sat} = A - \frac{B}{T + C},$$

with P in kPa and T in K, $A = 14.41$ kPa, $B = 2094$ kPa·K, and $C = -33.06$ K. The critical temperature is $T_c = 374.21$ K. Calculate the acentric factor. Is this molecule more or less acentric than that of water if the acentric factor of water is $\omega_{H_2O} = 0.3443$?

6.25 Suppose we want to store 100 kg of carbon dioxide in a 0.5 m^3 tank at 320 K and we want to calculate the required pressure using different thermodynamic models. Given that the critical temperature and pressure of carbon dioxide are 304.2 K and 7.382 MPa, respectively, calculate the pressure using:
(a) the ideal gas model;
(b) the generalized compressibility chart (corresponding state principle);
(c) the T–s chart in Appendix A.9.
Comment on the results obtained. Calculate the pressure using the cubic equation of state model implemented in STANMIX : is the percentage difference between this value and the one you obtained from the T–s diagram significant?

6.26 Compare the specific volume of steam at 40 MPa, 800 K as obtained with the ideal gas equation of state, the generalized compressibility chart, and the IF97 model. The critical temperature and pressure of water are equal to 647.096 K and 22.064 MPa, respectively.

6.27 Propane is to be stored at 4 MPa, 100 °C. The critical temperature and pressure of propane are equal to 369.8 K and 4.23 MPa, respectively. Estimate the specific volume at these conditions. How large is the tank needed to store 1150 kg of propane?

6.28 Given the speed of sound values for ethane from A. F. Estrada-Alexanders and J. P. M. Trusler, "The speed of sound and derived thermodynamic properties of ethane at temperatures between 220 K and 450 K and pressures to 10.5 MPa," *J. Chem. Thermodynamics*, **29**, (9), 991–1015, 1997, measured at low pressures:

T / K	measured c / m/s
220	273.340
250	289.263
275	301.550
295	310.542
315	319.322
330	325.977
345	332.479
365	340.899
385	349.095
415	361.051
450	374.524

Consider the speed of sound values as ideal gas speeds of sound and obtain the equation for $c_P^{IG}\{T\}$ according to (6.104).

6.29 Calculate the vapor pressure of water at 150 °C using the van der Waals equation of state. Repeat the calculation at 370 °C. How do these values compare with values obtained with the IF97 model?

6.30 Derive the expression for $\ln\phi$ for a simple compressible fluid if the thermodynamic model is given as the nondimensional specific Helmholtz energy

$$\frac{a\{T,\rho\}}{RT} = \alpha^{IG}\{\tau,\delta\} + \alpha^{R}\{\tau,\delta\}.$$

6.31 The vapor pressure of a simple compressible fluid can be roughly approximated as an exponential function of temperature. Show how such approximation is suggested by the Clapeyron equation.

6.32 Estimate the limiting metastable state for the liquid and the vapor of water at 100 °C using the van der Waals equation of state.

6.33 Obtain charts like those of Figure 6.23 in order to compare values for propane calculated with STANMIX and TPSI. What can you infer from these charts about the accuracy of the cubic equation of state model if compared to a multi-parameter thermodynamic model?

6.34 Use TPSI with a program of your choice and generate the T–s diagram of n-butane. The diagram should contain the saturation lines enclosing the vapor–liquid equilibrium region, the critical point and several correctly spaced iso-lines, namely isobars, isochores, isenthalps, and lines of constant vapor quality.

7 Energy Conversion Systems

CONTENTS

One of the most dazzling applications of thermodynamics is the study and design of devices to convert energy to obtain some useful effect. Modern societies are based on the extensive use of electricity, heating, and refrigeration for a broad variety of goals. These services are supplied by a system in which some sort of working substance, usually in the gas or liquid phase, flows through several devices at various temperatures. The same holds for engines, which provide the stunning mobility which is also a pillar of modern times.

Here we therefore describe and analyze systems such as stationary power plants, propulsion engines, refrigerators, heat pumps, and the like. Stationary power plants convert one kind of energy, like the chemical energy of a fuel, the thermal energy of a geothermal reservoir, or solar radiation into electricity. Propulsion engines do the same, but usually the primary energy source is a liquid fuel and the final objective is to obtain mechanical power or thrust to put something into motion. Refrigeration systems and heat pumps make use of electrical, mechanical, or thermal power in order to keep the temperature of a certain space at a value different from that of the natural environment.

7.1 Analysis of Thermodynamic Systems

Most energy systems involve a *working fluid*, like water or air, which is circulated through one or more components, often arranged in a *cycle*. The working fluid undergoes therefore a sequence of processes, which together form a thermodynamic cycle. In steam power plants (Figure 7.1a) the cycle is usually *closed*, but thermodynamic cycles can be *open* (closed by the atmosphere), as is the case for a turboprop engine (Figure 7.1b).

The basic methodology for the analysis of thermodynamic processes and systems has been laid down in Chapter 4, while in this chapter we will use that procedure, together with knowledge on the estimation of fluid properties (Chapter 6), in order to study some notable examples of energy conversion systems. In addition, we can make use of the many consequences of the second law of thermodynamics to discuss their performance and limitations.

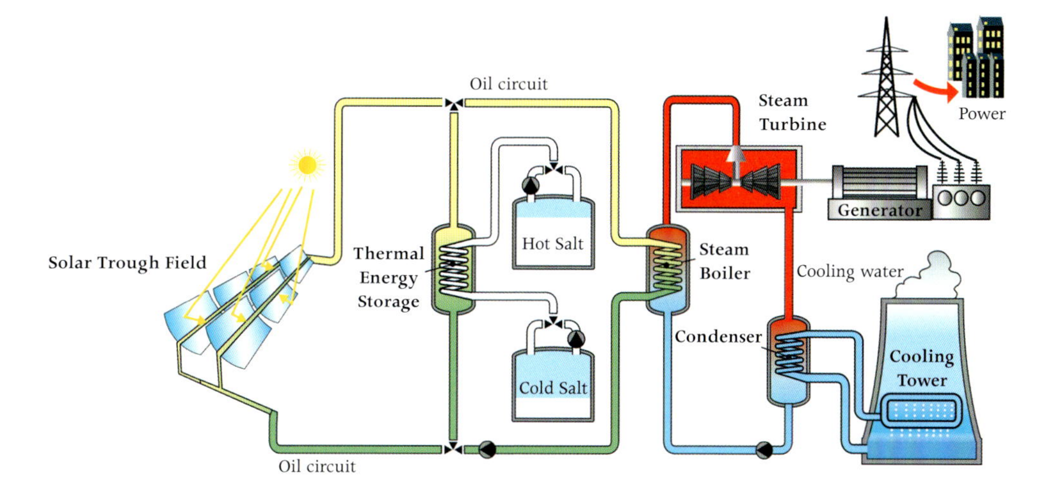

(a) A solar power station, using a Rankine-cycle plant to convert solar radiation into electricity.

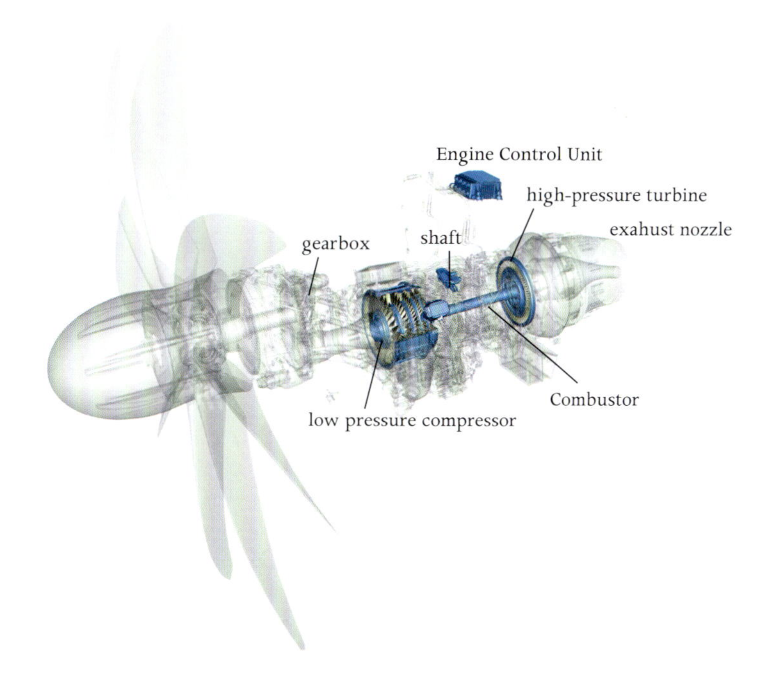

(b) A turboprop engine, the EPI TP400, Courtesy of Europrop International.

Figure 7.1 Examples of power plants for electricity generation and propulsion.

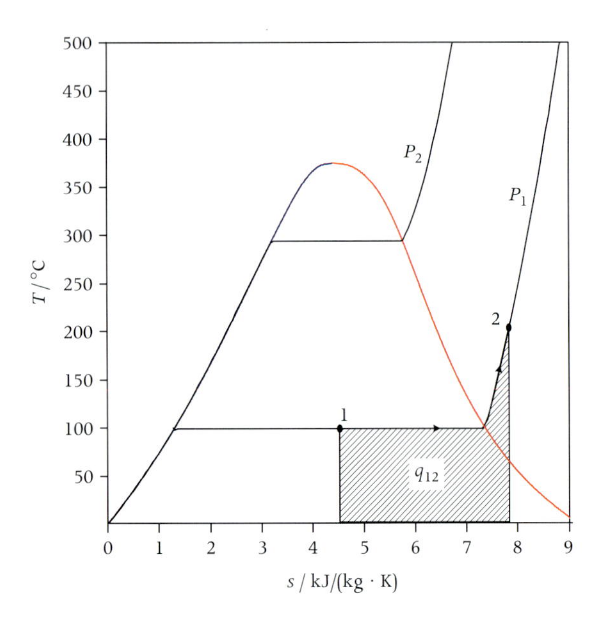

Figure 7.2 The *T*–*s* diagram of a fluid is particularly useful to show the amount of energy transfer as heat.

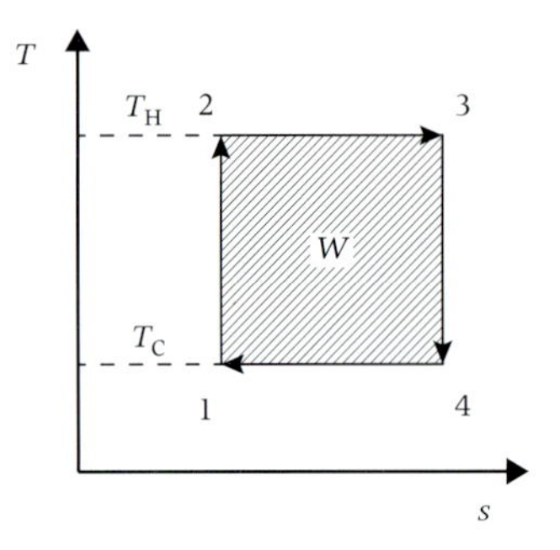

Figure 7.3 The Carnot engine of Figure 5.15 is characterized by a simple thermodynamic cycle in the *T*–*s* plane. (1-2) compression, (2-3) energy input as heat, (3-4) expansion, (4-1) energy output as heat. The work per unit mass of working fluid is proportional to the shaded area.

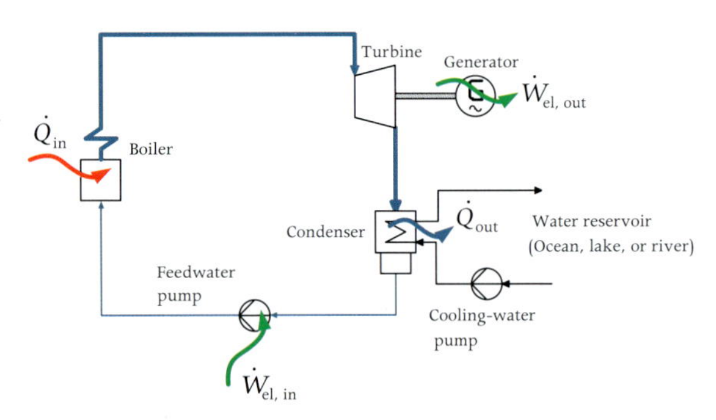

Figure 7.4 Process flow diagram of a simple steam power plant.

In Section 3.6 we have already stressed the importance of thermodynamic charts and of sketching relevant processes on such charts. Graphical representation of concepts is a very powerful tool for the understanding of thermodynamics. Here we will use these charts extensively and will show how you can use FLUIDPROP to effectively perform analyses and display processes on thermodynamic diagrams. Energy engineers are especially fond of the temperature–entropy thermodynamic plane for a given fluid. When some fluid undergoes a reversible process, the sequence of states experienced by the fluid is indicated by a line in the *T*–*s* plane (Figure 7.2). Since the process is reversible, the energy transferred as heat *to* a unit of mass of the fluid is represented by the area underneath the line indicating the process on the *T*–*s* diagram.

If the fluid undergoes a series of processes forming a thermodynamic cycle, its internal energy does not change over a cycle. Consequently the *net* energy transferred as heat *to* a unit mass of the fluid during one cycle must equal the *net* energy transfer as work from the fluid (work done), and both equal the area enclosed by the reversible path on the *T*–*s* plane (Figure 7.3).

Other thermodynamic charts like the *h*–*s*, *P*–*h*, and *P*–*v* diagrams are also very useful, as will be seen. Another important type of diagram in engineering thermodynamics is the *process flow diagram* or *flowsheet*. A simple example is given in Figure 7.4. These diagrams convey the functionality of a system, the relationship between components, and the flows of substances. Modern energy systems tend to be increasingly complex, therefore process flow diagrams are very helpful. The symbols and some rules to draw these diagrams are collected in documents provided by standardization bodies. These standards allow for a common graphical language which engineers can use to communicate about complex energy systems. Here we make use of simple exemplary flowsheets and they are sufficient to learn the basics in order to draw such diagrams.

When approaching a new system, it is generally a good idea to first draw the process flow diagram and then to sketch the process transformations on an appropriate thermodynamic plane. Working out the process representation in these two diagrams provides an understanding of how the system works and orients thinking as to how to proceed with first and second law analysis.

7.2 The Rankine Cycle

Thermodynamics developed as a result of nineteenth-century work with reciprocating steam engines. Present-day applications of thermodynamics go well beyond this single area, but steam power systems remain a major source of electricity and play an ever important role in the establishment of renewable energy technologies: solar concentrators can supply thermal energy to these systems, achieving a very high conversion efficiency.

Modern power plants employ rotating rather than reciprocating machinery to obtain mechanical power, for numerous practical reasons. Figure 7.4 shows the process flow diagram of a simple steam cycle plant, whose process representation in the T–s and P–h diagram is in Figure 7.5.

Let us first consider a simplified system (the thermodynamic cycle of actual steam power plants is more complicated, as we shall see). Liquid water is compressed by a pump and fed to a boiler. The boiler heats up the fluid until it evaporates, and continues the heating until there is no more liquid. The high-pressure steam is heated further (superheated) and it is delivered to the turbine where it expands thus converting the energy of the fluid into mechanical work. The expanded water vapor is condensed in an appropriate heat exchanger, which delivers liquid back to the pump.

It will be convenient to work with energy transfer as heat and work *per unit of mass flow*, because the analysis can then be made without regard to the system power or the mass flow rate. With reference to Figures 7.4, 7.5a, and 7.5b, the operating conditions are as follows:

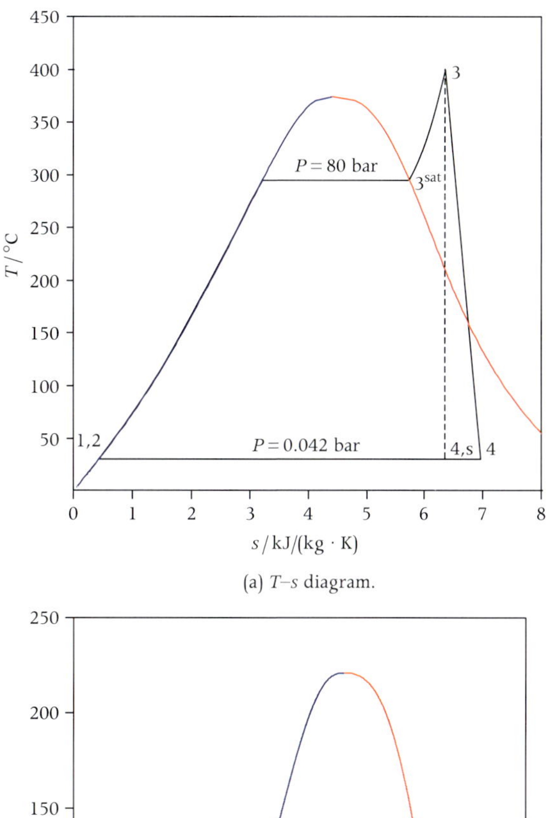

(a) T–s diagram.

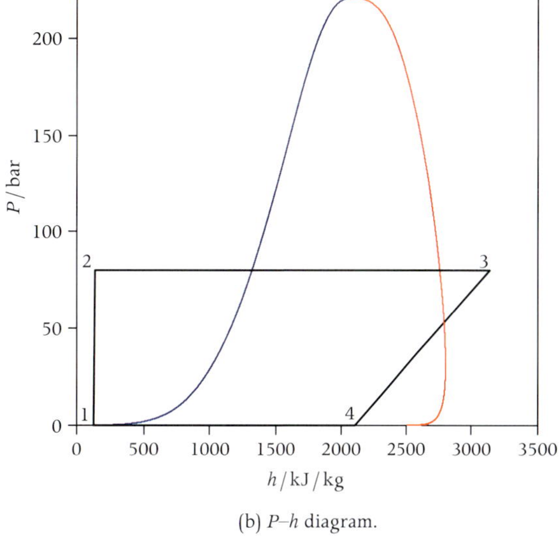

(b) P–h diagram.

Figure 7.5 The processes forming a simple Rankine steam cycle represented in useful thermodynamic planes. The numbering of the working fluid states is in accordance with the process flow diagram of Figure 7.4.

- State 1, liquid at 303.15 K (30 °C), in saturated conditions;
- State 2, 8.0 MPa (80 bar);

- State 3, superheated steam at 673.15 K (400 °C); and
- State 4, 303.15 K.

The isentropic efficiency of the machinery is:

- $\eta_{\text{pump},s} = 0.65$, and
- $\eta_{\text{turb},s} = 0.85$.

The following assumptions are made:

- kinetic and potential energy negligible at states 1, 2, 3, and 4;
- adiabatic pump and turbine;
- perfect electrical generator ($\eta_{\text{gen}} = 1$);
- the system is at steady state;
- water is at thermodynamic equilibrium at states 1, 2, 3, and 4; and
- there is no pressure drop along the heat exchangers and the connecting ducts.

Let us calculate the properties of water using IF97 (try it yourself using MATLAB or EXCEL). An energy balance on the pump yields

$$w_{\text{pump}} \equiv \dot{W}_{\text{pump}} / \dot{M} = h_2 - h_1.$$

w_{pump} is the specific shaft-work, that is the work per unit mass supplied by the pump to the fluid. The enthalpy of state 1 is

$$h_1 = h(T_1 = 303.15\ \text{K}, x = 0) = 126\ \text{kJ/kg}.$$

From the definition of isentropic efficiency we have that

$$h_2 = h_1 + \frac{(h_{2,s} - h_1)}{\eta_{\text{pump},s}}.$$

If the compression of liquid in the pump occurred without production of entropy, the enthalpy at the outlet of the pump would be

$$h_{2,s} = h(P_2, s_1).$$

We need therefore the entropy of state 1,

$$s_1 = s(T_1, x = 0) = 0.437\ \text{kJ}/(\text{kg} \cdot \text{K}),$$

thus $h_{2,s} = 134$ kJ/kg and

$$h_2 = \left(126 + \frac{134 - 126}{0.65}\right)\ \text{kJ/kg} = 138\ \text{kJ/kg}.$$

The specific shaft-work supplied by the pump is therefore

$$w_{\text{pump}} = h_2 - h_1 = (138 - 126)\ \text{kJ/kg} = 12\ \text{kJ/kg}.$$

Note the very small temperature rise, $T_2 = T(P_2, h_2) = 304.4$ K, due to the compression of the liquid. The enthalpy of state 3 can be found as

$$h_3 = h(P_2, T_3) = 3139\ \text{kJ/kg}.$$

Applying the energy conservation law to the boiler gives

$$\begin{aligned} q_{\text{boiler}} \equiv \dot{Q}_{\text{boiler}} / \dot{M} &= (h_3 - h_2) \\ &= (3139 - 138)\ \text{kJ/kg} \\ &= 3001\ \text{kJ/kg}, \end{aligned}$$

where q_{boiler} is the energy transfer as heat to the boiler per kg of water flowing through it. The working fluid entropy at 3 is

$$s_3 = s(P_3, h_3) = 6.366\ \text{kJ}/(\text{kg} \cdot \text{K}).$$

The calculation of the enthalpy corresponding to the end of the isentropic expansion from state 3 requires the computation of the condensing pressure,

$$\begin{aligned} P_1 = P^{\text{sat}}(T_1) = P(T_1, x = 0) &= 4247\ \text{Pa} \\ &= 0.042\ \text{bar} = 42\ \text{mbar}, \end{aligned}$$

which leads to

$$h_{4,s} = h(P_1, s_3) = 1923\ \text{kJ/kg}.$$

Finally the enthalpy at the outlet of the turbine is

$$h_4 = h_3 - \eta_{\text{turb},s}\left(h_3 - h_{4,s}\right) = 2106\ \text{kJ/kg},$$

and the specific shaft-work of the turbine is therefore

$$w_{\text{turb}} = h_3 - h_4 = (3139 - 2106)\ \text{kJ/kg} = 1033\ \text{kJ/kg}.$$

The energy-conversion efficiency of this system is

$$\begin{aligned} \eta = \frac{w_{\text{net}}}{q_{\text{boiler}}} &= \frac{w_{\text{turb}} - w_{\text{pump}}}{q_{\text{boiler}}} \\ &= \frac{(1033 - 12)\ \text{kJ/kg}}{3001\ \text{kJ/kg}} = 0.34. \end{aligned}$$

It is instructive to compare the energy conversion efficiency of this power plant to that of a Carnot cycle engine operating between the same maximum and minimum temperature:

$$\eta_{\max} = 1 - \frac{303.15\ \mathrm{K}}{674.15\ \mathrm{K}} = 0.55.$$

The Rankine cycle power plant therefore converts only about 62% of the maximum amount of mechanical power that could be generated by the ideal Carnot engine between the same maximum and minimum temperatures. Some of the difference is due to irreversibilities in the expansion process operated by the turbine, and some is due to the energy being transferred as heat to the working fluid at lower temperature if compared to the thermal energy input of the Carnot cycle engine.

The simple configuration and the low value of the isentropic efficiency of the machinery chosen for this example are typical of small steam plants, as for instance those of biomass-fueled power stations. If the plant must be designed such that the mechanical power output of the turbine is 30 MW, the required mass flow of water can be easily calculated as

$$\dot{M} = \frac{\dot{W}_{\mathrm{turb}}}{w_{\mathrm{turb}}} = \frac{30 \cdot 10^3\ \mathrm{kJ/s}}{1033\ \mathrm{kJ/kg}} = 29.0\ \mathrm{kg/s}.$$

The power used by the pump is

$$\dot{W}_{\mathrm{pump}} = \dot{M} \cdot w_{\mathrm{pump}} = 0.35\ \mathrm{MW},$$

the net power output is

$$\dot{W}_{\mathrm{net}} = \dot{W}_{\mathrm{turb}} - \dot{W}_{\mathrm{pump}} = 29.6\ \mathrm{MW},$$

and the thermal power transferred to the boiler is

$$\dot{Q}_{\mathrm{boiler}} = \dot{M} \cdot q_{\mathrm{boiler}} = 87.1\ \mathrm{MW}.$$

7.3 Vapor Power Plants

The configuration, and therefore the thermodynamic cycle, of real steam power plants is generally more complicated than that of the system we have just analyzed. Appropriate modifications to that basic plant configuration solve several technical problems and increase the energy conversion efficiency.

Superheating and reheating

The turbine operates over a large pressure range, therefore the density variation of the steam within the turbine is enormous. In our example the expansion ratio is $\beta_P \equiv P_3/P_4 = 1884$. Turbines of steam power plants condensing at low temperature are made of many stages (Figure 7.6), with each stage operating over a limited pressure ratio (say around 2). The stages of a large steam turbine can be grouped in separate machines sharing the same shaft, or are incorporated into a single large machine. The high-pressure stages are much smaller than the low-pressure stages, though the power converted by the various stages is typically about the same.

The vapor fraction at the outlet of the turbine (state 4 in Figure 7.5) is quite low,

$$x = x(P = 0.042\ \mathrm{bar}, h = 2106\ \mathrm{kJ/kg}) = 0.81,$$

which means that a mass corresponding to approximately 20% of the total mass of fluid flowing through the blades would form liquid droplets in the last turbine stage, and still a considerable amount in the preceding stages. These droplets would impact upon the blades, causing metal erosion and deterioration of the fluid dynamic performance. The last stages of modern steam turbines are therefore designed for at maximum $x \approx 0.9$. We conclude that superheating in condensing turbines is mandatory and the so-called *saturated configuration*, whereby the fluid at the turbine inlet is saturated (point 3^{sat} in Figure 7.5) is unfeasible. The comparison with a Carnot cycle engine operating between the same maximum and minimum temperatures tells us that increasing the pressure is beneficial because thermal energy is introduced in the thermodynamic cycle at comparatively higher temperature. Given that the vapor quality at the outlet of the turbine is a design constraint, we notice (Figure 7.7) that also the turbine inlet temperature must be increased. Superheating is often achieved with a separate heat exchanger, the *superheater*. Higher pressure and the addition of the superheater results in higher costs, therefore the maximum pressure/temperature of a power plant is usually proportional to its power output, and it is limited by the strength of materials. The effect of erosion due to the impact of liquid droplets depends on the peripheral velocity of the turbine rotor, which is greater in larger machines for several techno-economic reasons, therefore the minimum admissible vapor quality at the turbine outlet is higher in larger

(a) Cutout of the SST5-6000 steam turbine set (800 MW_e) made of several multi-stage units: high pressure, intermediate pressure, and two low pressure.

(b) Rotor of a SST-5000 multistage low-pressure unit.

Figure 7.6 Steam turbine. Courtesy of Siemens, www.siemens.com/press.

steam turbines, which can reach a capacity of up to 1000 MW. In large steam power plants (Figure 7.11), in order to decrease the amount of liquid droplets in the last turbine stage, *reheating* is another Rankine cycle modification normally adopted. It consists in reheating the steam at the discharge of the high-pressure turbine (or high-pressure turbines stages) by means of another heat exchanger before it is fed to the second turbine. Figure 7.12 shows that reheating increases the quality of the vapor at the turbine outlet. It can also be inferred that reheating increases the cycle efficiency with respect to a superheated cycle, as energy as heat is transferred to the working fluid at higher average temperature. Observe though, that reheating is very costly because the expanded steam at the high-pressure turbine outlet

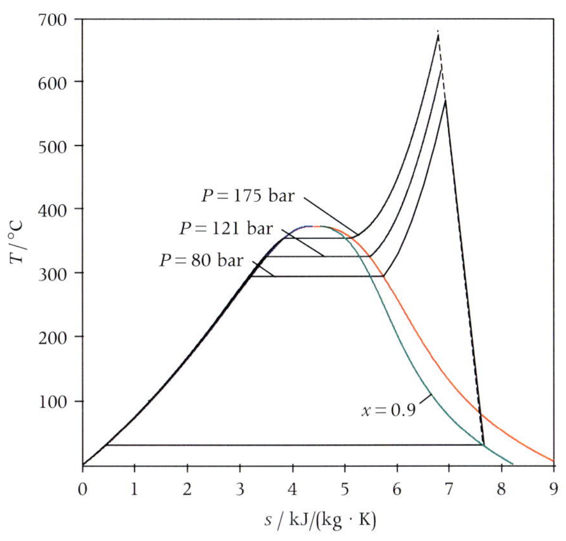

(a) Superheating at fixed vapor quality at the turbine outlet for the simple Rankine cycle of Figure 7.4 in the T–s diagram.

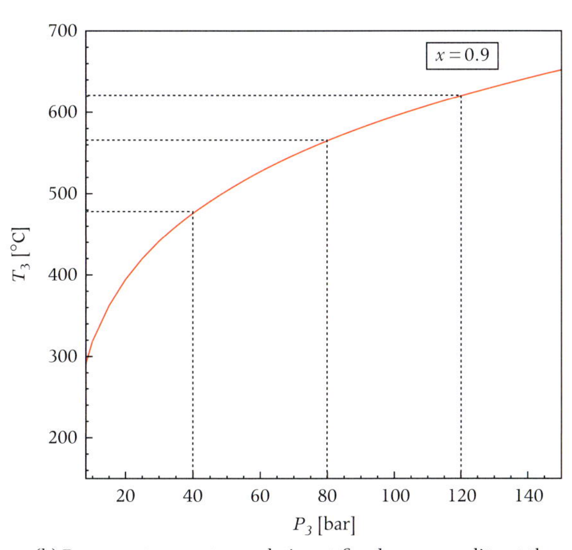

(b) Pressure–temperature relation at fixed vapor quality at the turbine outlet.

Figure 7.7 If the vapor quality at the turbine outlet is a design constraint to avoid blade erosion, increasing the cycle pressure to increase the cycle efficiency entails increasing the turbine inlet temperature.

already occupies a large volume, therefore the duct and the reheater are large and must withstand high temperature.

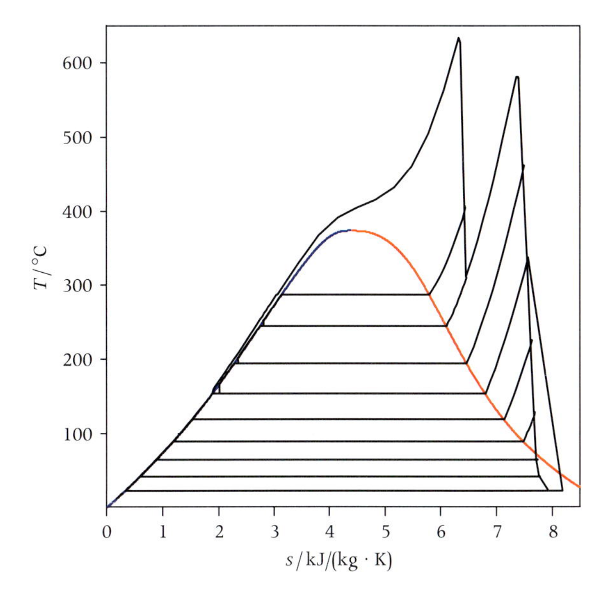

Figure 7.8 Representation of the processes of an ultra supercritical Rankine cycle power plant in the *T–s* diagram of water.

Supercritical cycle

The efficiency of a thermodynamic cycle is proportional to the difference between the average maximum and the minimum temperature at which thermal energy is exchanged with the environment. The maximum temperature is usually limited by the availability of suitable materials. Throughout the years the maximum temperature, and consequently pressure, of steam power plants have steadily increased, and nowadays the most modern steam power stations implement the so-called ultra-supercritical Rankine cycle (Figure 7.8). Liquid water is compressed in the pumping station at a pressure which is higher than the critical pressure of water, up to about 300 bar, and it is heated up to temperatures in the range 615 – 630 °C (up to more than 700 °C in the plants that are currently commissioned). In this process the fluid gradually transitions from the liquid state to a state of highly compressed vapor, without boiling.

Deaerator, regenerator, and economizer

The condensation temperature of a steam power plant depends on the temperature of the fluid that can

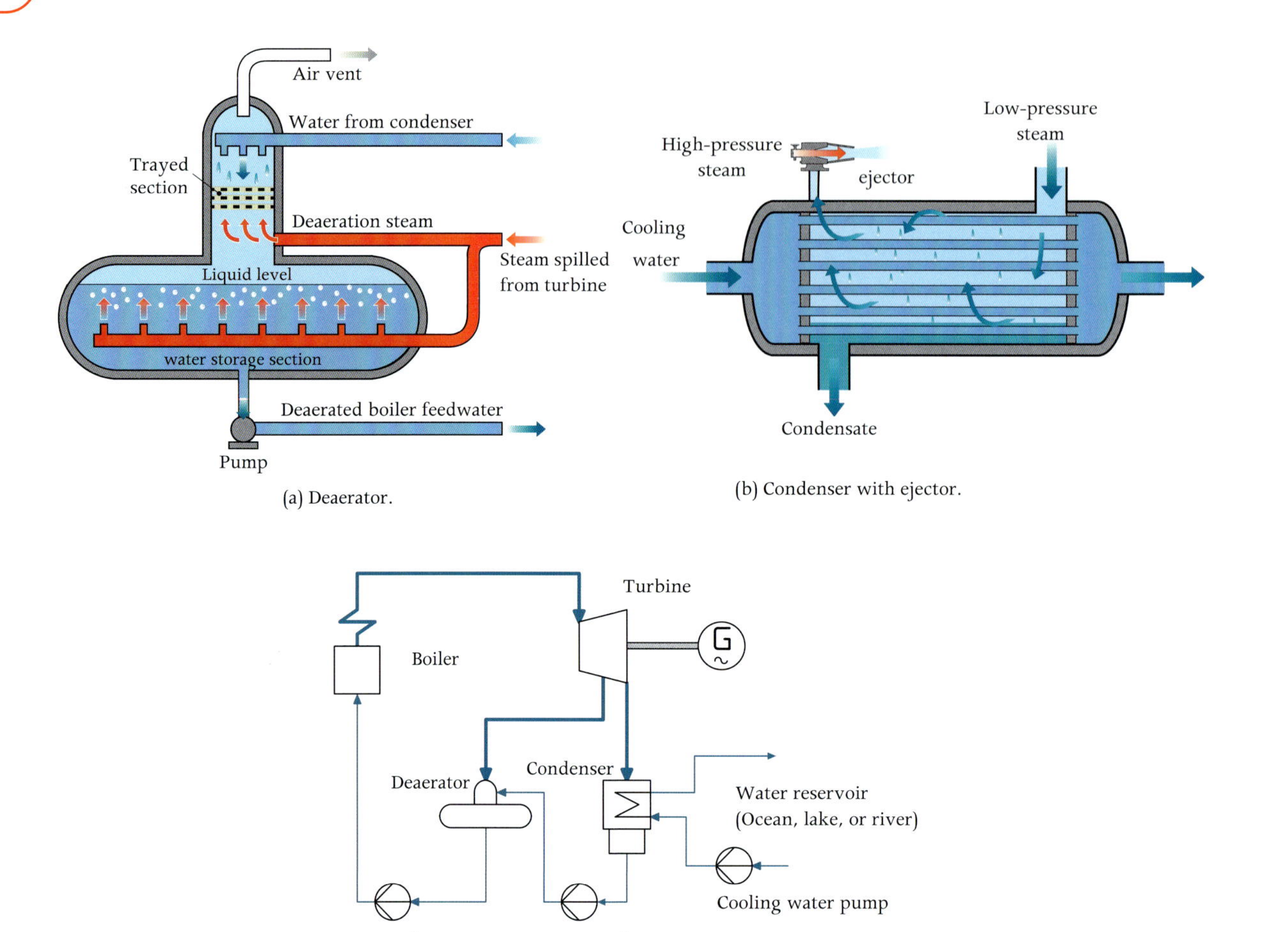

(a) Deaerator.

(b) Condenser with ejector.

(c) Process flow diagram of a simple steam power plant employing a deaerator. Note that the deaerator serves also as a regenerator, thus increasing the energy conversion efficiency of the system.

Figure 7.9 The deaerator is used to extract incondensible gases from the low-pressure section of a steam power plant by means of pressurized steam.

be used to cool the steam. This can be river or sea water. Often an intermediate water loop connects the condenser to the cooling sink. Environmental regulation often imposes the adoption of cooling towers, whereby a much smaller amount of water is evaporated on the surface of the packing of the intermediate loop, or directly on the surface of the condenser, thus, together with air, providing the cooling capacity. The condensation temperature varies therefore between close-to freezing temperature for colder climates to $40-60\ ^{\circ}C$. The vapor pressure for these temperatures is therefore much lower than atmospheric (42 mbar in our example), therefore any leak in the low-pressure part of the power plant permits air to contaminate the working fluid. Any impurity of the working fluid causes high corrosiveness in the high-temperature part of the power plant. This is why highly purified water is employed as the working fluid. Steam power plants condensing at sub-atmospheric pressure are therefore always equipped with vacuum pumps which initiate the low pressure when the plant is started and with a *deaerator* (Figure 7.9a) or otherdevices, like steam-driven *jet ejectors* (Figure 7.9b), to remove the small amount of air that inevitably leaks into the system. In these devices, a small amount of high-pressure steam is bled from the main line and fed through a Venturi-like device in which steam is accelerated until its pressure is lower than the condensation pressure. The air is therefore sucked by the *Venturi effect* into the ejector, and the air/steam mixture then slows down with a resulting pressure rise, which is enough to discharge it to the atmosphere.

By comparing the Carnot and the superheated Rankine cycle in the *T-s* plane, we see that one of the most efficiency-deteriorating processes is the heating of the liquid water emerging from the feed-water pump. The efficiency of a steam power plant considerably increases if some steam at the outlet of the high-pressure section of the turbine is extracted and used to increase the temperature of the compressed liquid at the pump outlet, before it enters the boiler. This process is called *regeneration* and such heat transfer is accomplished in a *feedwater heater*, where the hot steam is mixed with the cold water. In smaller and simpler power plants the deaerator also serves as a regenerator (Figure 7.9c). The regeneration process yields two counteracting effects: on the one hand the heat transfer internal to the thermodynamic cycle reduces the amount of energy that must be transferred from *outside* the system, on the other the steam that is spilled from the turbine does not expand in the turbine, thus it does not contribute to the conversion of energy into mechanical power. Which of the two effects is dominant cannot be stated a priori and requires an analysis (see the following example). In order to better understand the concept of regeneration, the analysis of a limiting case can help: if it were possible to realize an expansion with "continuous regeneration" in a saturated Rankine Cycle (Figure 7.10), the efficiency of such a system would be the same as that of a Carnot engine operating between the same maximum and minimum temperature. Regeneration, or *heat integration*, is common to almost all the energy conversion systems, and, even more generally, to any modern thermodynamic process aiming at increasing energy efficiency.

Steam power plants are the major source of electric power in the world. Extremely sophisticated technology is developed in order to achieve a very high efficiency in coal-fired stations and so called ultra-critical configurations, with dual-stage reheating are nowadays common. In these very large plants (Figure 7.11), multiple regeneration and preheating of the liquid and the combustion air with low-temperature flue gases from the furnace allow achieving approximately 48% net electric efficiency in optimal operating conditions, but the possibility of overcoming the 50% efficiency level seems feasible. Nuclear power plants for electricity generation or marine propulsion are also currently based on steam power plants.

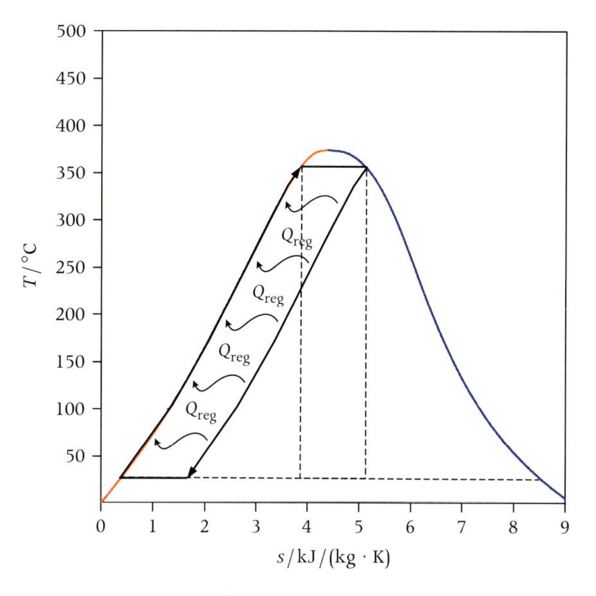

Figure 7.10 The idealized limit of continuous regeneration: the expansion process converting the energy of the working fluid into mechanical power occurs in the vapor–liquid region in very small steps, each followed by extraction of small amounts of steam that transfers energy as heat to the liquid. The efficiency of such a thermodynamic cycle is the same as that of a Carnot cycle operating between the same temperatures.

Steam technology is increasingly relevant for the conversion of sustainable energy sources. Small Rankine cycle power plants are adopted to generate electricity from the combustion of biomass, i.e., residues from forestry maintenance and agriculture, and material waste from the wood industry. The optimal capacity of such plants is small because the cost of gathering biomass is proportional to the distance from the plant. Solar concentrators of various types are increasingly used to transfer the sun's energy to the working fluid of a Rankine cycle plant,

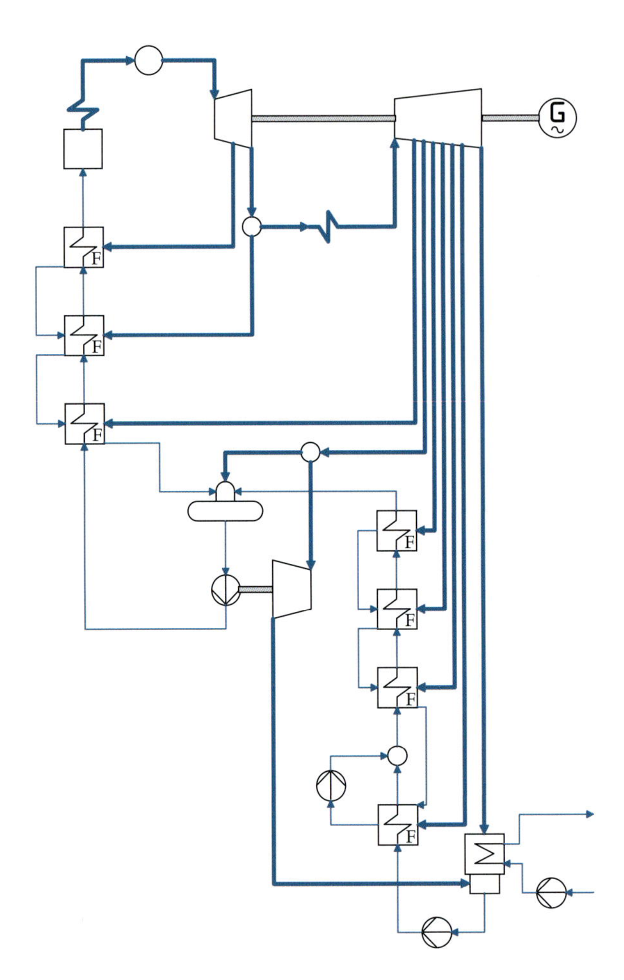

Figure 7.11 Process flow diagram of an actual steam power station (≈ 600 MW_e) displaying typical enhancements with respect to the simple Rankine cycle configuration, i.e., superheating, reheating, and regeneration. A deaerator is also present and one of the feedwater pumps is actuated by an auxiliary turbine deriving steam from the low-pressure stages of the main turbine.

thus achieving comparatively high energy conversion efficiency (Figure 7.13). Finally, the conversion of the energy content of geothermal reservoirs is another example of the use of the Rankine cycle principle to generate electricity from a sustainable energy source. In some cases geothermal steam can be expanded directly into the turbine, while, if the geothermal fluid is pressurized water or if it contains too many impurities, other more complex configurations are needed.

Example: efficiency of a Rankine cycle power plant. Consider the superheated Rankine cycle whose representation in the *T–s* thermodynamic diagram is in Figure 7.5a. A regenerated cycle configuration is derived from it by extracting steam from the turbine at 0.7 MPa = 7 bar. Such steam is mixed with compressed liquid delivered by the first feedwater pump, thus heating this liquid up to the saturation temperature at the extraction pressure (Figure 7.14). We want to calculate the efficiency of the regenerated cycle plant and compare it with that of the superheated cycle configuration without regeneration of the previous example. We need therefore to calculate states 3, 4, and 6 from the mass and energy balance over the control volume enclosing the deaerator. The amount of steam that is extracted from the turbine is what is needed to increase the temperature of the compressed liquid water at the outlet of the first feedwater pump up to the temperature of the extracted steam. The fraction of steam that must be spilled from the turbine can be obtained from the mass and energy balance equations applied to the control volume enclosing the deaerator:

$$\begin{cases} \dot{M}_2 + \dot{M}_{\mathrm{extr}} = \dot{M}_3 \\ \dot{M}_2 h_2 + \dot{M}_{\mathrm{extr}} h_6 = \dot{M}_3 h_3. \end{cases}$$

The mass fraction of extracted steam is $x_{\mathrm{extr}} \equiv \dfrac{\dot{M}_{\mathrm{extr}}}{\dot{M}_3}$. Solving for $\dot{M}_2$ the mass conservation equation, substituting for $\dot{M}_2$ in the energy balance and dividing by $\dot{M}_3$ gives

$$\frac{\dot{M}_3 - \dot{M}_{\mathrm{extr}}}{\dot{M}_3} h_2 + \frac{\dot{M}_{\mathrm{extr}}}{\dot{M}_3} h_6 = h_3,$$

which, by introducing x_{extr}, becomes

$$(1 - x_{\mathrm{extr}})\, h_2 + x_{\mathrm{extr}} h_6 = h_3.$$

This equation can be solved for the fraction of extracted steam and gives

$$x_{\mathrm{extr}} = \frac{(h_3 - h_2)}{(h_6 - h_2)}.$$

As already shown, the enthalpy of state 6 can be obtained by evaluating the enthalpy of the state having the same entropy as the state at the inlet of the turbine, and by applying the definition of isentropic efficiency for the steam expansion from state 5 to state 6 ($\eta_{\mathrm{turb},s} = 0.85$). The enthalpy of state 6 is therefore $h_6 = 2613$ kJ/kg. The mass fraction of extracted steam is thus

$$x_{\mathrm{extr}} = \frac{(697 - 127)\ \mathrm{kJ/kg}}{(2692 - 127)\ \mathrm{kJ/kg}} = 0.22.$$

The enthalpy of state 7 is calculated starting from state 6 in the same way. The total turbine specific mechanical work is therefore

$$w_{\mathrm{turb}} = h_5 - h_6 + (1 - x_{\mathrm{extr}})\,(h_6 - h_7) = 919\ \mathrm{kJ/kg}.$$

Analogously, the total specific compression work is

$$w_{\mathrm{pump}} = (1 - x_{\mathrm{extr}})\, w_{\mathrm{LP\ pump}} + w_{\mathrm{HP\ pump}} = 13\ \mathrm{kJ/kg}.$$

We can finally calculate the cycle efficiency as

$$\eta = \frac{w_{\mathrm{net}}}{q_{\mathrm{boiler}}} = \frac{w_{\mathrm{turb}} - w_{\mathrm{pump}}}{q_{\mathrm{boiler}}} = \frac{(919 - 13)\ \mathrm{kJ/kg}}{2430\ \mathrm{kJ/kg}} = 0.37.$$

The efficiency of the regenerated cycle is 3 percentage points higher than that of the homologous cycle without regeneration. The power engineer weighs such an increase in efficiency with the increase in cost and complexity due to the addition of extraction from the steam turbine and of the regenerating heat exchanger. Very importantly, nowadays, regulatory aspects related to local and global pollution more and more affect these decisions.

We also realize that the use of steam extracted from the turbine to pre-heat the liquid water before it enters the boiler implies two counteracting effects with respect to cycle efficiency: the positive effect of the decrease of thermal power input is diminished by the decrease of mechanical power produced by the turbine, due to the decrease of steam mass flow expanding in the low-pressure turbine section after the extraction. We suggest that enthusiastic students setup an EXCEL worksheet or a simple MATLAB program in order to calculate the extraction pressure that maximizes the cycle efficiency.

Organic Rankine cycle turbogenerator

For small power capacity or in the case where the heat source is at low temperature, water is not a suitable working fluid. The adoption of an organic substance as the working fluid solves a number of problems, and allows reaching a comparatively high conversion efficiency. The resulting organic Rankine cycle

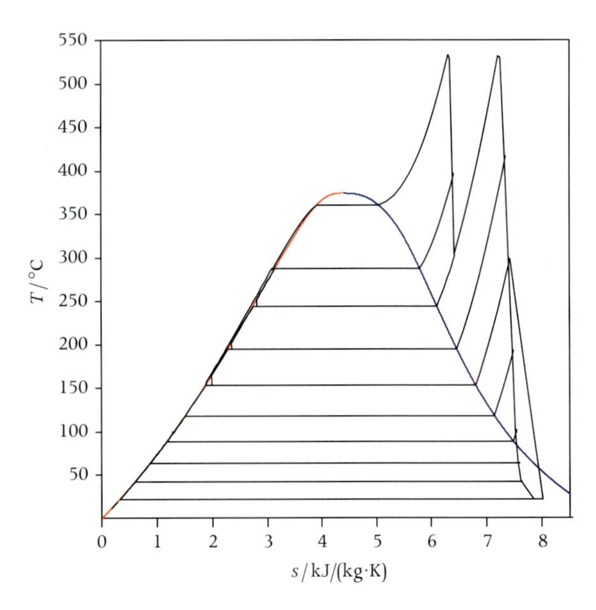

Figure 7.12 The representation in the T–s thermodynamic plane of the various thermodynamic processes undergone by water in the plant of Figure 7.11.

(ORC) power plant (Figure 7.15) is therefore a suitable technology for power outputs between a few kW_e up to several MW_e.[1]

The advantages of using an organic working fluid are mainly related to the achievable efficiency of the turbine and the possibility of choosing a fluid with optimal thermodynamic properties, depending on the temperature of the primary energy source and sink. The main fluid thermodynamic parameters affecting the optimal design of a turbine stage are the volumetric flow rate, $\dot{V} = \dfrac{\dot{m}}{\rho}$, and the specific enthalpy over the expansion, Δh_{turb}. If the power capacity is low, the mass flow rate $\dot{m}$ is small. Compared to organic fluids, the density of water vapor at typical turbine inlet conditions is large: at 380 °C and $P_r = 0.9$, water has a density of 119 kg/m^3, while toluene is 86 kg/m^3, and siloxane MDM 23.6 kg/m^3. If $\dot{m}$ is small and ρ is large, the volumetric flow rate $\dot{V}$ becomes small, which implies that the flow passages in the nozzle and the rotor of the turbines are too small. The use of an organic fluid solves the problem of the insufficient volumetric flow rate of low-power steam turbines. In addition, it can be demonstrated that the optimal rotational speed of a turbine stage is proportional to Δh_{turb}, that is the specific expansion work. For given turbine inlet and outlet conditions, the Δh_{turb} of water is much larger than that of organic fluids (try it yourself using FLUIDPROP!), therefore the optimal rotational speed of a small steam turbine is much greater than that of a turbine using an organic working fluid. In conclusion, the use of an organic fluid in small Rankine power plants enables the realization of a very efficient slow-rotating turbine, featuring a single stage or at the least very few stages, whose flow passages are large enough so that the fluid dynamic friction is low. Often the low rotational speed of ORC turbines allows for the direct coupling of the turbine rotor with that of the electrical generator. There is therefore no need of reduction gears or electronic devices to obtain grid-frequency current. Conversely, a steam turbine designed for low power output is very small, requires a much larger number of stages (which implies much higher costs), and rotates at a very high or even impractical speed.

The adoption of a fluid other than water for a Rankine power plant makes it possible to efficiently exploit low-temperature primary energy sources. A good example is a geothermal reservoir of pressurized hot water, say at 150 °C. The choice of a suitable organic fluid allows for the design of an optimal thermodynamic cycle and solves the problem of the excess of condensation at the turbine outlet, as shown in Figure 7.16b. One of the main distinguishing features of the organic Rankine cycle is that the expansion process is completely "dry": no vapor becomes liquid over the expansion. This can be noted in the *T–s* thermodynamic diagram of organic fluids, where the vapor saturation line, or *dew line*, has a positive slope, as opposed to water and other fluids made of simple molecules, which feature a dew line with a negative slope. Note also that the appropriate choice of the working fluid in low-temperature applications allows achieving a super-atmospheric phase change in the condenser, thus solving the problem of inward air-leaking.

[1] Extensive information on ORC technology is given in P. Colonna, E. Casati, C. Trapp, *et al.*, "Organic Rankine cycle power systems: From the concept to current technology, applications, and an outlook to the future," *J. Eng. Gas Turb. Power*, *137*, (10), 100 801–19, Oct. 2015.

(a) The steam turbine plant, with its intricate piping system.

(b) The parabolic trough collectors, concentrating the radiation on pipes where thermal oil is heated up.

Figure 7.13 A solar power plant (the 150 MW_e power station in Hassi R'Mel, northern Algeria). Courtesy of Siemens, www.siemens.com/press.

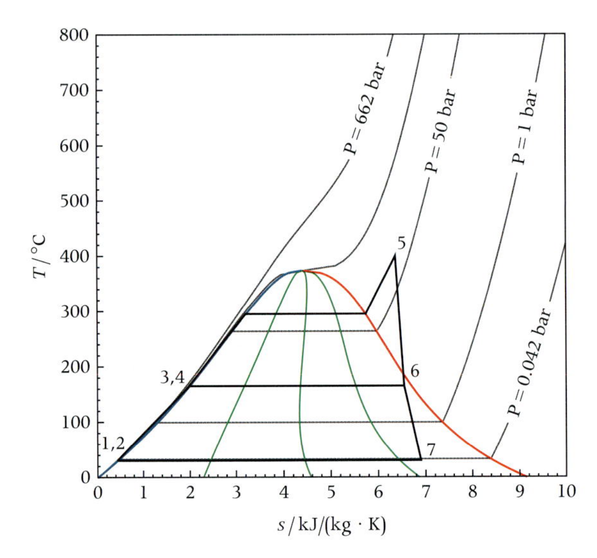

Figure 7.14 Representation in the water T–s thermodynamic plane of a simple regenerated steam Rankine cycle.

For this reason, the saturated cycle configuration can be adopted in ORC power plants, thus avoiding the additional cost and complication of superheating. In so-called high-temperature ORC systems, a regenerator is mandatory in order to achieve a high efficiency (Figure 7.17). High-temperature applications demand for a working fluid with a high critical temperature, which implies that the expanding vapor does not cool much. In order to recover the considerable amount of energy as heat available at the turbine outlet, a regenerator transfers this energy as heat to the liquid coming out of the pump, before entering the evaporator. This process is called *non-extractive desuperheating regeneration*. Superheated (Figure 7.18a) and supercritical (Figure 7.18b) cycle configurations are also possible, and the transfer of energy as heat from the primary source is easily achieved in a single heat exchanger. Also in the case of ORC plants, the supercritical cycle configuration allows reaching a higher efficiency. The critical pressure of organic working fluids for ORC power plants is an order of magnitude smaller than that of water. A supercritical ORC plant operates therefore at relatively low maximum pressure, and this is an advantage in terms of cost and safety. The supercritical cycle configuration is currently adopted in a few pilot plants for the conversion of geothermal energy or waste heat.

Common working fluids for low-temperature applications, like geothermal power and industrial heat recovery, are refrigerants and alkanes (e.g., propane, butane, etc.). ORC power plants for high-temperature applications, like biomass and solar conversion or heat recovery from recip engines and gas turbines, employ mostly toluene and siloxanes (light and low-viscosity silicon oils).

In summary, the advantages of ORC power plants can be listed as follows:

- the choice of the working fluid, thanks to its thermodynamic properties, and technological characteristics, is another degree of freedom in the thermodynamic design of the process. Cycle configurations that cannot be realized with water, become possible;
- the simplicity of the thermodynamic cycle, which at maximum requires a regenerator, matches very well the requirement of a small-capacity power plant, which often does not allow for complex and costly solutions;
- no condensation occurs in the expander, whose optimal design requires modest rotational speed;
- the maximum and minimum pressure within the plant can be chosen, to a certain extent, independently of the energy source/sink temperatures.

(a) 400 kW_e ORC power plant being transported to its site, a wood factory. The fuel for the plant is the residue from the manufacturing. The energy discharged as heat by the condenser of the ORC system is utilized to dry wood (cogeneration).

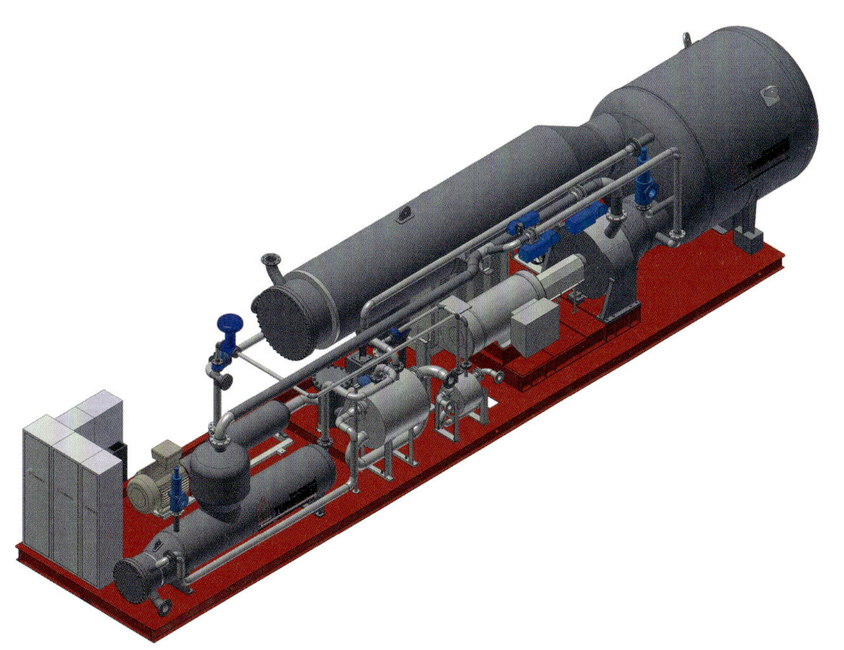

(b) A 600 kW_e heat recovery unit (TD 6 HR), capable of reaching an electrical efficiency in excess of 20%.

Figure 7.15 ORC power plants. Courtesy of Turboden.

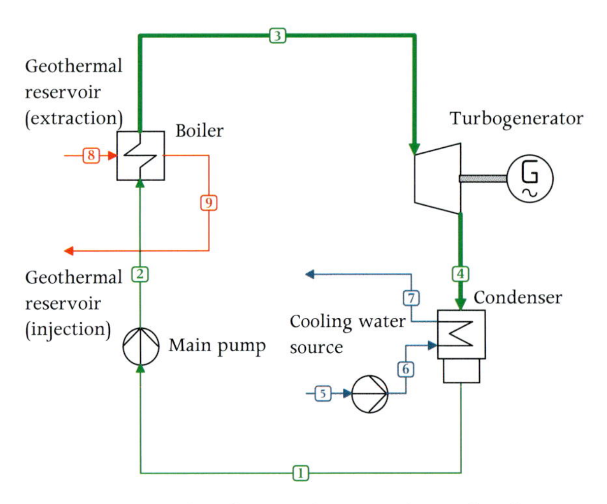

(a) The process flow diagram of a saturated ORC plant for low-temperature geothermal power production. The working fluid is R245fa, a so-called refrigerant.

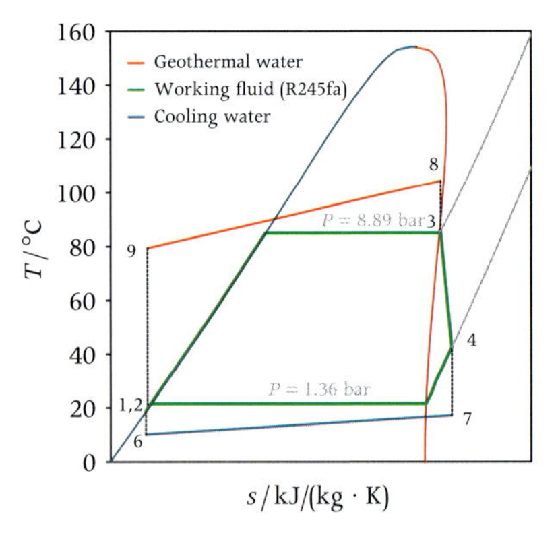

(b) The processes of the ORC power plant of (a) above in the T–s diagram of the working fluid.

Figure 7.16 Geothermal ORC power plant.

Low source/sink temperatures can be associated with high cycle pressures and high source/sink temperatures with low cycle pressures. The condensation of the fluid in super-atmospheric conditions is possible.

Points of concern in the design of an ORC system might be the toxicity, flammability, global warming potential, and cost of the working fluid. Conversely, some working fluids are good lubricants, thus saving on the complication and cost of a lubrication system for the turbine.

ORC turbogenerators are particularly suitable for the conversion of sustainable energy sources, as is the

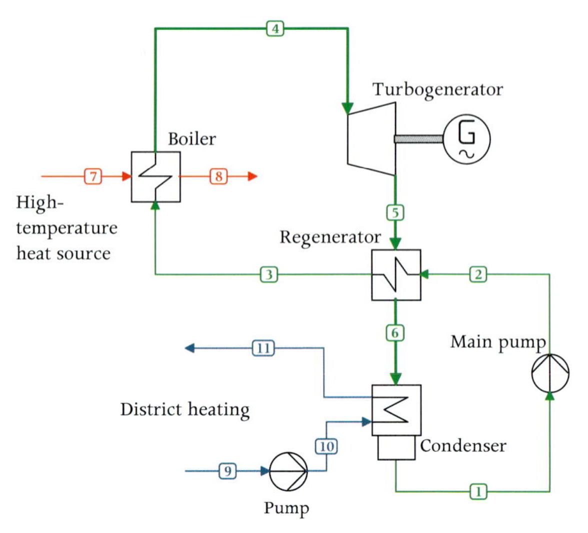

Figure 7.17 Process flow diagram of a high-temperature ORC power plant.

case for any "externally heated" system, i.e., a power system which does not require internal combustion as one of the processes of its thermodynamic cycle. ORC power plants have been widely employed for the conversion of geothermal heat sources over the past 50 years. More recently, ORC turbogenerators are being increasingly employed in biomass-fueled power plants and for heat-recovery from internal combustion engines, gas turbines, and industrial processes. Promising widespread applications for this technology are the conversion into electricity of solar radiation by means of solar mirrors, simultaneous generation of electricity and heat in buildings (the so-called domestic-cogeneration), and heat recovery from propulsion engines, like recip or gas turbine engines.

Example: efficiency and power of an ORC power plant. Consider a high-temperature ORC power plant featuring the process flow diagram of Figure 7.17, and employing MDM as working fluid. If the thermal energy source is given, and the maximum cycle temperature is fixed depending on the thermal stability of the working fluid, one can calculate the maximum cycle pressure P_4 which maximizes the net conversion efficiency. Figure 7.19 shows that the cycle efficiency increases with pressure until it reaches a

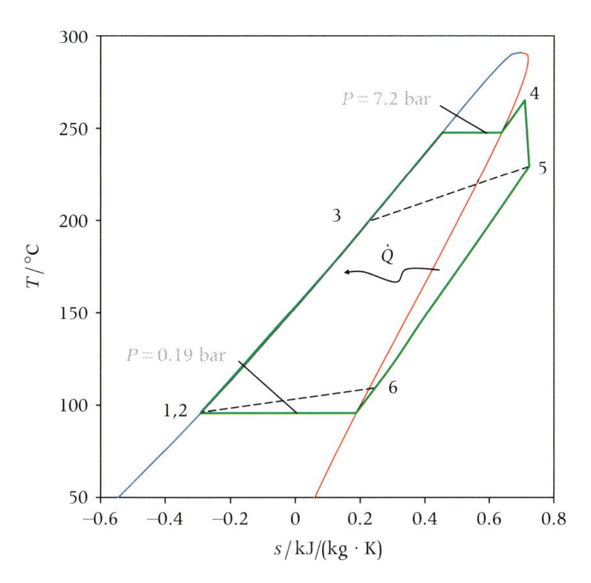

(a) Superheated ORC cycle, similar to that occurring in the plant of Figure 7.15a

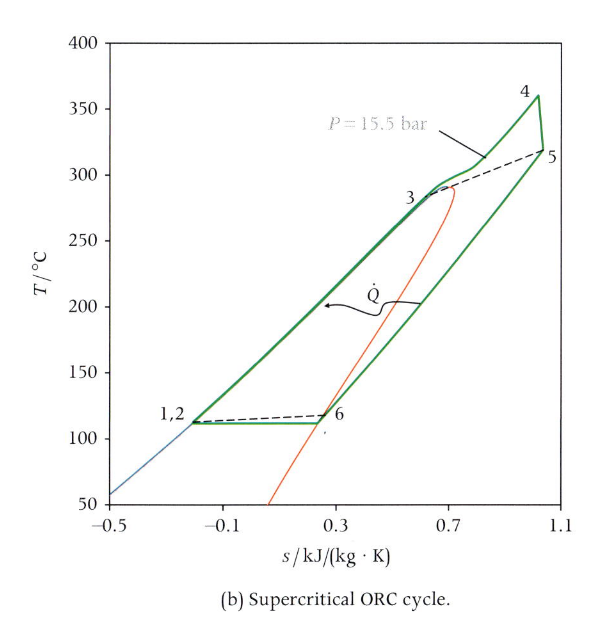

(b) Supercritical ORC cycle.

Figure 7.18 Processes of a high-temperature ORC power plant in the *T–s* diagram of the working fluid.

maximum value for slightly supercritical conditions. The power output increases steadily with the maximum cycle pressure, the only limiting factor being avoiding condensation throughout

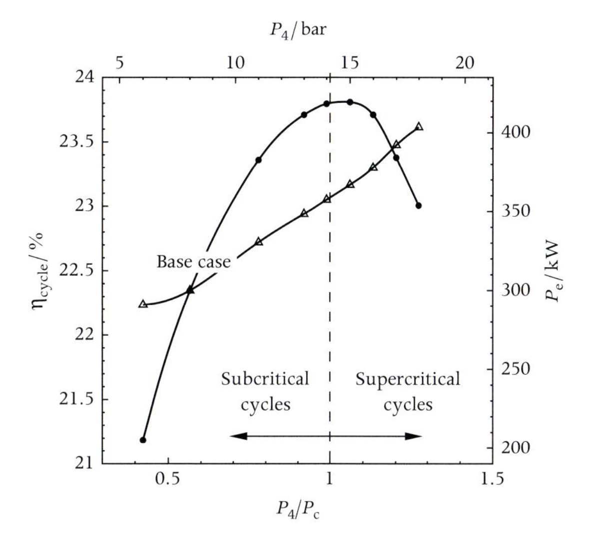

Figure 7.19 Calculated cycle efficiency (dots) and electric power output (triangles) of a waste heat recovery ORC plant as a function of the reduced pressure at the turbine inlet, from P. Colonna, J. Harinck, S. Rebay, *et al.*, "Real-gas effects in organic Rankine cycle turbine nozzles," *J. Propul. Power*, *24*, (2), 282–294, Mar. 2008.

the expansion. In waste heat recovery applications, the power output is the merit parameter.

7.4 Refrigeration

Refrigeration, the process of cooling a certain space or matter by putting it into contact with a substance at a temperature that is lower than the desired refrigeration temperature, is used for a large number of purposes. Refrigerators, small and large, are used to preserve food and medicines, air-conditioners keep the temperature at a comfort level in buildings, homes, and cars. Many oil, chemical, food and pharmaceutical processes make use of refrigeration to control the temperature to a certain low required value (which can also be very low, as in the case of air-liquefaction). The cooling effect can be obtained in several ways and here we only treat systems based on vapor-compression and absorption.

Vapor-compression refrigeration

A first example of the analysis of a simple heat pump thermodynamic cycle is presented in Section 4.7. A heat pump is a device that realizes the transfer of energy as heat from a source at lower temperature to an ambient at higher temperature. The thermodynamic cycle of a vapor-compression refrigerator is the same, but the desired effect is the transfer of energy as heat from a space to be refrigerated, to a warmer surrounding. In both cases the energy input is the mechanical power required by the compressor. Maybe, when you observed Figure 7.5b it looked familiar, because you have already seen a figure that is similar, Figure 4.5b. It is indeed quite comparable to the $P-h$ diagram of Figure 7.20, but the direction of the processes is reversed. The refrigeration and heat pump cycles are actually often called *inverse* Rankine cycles. In the refrigeration cycle, though, the expansion process, which is the counterpart of the compression in the Rankine power cycle, occurs in the vapor–liquid thermodynamic region, and it is realized by means of a throttling valve.

Analogously to steam power plants, refrigeration plants can also adopt more complex configurations. The double-stage configuration (Figure 7.21) allows achieving a higher COP, or makes it possible to generate the cooling effect at various temperatures (think of the home refrigerator and its freezing compartment, for example).

Similarly to what is done in case of ORC power plants, the working fluid, or *refrigerant*, is chosen depending on the operating conditions and other desired characteristics. Typical working fluids are hydrofluorocarbons (HFCs), several hydrocarbons, like butane or propane, carbon dioxide, and ammonia.

> Look in the on-line help of FLUIDPROP under REFPROP or STANMIX : all fluids having a name starting with "R" or "HFC" are used also as refrigerants in vapor-compression systems. Recently, international regulation has been put in place in order to phase out HFCs due to their negative impact on global warming.

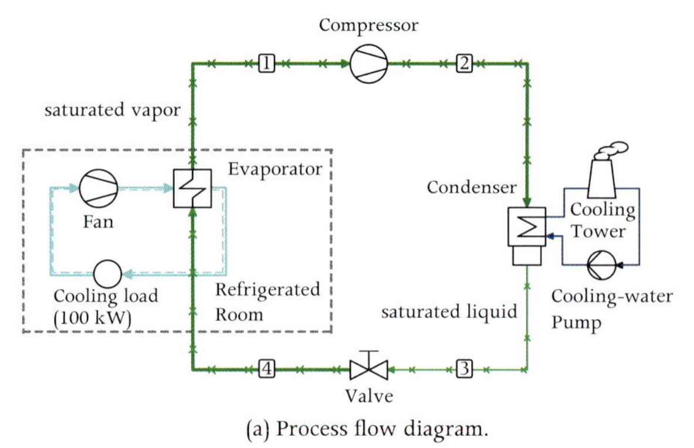

(a) Process flow diagram.

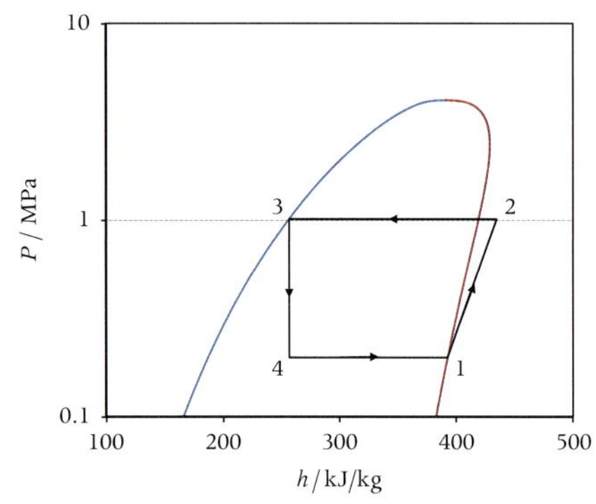

(b) The thermodynamic cycle in the $P-h$ diagram of the working fluid.

Figure 7.20 Vapor-compression refrigeration plant.

Water has many positive thermodynamic and technical features as a working fluid, but its vapor pressure at the temperatures of interest is extremely low, therefore the volume of the heat exchangers becomes large, they are prone to air leaks, and the pressure ratio of the compressor must also be very large. Nonetheless several vapor-compression systems utilizing water as a refrigerant have been realized.

Example: refrigeration plant enhanced by a turbocompressor. As we have seen, the transfer of energy as heat in a refrigeration plant is obtained at the cost of the mechanical power used by the compressor to increase the pressure of the working fluid. The fluid available at high pressure (state 3 in Figure 7.20b) could be put to a better use than just scrambled through a valve, similarly to what is done in the turboexpander of

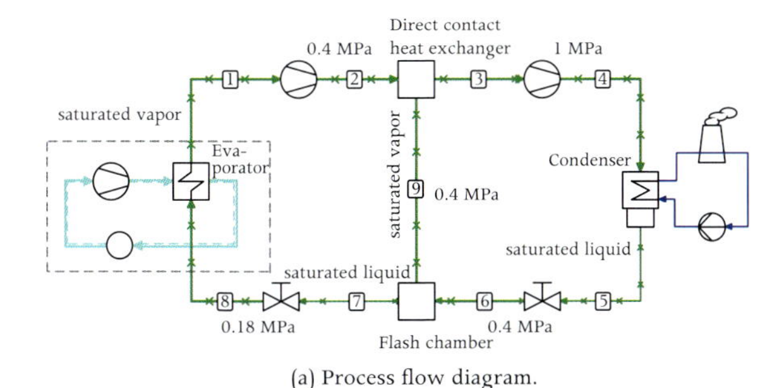

(a) Process flow diagram.

(b) The thermodynamic cycle in the P–h diagram of the working fluid.

Figure 7.21 Two-stage refrigeration plant using ammonia as the working fluid.

an automotive engine. This is not common practice, because efficiently extracting work from a vapor–liquid mixture implies technical challenges. Nonetheless, successful examples exist, therefore let us analyze what can be the efficiency gain of expanding the working fluid in a screw expander. Let refrigerant R134a be the working fluid, which enters the compressor as saturated vapor at $T_1 = -10$ °C (see Figure 7.20), and leaves the condenser as saturated liquid at $T_3 = 40$ °C. The isentropic efficiency of the compressor and of the expander are both 80% and we assume zero pressure loss in the evaporator and condenser. Using STANMIX, we can get the estimated saturation pressure at the condensing and evaporating temperatures, and these are

$$P_2 = P_3 = P(T_3) = 1.019 \text{ MPa},$$
$$P_1 = P_4 = P(T_1) = 0.201 \text{ MPa}.$$

Likewise, the enthalpy and entropy at the inlet of the compressor are

$$h_1 = h(T_1, q = 1) = -31.4 \text{ kJ/kg},$$
$$s_1 = s(T_1, q = 1) = -0.165 \text{ kJ/}\left(\text{kg} \cdot \text{K}\right).$$

The enthalpy of the working fluid at the outlet of the compressor is calculated with the outlet enthalpy of the isentropic process and the definition of isentropic efficiency as

$$h_{2,s} = h(P_2, s_1) = 2.8 \text{ kJ/kg},$$

and

$$h_2 = h_1 + \frac{h_{2,s} - h_1}{\eta_{\text{comp},s}} = -31.4 \text{ kJ/kg} + \frac{(2.8 - -31.4) \text{ kJ/kg}}{0.8} = 11.3 \text{ kJ/kg}.$$

Finally, because the expansion through the throttling valve is adiabatic, we have that

$$h_4 = h_3 = h(T_3, q = 0) = -168.6 \text{ kJ/kg},$$

which yields a coefficient of performance $\left(\text{COP} \equiv \dfrac{q_{\text{evap}}}{w_{\text{comp}}}\right)$

$$\text{COP} = \frac{h_1 - h_4}{h_2 - h_1} = 3.21.$$

If we employ a screw expander, instead of a throttling valve, and connect it to the compressor with a shaft, then the power generated by the expander can be used to help driving the compressor, therefore reducing the electric power input to the system. In this case the expansion is no longer isenthalpic (where do you expect state 4 to be located in the P–h diagram of Figure 7.20b?) and we can calculate h_4 as we did for the outlet of a Rankine cycle turbine. The entropy of state 3 follows from

$$s_3 = s(T_3, q = 0) = -0.712 \text{ kJ/}\left(\text{kg} \cdot \text{K}\right).$$

We hence have that

$$h_{4,s} = h(P_4, s_3) = -175.4 \text{ kJ/kg},$$

and

$$h_4 = h_3 - (h_3 - h_{4,s}) \times \eta_{comp}$$
$$= -168.6 \text{ kJ/kg}$$
$$- (-168.6 - -175.4) \text{ kJ/kg} \times 0.8$$
$$= -174.0 \text{ kJ/kg}.$$

The COP in this case is therefore

$$\text{COP} = \frac{q_{evap}}{w_{comp} - w_{turb}} = \frac{(h_1 - h_4)}{(h_2 - h_1) - (h_3 - h_4)}$$
$$= 3.82.$$

The COP of this system is 19 % better than that corresponding to the simple compression cycle. Note also that, for the same cooling load, the working fluid mass flow rate is smaller because q_{evap} is larger.

Absorption refrigeration

As strange as it may seem at first, refrigeration can be powered by energy transfer as heat from a high temperature source in place of the mechanical power needed by the compressor. These systems are based on the absorption refrigeration cycle, whose process flow diagram is shown in Figure 7.22. The working fluid in this case is a mixture of two substances,

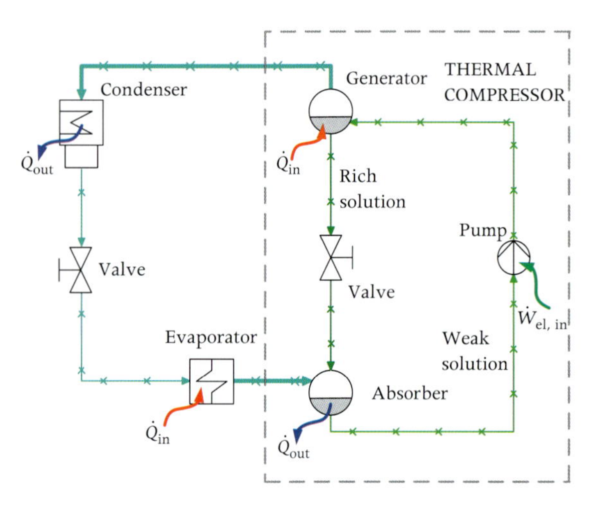

Figure 7.22 Process flow diagram of an absorption refrigeration plant.

usually water and ammonia, or water and lithium bromide (a salt). One is the refrigerant, that is the fluid that provides the refrigeration effect: by evaporating, it extracts energy as heat from the fluid or ambient to be cooled, and it rejects the energy as heat by condensing at a temperature higher than environment. The other fluid is called the absorbent. Lithium bromide acts as an absorbent with water, while water acts as an absorbent with ammonia. Lithium bromide/water systems are used at higher temperature levels than water/ammonia systems. As highlighted in Figure 7.22, the compression process in an absorption system is substituted by a subsystem, often called the "thermal compressor." In the thermal compressor the following processes take place:

- the low-pressure refrigerant mixes with the absorbent and condenses, thus requiring energy removal as heat;
- the liquid mixture is pumped at the maximum cycle pressure; and
- the refrigerant vapor is separated from the absorbent by means of energy transfer as heat at high temperature.

In this case the working fluid changes composition over the thermodynamic cycle, therefore the processes cannot be conveniently represented on one of the usual charts, like the T–s or the P–h state diagrams, which apply to fluids of fixed composition. After learning about thermodynamic properties of mixtures in the next chapter, you will also be able to perform the analysis of absorption refrigeration cycles.

The mechanical power required by an absorption refrigeration plant is much lower than that of a conventional vapor-compression refrigerator, because the working fluid is pressurized in the liquid phase rather than in the vapor phase. However, this system requires much larger heat transfer surfaces to reject energy as heat to the environment when compared to a vapor-compression refrigeration plant.

The fact that an absorption refrigeration plant or refrigerator can be powered by an external source of thermal energy at mild temperature is relevant in terms of sustainability. Such systems can be powered directly by solar collectors or can use waste heat from industrial processes and prime movers, like gas turbines and recip engines. We see once more that a much higher conversion efficiency can be obtained through smart integration of energy systems.

7.5 The Brayton Cycle

The concept of the Rankine power cycle is based on the use of a working fluid that, for some of the cycle processes, is in the liquid phase, while for others is in the vapor phase. The chief thermodynamic advantage is that the mechanical work required to pressurize the liquid is much smaller than the work that can be obtained by expanding the vapor. Therefore, even if the efficiency of the compression and of the expansion processes is low, the energy conversion efficiency of the entire system can still be sufficiently high. This advantage determined the success of steam engines and turbines in the early days. These power plants and engines could count on pumps and expanders with a very modest efficiency, and still they reached a remarkable state of advancement by the end of the nineteenth century, thus powering the industrial revolution.

If we choose to keep the working fluid in the gaseous phase throughout the cycle, the compression process drains off a large fraction of the work obtained in the expansion process. This problem initially hampered the development of machines based on all-gas cycles, like gas turbines. A simple example which introduces the Brayton cycle, the pattern cycle of gas turbines, clarifies this issue. Let us analyze a simple open-cycle system (Figure 7.23): air is compressed, heated, and expanded in a turbine.

We make the following idealizations:

- air, the working fluid, can be treated as a polytropic ideal gas with $\gamma = 1.4$ and $c_P = 1.04\ \mathrm{kJ/(kg \cdot K)}$,
- state 1, 20 °C and 1.013 bar (atmospheric pressure),
- state 2, 4 bar,
- state 3, 525 °C and 4 bar, and
- state 4, 1.013 bar.

The isentropic efficiency of the machinery is:

- $\eta_{\mathrm{comp},s} = 0.65$ for the compressor, and
- $\eta_{\mathrm{turb},s} = 0.85$ for the turbine.

Furthermore, we assume that:

- kinetic and potential energy are negligible at states 1, 2, 3, and 4,
- compressor and turbine are adiabatic,
- no pressure drop occurs in the heater,
- the system is in a steady state, and
- states at 1, 2, 3, and 4 are at equilibrium.

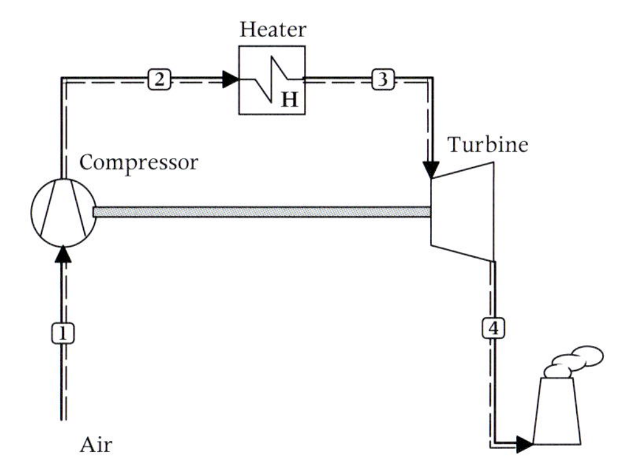

(a) Process flow diagram.

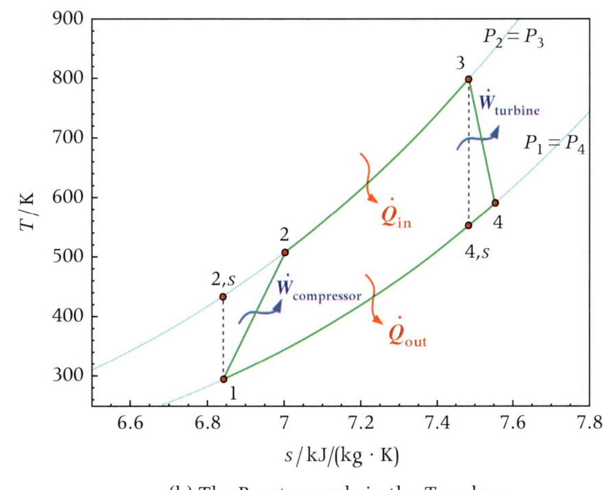

(b) The Brayton cycle in the T–s plane.

Figure 7.23 Simple open-cycle gas turbine power system.

Note that the minimum and maximum temperatures, as well as the efficiency of the components are similar to the Rankine cycle plant of Section 7.2.

Energy balances on the three components yield

$$w_{\text{comp}} = h_2 - h_1, \quad w_{\text{turb}} = h_4 - h_3, \quad q_{\text{heater}} = h_3 - h_2.$$

We already know how to compute the outlet temperature of the compressor from (6.25), and the work for the isentropic and real compression, and these are

$$\begin{aligned} T_{2,s} &= T_1 \left(\frac{P_2}{P_1}\right)^{(\gamma-1)/\gamma} \\ &= 293.15 \text{ K} \times 3.95^{0.286} = 434.0 \text{ K}, \\ w_{\text{comp},s} &= h_{2,s} - h_1 \\ &= c_P \left(T_{2,s} - T_1\right) \\ &= 1.04 \text{ kJ/}\left(\text{kg} \cdot \text{K}\right) (434.0 - 293.15) \text{ K} \\ &= 146.5 \text{ kJ/kg}, \\ w_{\text{comp}} &= \frac{w_{\text{comp},s}}{\eta_{\text{comp}}} = \frac{146.5 \text{ kJ/kg}}{0.65} = 225.4 \text{ kJ/kg}; \end{aligned}$$

we also have that

$$\begin{aligned} w_{\text{comp}} &= h_2 - h_1 = c_P (T_2 - T_1), \\ \Rightarrow T_2 &= T_1 + \frac{w_{\text{comp}}}{c_P} = 293.15 \text{ K} + \frac{225.4 \text{ kJ/kg}}{1.04 \text{ kJ/}\left(\text{kg} \cdot \text{K}\right)} \\ &= 509.8 \text{ K} \ (= 236.7 \ ^\circ\text{C}). \end{aligned}$$

The specific energy transfer as heat to the working fluid in the heater is

$$\begin{aligned} q_{\text{heater}} &= h_3 - h_2 = c_P (T_3 - T_2) \\ &= 1.04 \text{ kJ/}\left(\text{kg} \cdot \text{K}\right) (525 - 236.7) \text{ K} \\ &= 299.8 \text{ kJ/kg}, \end{aligned}$$

and, in analogy to the compression process, for the expansion we have that

$$\begin{aligned} T_{4,s} &= T_3 \left(\frac{P_3}{P_4}\right)^{(1-\gamma)/\gamma} \\ &= 798.15 \text{ K} \times 3.95^{-0.286} = 538.9 \text{ K}, \\ w_{\text{turb},s} &= h_3 - h_{4,s} = c_P \left(T_3 - T_{4,s}\right) \\ &= 1.04 \text{ kJ/}\left(\text{kg} \cdot \text{K}\right) (798.15 - 538.9) \text{ K} \\ &= 269.6 \text{ kJ/kg}, \\ w_{\text{turb}} &= w_{\text{turb},s} \times \eta_{\text{turb},s} = 269.6 \text{ kJ/kg} \times 0.85 \\ &= 229.2 \text{ kJ/kg}, \end{aligned}$$

and

$$\begin{aligned} w_{\text{turb}} &= h_3 - h_4 = c_P (T_3 - T_4), \\ \Rightarrow T_4 &= T_3 - \frac{w_{\text{turb}}}{c_P} = 798.15 \text{ K} - \frac{229.2 \text{ kJ/kg}}{1.04 \text{ kJ/}\left(\text{kg} \cdot \text{K}\right)} \\ &= 577.8 \text{ K}. \end{aligned}$$

The net work output is

$$\begin{aligned} w_{\text{net}} &= w_{\text{turb}} - w_{\text{comp}} = (229.2 - 225.4) \text{ kJ/kg} \\ &= 3.8 \text{ kJ/kg}, \end{aligned}$$

and the system energy-conversion efficiency is

$$\eta = \frac{w_{\text{net}}}{q_{\text{heater}}} = \frac{3.8}{299.8} = 0.013.$$

Only 1.3 % of the energy input is converted into mechanical work, compared to 34 % of the Rankine cycle plant. As we anticipated, we see that compressing a gas takes much more work than compressing a liquid, and this is the cause of the poorer performance of a machine based on the Brayton cycle, if compared to one based on the Rankine cycle *for comparable maximum and minimum temperature levels and efficiency of the components*.

On the other hand, it is much easier to achieve high maximum temperatures in a gas cycle because the working fluid can be burned with fuel *within* the system (*internal combustion*). Gas turbines therefore became practical when, starting from the 1940s, a large research and development effort brought the isentropic efficiency of large axial-flow compressors to the level of about 85 %. The compressor efficiency selected in this example was purposely low in order to emphasize this point. With a value of 85 % for the compressor efficiency in our example we have

$$\begin{aligned} w_{\text{comp}} &= 172.4 \text{ kJ/kg} \\ T_2 &= 458.9 \text{ K} \\ q_{\text{heater}} &= 352.9 \text{ kJ/kg} \\ w_{\text{net}} &= 56.7 \text{ kJ/kg} \\ \eta &= 0.16. \end{aligned}$$

The 16% conversion efficiency is still much lower than that of the Rankine cycle plant for the same temperature levels, but the improvement is large: this 15 percent points gain in compressor efficiency reduces the fuel consumption by a factor of almost 15.

The pertinent energy transfers for this system are summarized in Table 7.1, where they are compared to values of the ideal Brayton cycle for the same operating conditions. The values obtained for a cycle with isentropic compression and expansion show the effect on system efficiency due to irreversibilities in the compressor and the turbine.

A closer inspection of the ideal Brayton cycle is useful to introduce some concepts related to gas turbines. If the working fluid is treated as an ideal gas with constant specific heats, a simple expression for the cycle efficiency can be obtained as follows:

$$\eta = 1 - \frac{q_{out}}{q_{heater}} = 1 - \frac{c_P\,(T_4 - T_1)}{c_P\,(T_3 - T_2)} = 1 - \frac{T_4/T_1 - 1}{T_3/T_2 - 1} \cdot \frac{T_1}{T_2}.$$

We also have that $\frac{P_1}{P_2} = \frac{P_4}{P_3}$, thus $\left(\frac{T_1}{T_2}\right)^{\frac{\gamma}{\gamma-1}} = \left(\frac{T_4}{T_3}\right)^{\frac{\gamma}{\gamma-1}}$, therefore $\frac{T_4}{T_1} = \frac{T_3}{T_2}$. The ideal Brayton cycle efficiency is therefore

$$\eta = 1 - \frac{T_1}{T_2}.$$

It can also be calculated in terms of pressure ratio. Observing that $\frac{T_1}{T_2} = \frac{T_4}{T_3} = \left(\frac{P_2}{P_1}\right)^{(\gamma-1)/\gamma}$, with the definitions

$$\beta_P \equiv \frac{P_2}{P_1} = \frac{P_3}{P_4},$$

$$\vartheta \equiv \frac{(\gamma - 1)}{\gamma},$$

we also have that

$$\eta = 1 - \beta_P^{-\vartheta}. \tag{7.1}$$

We can therefore draw some conclusions that are *valid only for the ideal Brayton cycle*, that is:

- for a given working fluid, the efficiency depends only on the compression (or expansion) ratio;
- for a given pressure ratio, the efficiency depends on the working fluid (see Figure 7.24a). Fluids made of simpler molecules (higher γ) allow for higher efficiency at a given pressure ratio; and
- for a given maximum temperature of the cycle T_3, there exists a limiting pressure ratio, which is the one that gives the maximum energy conversion efficiency, but yields no net work (see Figure 7.24). This cycle is obtained for $T_3 \rightarrow T_2$, therefore it is an "infinitesimal" Carnot cycle operating between T_3 and T_2, with $\beta_{lim} = T_3/T_1$.

The energy-conversion efficiency and the net work output of the ideal Brayton cycle for a given T_3 (and T_1) are reported in Figure 7.24c. The pressure ratio for which the net work output is maximum is obtained for $T_4 = T_2$.

As a last remark, observe that the Brayton cycle can also be inverted to obtain the cycle of a refrigeration system.

7.6 Gas Turbines

In a real gas turbine power plant, the most relevant thermodynamic losses occur in the compressor and in the turbine. If one follows what is suggested by the analysis of the ideal cycle, and therefore designs a cycle with a very large pressure ratio, the result is a thermodynamic cycle with a very poor efficiency. The isentropic efficiency of the compressor in particular, but also of the turbine, tends to decrease with very large β_P. Only in the early 1940s did compressor efficiency became sufficiently high to provide positive net work from the systems that were realized. The efficiency of turbomachinery has increased enormously in the last 60 years, due largely to the perfecting of their fluid dynamic design. Hence, current turbofan engines (Figure 7.25a) feature compressors with pressure ratios that are close to the limit of what can be obtained with axial turbomachinery (the most modern axial compressors of large turbofan engines currently feature a β_P of approximately 50!), and still a very high isentropic efficiency. In these engines, and likewise in stationary gas turbine power plants (Figure 7.25b), the thermal energy input process is accomplished by combustion of fossil fuels, either liquid or gaseous, therefore the maximum temperature of the cycle is limited by the strength of materials at these high temperatures. Currently the turbine inlet temperature (TIT) of large gas turbines can reach up to approximately 1500 °C, by adopting special

Table 7.1 Comparison between ideal and real Brayton cycle, with $T_1 = 20$ °C, $P_1 = 1$ atm, $T_3 = 525$ °C, $\beta_P \approx 4$.

	w_{comp} /kJ/kg	w_{turb} /kJ/kg	w_{net} /kJ/kg	q_{heater} /kJ/kg	$\eta_{comp,s}$	$\eta_{turb,s}$	η
Ideal	146.5	269.4	122.9	378.7	1	1	0.32
Real	225.4	229.2	3.8	299.8	0.65	0.85	0.013

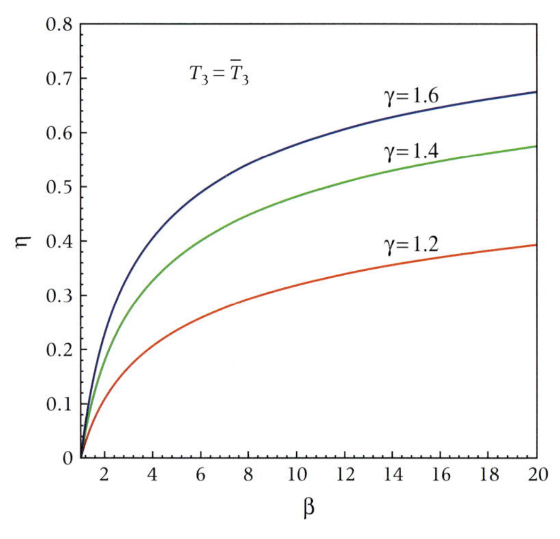

(a) Efficiency of the ideal Brayton cycle as a function of the pressure ratio and of the working fluid γ.

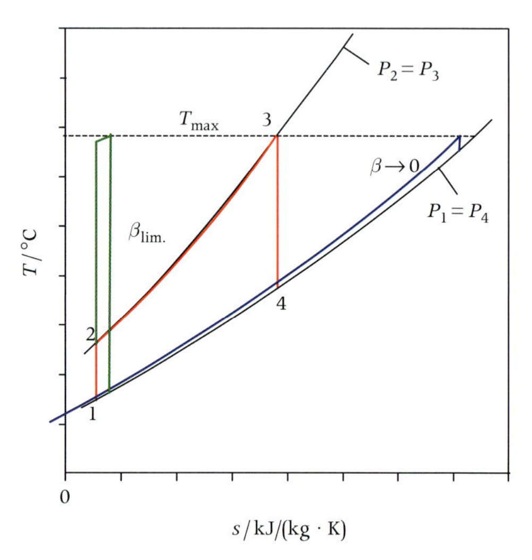

(b) Limiting cases of the idealized Brayton cycles in the T–s plane.

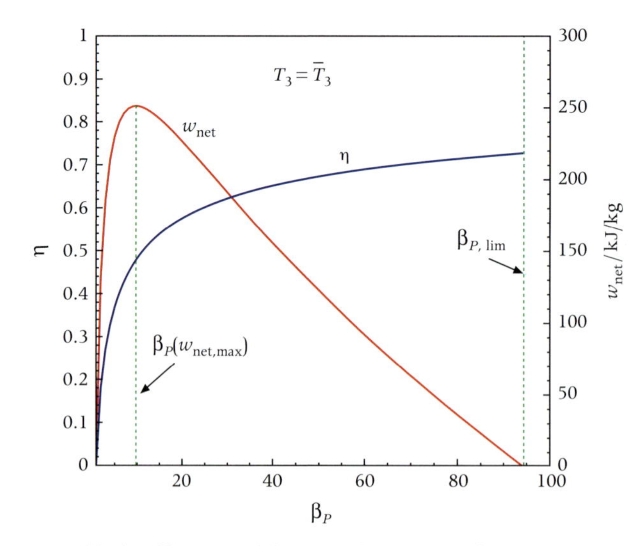

(c) The efficiency and the network output as a function of the pressure ratio.

Figure 7.24 The ideal Brayton cycle.

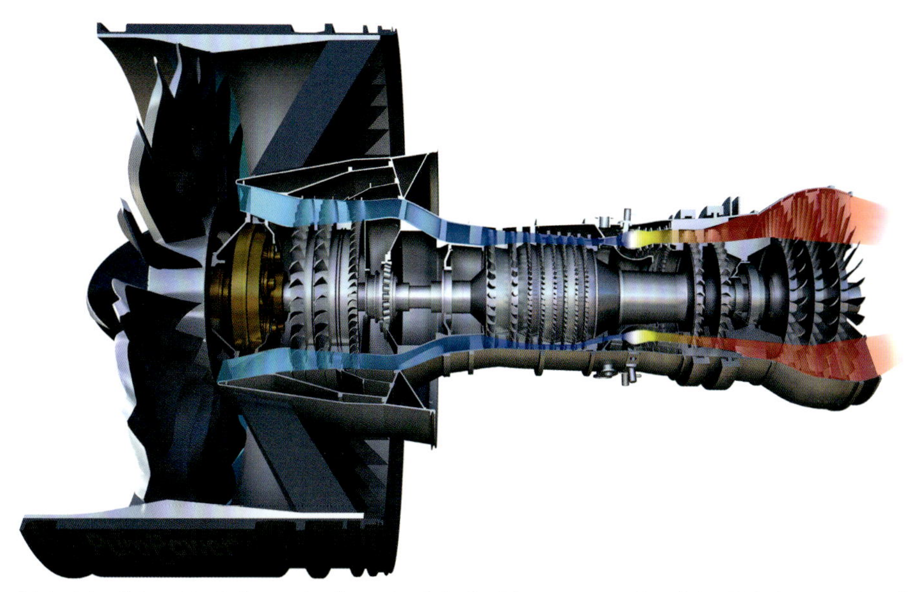

(a) A state-of-the-art turbofan engine for regional single-aisle passeneger aircraft (PW1000G, approximately 130 kN of thrust at take-off). Courtesy of Pratt&Whitney.

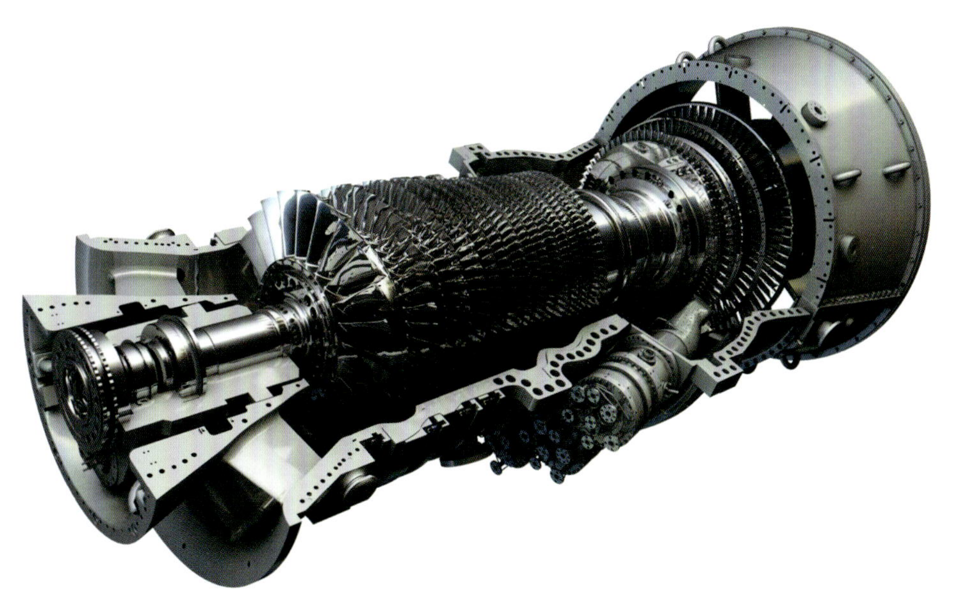

(b) A 230 MW_e heavy-duty gas turbine (GE 7F.05) for electricity production. Courtesy of General Electric.

Figure 7.25 Cut-outs showing modern gas turbines for different uses.

materials and sophisticated turbine-blade cooling. Air is extracted from the compressor and used to cool the turbine blades.

With this in mind, we see that, if gas-cycle systems do have some thermodynamic disadvantages, their practical advantages are numerous. The thermodynamic cycle can reach very high temperatures, but the maximum pressure is still quite low and heavy piping is not required. Gas turbines are therefore compact and their power-to-weight ratio is much larger than that of vapor power plants. Consequently they cost less and can be built more quickly. Their turbines are not beset by the erosive problems of wet vapor, and cavitation cannot occur in a compressor. However, the products of combustion must not damage the blades of the turbine, therefore open-cycle gas turbines must be operated with high-quality fuels, such as natural gas or kerosene, whose combustion does not produce particulate which would impact the blades. The combustion of coal and biomass is therefore unsuited for a gas turbine. Gas turbines are also used for ship propulsion and some attempts have been made to use mini-gas turbines as automotive engines. A mini gas turbine to be used as the electrical generator (or so-called *range extender*) for hybrid cars is also under scrutiny.

Let us now have a closer look at how β_P influences the performance of power plants based on the "real" Brayton cycle, together with the maximum cycle temperature (turbine inlet temperature, TIT). Also the thermodynamic cycle of a simple gas turbine can be easily calculated by applying mass and conservation balances to the components. Try and implement the cycle calculation in either EXCEL or MATLAB, using GASMIX (ideal gas) as the thermodynamic model for the fluid. The operational data typical of a medium-capacity industrial gas turbine and related assumptions are:

- temperature at the inlet of the compressor $T_1 = 298.15$ K,
- compressor inlet pressure $P_1 = 1.013$ bar,
- TIT, $T_3 = 1460$ K,
- pressure ratio of the compressor $\beta_P = 16.8$,
- the pressure losses in the combustor can be calculated as $\Delta P_{\text{comb}} = 0.05 \cdot P_3 = 2$ bar,
- the fuel is Dutch natural gas (Slochteren), thus the composition is

  CH_4 81.97 %,
  C_2H_6 2.87 %,
  C_3H_8 0.38 %,
  C_4H_{10} 0.15 %,
  C_5H_{12} 0.04 %,
  C_6H_{14} 0.05 %,
  CO_2 0.89 %,
  O_2 0.01 %,
  N_2 14.32 %,

- the fuel mass flow rate $\dot{m} = 0.6$ kg/s,
- the composition of 60% humid air at $P = 1.013$ bar and $T = 25$ °C is

  N_2 77.29 %,
  O_2 20.75 %,
  Ar 0.92 %,
  CO_2 0.03 %,
  H_2O 1.01 %, and

- the composition of the flue gas is

  N_2 74.83 %,
  O_2 13.23 %,
  Ar 0.88 %,
  CO_2 3.46 %,
  H_2O 7.60 %,

  the composition of the gas resulting from the combustion of fuel can be calculated by assuming that all chemical reactions reach equilibrium. You will be able to do it after studying Chapter 10.

- $\eta_{\text{comp},s} = 0.88$,
- $\eta_{\text{turb},s} = 0.91$,
- $\eta_{\text{gen}} = 0.98$.

With these data, the results of the Brayton cycle calculation are:

- net mechanical power output $w_{\text{net}} = 14.1$ MW,
- mechanical power transferred to the compressor $w_{\text{comp}} = 16.2$ MW,
- mechanical power obtained from the turbine $w_{\text{turb}} = 30.5$ MW,
- net energy conversion efficiency $\eta = 37.7$ %,
- temperature of the exhaust gases $T_4 = 830$ K.

In this calculation we have not taken into account that compressed air must be spilled from the compressor in order to pass it through the cooling passages of the turbine blades which are exposed to a very high temperature. This need can be appreciated by noting that the temperature of the gas within the first stage of large gas turbines is higher than the melting temperature of the metal of all parts in contact with the gas. The air taken from the compressor is used to generate a cold film around the metal surfaces to keep them

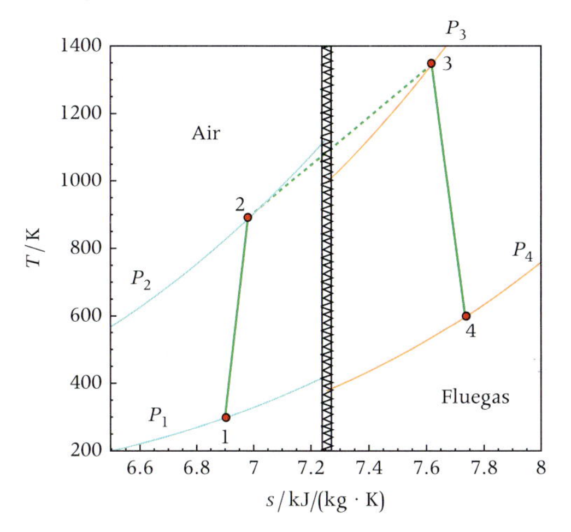

(a) The processes of a thermodynamic cycle of a gas turbine in the T–s plane.

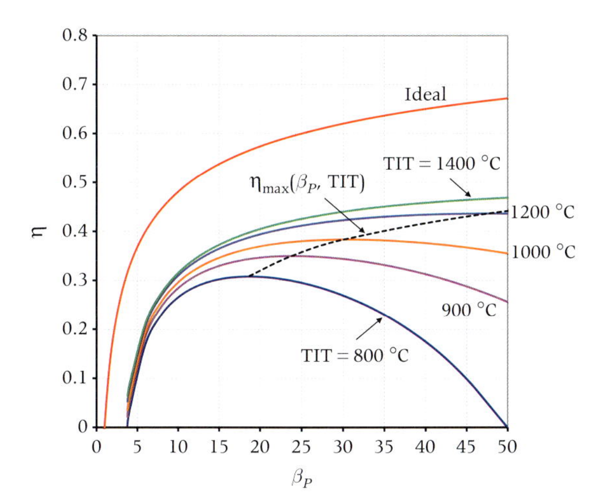

(b) The energy-conversion efficiency of a gas turbine cycle as a function of the pressure ratio (β_P) and of the turbine inlet temperature (TIT).

Figure 7.26 The Brayton thermodynamic cycle.

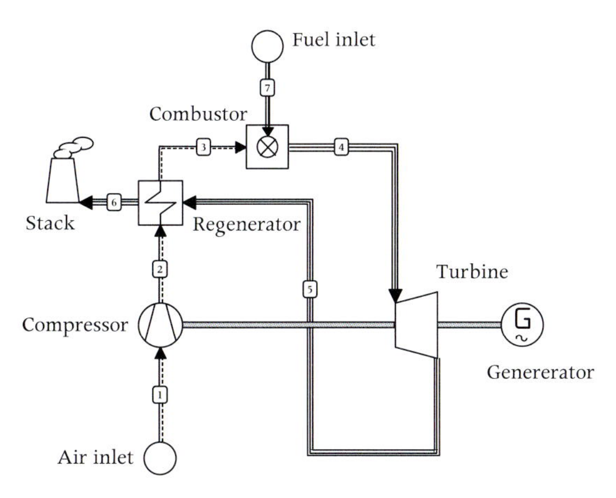

(a) The process flow diagram of are generative gas turbine power plant.

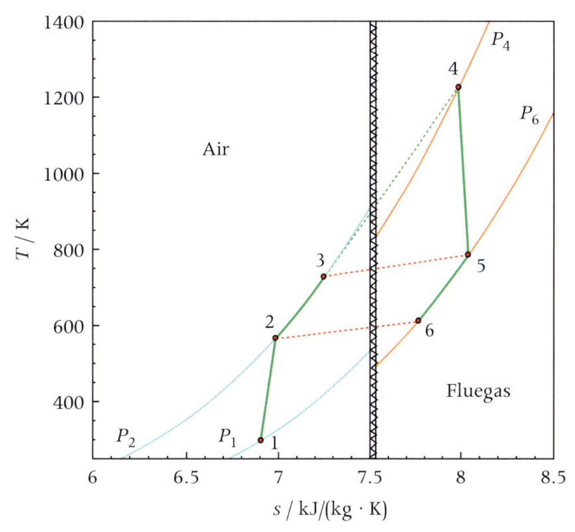

(b) The processes of a regenerative Brayton cycle in the T–s plane.

Figure 7.27 The regenerative gas turbine.

at a sustainable temperature. Blade cooling markedly decreases the efficiency we have calculated, and is neglected for simplicity.

The Brayton thermodynamic cycle in the T–s plane is shown in Figure 7.26a. If we use the simple EXCEL or MATLAB computer program you have just developed, we can calculate the net efficiency of the cycle as a function of the pressure ratio for several TITs. We thus obtain the chart of Figure 7.26b. We first notice that the curves related to "real" Brayton cycles are quite different from the curve of the ideal Brayton cycle, therefore the ideal cycle is not a good benchmark for a real gas turbine. The optimal pressure ratio in terms of cycle efficiency increases with increasing TIT, and this explains why the race to obtain ever more better performing gas turbines is focused on increasing both TIT and β_P at the same time.

Regeneration

The major source of thermodynamic losses for the simple Brayton cycle configuration is the energy transfer as heat, both to and from the working fluid. The energy input process occurs largely at low temperature, while the energy output process

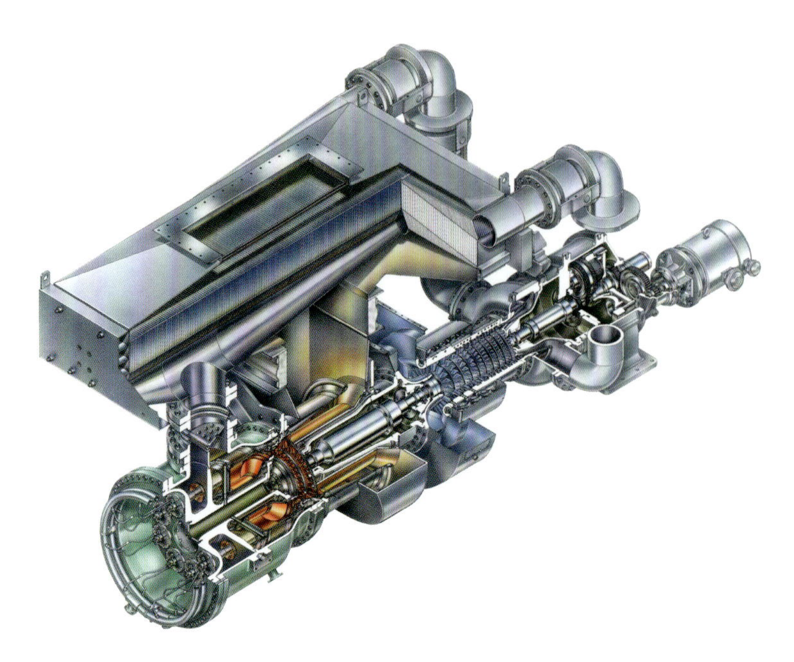

Figure 7.28 A 4.6 MW_e regenerative gas turbine. The regenerative Brayton cycle configuration makes a thermal efficiency as high as 38.5% in standard conditions possible for this relatively low power output. Courtesy of Solar Turbines Inc.

conversely occurs at comparatively high temperature. The previous example shows that the turbine exhaust temperature of a simple-cycle gas turbine is quite high. If it exceeds the compressor outlet temperature, regeneration becomes possible, as shown in Figure 7.27. In the regeneration process the high-temperature exhaust gas is used to preheat the air coming from the compressor, thus reducing the required energy input. With a counterflow regenerator it is possible to increase the temperature of the air at the compressor outlet T_2 up to almost T_5 and vice versa cool down the exhaust gases almost to T_2. Systems have been built in which up to 95% of the maximum regenerative effect ($T_3 = T_5$) has been obtained.

For a given TIT, the amount of regeneration is largest at low pressure ratio and, differently from the simple Brayton cycle, the regenerative cycle configuration features the highest efficiency for modest pressure ratio. The regenerative cycle is currently attractive for low-capacity units, for which the cost of high-efficiency and high-pressure ratio compressors is not affordable. A recuperator, the heat exchanger realizing the regeneration process, in turn introduces a series of technological problems, due to the large temperature and pressure differences between the heating and cooling flows, and its thermal inertia, which can hamper quick startup/shutdown and rapid power modulation.

By solving the technological problems of heat exchangers for high temperature gaseous streams, a very high efficiency can be obtained, even for gas turbine systems of relatively small capacity (Figure 7.28). Recuperated gas turbines usually hold the record for efficiency in their power class. Interesting developments are ongoing regarding very small recuperated gas turbine systems for cogeneration. Such appliances are suitable for providing electricity and heating at the same time to family dwellings (so-called domestic cogeneration).

Intercooling

In the simple Brayton cycle arrangement, a large portion of the power produced by the turbine must be used to power the compressor. It makes sense therefore to try to reduce the power demand of the compressor by cooling the compressed air at one or more intermediate stages (Figure 7.30a). This can be accomplished in a heat exchanger,

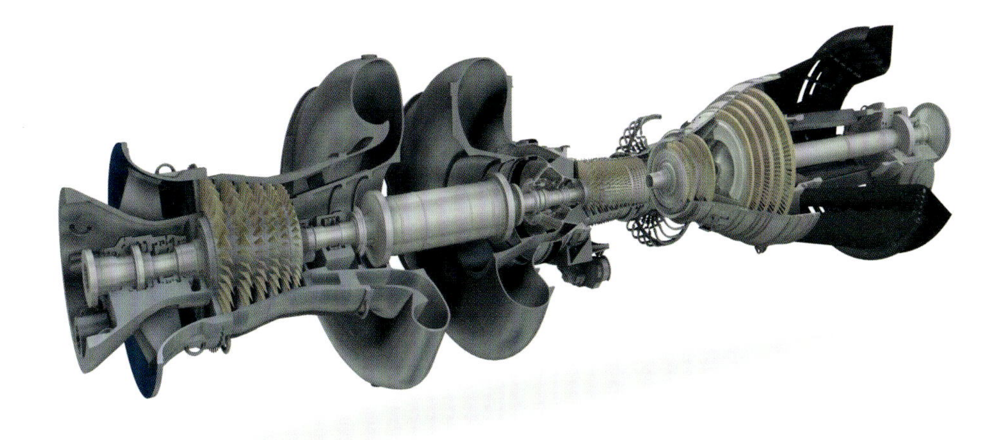

(a) Cutout of the intercooled gas turbine.

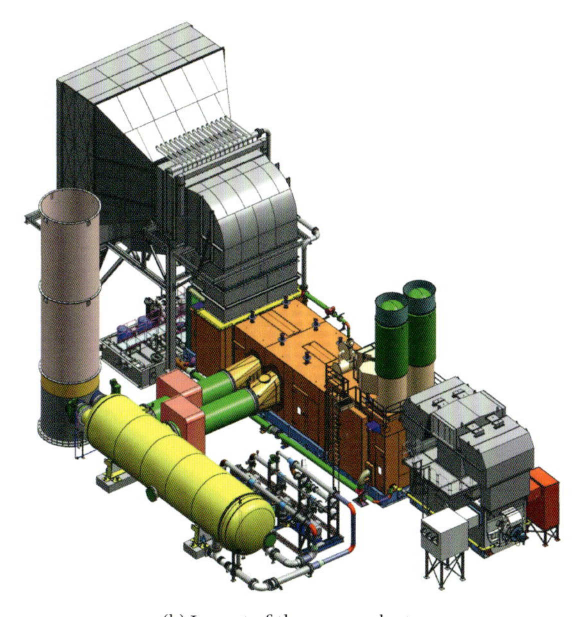

(b) Layout of the power plant..

Figure 7.29 A 100 MW$_e$ gas turbine power plant with intercooling, capable of achieving up to 46 % thermal efficiency. Courtesy of General Electric.

using water or air as coolant. The net mechanical work output increases, but also the energy input as heat increases, as the compressor outlet temperature decreases. Intercooling requires complex and bulky heat exchangers; therefore intercooling is applied in large gas turbines, for which high efficiency is a mandatory requirement. An advantage of intercooling is that air at comparatively low temperature is available for the cooling of the blades of the high-pressure stages of the turbine, thus allowing for a very high TIT. High efficiency can be obtained by means of a cycle analysis study aimed at optimizing cycle parameters. A current example of the implementation of the intercooling concept exists and it is a gas turbine system (100 MW$_e$) for stationary power production, which achieves approximately

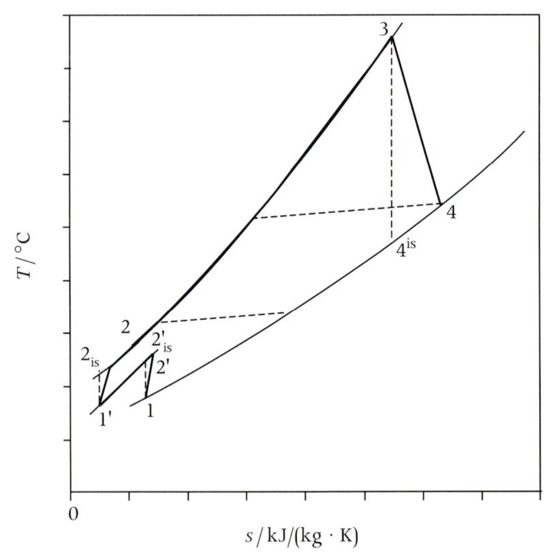

(a) The processes of the intercooled Brayton cyclein the T–s plane.

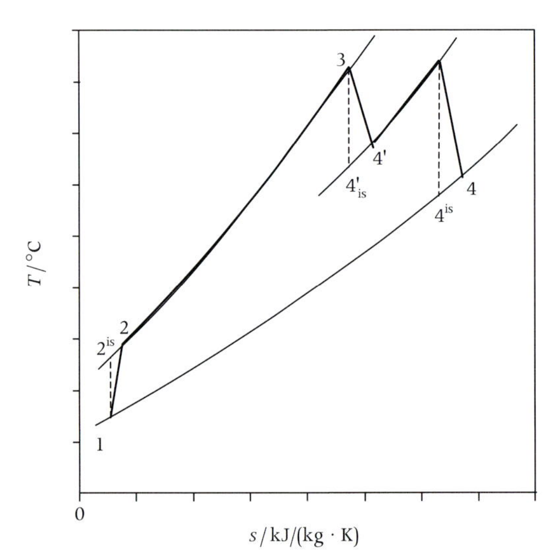

(b) The processes of the Brayton cycle with reheat in the T–s plane.

Figure 7.30 Modifications to the Brayton cycle.

46 % thermal efficiency (Figure 7.29) in nominal conditions.

Reheating

Analogously to intercooling, reheating between turbine stages (Figure 7.30b) also increases work output per unit of mass for a given pressure ratio and turbine inlet temperature. If compared to intercooling, reheating is less effective with respect to the net conversion efficiency: fluid dynamic losses are proportional to work output, therefore adding a process that produces more work, as in a Brayton cycle with reheating, increases the losses, as opposed to intercooling. Reheating is often utilized to increase the specific power output at the expense of fuel economy. A typical example of the implementation of the reheating concept is the afterburner of military aero-engines, which increases the thrust-to-weight ratio for short periods of time at the expense of a lower fuel efficiency. Gas turbines for power generation implementing the reheating concept are commercially available (Figure 7.31) and employed mostly in combined cycle power plants, Section 7.9. The high temperature at the gas turbine outlet positively affects the overall conversion efficiency as that thermal energy can be efficiently recovered by the integrated steam power plant, see Section 7.9.

Ultra-efficient gas turbine

The maximum benefit in terms of efficiency can be obtained by combining all the cycle variants so far described. The resulting cycle is the intercooled, reheated, and regenerated Brayton cycle. Looking at the cycle diagrams of Figure 7.30, we notice that intercooling and reheating reduce the energy-conversion efficiency if compared to the simple cycle, for more energy must be transferred as heat. Thinking graphically, the portions of the cycle that are "patched on" to the simple Brayton cycle are cycles with lower pressure ratio, and therefore less efficient. On the other hand, intercooling and reheating allow *increasing the potential for regeneration* by increasing the turbine outlet temperature and at the same time reducing the compressor outlet temperature. In theory, if an infinite number of reheat and intercooling stages is employed, together with regeneration, all the energy input as heat to the cycle occurs in the reheaters, therefore at the maximum cycle temperature, and all the energy rejection as heat takes place in the intercooler, where the working fluid is at its lowest temperature. This limiting case is the so-called Ericsson cycle, having the same efficiency as the Carnot cycle (Figure 7.32).

Everything comes at a price, therefore a gas turbine adopting intercooling, reheating, and regeneration at the same time is very complex and costly and no such turbine is currently in commercial operation,

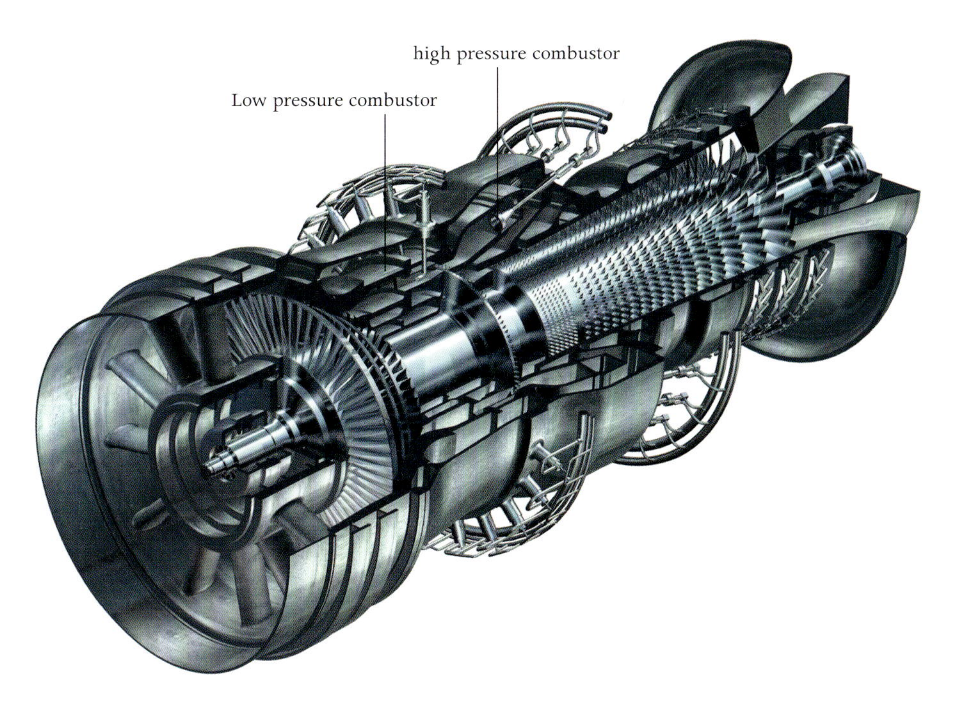

Figure 7.31 Cutout of a heavy-duty 345 MW_e gas turbine implementing reheating (Alstom GT 26). This model is mostly used for the topping cycle in combined cycle power plants. Courtesy of General Electric.

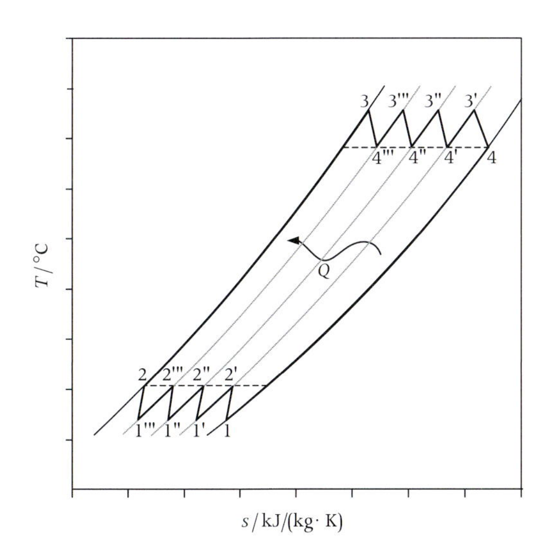

Figure 7.32 The Ericsson cycle in the *T*–*s* plane. The cycle is the limiting case of a regenerative Brayton cycle with a large number of intercooling and reheating stages and is characterized by the same efficiency as the Carnot cycle.

while the highest efficiency is obtained either with a simple cycle and very high TIT and pressure ratio, or by adopting only one or two of the Brayton cycle modifications (regeneration, intercooling, or intercooling plus regeneration).

Closed Brayton cycle gas turbine

The conversion into mechanical work of sustainable energy sources like concentrated solar radiation or biomass, or of thermal energy from a nuclear reactor, if done by means of a thermodynamic machine, leads to demands for configurations whereby the working fluid is kept separate from the energy source. An externally-heated gas turbine based on the closed Brayton cycle is a possible solution for the conversion of these sources. In this case, the limitation of available materials for heat exchangers constrains the maximum cycle temperature to values much lower than those of open-cycle gas turbines. Conversely, the use of a working fluid other than air/flue gas allows for a comparable efficiency to be achieved at lower TIT. As we have seen, the adoption of a low-molecular weight fluid like helium is beneficial in terms of efficiency, and indeed several helium closed-Brayton gas turbine power plants have been

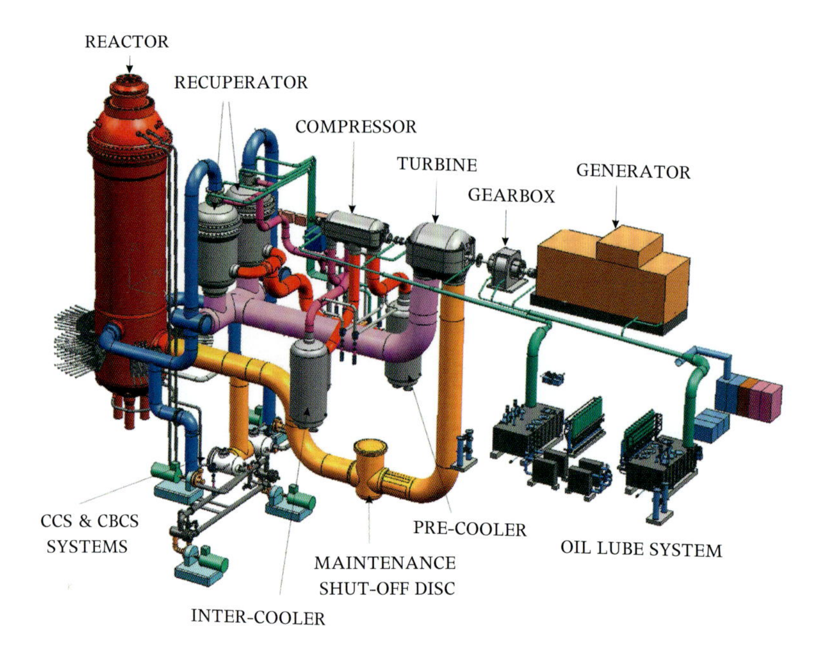

Figure 7.33 A new generation of nuclear reactors, the so-called pebble-bed reactor, can utilize a closed Brayton cycle gas turbine for power conversion. High conversion efficiency can be obtained by employing helium as the working fluid, together with regeneration. Courtesy of Pebble Bed Modular Reactor – PBMR.

developed throughout the years. A new generation of nuclear power plants based on the so-called pebble-bed reactor has been under development in South Africa, China and the United States. The conversion of heat from the nuclear fission into mechanical work is done by means of a gas turbine system adopting a closed regenerative Brayton cycle configuration, and helium as the working fluid (Figure 7.33).

Another promising concept that is currently being studied and actively developed is based on a closed Brayton cycle and supercritical CO_2 as the working fluid. The primary advantage is that the compression process occurs close to the critical point of carbon dioxide, thereby the compression work is much smaller if compared to traditional air-breathing or helium gas turbines. In addition, given the very high pressure of the working fluid, the power density of the system is very high. The processes of the supercritical CO_2 Brayton cycle are shown in Figure 7.34a, while the process flow diagram of the power plant is in Figure 7.34b. The value of the maximum cycle temperature is currently limited to $\approx$ 700 °C by the need for primary heat exchanger materials capable of withstanding the very high pressure and temperature. The closed Brayton cycle CO_2 turbine is suitable for both nuclear and concentrated solar power generation.

7.7 Other Gas Power Cycles and Engines

Besides the Brayton cycle, several other gas power cycles exist, whereby mechanical power is obtained by an alternating motion, instead of rotation. Similarly to the gas turbine, useful work is obtained by making the specific volume of the working fluid during the expansion process much larger than that characterizing the compression process. The Otto and the Diesel open gas cycles are at the basis of reciprocating or internal combustion engines used massively for automotive propulsion, but also for naval propulsion, stationary power and many other applications (Figure 7.35).

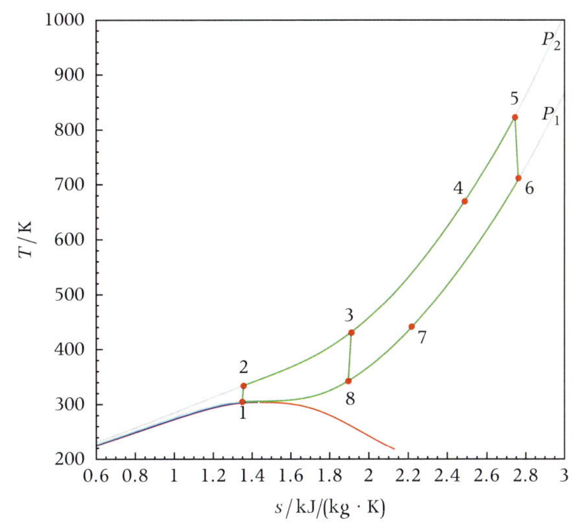

(a) Thermodynamic cycle of a supercritical CO_2 gas turbine in the T–s plane.

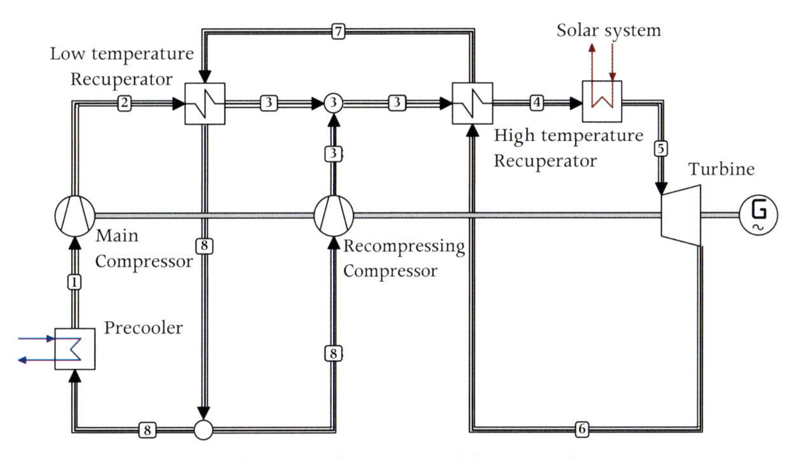

(b) Process flow diagram of the power plant.

(c) The 3D-printed model of the rotor of a10MW$_e$ supercritical CO_2 turbine under development. Courtesy of General Electric.

Figure 7.34 The closed Brayton cycle gas turbine, utilizing supercritical carbon dioxide as the working fluid.

Figure 7.35 Example of a current long-haul truck diesel engine. This is a 10.8 liter engine, with six cylinders in line, turbocharged with intercooling. Its maximum power is 320 kW, its compression ratio is 17.5:1. It complies with Euro 6 regulations, thus it features an Exhaust Gas Recirculation (EGR) system, a Diesel Particular Filter (DPF) and Selective Catalytic Reduction (SCR) after-treatment. The thermal efficiency of modern heavy-duty truck diesel engines in nominal conditions is usually in excess of 45%. Courtesy of DAF Trucks.

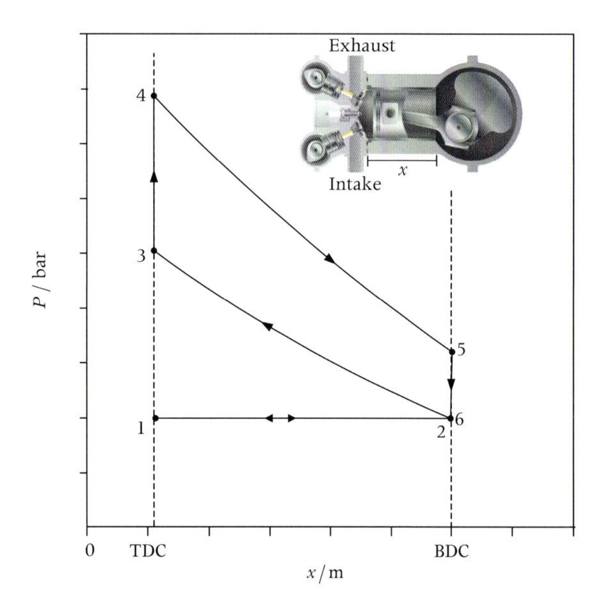

Figure 7.36 Evolution of pressure within the cylinder as a function of piston position in an idealized spark-ignition engine.

Otto cycle

This is the model cycle for spark-ignition engines. Figure 7.36 shows the pressure of the gas within the cylinder of an idealized spark-ignition engine as a function of the piston position. With the piston at top dead center (TDC), the intake valve opens and a fresh charge of fuel-air mixture is sucked in (1–2). At bottom dead center (BDC) the intake valve closes, and the return stroke causes the gas to be compressed (2–3). In the idealized system, ignition occurs instantaneously at TDC, causing a rapid rise in temperature and pressure (3–4). The gas is then expanded on the outstroke (4–5), until at BDC the exhaust valve opens, and the gas "blows down" through the exhaust port (5–6). With a fourth stroke the gases are purged out. In the idealized Otto cycle the compression and expansion processes are considered reversible and adiabatic, that is, isentropic, and it is assumed that the pressure within the cylinder during the intake and exhaust strokes is equal to the atmospheric pressure. The work done by the piston on the gas inside the cylinder during the exhaust stroke is exactly equal to the work done on the piston by the gas during the intake stroke, so that useful work output results only from the excess of the work done by the gas during the expansion stroke over that on the gas during the compression stroke.

The process representation for the fluid during the compression, ignition, and expansion parts of the cycles is shown in Figure 7.37a. The combustion process is idealized in terms of simple energy addition as heat, and the changes in the chemical composition of the mixtures are neglected. Similarly to what we have seen for the Brayton cycle, by assuming that the working fluid is a polytropic perfect gas, the usual thermodynamic analysis (energy balances over the processes) leads to a simple algebraic expression for the efficiency of the ideal Otto cycle in terms of the compression ratio, namely

$$\eta = \frac{w_{\rm net}}{q_{\rm in}} = 1 - \beta_v^{(1-\gamma)}, \qquad (7.2)$$

where the volumetric compression ratio β_v is

$$\beta_v = \frac{v_2}{v_3} = \frac{v_5}{v_4}.$$

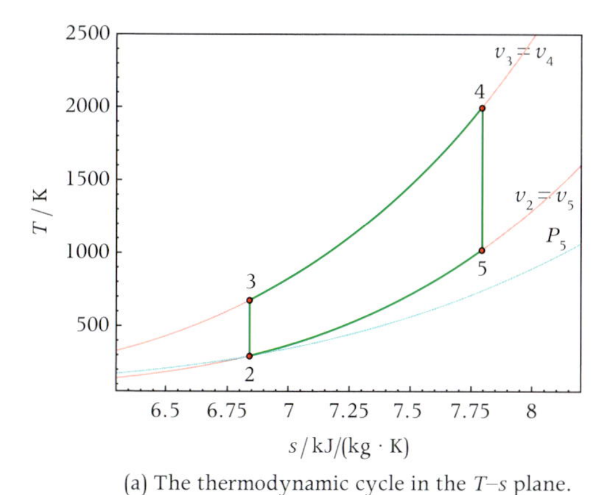

(a) The thermodynamic cycle in the T–s plane.

(b) The cycle efficiency as a function of the compression ratio ρ and of the perfect gass pecific heat ratio γ of the working fluid.

Figure 7.37 The ideal Otto cycle.

This relationship is shown in Figure 7.37b for various values of γ. Fluid dynamic losses in this case have little influence on the net conversion efficiency, as opposed to what happens in the Brayton cycle, because they are proportional to the kinetic energy of the fluid, which is orders of magnitude smaller than in the compressor and the expander of a gas turbine. For this reason the efficiency steadily increases with the compression ratio, while the efficiency of the Brayton cycle displays a maximum value (Figure 7.26b). The efficiency of the ideal Otto cycle is therefore a good indicator for comparison with the performance of the real cycle. Another important observation is that a working fluid with a higher γ yields higher efficiencies. The working fluid is a mixture of air and gasoline first and flue gases thereafter. A mixture with a larger amount of air has a higher value of γ, therefore currently large efforts are made to keep the air/fuel ratio of the charge to the lowest possible value in order to maximize fuel economy.

The cycle of a spark-ignition engine is affected by a number of non-idealities, which make the real thermodynamic cycle radically different from the ideal cycle. Combustion takes time, and for this reason it is initiated before the piston reaches TDC during the compression stroke; the piston must do work on the air to get out, and this is more than the work done on the piston by the cylinder gases during the intake stroke. Heat transfer is involved, so the compression and expansion processes are not isentropic. The pressure-displacement diagram of a realistic spark-ignition engine is shown in Figure 7.38.

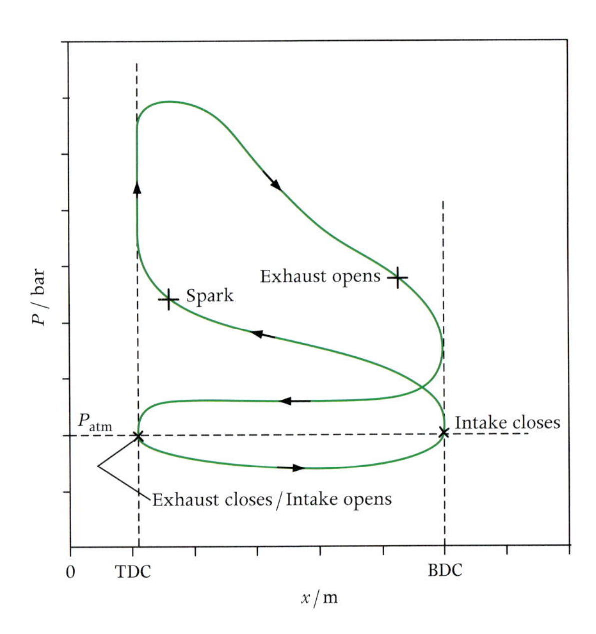

Figure 7.38 Evolution of pressure within the cylinder as a function of the position of the piston in a spark-ignition engine.

The peak temperature in such an engine is of the order of 1500 to 2000 K, and peak pressures are in the range 4–5 MPa and are lower than in the ideal cycle. The pressure at the engine exhaust must be slightly super-atmospheric, and the temperature is about 1000 K (not at the tailpipe).

The curve displaying the efficiency versus compression ratio clearly shows that one would like to operate at high compression ratio ρ. However, in the spark-ignition engine this is limited by the problem of *detonation*: at high pressure the fuel begins to burn beyond the area ignited by the spark. This causes shock waves in the combustion chamber ("knocking"), which may cause severe damage if allowed to continue over a period of time.

Nowadays, modern engines like those used on board of hybrid vehicles operate according to a modification of the Otto cycle, named the Atkinson cycle. The intake valve stays open longer than in the Otto cycle, such that the flow of air in the manifold gets reversed, because the piston expels some of the fresh fuel mixture back out of the intake port. This reduces the compression ratio. However, the expansion ratio remains the same, as the pressure inside the cylinder at the end of the expansion is the same as the atmospheric pressure. The net result is that the Akinson cycle allows reaching higher efficiencies than those of Otto engines. The trade-off is that the power density of an Atkinson engine is lower than that of an equivalent Otto engine.

Diesel cycle

One method of avoiding detonation during compression, with its limitation on the compression ratio, is to inject the fuel *after* the compression process. Since there is no fuel in the cylinder during compression, higher compression ratios (and therefore higher cycle temperatures) can be reached without detonation. Upon injection, the fuel ignites spontaneously owing to the high air temperature.

The Diesel cycle is the model cycle for reciprocating auto-ignition engines. In the ideal thermodynamic cycle, air is compressed to the TDC, at which time fuel is injected, and the combustion process is isobaric for part of the expansion stroke. The rest of the expansion stroke and the compression stroke are assumed isentropic. Figure 7.39a shows the evolution of pressure with the position of the piston, while the ideal Diesel cycle in the T–s plane is shown in Figure 7.39b.

Thermodynamic analysis of the ideal cycle allows us to write the energy conversion efficiency in terms of compression ratio and cutoff ratio ρ_{cutoff} as

$$\eta = 1 - \frac{1}{\rho^{(\gamma-1)}} \cdot \frac{(\rho_{\mathrm{cutoff}}^{\gamma} - 1)}{\gamma(\rho_{\mathrm{cutoff}} - 1)}, \tag{7.3}$$

where $\rho \equiv \dfrac{v_2}{v_3}$ is the compression ratio and $\rho_{\mathrm{cutoff}} \equiv \dfrac{v_4}{v_3}$.

Figure 7.40 shows the efficiency of the ideal Diesel cycle as a function of the compression and cutoff ratios. Note that the efficiency for $\rho_{\mathrm{cutoff}} = 1$ is the same as for the Otto cycle. Figure 7.41 shows a comparison of the ideal Brayton, Otto and Diesel cycles. For the same amount of energy input as heat, the Diesel cycle produces less net work. For this reason the efficiency of the Diesel cycle is lower than that of the Otto and Brayton cycle *operating at the same*

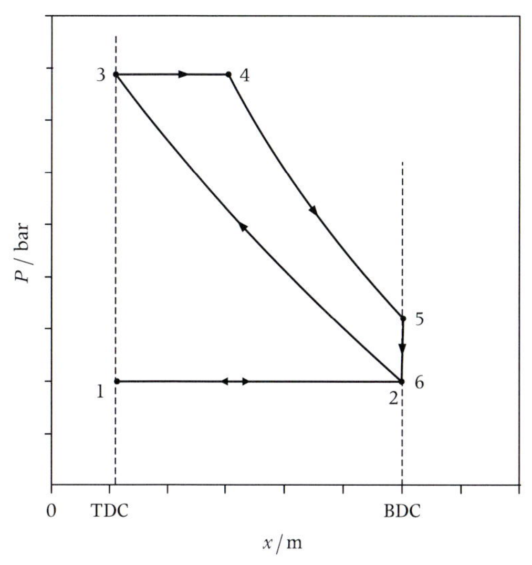

(a) Evolution of pressure with in the cylinder as a function of piston position in the ideal auto-ignition engine.

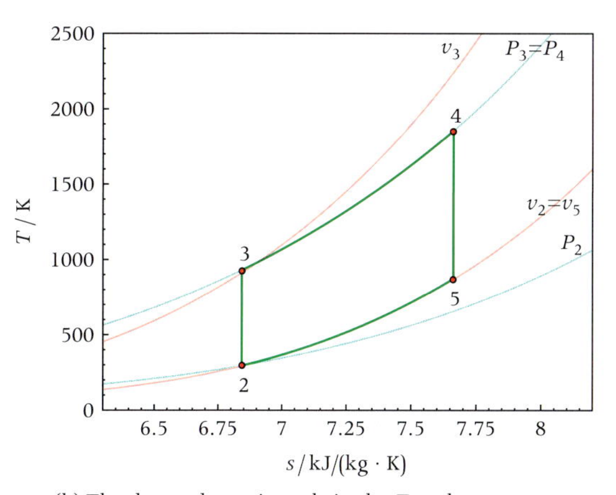

(b) The thermodynamic cycle in the T–s plane.

Figure 7.39 The ideal Diesel cycle.

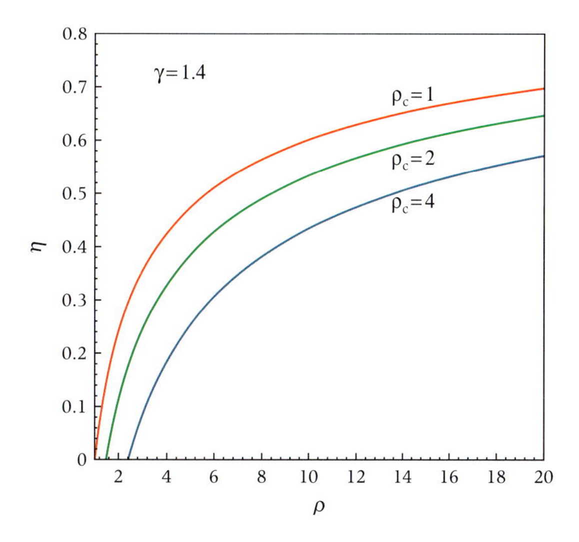

Figure 7.40 The ideal Diesel cycle efficiency as a function of the compression ratio, the cutoff ratio and the perfect gas specific heat ratio of the working fluid.

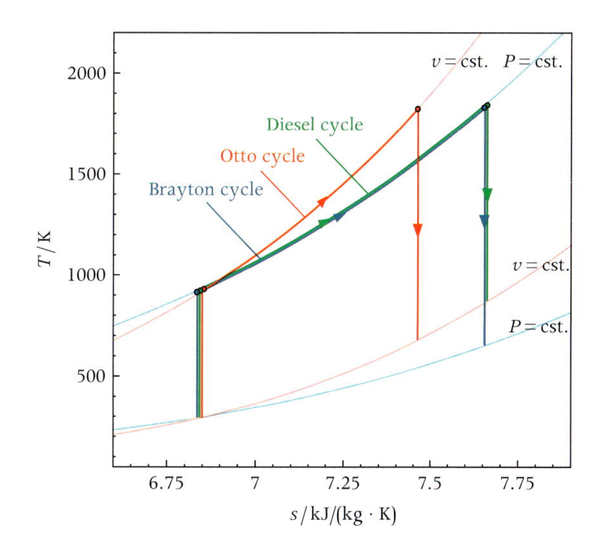

Figure 7.41 The ideal Brayton, Otto, and Diesel cycles in the T–s plane.

pressure ratio and between the same minimum and maximum temperatures. Compression ratios of practical engines are limited by the pressures and temperatures which can be tolerated within the cylinders. Compression ratios in Diesel engines are much higher than in gas turbines of the same capacity, which is one of the reasons for the success of reciprocating engines. The discontinuous nature of the process allows for easier cooling and therefore for fewer problems with the strength of materials at high pressure. In the spark-ignition engine, combustion occurs as the gas is being compressed, while in the auto-ignition engine combustion begins as the gas is being expanded. As a result, Diesel engines operate at much higher compression ratios (up to 25:1) than spark-ignition engines (up to almost 15:1), and consequently can achieve a higher efficiency.

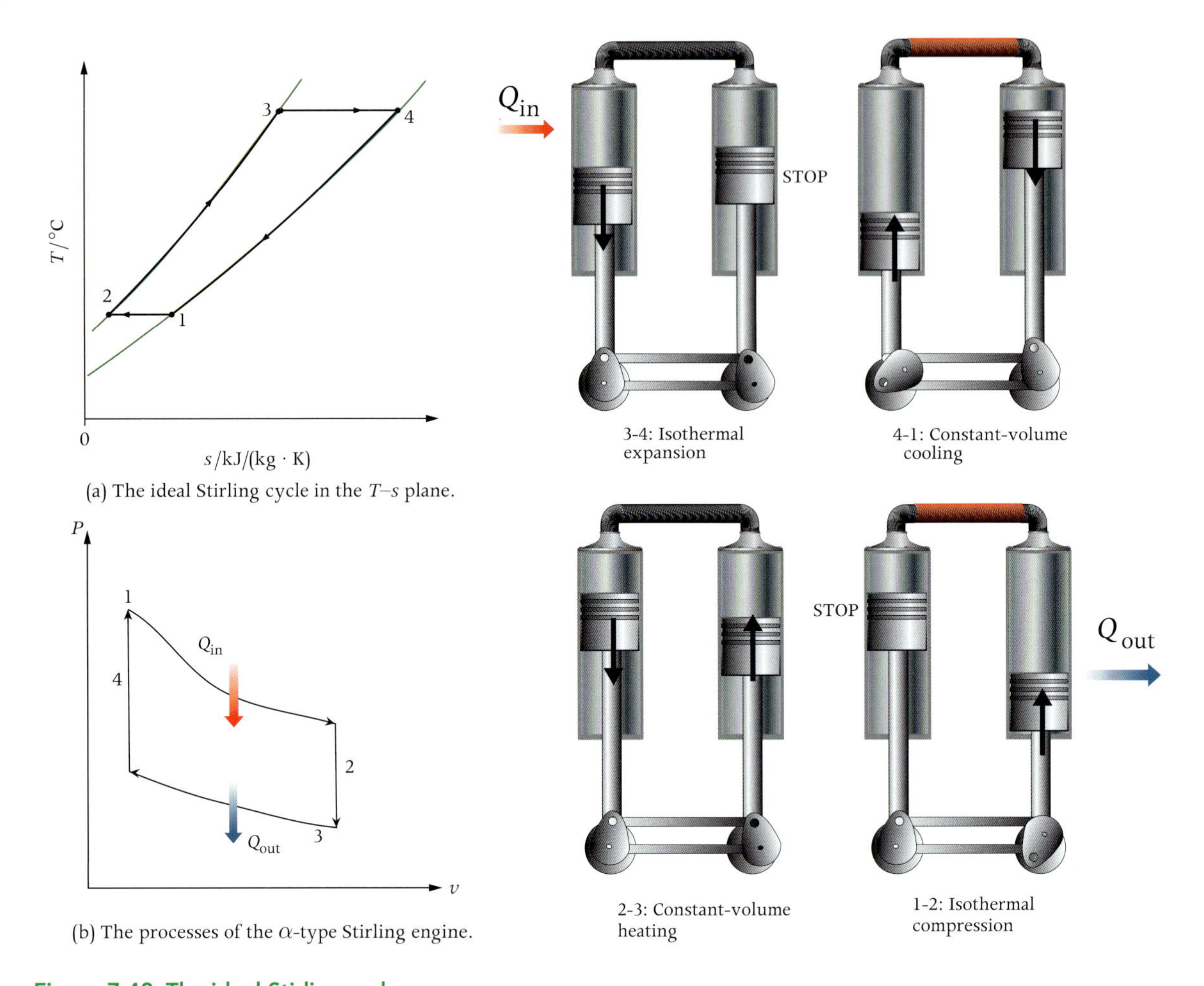

Figure 7.42 The ideal Stirling cycle.

Stirling cycle

A reciprocating engine suitable for sustainable energy sources is the Stirling engine. The ideal Stirling cycle is a closed cycle, therefore suitable for the conversion of external energy sources. Figure 7.42a shows the ideal cycle in the T–s plane, while the strokes of the so-called α-type Stirling engine are depicted in Figure 7.42b. The ideal cycle consists of an isothermal compression (1–2), an isochoric heating (2–3), an isothermal expansion (3–4), and an isochoric cooling (4–1). Note that regeneration is possible and ideally the energy rejected as heat in process 4–1 can be transferred to process 2–3. Such an idealized system would have an efficiency equal to the Carnot efficiency for the two temperature extremes.

Stirling engines can be realized with several different and fairly complex mechanisms. The working fluid can be air, helium, or some other light gas in a closed cycle. Another option would be to utilize an organic fluid which allows for the compression to take place in the thermodynamic region close to the critical point, thereby requiring far less compression work, much as the closed CO_2 Brayton cycle turbine. Such an organic Stirling engine would however be limited by the thermal decomposition temperature of the working fluid, which is far lower than that of conventional working fluids, and by the maximum pressure, which would be much higher than that of light gases.

Stirling engines have a long history, but the complexity of the mechanisms and of the required moving regenerator have so far hampered their widespread utilization. They are suitable for small capacity power generation and they are being developed, for example, for concentrated solar power (Figure 7.43) and biomass combustion applications. As is the case for many power cycles, the Stirling cycle can also be reverted, thus becoming an appropriate thermodynamic cycle for refrigeration or heat pump systems.

As a final remark regarding gas power cycles, beware that all these model cycles are highly idealized. Idealization certainly helps the understanding of the operating principle, and can be used to study real machines, but a good engineer should never lose sight of the hypotheses and assumptions that have been made. For example, the thermal energy input process of a spark-ignition engine is not exactly isochoric, and the combustion in the Diesel engine is not isobaric. The same holds for the processes occurring in a real Stirling engine. The working fluid of the ideal Otto and Diesel cycles is air, while in the analysis of the thermodynamic cycle of real systems, combustion plays a very important role, and the composition of the working fluid changes (see Chapter 10).

7.8 Fuel Cells

The conversion efficiency of chemical energy into mechanical work can reach a very high level if it occurs directly, without the need of a thermodynamic cycle. The fuel cell is a device in which a chemical reaction is harnessed directly to obtain electric power. Unlike a chemical battery, in which an electrolyte is decomposed into its basic components, a fuel cell utilizes a controlled reaction between a fuel and an oxidizer. Since the feeding of the fuel and of the oxidizer can be continuous, the fuel cell does not run down as a chemical battery does. As an example, schematics of typical fuel cells are shown in Figure 7.44. A gaseous fuel (usually hydrogen) and an oxidizer (usually oxygen) enter the cell and are brought into contact with porous electrodes. The electrodes are separated by an electrolyte (a substance with special properties, either liquid or solid, that can ionize), which serves to limit the reaction rate. The fuel diffuses through the porous anode, is absorbed on the surface, and reacts with ions in the electrolyte, yielding free electrons. The oxidant diffuses through the cathode, is absorbed by the surface, and reacts to form ions. The product of the fuel reaction (i.e., water in the case of hydrogen as fuel) is continually being formed at the anode and the oxidant continually decomposes at the cathode. Electrons which are formed at the cathode as a result of the electrochemical reactions provide the current to connected electrical loads. Several combinations of fuels/oxidizers/electrolytes are possible and, depending on that, fuel cells must be operated at different temperatures. Table 7.2 contains an overview of different types of fuel cells and their operating temperatures.

The fuel cell does not convert energy as heat into electricity (in fact the electrochemical process is almost isothermal), and consequently the Carnot efficiency is irrelevant with respect to the evaluation of its performance. Given a supply of oxygen and hydrogen, more useful work can be obtained with a fuel cell than if the gases were allowed to react spontaneously and a Carnot cycle were run from between the flame temperature and the environment. The maximum theoretical efficiency of a fuel cell, based on the conversion of chemical to electric energy, is 100 %. High temperature fuel cells reach electrical efficiencies of the order of 55–60 %.

Fuel cells have been initially developed for space applications, but their development is now forcefully pursued for both stationary and automotive applications, because of the high conversion efficiency that can be achieved. The cost of materials and of the total system, together with the production cost are the major hampering issues. Fuel cells are easily scalable, so applications range from power supply of laptop computers up to car/bus/truck engines and multi-megawatt stationary power plants (see Figure 7.45).

7.9 Combined, Cogenerating, and Binary Cycle Power Plants

The transfer of energy as heat at high temperature to the environment is such a thermodynamic squandering that it goes simply by the name of *waste heat*.

Table 7.2 Different types of fuel cells and their operating temperatures.

Electrolyte	Fuel cell type, abbreviation	Transported ions	Operating temperature / K
Solid oxide	SOFC	O^{2-}	1100–1300
Molten carbonate	MCFC	CO_3^{2-}	875–925
Phosphor acid	PAFC	H^+	470
Polymer[a]	PEFC	OH^-	350
Alkaline	AFC	O^{2-}	350

[a] Also known as proton exchange membrane (PEM), or solid polymer fuel cell (SPFC).

Figure 7.43 A solar Stirling power plant (Maricopa Solar in Peoria, AZ): the 25 kW_e SunCatcher engine is mounted in the focus of a parabolic mirror. Courtesy of Lars Jacobsson, United Sun Systems.

> The implications of wasting a mind-numbing amount of thermal energy all over the world go well beyond the effect it has on the efficiency of energy conversion systems. The authors strongly advocate regulation favoring technology for the recovery of wasted thermal energy. These technologies should be supported by considering them equivalent to those for the conversion of renewable energy sources.

Very often the key to achieving very high energy conversion efficiency is to integrate into one single energy plant systems driven by the energy discarded from the plant working at the highest temperature. Nowadays the power plants achieving the highest energy-conversion efficiency are combined-cycle power plants, most often operated on natural gas. One or more large and highly efficient gas turbines discharge their exhaust gas into a heat recovery steam generator (HRSG) powering a steam power plant (Figure 7.46). This way an energy conversion efficiency in excess of 60% has become possible. If the waste heat of the gas turbine can be efficiently recovered by a steam power plant, the optimization of the thermodynamic system may lead to choosing gas turbines with lower-than-maximum efficiency, therefore higher exhaust temperature, because this increases the power that is efficiently converted by the bottoming Rankine-cycle power plant. For this reason large gas turbines with reheat are a viable

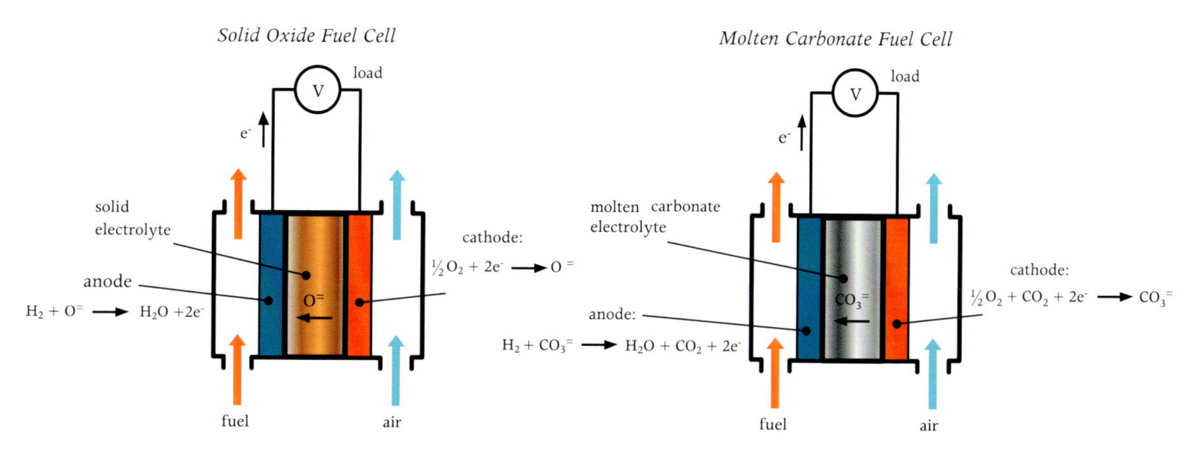

Figure 7.44 Schemes of typical current fuel cells.

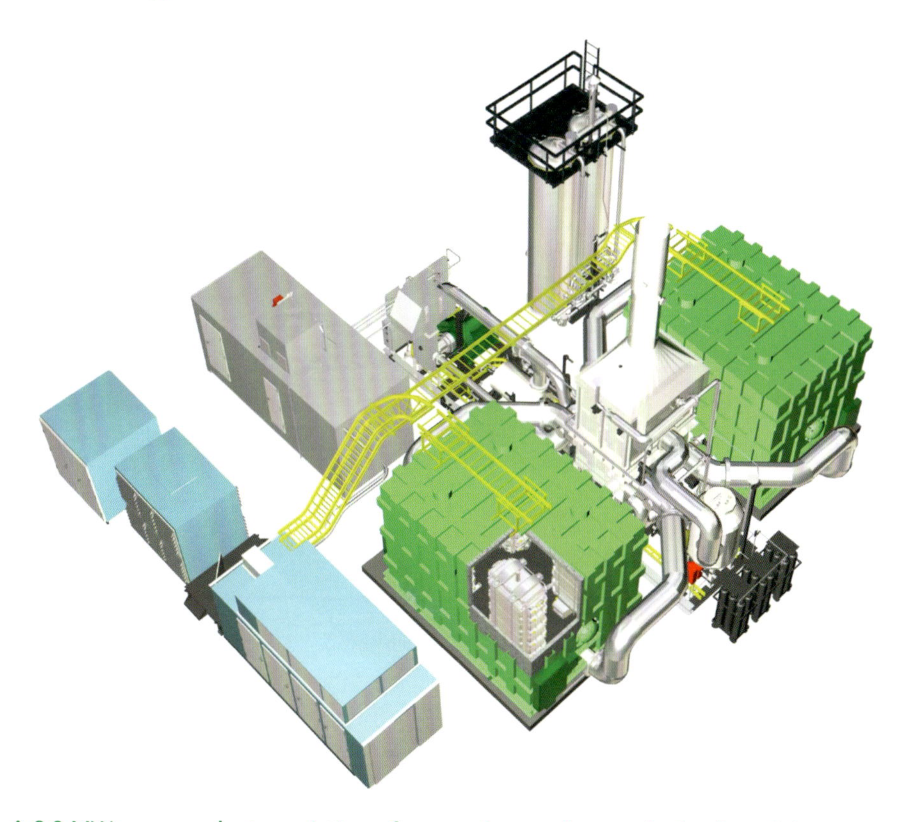

Figure 7.45 A 2.8 MW$_e$ power plant consisting of two molten carbonate fuel cell modules (green). One of four fuel cell stacks within each of the modules is visible in the cutaway. The incoming fuel is processed by the mechanical balance of plant (grey). The electrical output is processed by the electrical balance of plant (blue). Courtesy of FuelCell Energy.

option in combined-cycle power plants. Note also that concentrating solar radiation permits achieving very high temperatures, comparable to the turbine inlet temperatures of gas turbines, therefore solar combined-cycle technology with open or closed-cycle gas turbines is currently beging studied.

The heating of buildings or heating needed by industrial processes can often be accomplished by

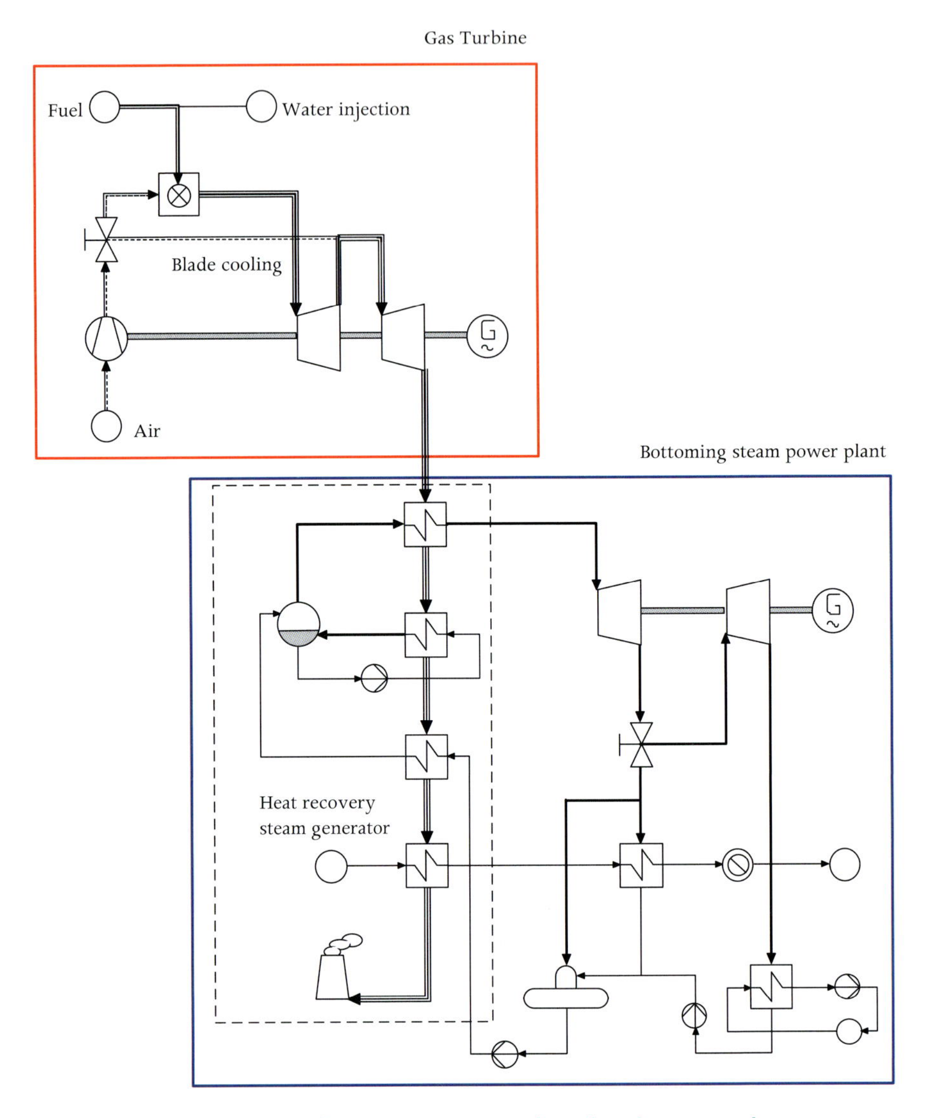

Figure 7.46 The process flow diagram of a cogenerative combined-cycle power plant.

utilizing the energy discarded by a power plant. Combined heat and power (CHP), also known as cogeneration, allows energy conversion efficiencies as high as 90 % and more. In this case the energy conversion efficiency is defined as the sum of the energy converted into a useful form (electricity and thermal energy utilized in buildings or production processes) divided by the energy content of the primary source. The energy discarded as heat from the power plant can also drive an absorption refrigeration plant, therefore plants converting a primary energy source into electricity, heating, and cooling are called *trigeneration* systems. The cost of the infrastructure to distribute warm water or steam used for heating purposes is often one of the issues to be solved. In many cold regions of the world district heating has been successfully realized, but the potential for cogeneration is far from being exhausted and the advent of distributed power generation would facilitate the adoption of this technology.

Another example of the utilization of energy discharged as heat from one energy system to another is the so-called binary-cycle power plant (Figure 7.47). The energy conversion efficiency can be enhanced by increasing the temperature at which energy is transferred as heat into the conversion system. Several

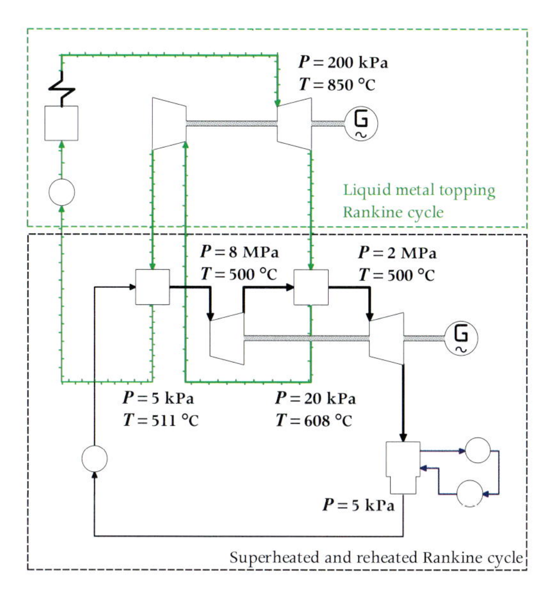

(a) The process flow diagram.

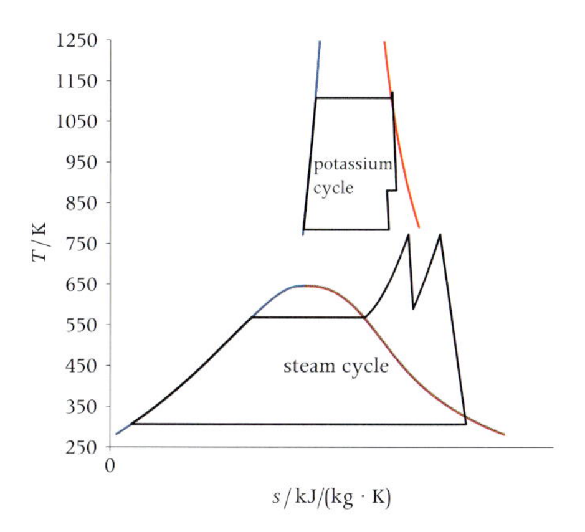

(b) The thermodynamic cycles in the T–s plane.

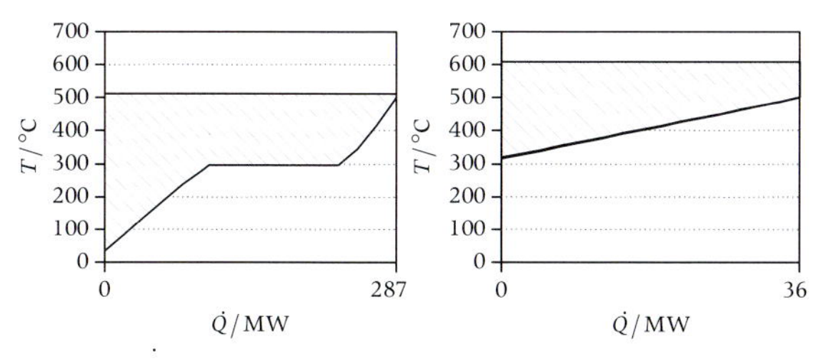

(c) The $Q - T$ diagrams of the heat recovery steam generators.

Figure 7.47 A topping Rankine power plant with an alkali metal (potassium) as working fluid and a conventional bottoming steam power plant with reheating.

alkali metals, like potassium, rubidium, and cesium, can be vaporized at temperatures in the 900–1300 °C range and condensed at 300–600 °C. Rankine cycle power plants using an alkali metal as working fluid have been studied for the generation of electricity in space, and several prototypes have been realized in the United States and Russia. The binary Rankine cycle configuration is a very attractive option for solar energy conversion, once the problems connected with sun radiation concentration and material at high temperature have been solved.

7.10 Thermodynamic Design, Sustainability, and Other Criteria

Energy is often in the news because the conversion of the energy content of fossil fuels into mechanical power and electricity is related to grave pollution problems and to global climate change. The study and development of new energy conversion technologies which do not permanently disrupt the Earth's equilibrium has therefore enormous societal relevance. The few examples that are treated in this chapter show that indeed several sustainable technologies are readily available, even if more can be done to reduce their cost and increase their efficiency (join the team and become an energy engineer!). The reduction of the humongous waste of energy resources affecting the most developed parts of the world has a prominent role in tackling the challenge of achieving sustainable development. Thermodynamics is at the basis of studies to implement the rational use of energy resources. The ever increasing waste has been generated by the very low economic value of energy, whose price has not included disguised costs, like pollution, its connection to health costs, negative effects on the environment and on international political equilibrium.

It is desirable and likely that *sustainability* will become one of the main design criteria for energy systems. A sustainable energy system must not permanently deplete Earth resources, therefore its life-cycle must also be addressed, and measures be taken to recycle its materials, once it reaches end-of-operation. Other common design criteria include efficiency, cost, weight, volume, aesthetics. Some of these criteria are addressed in the preliminary design phase of an energy system, which is usually performed by means of a steady-state thermodynamic analysis, much like those illustrated here, though often more detailed. If the requirements include a certain dynamic performance, the same principles are applied to obtain a time-dependent model of the system, which allows for its simulation under transient conditions (think of the study of the power-modulation of a car engine, or of the analysis of the response of a solar converter to the passage of clouds). Once the thermodynamic design is finalized, the components of the system are designed/sized or selected among available products, using the results of the thermodynamic analysis as specifications. The design process is usually completed by the design of the control system and the necessary instrumentation. The design of a system is almost always iterative, as the following phases usually highlight issues that can be solved by iterating on the design procedure.

The new requirements on sustainable energy conversion and rational use of energy resources might change the current paradigm, whereby: i) electricity is generated by large centralized power stations utilizing fossil fuels as primary sources; and ii) ground transportation is largely performed with private gasoline- or diesel-fueled vehicles. Given the localized and variable nature of sustainable energy resources like wind, sun radiation, geothermal reservoirs, and biomass, power stations could be more numerous, smaller and distributed. Sustainability could be achieved by locally exploiting and utilizing the most favorable renewable resource: wind where it is abundant and constant, solar radiation in sunny regions, etc. A wide variety of technologies are therefore necessary. Energy storage is in demand wherever solar radiation or wind are the primary source. Public transportation must become much more widespread, and electric vehicles will substitute the current generation of cars and trucks when electrical batteries have reached a suitable technological level at affordable cost. The batteries of the next generation will also affect the widespread application of solar and wind energy technology. Several studies indicate as a possibility the use of automotive batteries (vehicles are still most of the time) for storing the electricity produced by sustainable power plants.

EXERCISES

If applicable, unless explicitly stated, the following assumptions are made:

- kinetic and potential energy are negligible at the inlet and outlet of components;
- turbomachinery is adiabatic;
- the electrical generator is perfect ($\eta_{gen} = 1$);
- the system is at steady state;
- working fluids are at thermodynamic equilibrium; and
- there is no pressure drop along heat exchangers and ducts.

7.1 What is the value of the efficiency of the *adjacent* reversible cycle displayed here in the T–s diagram of its working fluid?

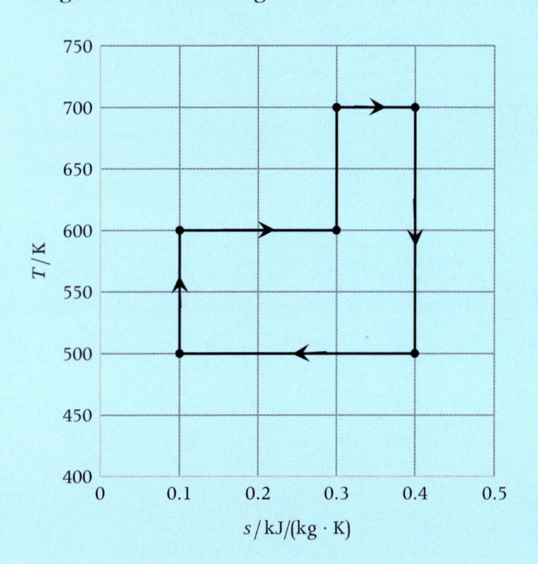

7.2 Technicians of a steam turbine manufacturing company have measured the following data for a simple 500 kW turbine:

- inlet state at $T = 673$ K and $P = 40$ bar, and
- outlet state at $T = 378$ K and $P = 1.0$ bar.

Draw the steam expansion process in the T–s diagram of water (you can use the chart in Appendix A.3 or make your own diagram with relevant isobars and state points using EXCEL or MATLAB) and determine the turbine efficiency.

7.3 Steam flows through a condenser at 2×10^6 kg/hr, entering as saturated vapor at 40 °C, and leaving at the same pressure as subcooled liquid at 30 °C. Cooling water enters the condenser pipes at 18 °C and 1.5 bar. Assume no pressure loss occurs in the pipes. Environmental regulations require that the exit temperature is at maximum 20 °C. Determine the minimum cooling water flow rate.

7.4 The booster pump of a large steam power plant raises the pressure from 3×10^6 Pa at 313 K to 6×10^6 Pa. If the flow rate is 20 000 kg/ min, and the isentropic efficiency is 0.75, calculate the required power in kW.

7.5 Calculate the thermal efficiency of a simple steam power plant implementing the superheated cycle configuration. Superheated steam enters the turbine at 10 MPa and 680 K. The pressure in the condenser is kept at 0.1 MPa, as the discharged thermal energy is used for district heating. Assume that the water state at the feedwater pump inlet is saturated. The isentropic efficiency of the pump is 0.6, while that of the turbine is 0.87.

7.6 Calculate the increase in thermal efficiency due to the introduction of a regenerating dearator in the Rankine-cycle power plant of Exercise 7.5. The plant operates at the optimal pressure (the extraction pressure that makes the thermal efficiency maximum). Figure 7.9c shows the process flow diagram of the system. The second feedwater pump has the same isentropic efficiency as the first. Assuming that the nominal net power output of the power plant is 120 MW and that it operates at nominal power for 8000 hours per year, how many more kWh per year does it deliver if compared to the power plant of Exercise 7.5? With a value of the kWh that is typical of your location, what is the value of the extra electrical energy due to the more efficient cycle configuration?

7.7 A Rankine-cycle power plant has one stage of reheat. The steam at the turbine inlet is at 500 °C and 3×10^6 Pa. After expansion to 0.5×10^6 Pa, the steam is reheated to 500 °C, and expanded in a second turbine to a condenser pressure of 0.007×10^6 Pa. The steam leaves the condenser as saturated liquid. Calculate the cycle efficiency, using a pump isentropic efficiency of 0.65 and a turbine isentropic efficiency of 0.90 for both turbines.

7.8 Determine the energy conversion efficiency of a supercritical steam power plant whose steam condenses at 310 K, and reaches 30 MPa in the primary heater. Assume that the inlet pump state is saturated for simplicity, that the maximum system temperature is 923 K and that the overall isentropic efficiency of the turbines is 0.93 (several machines in series). The overall efficiency of the pumps is 0.75. Calculate also the thermal power input and steam-flow rate for 1000 MW of power output.

7.9 A power system for space applications is based on a Rankine cycle using potassium as the working fluid. The working fluid at the turbine inlet is at 1200 K and 2.0 bar, and is fixed. The pump inlet is saturated liquid and the pump work is negligible. The turbine efficiency is 0.7. In space the condenser must be a radiator, and weight is of crucial importance. Determine the operating conditions corresponding to minimum-weight radiator for a given turbine power. Calculate and plot the ratio of radiator area to power output (m^2/kW) versus the condenser temperature. Try a condenser temperature range from 800 to 1100 K. The thermal power discharged by the condenser is radiated according to the equation $Q_c = \sigma T^4 A$, where A is the radiator area, and σ is the physical constant $\sigma = 5.68 \times 10^{-8}$ W/($m^2 \cdot K^4$).

7.10 The fuel efficiency of reciprocating engines could be greatly increased (and thus pollution decreased) if the thermal energy that is wasted with the exhaust gases and by cooling the engine jacket and the lube oil were converted into additional work. As a first step in this direction, the heat recovery from the exhaust of truck engines by means of mini-Organic Rankine Cycle (ORC) power systems is currently being pursued. Calculate the net power output of a mini-ORC turbogenerator recovering heat from the flue gases of a truck diesel engine in cruise conditions, thus delivering 150 kW. The working fluid is siloxane MDM, and the system implements the superheated, regenerated cycle configuration. The mass flow rate of the engine exhaust is 0.18 kg/s and the temperature of the gases at the inlet of the primary heat exchanger is 590 K. The isobaric specific heat of the gases can be assumed constant and is $c_P = 1.1$ kJ/(kg · K). The temperature of the cooling water of the radiator that can be used to condense the ORC working fluids is maintained by the control system around 358 K independently of the engine load. A condensing pressure of 0.2 bar is thus deemed to be feasible under all operating conditions. The minimum temperature difference in the primary heat exchanger is 20 K and in the regenerator is 10 K. The pressure in the evaporator is 5 bar, while the degree of superheating is 5 K. All pressure losses within the ORC system can be assumed negligible. The turbine efficiency is 0.70, while that of the pump is 0.6.

7.11 A renewable energy system for the (co)generation of electricity and cooling in sunny areas of the world consists of a solar Organic Rankine Cycle (ORC) power system, whose discharged thermal power is used to power an absorption refrigerator. The thermal power collected by the solar troughs and transferred to the working fluid (MDM) is 100 kW. The ORC power plant implements the supercritical and regenerated cycle configuration. The reduced pressure of the working fluid in the primary heat exchanger is $P_r = 1.15$, the temperature at the inlet of the turbine is 305 °C, the temperature in the condenser is 100 °C. The minimum temperature difference in the regenerator is 20 K. The isentropic efficiency of the turbine and the pump are 0.85 and 0.65, respectively. The auxiliaries' consumption, including the solar field recirculation pump but excluding the main ORC pump, is 10% of the turbine gross power. The COP of the absorption refrigerator is 0.5. Calculate the electric power and the cooling power delivered by this system in these conditions. Calculate also the mass flow rate of diesel fuel if the same electric and cooling power were obtained with an engine generating both the electricity and the electricity for a compression refrigerator. The diesel engine efficiency is 0.35, while the coefficient of performance of the vapor-compression refrigerator is COP = 4.5.

7.12 Determine the COP of a vapor-compression refrigerator implementing the simple inverse Rankine cycle, employing R134a as refrigerant, with saturated vapor leaving the evaporator at 278 K and saturated liquid leaving the condenser at 300 K. The compressor has an isentropic efficiency of 0.87. Derive the Carnot efficiency for a refrigerator and compare the calculated COP with that of a Carnot refrigerator operating between the same temperatures.

7.13 Compare the COP of the ammonia refrigeration system of Figure 7.21 to that of the single-stage system of Figure 7.20, operating with the same working fluid between the same minimum and maximum cycle pressures, and given that the temperature at the outlet of the direct contact heat exchanger equals 40 °C. Assume that the isentropic efficiency of all the compressors is the same and is 0.8.

7.14 Consider an externally heated air gas turbine implementing the thermodynamic cycle described in Section 7.5. Using a compressor isentropic efficiency 0.85, determine the conversion efficiency of the system as a function of the pressure ratio, keeping the turbine inlet temperature fixed. What pressure ratio must be reached before the system can be started? Specify the mass flow rate, the thermal power input and output of the system, the compressor and turbine power, and pressure ratio at (a) the maximum efficiency operating point, and (b) the minimum flow rate for 225 kW.

7.15 In the regenerative gas turbine power plant whose process flow diagram is shown in Figure 7.27a, the mass flow of air entering the compressor at 1 bar and 25 °C is 17.9 kg/s. The pressure ratio of the compressor is 9.9. The regenerator effectiveness is 0.9. The effectiveness of a counter flow heat exchanger is defined as

$$\epsilon = \frac{\dot{m}_h c_{P_h}\left(T_{h,in} - T_{h,out}\right)}{(\dot{m}c_P)_{min}\left(T_{h,in} - T_{c,in}\right)} = \frac{\dot{m}_c c_{P_c}\left(T_{c,out} - T_{c,in}\right)}{(\dot{m}c_P)_{min}\left(T_{h,in} - T_{c,in}\right)},$$

where subscripts h and c stand for the hot and the cold stream, respectively, and $(\dot{m}c_P)_{min}$ is the smallest heat capacitance of these two streams. The isentropic efficiency of the compressor is 0.89 and that of the turbine is 0.92. The turbine inlet temperature is 1160 °C. The mass flow of fuel entering the combustor is 0.36 kg/s and the lower heating value of the fuel is 38 MJ/kg (Dutch Slochteren natural gas). The air and exhaust gas composition are[2]

[2] Usually the fuel mass flow is calculated from the fuel composition and the energy balance of the combustor. However, given that you will learn how to carry out chemical equilibrium calculations only once you reach Chapter 10, here the fuel mass flow and exhaust gas compositions are specified.

	Air	Fluegas
N_2	77.29 %	75.34 %
O_2	20.75 %	14.79 %
H_2O	1.01 %	6.23 %
CO_2	0.03 %	2.75 %
Ar	0.92 %	0.89 %

Assume that there are no pressure or thermal losses anywhere in the system and that no air is spilled from the compressor for turbine blade cooling. Determine the net mechanical power output in kW and the conversion efficiency of the system.

7.16 Consider a two-stage gas compressor with inter-cooling.

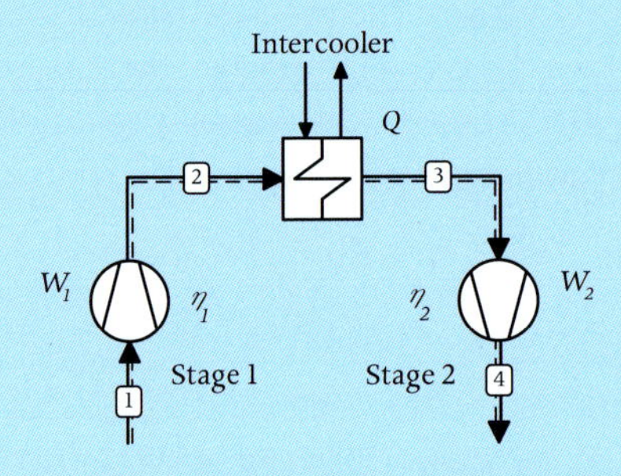

Assume the gas can be treated as ideal with constant specific heat. Derive an expression for the specific work input in terms of $P_1, P_2, P_4, \eta_1, \eta_2, T_1$ and T_3. Then, for all the other parameters fixed, find the value of P_2 which minimizes the work requirement. Show that, if $T_3 = T_1$ and $\eta_1 = \eta_2$, the work is minimum for $P_2/P_1 = \sqrt{P_4/P_1}$.

7.17 Determine the efficiency and the net mechanical power output of the system of Exercise 7.15 in the case where the compressor includes one inter-cooling stage, the isentropic efficiency of the compressors is the same as in Exercise 7.15, and the pressure ratio of the two stages is optimal. Assume an increase in fuel consumption of 5 % with respect to the fuel consumption in Exercise 7.15. What considerations can you make?

7.18 Combustion gas with the composition of Exercise 7.15 at the inlet of the reheated stages of a gas turbine is at 1400 K and 40 bar. The gas expands to atmospheric pressure. Between the high pressure and low pressure turbines, the gas is reheated at a constant pressure of 10 bar up to 1400 K. Assume that the efficiency of both turbines is 0.90, and that there are no pressure losses in the reheater. Calculate:

(a) the specific expansion work for each turbine stage,
(b) the specific energy as heat that must be supplied through the reheater,
(c) the increase in specific work if compared to a single expansion stage with no reheat, and
(d) the pressure ratio that maximizes the power output of the turbine.

7.19 Consider a closed recuperative Brayton cycle turbine power plant employing supercritical CO_2 as the working fluid. The working fluid at the compressor inlet is at $T = 32$ °C and $P = 7.7$ MPa. CO_2 pressure in the heater is 20 MPa and the turbine inlet temperature is 550 °C. The regenerator effectiveness is 0.95, see Exercise 7.15. The isentropic efficiency of the compressor is 0.89 and that of the turbine is 0.90. Calculate the mass flow rate of CO_2 if the power plant must deliver 50 MW, and its conversion efficiency. All pressure and thermal losses can be assumed as zero.

7.20 An ideal diesel cycle using air as the working fluid has a compression ratio of 15. The energy input is idealized as a transfer of energy as heat to the working fluid of 1650 kJ/kg. The inlet conditions are 20 °C and 1 bar. Calculate the temperature and pressure of all remaining states in the cycle, and determine the cycle efficiency.

7.21 The compression ratio of an ideal Otto cycle using air as the working fluid is 8. The energy input is idealized as a transfer of energy as heat to the working fluid of 1650 kJ/kg. The inlet conditions are 20 °C and 1 bar. Calculate the temperature and pressure of all remaining states in the cycle, and determine the cycle efficiency.

7.22 Discuss why the idealized cycles of Exercises 7.20 and 7.21 are inadequate for the analysis of real engines. Which idealizations are likely to be the weakest? What effects will they have on the predicted performance?

7.23 An idealized recip engine implementing the ideal Otto cycle operates with a compression ratio of 9:1 at intake conditions of 100 kPa and 298 K. The cylinder volume is initially 2000 cm^3. If 4 kJ of energy is transferred as heat to the working fluid during an isochoric process, calculate the cycle efficiency, and the pressure and temperature at the end of each process of the cycle.

7.24 The compression ratio of an idealized recip engine implementing the ideal diesel cycle is 15:1. The air intake conditions are 100 kPa and 298 K. The cylinder volume is initially 2000 cm^3. If 4 kJ of energy is transferred as heat to the working fluid during an isobaric process, calculate the cycle efficiency, and the pressure and temperature at the end of each process of the cycle.

8 Thermodynamic Properties of Multicomponent Fluids

CONTENTS

Fluids made of several substances are relevant in many energy conversion systems and other technical processes. The working fluid of a compression or absorption refrigeration system can be a mixture; the fuel of internal combustion engines and gas turbines is formed by various hydrocarbons blended together. Using mixtures as working fluids is currently being studied in the case of non-conventional thermal power systems, like those based on the organic Rankine cycle and the closed Brayton cycle concepts. Many other examples can be formulated. For all these applications, the estimation of fluid thermodynamic properties is essential.

8.1 Simple Mixtures

Here we treat thermodynamic properties of *simple fluid mixtures*: these are mixtures of pure fluids that can blend with each other in the liquid phase without separating, as water does from gasoline. In addition, they do not chemically react, nor dissociate. Still, the physical behavior of simple mixtures in vapor–liquid equilibrium is rather different from that of pure fluids.

We start by describing these differences using thermodynamic diagrams because this facilitates the understanding of processes involving mixtures and the theoretical treatment that follows. The first major difference can be noted in the T–s and T–z diagrams of Figure 8.1 related to a *binary* (two-component) mixture of propane (1) and n-pentane (2). We indicate the overall mole fraction composition, the composition of the fluid when it is either only liquid or only vapor, with z. In this case the fluid composition is *equimolar*, that is $z_1 = 0.5$ and $z_2 = 0.5$. The T–s

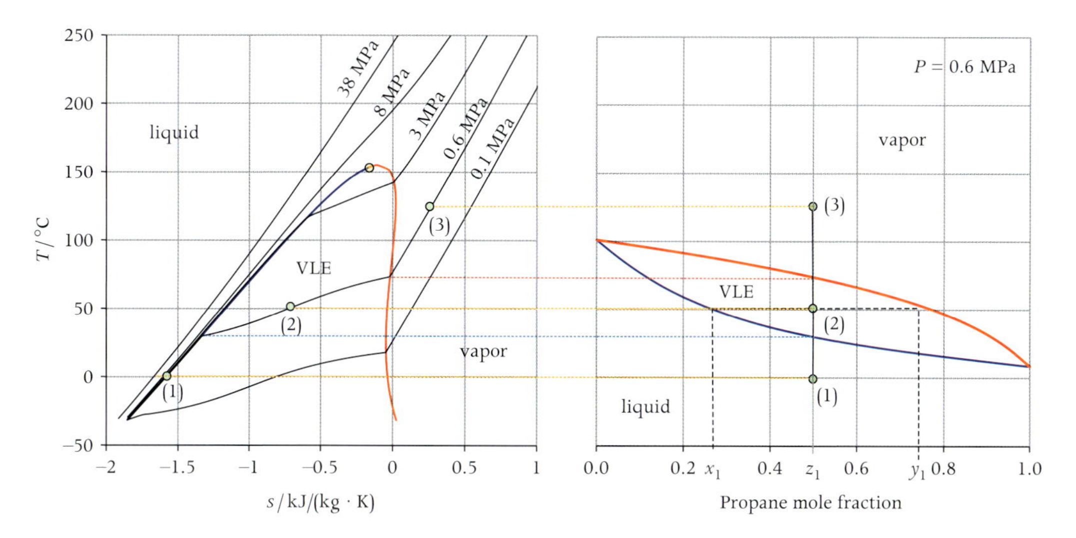

Figure 8.1 Example thermodynamic diagrams for a simple binary mixture (propane/n-pentane). The *T–s* diagram (left) is for the equimolar composition. The bubble lines are blue, while the dew lines are red. The solid circle is the critical point. Line (1)–(2)–(3) indicates an isobaric heating process for a mixture with equimolar overall composition. The same process is indicated also in the *T–z* diagram (right), which gives the composition of the liquid and vapor phase during vaporization. These charts are generated with the StanMix library. StanMix implements the mixture thermodynamic model described in Section 8.10.

diagram of Figure 8.1 shows the saturation line which separates subcooled liquid states and the superheated vapor states from the region of vapor–liquid equilibrium (VLE) states. This line in the T–s diagram is very similar to the saturation line of a pure fluid, but there are important differences in the VLE region that can be highlighted by describing the process of slowly heating the mixture from a state of subcooled liquid (1) to that of superheated vapor (3) at constant pressure $P = \bar{P}$.

By transferring energy as heat to the mixture, its temperature rises until it reaches its boiling point at that pressure. In this particular state, a first infinitesimal bubble of vapor is formed. This thermodynamic state is called the *bubble point*. The liquid composition, given by the mole fractions of liquid in saturated state that we indicate with x_i, must be the same as the overall composition given by z_i. However, the composition of the small vapor bubble is different and contains a larger amount of the most volatile compound, propane in this case. We indicate the composition of the vapor in the saturated state with y_i. During the isobaric phase change caused by the steady slow energy transfer to the mixture, the temperature continues to increase, differently from a pure fluid, as an effect of the more complex molecular interactions between the two substances. Each of the saturated states along the isobaric heating line is characterized by different liquid and vapor compositions, which can be read on the corresponding T-z diagram in Figure 8.1. The composition of the vapor phase corresponding to any saturated state (2) on the isobar is displayed on the upper line (red) of the so-called *lens diagram*, while the composition of the liquid phase is given by the lower line (blue). The state at which all the liquid is vaporized is called *dew point*. This is because, if one imagines the process in the opposite direction (slow isobaric cooling), the dew point occurs when the first infinitesimal droplet is formed while lowering the temperature of the mixture. At the dew point, the composition of the vapor phase and the overall composition are the same ($y_i = z_i$), while the composition of the droplet (x_i) is different and it contains a larger amount of the less volatile compound. The line that is the locus of all the bubble points is called the *bubble line*, while

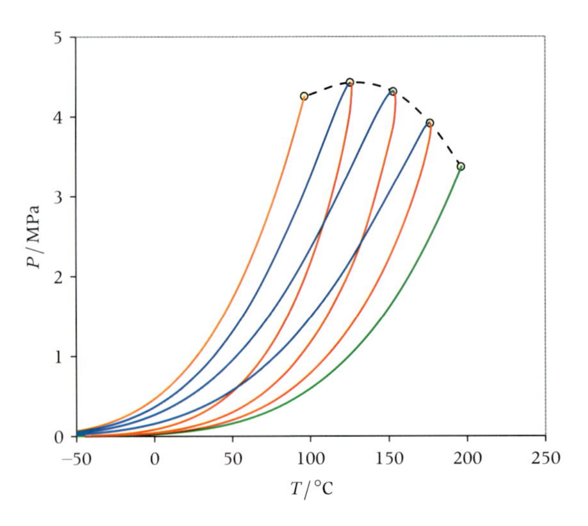

Figure 8.2 ***P–T*** **thermodynamic diagram of propane/n-pentane mixtures showing the saturation lines of – from left to right – propane (orange), propane/n-pentane mixtures with propane mole fractions of 0.75, 0.5, and 0.25 (blue: bubble lines, red: dew lines), and n-pentane (green). Circles indicate the mixture critical points, and the dashed line that connects them is the critical locus.**

that formed by all the dew points is called the *dew line*.

The *P–T* diagram of Figure 8.2 is useful to illustrate another difference between pure fluids and mixtures. The orange line represents the saturation line of pure propane, while the green line that of pure n-pentane. Blue and red lines refer to the bubble and dew lines of mixtures of various compositions. The line in the *P–T* diagram formed by the bubble and the dew line for a certain composition and enclosing all the vapor–liquid equilibrium states is called the *phase envelope*. Observe that the vapor–liquid critical point of a mixture is not the state at the maximum temperature and pressure at which vapor and liquid can co-exist, like in a pure fluid, but its general definition, valid for any mixture and any combination of phases, comes from a more theoretical approach involving stability analysis of equilibrium states. A vapor–liquid critical point can be defined for any simple mixture as the state at which there is no difference between the properties of the two co-existing phases, therefore they feature identical composition.

The great scientist Josiah Willard Gibbs was the first who provided a rigorous definition of the critical points of multicomponent fluids. It can be demonstrated that for a simple mixture the vapor–liquid critical point is where the saturation and the spinodal lines in the *P–v* diagram are tangent. For a general treatment of the critical point of mixtures the interested student is referred, for example, to R. J. Sadus, "Calculating critical transitions of fluid mixtures: Theory vs. experiment," *AIChE J.*, **40**, (*8*), 1376–1403, 1994.

The line that connects all vapor–liquid critical points is called the *critical locus*. All saturation lines are tangent to the vapor–liquid critical locus. We also notice that there can be vapor–liquid equilibrium states at temperatures that are higher than the critical temperature and states at pressures that are higher than the critical pressure. The state at the highest temperature on the phase envelope line is called *cricondentherm*, while that at the highest pressure on the phase envelope is called *cricondenbar*. The phase behavior of multicomponent mixtures can be far more complicated than this, with multiple liquid phases, discontinuous phase envelopes, and multiple or no critical points, however, the general treatment of phase equilibria is outside the scope of this book.

8.2 Extension of Thermodynamic Relations to Mixtures

Estimating thermodynamic properties of simple multicomponent fluids is at the basis of the analysis and design of many energy conversion systems and other technical processes, and the theory that makes it possible is an extension of what is described in Chapter 6. Equations of state are still the most widespread type of model for mixture property calculations, while more fundamental models based on statistical thermodynamics and the description of molecular forces are actively researched, but are currently too computationally expensive for most engineering needs.

The large variety of ways in which molecules can interact with each other makes the problem of modeling thermodynamic properties of mixtures more complicated, if compared to that of pure fluids. In general, mixture thermodynamic properties are not additive with respect to the pure constituents, though their calculation depends in complex ways on the properties of each component. The mutual interaction of different molecules must be accounted for, and this information is at the basis of the methods used to obtain suitable functions for the parameters of mixture equations of state. As always, the correct application of the First and Second Laws of Thermodynamics are at the basis of the framework allowing for the study of equilibrium states, hence the calculation of the associated fluid properties.

8.3 The Perfect Gas Mixture

The definition of *ideal gas* does not depend on the number and type of interacting molecules. As we have seen in Sections 3.8 and 3.9, the volumetric properties of a gas are described by the equation $PV = N\hat{R}T$ as long as the molecules are so distant that their interaction is similar to elastic collisions among spherical particles, and their volume can be considered negligible. In addition, properties u, h, c_P, and c_v depend solely on temperature.

Let us now show how all thermodynamic properties of a perfect gas mixture can be obtained from the properties of the constituents together with the composition of the mixture.

In analogy with (3.20), the volumetric equation of state for a multicomponent fluid in the ideal gas state is

$$PV = (N_1 + N_2 + \cdots + N_{nc})\,\hat{R}T = \left(\sum_{i=1}^{nc} N_i\right)\hat{R}T, \tag{8.1}$$

where nc is the total number of components in the mixture. Let us denote with N and M the total number of moles and the total mass of the mixture.

The *composition* of a mixture can be given either by assigning the mole fraction $\chi_i \equiv \dfrac{N_i}{N}$ of each component, or its mass fraction $\xi_i \equiv \dfrac{M_i}{M}$. It follows that $\sum_{i=1}^{nc} \chi_i = 1$ and $\sum_{i=1}^{nc} \xi_i = 1$. In addition, $\hat{M} \equiv \dfrac{M}{N} = \sum_{i=1}^{nc} \chi_i \hat{M}_i$ is the average molar mass of the mixture.

> We thus use χ for mole fraction and ξ for mass fractions, if the fluid behaves as an ideal gas. For the mole fractions of vapor and liquid in equilibrium we use y and x, respectively, while for the overall mole fractions, that is if the fluid is either entirely liquid or entirely vapor, we use z.

With these definitions we can write the perfect gas equation of state of a mixture as

$$Pv = RT = \frac{\hat{R}}{\hat{M}}T = \frac{\hat{R}}{\sum_{i=1}^{nc} \chi_i \hat{M}_i}T.$$

The gas constant of an ideal gas mixture is obtained from the molar masses of the constituents and the mole composition with

$$R = \frac{\hat{R}}{\sum_{i=1}^{nc} \chi_i \hat{M}_i}. \tag{8.2}$$

Moles and molar quantities are most often adopted when dealing with mixture thermodynamic properties or chemical reactions because in these cases we need to describe mathematically the interactions between elementary entities, i.e., molecules and atoms, and therefore we need to account for their amounts (the number of constituent particles in the thermodynamic system).

The molar volume is defined as

$$\hat{v} \equiv \frac{V}{N} = \frac{\hat{R}T}{P},$$

and it is our first example of a molar thermodynamic property.

A concept that is sometimes useful in technical applications is that of *partial pressure* P_i of one of the components of the ideal gas mixture, which is defined as

$$P_i \equiv \chi_i P. \tag{8.3}$$

For a perfect gas mixture we have therefore

$$P_i = \frac{N_i}{\sum\limits_{k=1}^{nc} N_k} P = \frac{N_i}{\sum\limits_{k=1}^{nc} N_k} \left(\sum_{k=1}^{nc} N_k \frac{RT}{V} \right) = \frac{N_i RT}{V},$$

which shows that the partial pressure of species i in the mixture is equal to the pressure that would act on the container of volume V, if the same number of moles of that constituent (N_i) were alone and maintained at the same temperature of the mixture. This is a consequence of the absence of molecular interaction that is assumed for the ideal gas.

All other molar properties, namely $\hat{c}_v$, $\hat{u}$, $\hat{s}$, etc., can be obtained as a function of the corresponding molar properties of the mixture constituents and of the mole fraction composition. If we have an equation of state in the form $P = P(T, v)$, or $P = P(T, \hat{v})$, in order to complete the fluid thermodynamic model, we need in addition a relation for the isobaric specific heat capacity. The total internal energy of a fluid is an extensive property, and in the case of a perfect gas it is independent of pressure, therefore for a gas mixture we have

$$U = U_1 + U_2 + \cdots + U_{nc} = \sum_{i=1}^{nc} U_i.$$

The internal energy of each of the constituents depends only on the temperature and on its amount, independently of the mutual interactions between different molecules, therefore the total energy of the system at a certain temperature is obtained by summation of the contributions of each component. On a molar basis we therefore have that

$$N\hat{u} = N_1\hat{u}_1 + N_2\hat{u}_2 + \cdots + N_{nc}\hat{u}_{nc} = \sum_{i=1}^{nc} N_i\hat{u}_i, \quad (8.4)$$

where $\hat{u}_i$ is the molar internal energy of the ith constituent of the mixture, that is the internal energy of component i per mole of i. Dividing (8.4) by the total number of moles N, yields

$$\hat{u} = \sum_{i=1}^{nc} \chi_i \hat{u}_i.$$

Also the enthalpy of a perfect gas mixture is independent of pressure, therefore the similar relation

$$\hat{h} = \sum_{i=1}^{nc} \chi_i \hat{h}_i \quad (8.5)$$

holds. We can then use the definition of molar isobaric specific heat capacity $\hat{c}_P \equiv \left(\frac{\partial \hat{h}}{\partial T} \right)_P$ and apply it to $\hat{h} = \sum_{i=1}^{nc} \chi_i \hat{h}_i$, thus obtaining a relation between the molar isobaric specific heat capacity of the ideal gas mixture and that of its constituents in the form

$$\hat{c}_P = \sum_{i=1}^{nc} \chi_i \hat{c}_{P_i}.$$

Example: entropy change of mixing of an ideal gas mixture. The dependence of the entropy of an ideal gas mixture on the contribution of its constituents is different from what we have seen for enthalpy or internal energy, as the entropy of the ideal gas does depends on pressure. By recalling (5.18), we have that

$$d\hat{s} = \frac{\hat{c}_v}{T} dT + \hat{R} \frac{d\hat{v}}{\hat{v}} = \hat{c}_P \frac{dT}{T} - \frac{\hat{R}}{P} dP.$$

Let us evaluate the difference in entropy for species i at temperature T, but at the pressure of the mixture P and at its partial pressure P_i. At constant T we have that

$$d\hat{s} = -\hat{R} \times d \ln P \qquad (T = \text{const.}),$$

therefore

$$\hat{s}_i(T, P) - \hat{s}_i(T, P_i) = -\hat{R} \ln \frac{P}{P_i} = \hat{R} \ln \chi_i,$$

and hence

$$\hat{s}_i(T, P_i) = \hat{s}_i(T, P) - \hat{R} \ln \chi_i.$$

For an ideal gas mixture

$$\hat{s}(T, P) = \sum_{i=1}^{nc} \chi_i \hat{s}_i(T, P_i),$$

because in a perfect gas the property contribution of a mixture constituent to the total mixture property must be evaluated at its partial pressure. This is because the definition of ideal gas implies that the molecules of any constituent do not interact with the molecules of the other compounds, therefore the contribution to the entropy of one constituent is given by the amount that the constituent supplies as if it were alone in the volume at the given temperature. We can therefore conclude that the entropy of the ideal gas mixture is

$$\hat{s}(T, P) = \sum_{i=1}^{nc} \chi_i \hat{s}_i(T, P) - \hat{R} \sum_{i=1}^{nc} \chi_i \ln \chi_i. \quad (8.6)$$

Equation (8.6) shows that a mixture of perfect gases features more entropy than the contribution of each constituent at the given temperature and pressure, and this is due to the pressure effect: a constituent contributes to the mixture molar entropy at its partial pressure, which is always lower than the mixture pressure, thus associated with a larger entropy (more uncertainty and disorder). This fact is expressed by the molar *entropy change of mixing* of a perfect gas mixture

$$\Delta\hat{s}_{\text{mixing}}(T, P) \equiv \hat{s}(T, P) - \sum_{i=1}^{nc} \chi_i \hat{s}_i(T, P)$$
$$= -\hat{R} \sum_{i=1}^{nc} \chi_i \ln \chi_i,$$

which is always positive because $\chi_i < 1$. Note that the entropy pressure-dependence of an ideal gas can be appreciated on a fluid T–s diagram by following an isotherm at low reduced pressures, see, e.g., Figure 6.12a.

The enthalpy and internal energy of a perfect gas mixture do not depend on pressure, therefore their value is not affected by any change due to mixing. The molar Gibbs function of a perfect gas mixture on the other hand is influenced by a similar pressure effect. We can obtain its expression from its definition $\hat{g} \equiv \hat{h} - T\hat{s}$, by substituting in it (8.5) and (8.6), which gives

$$\hat{g}(T, P) = \sum_{i=1}^{nc} \chi_i \hat{g}_i(T, P) + \hat{R}T \sum_{i=1}^{nc} \chi_i \ln \chi_i. \quad (8.7)$$

The conversion from molar properties to (mass-) specific properties is readily made with

$$u = \frac{\hat{u}}{\hat{M}}, \quad h = \frac{\hat{h}}{\hat{M}}, \quad s = \frac{\hat{s}}{\hat{M}}, \quad c_v = \frac{\hat{c}_v}{\hat{M}}, \ldots$$

Example: condensing home boiler. Modern home boilers achieve higher conversion efficiency by using the thermal energy that can be obtained by condensing the water vapor contained in the flue gas in order to pre-heat the warm water supply, which then receives thermal energy from the combustion gases, as shown in the scheme of Figure 8.3. Let us calculate the condensing pressure of the flue gas, that is the pressure at which water vapor can be condensed.

A valid assumption is that the flue gas can be modeled as an ideal gas mixture, as its pressure is low (combustion occurs at atmospheric pressure) and the temperature is relatively high. The natural gas used as fuel mainly consists of methane and burning it with a little excess air generates flue gas containing a mole fraction of H_2O of about 0.15 (you will learn how to compute the composition of combustion products in Chapter 10). The water vapor in the ideal gas mixture is unaffected by the presence of the other constituents (e.g., CO_2), therefore it condenses at a partial pressure corresponding to the saturation pressure of water at the temperature of the flue gas. If we assume a temperature of the flue gas at the outlet of the heat exchanger of 325 K, then the saturation pressure of water vapor can be calculated with the IF97 library, yielding

$$P^{\text{sat}}_{H_2O}(\bar{T} = 325\ \text{K}) = 13.5\ \text{kPa}.$$

In order to condense the water vapor content in the flue gas, the water partial pressure in the heat exchanger must be greater than or equal to the saturation pressure of water at the flue gas temperature, meaning that

$$P_{H_2O}(\bar{T}) = \chi_{H_2O} \times P_{\text{fluegas}}(\bar{T}) \geq P^{\text{sat}}_{H_2O}(\bar{T})$$
$$= 13.5\ \text{kPa}.$$

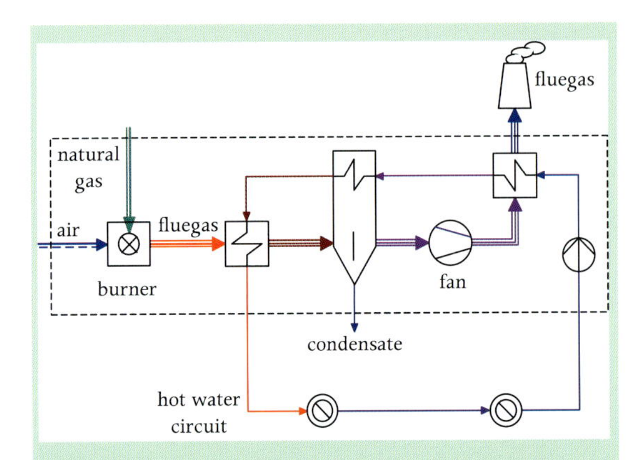

Figure 8.3 Simple process flow diagram of a condensing home boiler.

The pressure of the flue gas must therefore be at least 13.5 / 0.15 = 90.5 kPa to condense the water vapor content of the flue gas. Such pressure is slightly lower than atmospheric: a fan that moves the flue gases through the heat exchangers is thus always adopted.

8.4 Partial Molar Properties

The properties of an ideal gas mixture depend only on the properties of its constituents, while in states for which the ideal gas assumption does not hold, interaction between different molecules must be taken into account. *Partial* molar properties allow extending all the thermodynamic concepts seen so far to this case.

In order to get to the definition of partial molar property, let us start with an example that is intuitively easier to understand. At a given T and P, the volume occupied by a mixture depends on the volume occupied by its components and on their amounts. The total volume of a mixture is therefore a function of the number of moles of each of the nc components, that is

$$V = V(T, P, N_1, N_2, \ldots, N_{nc}).$$

The volume occupied by a molecule surrounded by molecules of the same kind is in general different from that occupied by the same molecule surrounded by molecules of a different kind, as it strongly depends on the interaction forces, which can be widely different. This can be easily verified: if water is mixed with a liquid alcohol, the volume of the mixture is always smaller than the sum of the volumes of the same amounts of water and alcohol, if measured separately, unmixed. This is a result of the different forces acting between pure water and alcohol molecules, if compared to the water–water or to the alcohol–alcohol interaction: water and alcohol molecules attract each other more strongly, thus the volume they occupy when mixed at a given temperature and pressure is smaller. This is shown for example in Figure 8.4, displaying the *excess* molar volume (or *volume change of mixing*) of water–ethanol mixtures for all possible compositions. The excess molar volume is the difference between the volume of the mixture and the summation of the volumes of the same amount of constituents "unmixed." The excess molar volume of the liquid ethanol–water mixture at

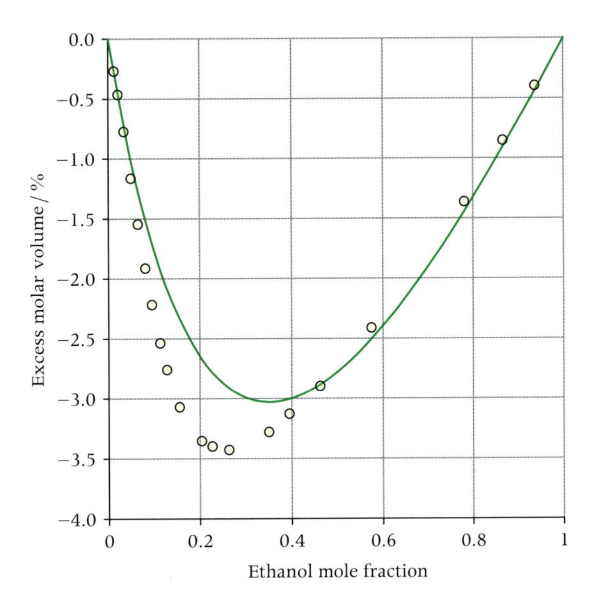

Figure 8.4 Excess molar volume of water – ethanol liquid mixtures at atmospheric pressure and $T = 298.15$ K. The green line shows values calculated with the PCP-SAFT thermodynamic model, while circles are experimental data taken from J.-P.E. Grolier and E. Wilhelm, "Excess volumes and excess heat capacities of water + ethanol at 298.15 K," *Fluid Phase Equilib.*, 6, (3–4), 283–287, 1981.

atmospheric pressure and $T = 298.15$ K is negative for any composition. Note that the volume difference changes with the composition, as the surrounding of a molecule does depend on how many molecules of a certain type are in the surrounding. Observe also that the values predicted by the model are somewhat different from accurate experimental values: modeling the interaction between strongly polar molecules is challenging even for state of the art mixture models!

Now, let us assume that we have a mixture of two components, and that we double the amount of the two components at the same temperature and pressure. The volume will double, because the interaction between the molecules remains on average the same, therefore

$$V(2 \times N_1, 2 \times N_2) = 2 \times V(N_1, N_2).$$

The mixture volume is a function of the mole numbers of its constituents such that, if its arguments scale with a certain factor, its value scales by the same factor. In mathematics a function which satisfies $f(\alpha x, \alpha y) = \alpha^n f(x, y)$ is called a *homogeneous function of degree n*. The volume of a mixture is therefore a homogeneous function of first degree, and this is important because first-degree homogeneous functions feature several characteristics that are useful here. For instance, for the extensive property volume we have that

$$V = \sum_{i=1}^{nc} N_i \times \bar{v}_i, \tag{8.8}$$

where

$$\bar{v}_i \equiv \left(\frac{\partial V}{\partial N_i}\right)_{T,P,N_{j \neq i}}. \tag{8.9}$$

Here $\bar{v}_i$ is called the *partial molar volume* of component i in the mixture. It is the change in volume per mole, if component i is added to the mixture, and it is therefore an intensive property. Relation (8.8) between the total volume and the partial molar volume of the constituents is derived using Euler's theorem on homogeneous functions of degree n as follows. If

$$f(\alpha x, \alpha y) = \alpha^n f(x, y),$$

and we define $x' \equiv \alpha x$ and $y' \equiv \alpha y$, we have that

$$\begin{aligned} n\alpha^{n-1} f(x, y) &= \frac{\partial f}{\partial x'}\frac{\partial x'}{\partial \alpha} + \frac{\partial f}{\partial y'}\frac{\partial y'}{\partial \alpha} \\ &= \frac{\partial f}{\partial x'} x + \frac{\partial f}{\partial y'} y \\ &= x\frac{\partial f}{\partial (x\alpha)} + y\frac{\partial f}{\partial (y\alpha)}. \end{aligned}$$

This equation is valid for any α, therefore also for $\alpha = 1$, which gives

$$nf(x, y) = x\frac{\partial f}{\partial x} + y\frac{\partial f}{\partial y}. \tag{8.10}$$

If f is the volume V, x is N_1, y is N_2 and $n = 1$, and we generalize (8.10) to the case of many arguments, we find that at given T and P

$$V(N_1, \ldots, N_{nc}) = \sum_{i=1}^{nc} N_i \left(\frac{\partial V}{\partial N_i}\right)_{T,P,N_{j \neq i}} = \sum_{i=1}^{nc} N_i \bar{v}_i. \tag{8.11}$$

According to our reasoning, the volume of a mixture is different from the ideal volume that would be obtained if there were no difference in the interaction between the different molecules, and this can be made explicit, in the case of a binary mixture, by writing

$$\begin{aligned} \Delta V_{\text{mixing}} &= V - V_{\text{ideal}} \\ &= (N_1\bar{v}_1 + N_2\bar{v}_2) - (N_1\hat{v}_1 + N_2\hat{v}_2), \end{aligned}$$

which shows the difference between the partial molar volume $\bar{v}$, which refers to the molar volume of a substance *in a mixture*, and the molar volume $\hat{v}$, which in turn refers to the molar volume of a certain pure substance. This relation can be generalized and it becomes

$$\Delta V_{\text{mixing}} = V - V_{\text{ideal}} = \sum_{i=1}^{nc} N_i (\bar{v}_i - \hat{v}_i). \tag{8.12}$$

In an ideal gas the molecules are so far apart that their interaction does not depend on the type of molecules in the mixtures, therefore the partial molar volume and the molar volume are the same, and the volume change for mixing is zero.

Furthermore, observe that the partial molar volume $\bar{v}_i$ is also a homogeneous function, but of degree 0, because

$$\bar{v}_i(\alpha N_1, \ldots, \alpha N_{nc}) = \bar{v}_i(N_1, \ldots, N_{nc}), \tag{8.13}$$

which shows the physical fact that the molar volume of a substance *in the mixture* depends on the relative amounts of the other compounds in the mixture, therefore equally scaling all the amounts of the constituents has no effect on it. If we take $\alpha = \frac{1}{N_{\text{tot}}}$, (8.13) becomes

$$\bar{v}_i(z_1, z_2, \ldots, z_{nc}) = \bar{v}_i(N_1, N_2, \ldots, N_{nc}). \tag{8.14}$$

Note though that since $\sum_{i=1}^{nc} z_i = 1$, the partial molar volume is a function only of $nc - 1$ molar concentrations, i.e., $\bar{v}_i = \bar{v}_i(z_1, z_2, \ldots, z_{nc-1})$.

Finally, if we divide (8.8) by N_{tot} and use equality (8.14), the molar volume of a mixture is given by

$$\hat{v} = \sum_{i=1}^{nc} z_i \bar{v}_i. \tag{8.15}$$

As we said, the volume change of mixing is also called excess volume, thus, in terms of molar properties, we have that

$$\hat{v}^{\text{E}} = \Delta\hat{v}_{\text{mixing}} = \hat{v} - \sum_{i=1}^{nc} z_i \hat{v}_i.$$

All that has been shown for a property with an-easy-to-understand physical meaning like the mixture volume, is valid for any other mixture thermodynamic property. It can be concluded that any extensive mixture property is given by the sum of the partial molar properties of each component weighted by its amount, expressed in number of moles. For example, we have that

$$H = \sum_{i=1}^{nc} N_i \times \bar{h}_i, \tag{8.16a}$$

$$S = \sum_{i=1}^{nc} N_i \times \bar{s}_i, \tag{8.16b}$$

$$G = \sum_{i=1}^{nc} N_i \times \bar{g}_i. \tag{8.16c}$$

All partial molar properties are defined as partial derivatives of a given property with respect to N_i at constant $T, P, N_{T,P,N_{j \neq i}}$.

Similarly, the molar properties of the mixture can be obtained by summing the partial molar properties of each constituent weighted by the corresponding molar fraction, as in

$$\hat{h} = \sum_{i=1}^{nc} z_i \times \bar{h}_i, \tag{8.17a}$$

$$\hat{s} = \sum_{i=1}^{nc} z_i \times \bar{s}_i, \tag{8.17b}$$

$$\hat{g} = \sum_{i=1}^{nc} z_i \times \bar{g}_i. \tag{8.17c}$$

Before we dive into the treatment of the all-important Gibbs function for the calculation of vapor–liquid equilibria, it is worth pointing out that nomenclature and its consistency is particularly important when studying mixture thermodynamics. The notation of Table 8.1 is therefore particularly important, and you should pay attention to the difference between:

- a total or intensive property (uppercase),
- a specific property (lowercase, property per kilogram),
- a molar property (circumflexed lower case, a property per mole of an *unmixed* or pure substance),
- a partial molar property (over-barred lowercase, a property per mole of a *mixed* substance obeying the partial molar property relations), and
- a property of the ith component *in the mixture* (lower case with a tilde, a mixture molar property, but *not* a partial molar property).

8.5 Vapor–Liquid Equilibrium

As you may expect given what you have learned from Sections 5.4 and 6.11, vapor–liquid equilibrium conditions for a mixture at fixed P and T can be obtained by applying the first and second laws of thermodynamics.

Here again we show how equations that allow for the computation of the vapor–liquid equilibrium state of a mixture can be obtained from these laws: we apply them to a control volume where pressure P

Table 8.1 Notation for thermodynamic properties of mixtures.

Property	Typeface	Example
Intensive and total	uppercase	P / MPa, T / K, V / m^3, H / J, ...
Specific	lowercase	v / m^3/kg, h / J/kg, s / J/ (kg · K), ...
Ideal gas	with IG superscript	v^{IG}, h^{IG}, s^{IG}, c_P^{IG}, ...
Molar, pure, or *unmixed* fluid	lowercase, over-careted	$\hat{v}$ / m^3/mol, $\hat{h}$ / J/mol, $\hat{s}$ / J/ (mol · K), ...
Partial molar, fluid *in the mixture*	lowercase, over-barred	$\bar{v}$ / m^3/mol, $\bar{h}$ / J/mol, $\bar{s}$ / J/ (mol · K), ...
Of fluid i in the mixture	lowercase, over-tilded	$\tilde{f}_i$, ...
Ideal solution	with IS superscript	v^{IS}, h^{IS}, s^{IS}, c_P^{IS}, ...

and temperature T are fixed. This is a paradigmatic case of interest in engineering and scientific applications, because temperature and pressure can be easily measured and controlled. Devices to measure vapor–liquid equilibrium states (see Figure 8.9) whose data are needed as input to thermodynamic models are a good example. As we shall see in the following, other equilibrium conditions can be obtained in a similar way by considering other couples of fixed variables.

Let us therefore consider the piston–cylinder system of Figure 8.5. The isolated inner volume contains a fluid, and we assume that T and P are kept constant. Let us assume that small energy transfers as heat and work perturb this system, therefore we can write the first and second law equations (energy and entropy balance) for this system as

$$-dU = đQ + đW,$$

$$0 + đ\mathcal{P}_S = \frac{đQ}{T} + dS.$$

The small amount of energy transfer as work due to the infinitesimal expansion perturbation is

$$đW = PdV,$$

while the small amount of energy transfer as heat can be obtained from the entropy balance as

$$đQ = T\left(đ\mathcal{P}_S - ds\right).$$

We can now substitute these expressions for $đW$ and $đQ$ in the energy balance, which yields

$$0 = T\left(đ\mathcal{P}_S - ds\right) + PdV + dU,$$

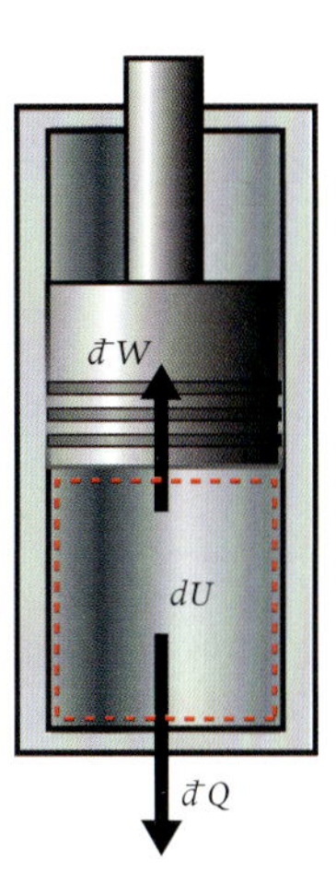

Figure 8.5 Piston–cylinder system: the fluid inside the inner volume is kept in thermodynamic equilibrium at T and P, and is subjected to small energy transfers as heat and work.

and rearrange it to obtain

$$\underbrace{-Tđ\mathcal{P}_S}_{\leq 0} = dU + PdV - TdS. \qquad (8.18)$$

As the small entropy production must be positive or zero and the temperature is also positive by definition, the right-hand side of this equation must also be lower than or equal to zero.

The Gibbs function is defined as $G \equiv H - TS$, therefore its total differential is

$$dG = dH - d(TS) = dU + PdV + vdP - TdS - sdT,$$

which in this case, given that temperature and pressure are kept constant, is

$$dG = dU + PdV - TdS, \qquad T = \text{const.}, P = \text{const.}$$

Equation (8.18) therefore prescribes that for any small perturbation

$$dG \leq 0, \qquad T = \text{const.}, P = \text{const.}$$

This relation expresses a fundamental concept for isolated thermodynamic systems in equilibrium at a given pressure and temperature: any perturbation of the system determines either a zero or an infinitesimally negative variation of its Gibbs function, i.e., the Gibbs function of the system must be at its minimum. Concisely, we have obtained that:

$$\begin{gathered}\text{Isolated system,}\\ T, P = \text{const.}\\ \Downarrow\\ G \text{ is minimum @ equilibrium.}\end{gathered}$$

This concept can be graphically expressed with plots of the Gibbs function as a function of time (Figure 8.6a), of the volume (Figure 8.6b), and of the internal energy (Figure 8.6c).

For such a system in equilibrium G is at a stationary point, therefore we can also write:

$$\begin{gathered}\text{Isolated system in equilibrium,}\\ T, P = \text{const.}\\ \Downarrow\\ dG = 0.\end{gathered} \tag{8.19}$$

In a similar way it can be shown that if in a system V and U are kept constant, the condition for equilibrium is that S is maximum, and if V and T are kept constant, A must be minimum.

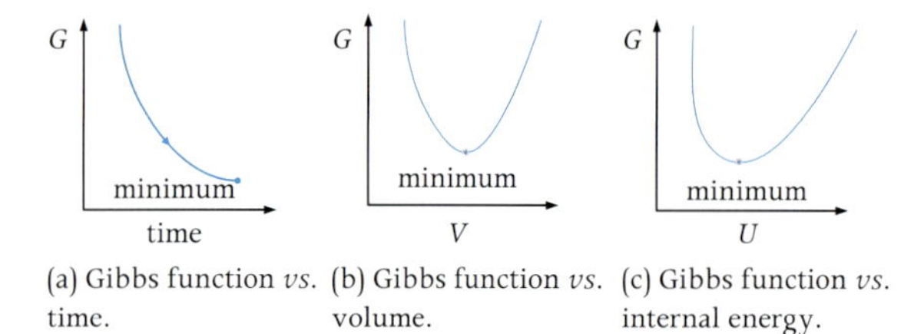

(a) Gibbs function *vs.* time. (b) Gibbs function *vs.* volume. (c) Gibbs function *vs.* internal energy.

Figure 8.6 Gibbs function of a fluid at a given temperature and pressure.

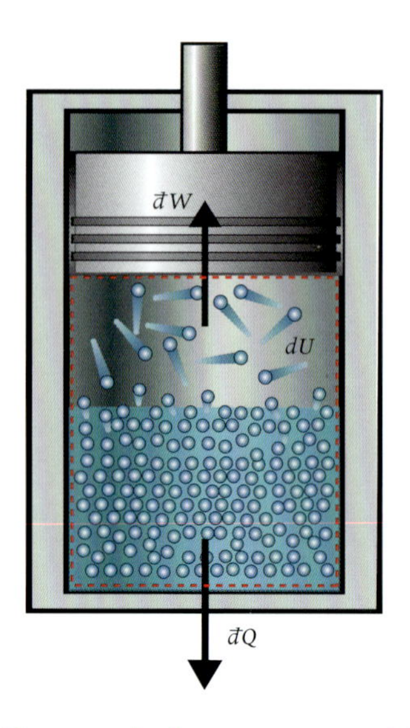

Figure 8.7 Piston–cylinder system containing a mixture in vapor–liquid equilibrium at *T* and *P*.

In the following, we want to exploit the general equilibrium condition for an isolated thermodynamic system at fixed pressure and temperature (Figure 8.7) in order to obtain a set of equations valid for the vapor–liquid equilibrium of a simple, non-reacting fluid mixture. These equations allow calculation of saturated states from a thermodynamic model based on an equation of state. Note that this condition is applicable even to mixtures forming more than two phases, for example those forming two liquid phases together with the vapor phase. A typical example is a mixture of water and gasoline, so-called immiscibile liquids. However, we treat only the simplest case of vapor–liquid equilibrium and leave the general case to more specialized textbooks.

The equilibrium of two gas phases has been predicted and actually experimentally verified. Van der Waals first speculated on its existence in 1894, while it was first experimentally discovered in a mixture of helium and xenon at very high pressure (thousands of bar) by a research group at the Delft University of Technology, see, J. D. S. Arons and G. A. M. Diepen, "Gas–gas equilibria," *J. Chem. Phys.*, **44**, (6), 2322–2300, 1966. There is debate though about what an extremely dense gas should be called.

The total differential of the Gibbs function for a multicomponent fluid is

$$dG(T, P, N_{i=1,\ldots,nc}) = \left(\frac{\partial G}{\partial T}\right)_{P,\mathrm{N}} dT + \left(\frac{\partial G}{\partial P}\right)_{T,\mathrm{N}} dP + \sum_{i=1}^{nc} \left(\frac{\partial G}{\partial N_i}\right)_{T,P,N_{j\neq i}} dN_i. \tag{8.20}$$

In analogy to what we have seen for the partial molar volume, the partial molar Gibbs function of substance i in the mixture is defined as

$$\bar{g}_i \equiv \left(\frac{\partial G}{\partial N_i}\right)_{T,P,N_{j\neq i}}. \tag{8.21}$$

This quantity is sometimes given the name of *chemical potential* μ_i, but we do not see a good reason for introducing a new function.

From (8.20) and (8.21), at fixed T and P we have that

$$dG_{T,P} = \sum_{i=1}^{nc} \left(\frac{\partial G}{\partial N_i}\right)_{T,P,N_{j\neq i}} dN_i = \sum_{i=1}^{nc} \bar{g}_i dN_i.$$

The Gibbs function is an extensive property, therefore

$$G = G^{\mathrm{V}} + G^{\mathrm{L}} \qquad \text{and} \qquad dG = dG^{\mathrm{V}} + dG^{\mathrm{L}}.$$

The total number of moles is fixed (isolated system) and $N_i = N_i^{\mathrm{L}} + N_i^{\mathrm{V}}$. Mass conservation and the absence of chemical reactions imply that $dN_i = 0$, hence

$$dN_i^{\mathrm{L}} = -dN_i^{\mathrm{V}},$$

which tells us that if some amount of liquid i disappears, it becomes vapor and vice versa. We can thus write

$$dG = \sum_{i=1}^{nc} \bar{g}_i^{\mathrm{L}} dN_i^{\mathrm{L}} + \sum_{i=1}^{nc} \bar{g}_i^{\mathrm{V}} dN_i^{\mathrm{V}} = \sum_{i=1}^{nc} \left(\bar{g}_i^{\mathrm{L}} - \bar{g}_i^{\mathrm{V}}\right) dN_i^{\mathrm{L}}.$$

The equilibrium condition (8.19) applied to the *vapor–liquid equilibrium of a simple non-reacting multicomponent fluid that cannot exchange mass with the environment at a given pressure and temperature* implies that

$$\bar{g}_i^{\mathrm{V}} = \bar{g}_i^{\mathrm{L}} \qquad i = 1, 2, \ldots, nc, \tag{8.22}$$

because (8.19) must be satisfied for any variation dN_i.

The partial molar Gibbs function of each component of the mixture in both the liquid and the vapor phase can be related to the fugacity function. The fugacity of components in the mixture can be obtained from a thermodynamic model in a way similar, albeit a bit more complicated, to what we have described in Section 6.11 for a pure fluid. In order to compute a vapor–liquid equilibrium state for a mixture we need to solve a system of equations, while in the case of a pure fluid this can be done by solving one single equation. However, before plunging into it, let us treat some simpler cases of vapor–liquid equilibria in mixtures.

Example: Clapeyron equation for mixtures. As an application of the mixture vapor–liquid equilibrium concept, let us obtain a version of the Clapeyron equation

$$\frac{dP^{\mathrm{sat}}}{dT^{\mathrm{sat}}} = \frac{h^{\mathrm{V}} - h^{\mathrm{L}}}{T\left(v^{\mathrm{V}} - v^{\mathrm{L}}\right)}$$

which is valid for mixtures.

We consider a simple mixture of given composition at its bubble point, see Figure 8.1. In this state, the composition of the liquid at the limit of incipient boiling is the same as the composition of the liquid when the temperature is below its saturation value ($x_i = z_i$), while the composition of the vapor starts to change.

Let us look at the infinitesimal change of the Gibbs function in the vapor phase as the temperature is changed by an infinitesimal value dT along the bubble line. If equilibrium exists we have

$$\bar{g}_i^{\mathrm{V}} = \bar{g}_i^{\mathrm{L}}.$$

An infinitesimal change of temperature at equilibrium implies that also

$$d\left(\bar{g}_i^{\mathrm{V}} - \bar{g}_i^{\mathrm{L}}\right) = 0.$$

Multiplying these equations by y_i and summing over all the constituents yields

$$\sum_{i=1}^{nc} y_i d\bar{g}_i^{\mathrm{V}} = \sum_{i=1}^{nc} y_i d\bar{g}_i^{\mathrm{L}}. \tag{8.23}$$

Analogously to the procedure that we followed in order to obtain the Clapeyron equation for pure fluids, we need to get to an expression that contains enthalpy and volume differences (partial molar in this case), therefore we start from the total differential of the Gibbs function, see (6.111) and (8.20),

$$dG = -SdT + VdP + \sum_{i=1}^{nc} \left(\frac{\partial G}{\partial N_i}\right)_{T,P,N_{j\neq i}} dN_i,$$

where therefore $-S = \left(\frac{\partial G}{\partial T}\right)_P$, and we bring in the enthalpy. To do so, we consider in addition the total differential of the $\frac{G}{T}$ function

$$d\left(\frac{G}{T}\right) = \left(\frac{\partial\,(G/T)}{\partial T}\right)_P dT + \frac{V}{T}dP + \frac{1}{T}\sum_{i=1}^{nc}\left(\frac{\partial G}{\partial N_i}\right)_{T,P,N_{j\neq i}} dN_i.$$

We thus have that

$$\left(\frac{\partial\,(G/T)}{\partial T}\right)_P = \frac{1}{T}\left(\frac{\partial G}{\partial T}\right)_P + G\frac{\partial}{\partial T}\left(\frac{1}{T}\right) = -\frac{S}{T} - \frac{G}{T^2} = -\frac{H^2}{T}, \quad (8.24)$$

which is known as the *Gibbs Helmholtz* equation. This allows us to write

$$d\left(\frac{G}{T}\right) = -\frac{H^2}{T}dT + \frac{V}{T}dP + \frac{1}{T}\sum_{i=1}^{nc}\left(\frac{\partial G}{\partial N_i}\right)_{T,P,N_{j\neq i}} dN_i. \quad (8.25)$$

At constant P and T and at equilibrium, $dG_{T,P} = \sum_{i=1}^{nc} \bar{g}_i dN_i = 0$, therefore the last term in the right-hand side of Equation (8.25) is zero. Furthermore

$$d\left(\frac{G}{T}\right) = \sum_{i=1}^{nc} N_i d\left(\frac{\bar{g}_i}{T}\right), \quad (8.26)$$

therefore (8.25) becomes

$$\sum_{i=1}^{nc} N_i d\left(\frac{\bar{g}_i}{T}\right) = \frac{V}{T}dP - \frac{H^2}{T}dT.$$

We can divide this equation by N_{tot} and state that for the vapor at the bubble point

$$\sum_{i=1}^{nc} y_i d\left(\frac{\bar{g}_i^{\text{V}}}{T}\right) = \frac{\hat{v}^{\text{V}}}{T}dP - \frac{\hat{h}^{\text{V}}}{T^2}dT = \frac{\sum_{i=1}^{nc} y_i \bar{v}_i^{\text{V}}}{T}dP - \frac{\sum_{i=1}^{nc} y_i \bar{h}_i^{\text{V}}}{T^2}dT. \quad (8.27)$$

In addition, given that at the bubble point the liquid composition does not change for an infinitesimal change of temperature along the bubble line, we have that

$$d\left(\frac{\bar{g}_i^{\text{L}}}{T}\right) = \frac{\bar{v}_i^{\text{L}}}{T}dP - \frac{\bar{h}_i^{\text{L}}}{T^2}dT. \quad (8.28)$$

Now, by considering that if (8.23) holds, also

$$\sum_{i=1}^{nc} y_i d\left(\frac{\bar{g}_i^{\text{V}}}{T}\right) = \sum_{i=1}^{nc} y_i d\left(\frac{\bar{g}_i^{\text{L}}}{T}\right)$$

must hold, and by substituting (8.27) and (8.28) in it, and by gathering $\frac{dP}{dT}$ at the left-hand side, we obtain

$$\left(\frac{dP}{dT}\right)^{\text{bubble}} = \frac{\sum_{i=1}^{nc} y_i \bar{h}_i^{\text{V}} - \sum_{i=1}^{nc} y_i \bar{h}_i^{\text{L}}}{T\left(\sum_{i=1}^{nc} y_i \bar{v}_i^{\text{V}} - \sum_{i=1}^{nc} y_i \bar{v}_i^{\text{L}}\right)}. \quad (8.29)$$

This equation shows that the slope of the bubble line depends on the partial molar enthalpies and volumes of the constituents, and on the composition of the vapor phase. As can be expected, this dependence is more complicated than in pure fluids, as it also involves the composition.

An interesting use of the Clapeyron equation for mixtures occurs with multicomponent fluids forming an *azeotrope*, see Section 8.11. Also in this case the overall composition and that of the liquid are the same. The Clapeyron equation for a binary mixture can be used to correlate the states forming azeotropes. This is nicely shown for example in J. P. O'Connell and J. M. Haile, *Thermodynamics: Fundamentals for Applications*. Cambridge University Press, May 2005, page 387.

Gibbs–Duhem equation

A relation that is often useful in phase equilibria problems can be derived by observing that, at constant T and P, $G = \sum_{i=1}^{nc} N_i \bar{g}_i$, therefore the definition of total differential implies that

$$dG_{T,P} = \sum_{i=1}^{nc} N_i d\bar{g}_i + \sum_{i=1}^{nc} \bar{g}_i dN_i.$$

However, we have demonstrated using properties of homogeneous functions that (8.11) is valid if P and T do not vary, thus

$$dG_{T,P} = \sum_{i=1}^{nc} \bar{g}_i dN_i,$$

also holds. Consequently, as these two expressions for the total differential of the Gibbs function are both valid, it must be that

$$\sum_{i=1}^{nc} N_i d\bar{g}_i = 0. \tag{8.30}$$

Equation (8.30) is known as the *Gibbs–Duhem equation*, which can also be written in terms of mole fractions as

$$\sum_{i=1}^{nc} z_i d\bar{g}_i = 0. \tag{8.31}$$

Phase rule for mixtures

With this rule we gain a way of knowing how many intensive thermodynamic variables must be specified in order to determine the equilibrium state of a substance if we know how many phases are present at equilibrium. The phases of a mixture can be gaseous, liquid, and solid. The rule descends from the equilibrium condition, therefore from the First and Second Laws of Thermodynamics.

Consider a mixture in vapor–liquid equilibrium made of nc components. According to (8.22) we have that

$$\bar{g}_i^{\mathrm{L}}(T, P, x_1, x_2, \ldots, x_{nc-1}) = \bar{g}_i^{\mathrm{V}}(T, P, y_1, y_2, \ldots, y_{nc-1})$$
$$\text{where } i = 1, \ldots, nc.$$

Let us call π the number of phases, and provide a general accounting of the number of independent equations as

$$\#\mathrm{E} = (\pi - 1)\, nc.$$

The dependence of $\bar{g}_i^{\mathrm{L}}$ is only from x_1 to x_{nc-1} because $\sum_{i=1}^{nc} x_i = 1$. The same holds for $\bar{g}_i^{\mathrm{V}}$. The independent variables of the problem are

$$T, P, x_1, x_2, \ldots, x_{nc-1}, y_1, y_2, \ldots, y_{nc-1}.$$

Let us now count the total number of variables in the general case of π phases and we get

$$\#\mathrm{V} = \underbrace{(nc - 1)\,\pi}_{x_1,\ldots,x_{nc-1},y_1,\ldots,y_{nc-1}} + \underbrace{2}_{T,P}.$$

The independent variables, or as they are also called, the *degrees of freedom* of the problem, come from subtracting the number of equations (constraints) from the number of independent variables, that is

$$\#\mathrm{DG} = \#\mathrm{V} - \#\mathrm{E} = (nc - 1)\,\pi + 2 - (\pi - 1)\, nc.$$

The *phase rule* states therefore that the number of independent intensive variables that must be fixed in order to determine the equilibrium state of a multiphase system is

$$\#\mathrm{DG} = nc - \pi + 2. \tag{8.32}$$

Example: fixing the state of a binary mixture. The number of thermodynamic variables that must be known in order to determine the equilibrium state of a simple compressible substance if it is single-phase (for example gaseous) is $\#\mathrm{DG} = 1 - 1 + 2 = 2$. A possible choice is therefore T and P. If the fluid is in vapor–liquid equilibrium, $\#\mathrm{DG} = 1 - 2 + 2 = 1$, and we know that if the vapor is saturated, we can specify either the temperature or the pressure, but not both. At the triple point $\pi = 3$, hence it is not possible to change any variable without changing the others, which means that changing one variable causes one of the phases to vanish, see Figure 6.18.

Let us now consider a binary mixture of nitrogen and oxygen in thermodynamic equilibrium. The independent variables are T, P, and z_{O_2} (or z_{N_2}) because $z_{N_2} = 1 - z_{O_2}$. If we have one single phase, for example either gas or liquid, the application of the phase rule yields that in order to fully specify an equilibrium state we have to fix $\#DG = 2 - 1 + 2 = 3$ variables and these are for example T, P, and z_{O_2}. If the mixture is in vapor–liquid equilibrium, the phase rule tells us that $\#DG = 2 - 2 + 2 = 2$ variables must be specified, and these could be P and x_{O_2}, or T and y_{N_2}, or any of the other possible combinations. The charts of Figure 8.1 can further help the understanding of these concepts.

8.6 The Ideal Solution

When treating thermodynamic properties of substances, we often resort to defining first an ideal case, and then obtaining models and relations applicable to real substances by focusing on the deviation or departure from the model describing the idealization. Ideal models are also very useful in a number of practical cases, those for which the idealization implies an acceptable error for a given purpose.

For this reason, when dealing with real mixtures that can exhibit a vapor and a liquid phase, it is useful to define an *ideal solution* as the liquid mixture in which molecular interactions are all the same, independently from the type of molecules in the mixture: by ideally mixing two or more liquids we therefore assume that the total volume of the mixture is given by the weighted summation of the volumes of the constituents, as there are no different molecular interactions (*e.g.*, attraction and repulsion) that can alter this result. The molar volume of the ideal solution, or ideal liquid mixture, is therefore

$$\hat{v}^{IS} \equiv \sum_{i=1}^{nc} x_i \hat{v}_i. \tag{8.33}$$

All molecular interactions are the same: if we mix the liquids A and B, the interactions A $\Leftrightarrow$ A, B $\Leftrightarrow$ B, and A $\Leftrightarrow$ B are all the same. As a consequence, for an ideal solution $\Delta V_{\text{mixing}} = 0$. We note that a mixture of ideal gases also satisfies the ideal solution condition, as in that case also molecules do not interact, therefore $\hat{v}^{IG} = \sum_{i=1}^{nc} \chi_i \hat{v}_i$. In the case of an ideal solution, however, the volume of the liquid of species i CANNOT be computed with the perfect gas equation as $\hat{v}_i = \dfrac{\hat{R}T}{P}$, but it must be computed with another model which is appropriate for liquids.

The absence of difference in molecular interactions not only has consequences on the volume but also on internal energy. Thus again, similarly to an ideal gas mixture, the molar internal energy of an ideal solution is given by

$$\hat{u}^{IS} \equiv \sum_{i=1}^{nc} x_i \hat{u}_i.$$

It follows also that the molar enthalpy of an ideal solution is $\hat{h}^{IS} = \sum_{i=1}^{nc} x_i \hat{h}_i$ and $\Delta H_{\text{mixing}} = 0$. Statistical mechanics, that is the study of the problem at molecular level, taking into account the interaction between molecules, allows us to prove that the mixing of different molecules at a given pressure and temperature and under the assumption of an ideal solution, is such that the molar entropy generation due to mixing is equal to $-\hat{R} \sum_{i=1}^{nc} x_i \ln x_i$.[1] The molar entropy of an ideal solution is therefore given by

$$\hat{s}^{IS} = \sum_{i=1}^{nc} x_i \hat{s}_i - \hat{R} \sum_{i=1}^{nc} x_i \ln x_i. \tag{8.34}$$

Not surprisingly, this expression is analogous to (8.6), which is valid for the perfect gas, though its derivation is conceptually less straightforward.

[1] See, for example, S. I. Sandler, *Chemical, Biochemical and Engineering Thermodynamics*, 4th ed. New York: John Wiley & Sons, 2006, Section 9.1, p. 476.

The definition of Gibbs function then allows us to obtain the equation valid for an ideal solution as

$$\hat{g}^{IS} \equiv \hat{h}^{IS} - T\hat{s}^{IS} = \sum_{i=1}^{nc} x_i\hat{g}_i + \hat{R}T\sum_{i=1}^{nc} x_i \ln x_i. \quad (8.35)$$

Vapor–liquid equilibrium of an ideal mixture

Now we have the elements to obtain the thermodynamic model of a simple, though sometimes applicable, vapor–liquid equilibrium of a mixture. There are mixtures of liquids whereby the interaction between different molecules is so similar to that of molecules of the same type that the ideal solution model provides satisfactory results for engineering applications. If the mixture is in equilibrium with its vapor at low pressure, the vapor closely obeys the perfect gas law. This is the case for example of a mixture of propane and n-pentane at low reduced pressure and temperature. In addition, at that thermodynamic state, the change of properties of the liquid is almost independent of pressure.

The vapor–liquid equilibrium equations for a mixture at fixed T and P (8.22) in the case where the liquid phase is an ideal solution and the vapor phase a perfect gas mixture become, see (8.7) and (8.35),

$$\hat{g}_i^{\,V}(T,P) + \hat{R}T\ln y_i = \hat{g}_i^{\,L}(T,P) + \hat{R}T\ln x_i, \quad (8.36)$$

which can be written as

$$\hat{g}_i^{\,V}(T,P) - \hat{g}_i^{\,L}(T,P) = \hat{R}T\ln\frac{x_i}{y_i}. \quad (8.37)$$

If the assumption of low pressure holds, the pressure has negligible influence on the properties of the liquid, therefore

$$\hat{g}_i^{\,L}(T,P) = \hat{g}_i^{\,L}(T,P_i^{sat}). \quad (8.38)$$

Let us find a way to write $\hat{g}_i^{\,V}(T,P)$ as a function of the saturation pressure of species i such that we can exploit the fact that for a pure species i in saturated conditions $\hat{g}_i^{\,L}(T,P_i^{sat}) = \hat{g}_i^{\,V}(T,P_i^{sat})$. In order to do this, we can find the difference between $\hat{g}_i^{\,V}(T,P_i^{sat})$ and $\hat{g}_i^{\,V}(T,P)$ by integrating the function $\hat{g}_i^{\,V}$ at constant temperature. For the perfect gas we have that at constant temperature $d\hat{g}_i^{\,V} = \hat{v}dP$, therefore

$$\hat{g}_i^{\,V}(T,P_i^{sat}) - \hat{g}_i^{\,V}(T,P) = \int_P^{P_i^{sat}} \frac{\hat{R}T}{P}dP = \hat{R}T\ln\frac{P_i^{sat}}{P},$$

and

$$\hat{g}_i^{\,V}(T,P) = \hat{g}_i^{\,V}(T,P_i^{sat}) - \hat{R}T\ln\frac{P_i^{sat}}{P}. \quad (8.39)$$

Let us now plug (8.38) and (8.39) into (8.37) and we obtain

$$\cancel{\hat{g}_i^{\,V}(T,P_i^{sat})} - \hat{R}T\ln\frac{P_i^{sat}}{P} - \cancel{\hat{g}_i^{\,L}(T,P_i^{sat})} = \hat{R}T\ln\frac{x_i}{y_i},$$

which reduces to

$$x_i P_i^{sat} = y_i P \qquad i = 1, 2, \ldots, nc. \quad (8.40)$$

These equations are known as *Raoult's law*. Its application allows us to easily compute the vapor–liquid equilibrium state of an ideal mixture if the properties of the components are known.

Example: P-xy chart using Raoult's law. A fluid consisting of nitrogen (N_2) and oxygen (O_2) at low reduced temperature and pressure is an example of an ideal binary mixture. We want to obtain a P–xy chart for these mixtures at $T = 100$ K using Raoult's law. The saturation pressures at this temperature, calculated with STANMIX, are

$$P_{N_2}^{sat}(T = 100\text{ K}) = 0.779\text{ MPa},$$
$$P_{O_2}^{sat}(T = 100\text{ K}) = 0.254\text{ MPa}.$$

The application of Raoult's law to this mixture results in

$$x_{N_2}P_{N_2}^{sat} = y_{N_2}P,$$
$$x_{O_2}P_{O_2}^{sat} = y_{O_2}P,$$

and because it is a binary mixture, we have that $x_{O_2} = 1 - x_{N_2}$ and $y_{O_2} = 1 - y_{N_2}$. Let us choose x_{N_2} as the independent variable of the P–xy chart at $T = 100$ K. The bubble line is obtained by plotting P as a function of x_{N_2}, while the dew line is obtained by plotting y_{N_2} at the pressure corresponding to x_{N_2}. We need therefore two independent equations. The equation for the bubble line is

$$P = x_{N_2} P_{N_2}^{sat} + \left(1 - x_{N_2}\right) P_{O_2}^{sat},$$

while that for the dew line is

$$y_{N_2} = x_{N_2} \frac{P_{N_2}^{sat}}{P}.$$

The result is the P–xy diagram of Figure 8.8, which you can easily replicate with, for example, EXCEL or MATLAB. The comparison with experimental data shows you that the ideal mixture model provides good estimates, though a certain deviation can be observed. You might want to check if one of the FLUIDPROP mixture models predicts values that are even closer to measurements. Observe that the bubble line is straight, given the linear dependence of P^{bubble} on the mole fraction, a feature of ideal binary mixtures. The deviation from such a straight line can be used to visually evaluate the deviation of the vapor–liquid equilibrium behavior of mixtures from ideality (*non-ideality* of a mixture).

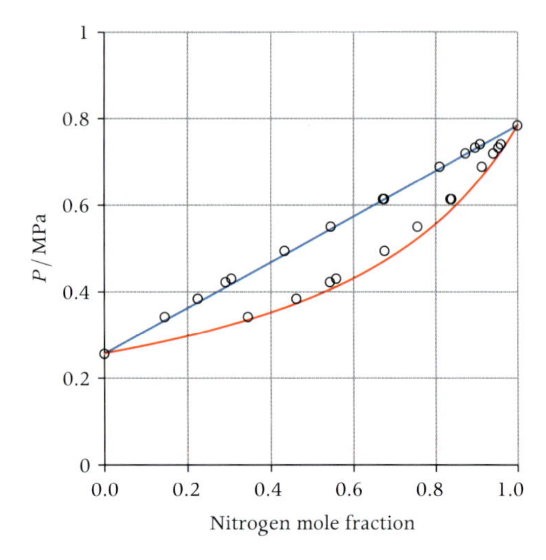

Figure 8.8 The P–xy chart of nitrogen/oxygen mixtures at $T = 100$ K. The circles are experimental data of A. Baba-Ahmed, P. Guilbot, and D. Richon, "New equipment using a static analytic method for the study of vapour–liquid equilibria at temperatures down to 77 K," *Fluid Phase Equilib.*, 166, (2), 225–236, 1999. The lines (blue: bubble, red: dew line) are calculated with the help of Raoult's law.

8.7 Fugacity and Fugacity Coefficient

The fugacity function introduced in Section 6.11, page 131, is equally useful in the calculation of mixture vapor–liquid equilibria. Recall that fugacity is an auxiliary function brought in to avoid numerical problems due to values of the specific Gibbs function that become very large at low pressure.

For a pure fluid at a given P and T the fugacity is defined as $f \equiv P \exp\left(g - \frac{g^{\text{IG}}}{RT}\right)$, or, implicitly, by the relation $\ln \frac{f}{P} = \frac{g - g^{\text{IG}}}{RT}$. The fugacity of component i *in the mixture* is bound to be different from the fugacity of pure i, because of molecular interactions affecting the value of all thermodynamic properties. For this reason the fugacity of a fluid in the mixture is defined for convenience via an appropriate relation to the partial molar Gibbs function as

$$\tilde{f}_i \equiv \chi_i P \exp\left(\frac{\bar{g}_i - \bar{g}_i^{\text{IG}}}{\hat{R}T}\right). \tag{8.41}$$

Note that here the tilde stands for a property of a fluid when mixed with other substances, but that the fugacity is not a partial molar property, as $\hat{f} \neq \sum_{i=1}^{nc} z_i \bar{f}_i$. The notation for the fugacity of component i is therefore different from that of a partial molar property. The definition of $\tilde{f}_i$ can also be obtained implicitly from the relation

$$\bar{g}_i - \bar{g}_i^{\text{IG}} = \hat{R}T \ln\left(\frac{\tilde{f}_i}{\chi_i P}\right). \tag{8.42}$$

In some textbooks this equation is given in the form $\bar{g}_i = \hat{R}T \ln \tilde{f}_i + \mathcal{F}_i(T)$, where $\mathcal{F}_i(T)$ comes from grouping the ideal gas terms. We do not like this form of the relation defining the fugacity of a mixture constituent because it is incorrect with respect to dimensions, the unit of fugacity being Pa.

Equilibrium using the fugacity and fugacity coefficient functions

The fugacity function is useful to numerically solve the vapor–liquid equilibrium conditions for a mixture, namely $\bar{g}_i^{\mathrm{L}} = \bar{g}_i^{\mathrm{V}}$, for $i = 1, \ldots, nc$. If we substitute (8.42) for both the liquid and the vapor partial molar Gibbs functions of the ith mixture component, the ideal gas terms cancel out, and we obtain the equivalent conditions for equilibrium

$$\tilde{f}_i^{\mathrm{L}} = \tilde{f}_i^{\mathrm{V}} \qquad i = 1, \ldots, nc. \tag{8.43}$$

Let us now introduce the fugacity coefficient of constituent i in the mixture $\tilde{\phi}_i$ in analogy with what was done for a pure substance. The partial residual molar Gibbs function of component i in the mixture is

$$\bar{g}_i^{\mathrm{R}} = \bar{g}_i - \bar{g}_i^{\mathrm{IG}} = \bar{g}_i - \bar{g}_i^{\mathrm{IG}} = \hat{R}T \ln\left(\frac{\tilde{f}_i}{\chi_i P}\right).$$

We define the fugacity coefficient of component i in the mixture as

$$\tilde{\phi}_i \equiv \frac{\tilde{f}_i}{y_i P}, \tag{8.44}$$

so that it allows us to write the residual of the partial molar Gibbs function of component i in the mixture as

$$\bar{g}_i^{\mathrm{R}} = \hat{R}T \ln \tilde{\phi}_i. \tag{8.45}$$

The relation between the fugacity coefficient of component i in the mixture and the residual of its partial molar Gibbs function can therefore be conveniently expressed as

$$\ln \tilde{\phi}_i = \frac{\bar{g}_i^{\mathrm{R}}}{\hat{R}T}. \tag{8.46}$$

Finally, the equilibrium conditions in terms of partial molar Gibbs function can be rewritten in terms of the fugacity coefficient of component i in the mixture as illustrated in the following steps. At constant T and P we have

$$\begin{aligned} \bar{g}_i &= \bar{g}_i^{\mathrm{IG}} + \bar{g}_i^{\mathrm{R}}, \\ \bar{g}_i^{\mathrm{IG}} &= \hat{g}_i + \hat{R}T \ln \chi_i, \\ \bar{g}_i^{\mathrm{R}} &= \hat{R}T \ln \tilde{\phi}_i. \end{aligned}$$

If we now indicate with y_i the mole fraction of component i in the vapor phase and with x_i the mole fraction in the liquid phase, the equilibrium conditions $\bar{g}_i^{\mathrm{L}} = \bar{g}_i^{\mathrm{V}}$ at given T and P become

$$\bar{g}_i^{\mathrm{IG}} + \hat{R}T \ln x_i + \hat{R}T \ln \tilde{\phi}_i^{\mathrm{L}} = \bar{g}_i^{\mathrm{IG}} + \hat{R}T \ln y_i + \hat{R}T \ln \tilde{\phi}_i^{\mathrm{V}}.$$

The molar Gibbs function of species i in the ideal gas and unmixed state cancel out, therefore the equilibrium conditions at fixed T and P reduce to

$$y_i \tilde{\phi}_i^{\mathrm{V}} = x_i \tilde{\phi}_i^{\mathrm{L}} \qquad i = 1, \ldots, nc. \tag{8.47}$$

This system of nonlinear equations is solved when computing the vapor–liquid equilibrium state of a multicomponent fluid using a fluid model based on an equation of state, as we shall show in Section 8.10. The fugacity function $\tilde{\phi}_i$ can be obtained for both the vapor and the liquid phase from an equation of state, similarly to what was shown for a pure fluid in the example at page 132 of Section 6.11. It is well-behaved from a numerical point of view, as it does not diverge at low pressure and it features values that are between 0 and a few units.

Observe that the fugacity coefficient is related to the non-ideality of a mixture because $\bar{g}_i^{\mathrm{R}}$ is a measure of the deviation from ideal behavior of component i in the mixture. Such a relation is shown by

$$\ln \tilde{\phi}_i = \frac{\bar{g}_i^{\mathrm{R}}}{\hat{R}T} = \int_0^P \frac{\bar{v}_i^{\mathrm{R}}}{\hat{R}T} dP \qquad (T = \text{const.}).$$

If the molecules strongly interact, the partial molar volume of constituent i is very different from the value of the volume that the fluid i would have if unmixed, consequently the residual molar volume is large, and the fugacity coefficient is thus very different from its ideal value of 1. The deviation of $\tilde{\phi}_i$ from 1 provides a measure of the non-ideality of the interaction of i with other molecules in a mixture.

Example: fugacity of a species in a mixture of perfect gases. Let us see what happens if we apply the definition of fugacity and fugacity coefficient to a component in a mixture of perfect gases. The residual partial molar Gibbs function of a substance i that behaves as a perfect gas must be zero, therefore $\tilde{\phi}_i^{\mathrm{IG}} = 1$ and

$$\tilde{f}_i^{\mathrm{IG}} = y_i P, \tag{8.48}$$

which shows that the fugacity of constituent i is equal to the partial pressure of i if the mixture behaves in accordance with the perfect gas model.

8.8 Activity Coefficient Models for the Liquid Phase

In order to obtain thermodynamic models that are applicable to real mixtures, therefore accounting for the interactions between the various molecules, we need mathematical models at microscopic or macroscopic level that can describe the sheer complexity and variety of the forces acting between molecules, i.e., of their molecular potentials, as we have seen in Section 6.9, page 126. Mixtures of molecules of different dimensions, generating dipole or quadrupole interactions, affected by hydrogen bonding and other types of interaction reach vapor–liquid equilibrium states that are very different from those that can be modeled according to the ideal solution assumption.

Similarly to what is done to model the equilibrium states of pure fluids, two paths are possible. Modern research on the modeling of thermodynamic properties of fluids often focuses on first-principle models centered on the description of molecular potential. Statistical mechanics and heavy computations then allow obtaining mixture properties. The other more practical approach, often used in engineering, is based on carefully selected measurements of mixture equilibria states and correlation of measurements with equations that are then incorporated into the complete thermodynamic model. The models based on these correlations have predictive capabilities (accurate computation of states outside the range for which measurements are available), if they are based on thermodynamically-based functions, and if they comply with thermodynamic constraints. Vapor–liquid experiments aimed at measuring the pressure, temperature, and compositions of the vapor and of the liquid phase are performed with devices like the one depicted in Figure 8.9.

There exist large databases collecting the results of vapor–liquid equilibria experiments, which make it possible to obtain models of various degree of accuracy for a large variety of mixtures.

Well-known databases collecting mixture vapor–liquid equilibrium data are:

- DECHEMA Chemistry data series (www.dechema.de/en/CDS.html)
- Dortmund Data Bank (www.ddbst.com)
- Infotherm Database of Thermophysical Properties (www.infotherm.com),
- Korean Thermophysical Properties DataBank (www.cheric.org/research/kdb/)
- NIST ThermoData Engine (http://trc.nist.gov/tde.html).

An approach not depending on actual vapor–liquid measurements, but able to predict how molecules interact by knowing their potential would be general, and its benefit apparent, but it is still a topic driving research in this field of science.

We treat here models describing the non-ideal interaction between molecules in a mixture from a macroscopic point of view. These models are based on experimental data, in particular related to non-ideal molecular interaction within the liquid phase, and their development starts from the usual trick of mathematically describing the non-ideal interaction as a deviation from ideal behavior. The *activity coefficient* is another thermodynamic function traditionally used in models capable of correlating non-ideal vapor–liquid equilibrium data at low reduced pressure. The description of a general and complete model for mixtures that is not limited to low reduced pressure VLE, and that extends activity coefficient models is given in Section 8.10.

Activity coefficient

The vapor–liquid equilibrium condition for a mixture at given T and P can be expressed in terms of a partial molar Gibbs function of its constituents (8.22). Let us focus on the liquid phase, and define the deviation of

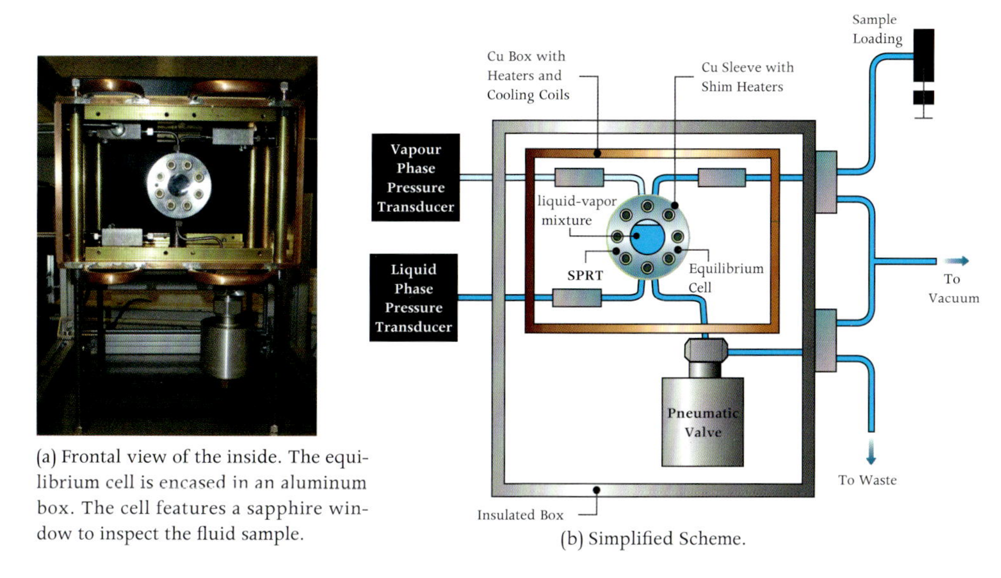

(a) Frontal view of the inside. The equilibrium cell is encased in an aluminum box. The cell features a sapphire window to inspect the fluid sample.

(b) Simplified Scheme.

Figure 8.9 Apparatus for the measurement of vapor–liquid equilibria at the National Institute of Standards and Technology in Boulder, CO, USA. Temperature is controlled with a two stage thermostat consisting of a heated copper sleeve and a temperature controlled copper box. The specifications are: 270 K $< T <$ 370 K, $P <$ 7 MPa. **From S.L. Outcalt, B.C. Lee, "A Small-Volume Apparatus for the Measurement of Phase Equilibria," J. Res. Natl. Inst. Stand. Technol. 2004, 109, 525–531.**

G from the value it would have if the solution were ideal as the *excess Gibbs function*

$$G^{\mathrm{E}} = G - G^{\mathrm{IS}}. \tag{8.49}$$

All other extensive properties can be treated in the same way, and excess properties similarly defined. At constant T and P and recalling (8.21) we therefore have that

$$\bar{g}_i^{\mathrm{E}} = \bar{g}_i - \bar{g}_i^{\mathrm{IS}},$$

which we call the partial molar excess Gibbs function of component i in the mixture.

Let us write this quantity as a function of fugacities, by making explicit the dependence of the partial molar Gibbs function of component i in the perfect gas state. This results in

$$\bar{g}_i^{\mathrm{E}} = \bar{g}_i - \bar{g}_i^{\mathrm{IS}} = \left(\bar{g}_i - \bar{g}_i^{\mathrm{IG}}\right) + \left(\bar{g}_i^{\mathrm{IG}} - \bar{g}_i^{\mathrm{IS}}\right). \tag{8.50}$$

Recalling that for the ideal gas $\bar{g}_i^{\mathrm{IG}} = \hat{g}_i^{\mathrm{IG}} + \hat{R}T\ln x_i$ and for the ideal solution $\bar{g}_i^{\mathrm{IS}} = \hat{g}_i + \hat{R}T\ln x_i$, and substituting these expressions in (8.50) gives

$$\bar{g}_i^{\mathrm{E}} = \left(\bar{g}_i - \bar{g}_i^{\mathrm{IG}}\right) + \left(\hat{g}_i^{\mathrm{IG}} + \cancel{\hat{R}T\ln x_i} - \hat{g}_i - \cancel{\hat{R}T\ln x_i}\right). \tag{8.51}$$

We thus recognize that

$$\bar{g}_i - \bar{g}_i^{\mathrm{IG}} = \bar{g}_i^{\mathrm{R}} = \hat{R}T\ln\tilde{\phi}_i = \hat{R}T\ln\frac{\tilde{f}_i}{x_i P},$$

by combining (8.51) with (8.44) and (8.46). Similarly, from (6.112) we see that for a pure fluid

$$\hat{g}_i - \hat{g}_i^{\mathrm{IG}} = \hat{g}_i^{\mathrm{R}} = \hat{R}T\ln\frac{f_i}{P}.$$

Equation (8.51) then becomes

$$\bar{g}_i^{\mathrm{E}} = \hat{R}T\ln\frac{\tilde{f}_i}{x_i P} - \hat{R}T\ln\frac{f_i}{P} = \hat{R}T\ln\frac{\tilde{f}_i}{x_i f_i},$$

where $\tilde{f}_i$ is the fugacity of component i in the solution, while f_i is the fugacity of i when it is not mixed. By analogy with the fugacity coefficient and the partial residual Gibbs function of a vapor, it is convenient in the case of the partial excess Gibbs function of a liquid to define the *activity coefficient* of component i in the solution as

$$\gamma_i \equiv \frac{\tilde{f}_i}{x_i f_i}, \qquad (8.52)$$

such that we have

$$\ln \gamma_i = \frac{\bar{g}_i^{\mathrm{E}}}{\hat{R}T}. \qquad (8.53)$$

The ideal solution is characterized by $\bar{g}_i^{\mathrm{E}} = 0$, therefore in an ideal solution all activity coefficients are unity. Given that $\bar{g}_i^{\mathrm{E}}$ is a partial molar property, we have some other useful relations, namely

$$\hat{R}T \ln \gamma_i = \left(\frac{\partial G^{\mathrm{E}}}{\partial N_i}\right)_{T,P,N_{j\neq i}}, \qquad (8.54)$$

and

$$\frac{\hat{g}^{\mathrm{E}}}{\hat{R}T} = \sum_{i=1}^{nc} x_i \ln \gamma_i. \qquad (8.55)$$

Moreover, note that the Gibbs–Duhem equation (8.31) is valid for any partial molar property, and the activity coefficient is a partial molar property, therefore at constant T and P we also have that

$$\sum_{i=1}^{nc} x_i d \ln \gamma_i = 0. \qquad (8.56)$$

These equations are useful because we shall see that it is possible to determine γ_i from vapor–liquid equilibria (VLE) measurements, which in turn allow for the computation of the VLE states of non-ideal mixtures.

Example: activity coefficients from experimental data. Let us learn how activity coefficients can be obtained from experimental VLE data with an example. Consider the VLE measurements related to a binary mixture of methanol and acetone listed in Table 8.2, which are also reported on the P–xy chart of Figure 8.10. The measurements were taken at low reduced temperature and pressure, therefore the assumption that the vapor phase can be modeled as a perfect gas is valid. Now we want to take into account that the liquid phase deviates from the ideal solution behavior in order to model the vapor–liquid equilibrium states of this mixture more accurately. We can write the fugacity conditions for vapor–liquid equilibrium as

$$y_i P = x_i f_i \gamma_i, \qquad i = 1, 2, \ldots, nc, \qquad (8.57)$$

having noticed that for the fugacity of the ideal gas we can use (8.48), while for the fugacity of the non-ideal liquid we can use the definition of activity coefficient (8.52). The fugacity of liquid i at T and P unmixed (f_i) is rather easy to evaluate. The fugacity of liquid i at temperature T can be evaluated starting from the fugacity of i in saturated conditions f_i^{sat}. At constant T we have from (6.112) and (6.111) that

$$d\hat{g}_i = \hat{v}_i dP = \hat{R}T d \ln f_i \qquad T = \text{const.},$$

hence

$$d \ln f_i = \frac{\hat{v}_i}{\hat{R}T} dP \qquad T = \text{const.} \qquad (8.58)$$

We can now integrate (8.58) from a subcooled state at temperature T to the saturated state and we obtain

$$\ln \frac{f_i}{f_i^{\mathrm{sat}}} = \frac{1}{\hat{R}T} \int_{P_i^{\mathrm{sat}}}^{P_i} \hat{v}_i dP. \qquad (8.59)$$

At low reduced pressure the compressibility of the liquid is negligible, therefore it can safely be assumed that $\hat{v}_i = \hat{v}_i^{\mathrm{sat}} = \text{const.}$, therefore

$$\ln \frac{f_i}{f_i^{\mathrm{sat}}} = \frac{\hat{v}_i^{\mathrm{sat}} \left(P_i - P_i^{\mathrm{sat}}\right)}{\hat{R}T}.$$

Introducing the definition of fugacity coefficient and solving for the fugacity of the liquid at T we have that

$$f_i = \phi_i^{\mathrm{sat}} P_i^{\mathrm{sat}} \exp \frac{\hat{v}_i^{\mathrm{sat}} \left(P_i - P_i^{\mathrm{sat}}\right)}{\hat{R}T}. \qquad (8.60)$$

At low reduced pressure the ideal gas assumption applies to the vapor phase, therefore $\phi_i^{\mathrm{sat,L}} = \phi_i^{\mathrm{sat,V}} = 1$, moreover it can be verified numerically that

$$\exp \frac{\hat{v}_i^{\mathrm{sat}} \left(P_i - P_i^{\mathrm{sat}}\right)}{\hat{R}T} \approx 1.$$

We can thus conclude that at low reduced pressure for the liquid phase $f_i \approx P_i^{\mathrm{sat}}$, therefore

$$y_i P = x_i P_i^{\mathrm{sat}} \gamma_i, \qquad i = 1, 2, \ldots, nc. \qquad (8.61)$$

This development shows that the choice of expressing f_i in terms of f_i^{sat} is quite natural, and the resulting Equation (8.61) is practical. However, this choice limits the applicability of the activity-coefficient approach to substance i whose temperature is lower than the critical temperature (and greater than the triple-point temperature), so that P_i^{sat} is defined. Conversely the pressure P can very well exceed the critical pressure of substance i.

Equation (8.61) makes the calculation of γ_1 and γ_2 for all the measurement points straightforward, and the values are reported in Table 8.2.

It thus becomes apparent that in order to make use of this information in a general model, which provides values of the activity coefficients for any composition at the given temperature, we need to correlate these data in order to obtain a suitable equation. In addition, the activity coefficient model must comply with constraints that come from thermodynamics. This is shown in the following for the case of the simplest activity coefficient model applicable to non-ideal binary mixtures.

Table 8.2 Experimental VLE data for methanol/acetone mixtures at 298.15 K taken from A. Tamir, A. Apelblat, and M. Wagner, "An evaluation of thermodynamic analyses of the vapor–liquid equilibria in the ternary system acetone + chloroform + methanol and its binaries," *Fluid Phase Equilib.*, *6*, (1–2), 113–139, 1981, together with the corresponding calculated activity coefficients whereby the vapor is considered as a perfect gas and the saturation pressures of methanol and acetone are calculated with StanMix.

P / kPa	$x_{methanol}$	$y_{methanol}$	$\gamma_{methanol}$	$\gamma_{acetone}$
30.73	0	0		1.007
30.76	0.122	0.145	2.171	0.982
30.82	0.188	0.190	1.850	1.008
30.80	0.214	0.193	1.649	1.036
29.96	0.339	0.259	1.359	1.101
29.44	0.418	0.305	1.275	1.152
28.16	0.540	0.377	1.167	1.250
27.36	0.598	0.402	1.092	1.334
24.45	0.764	0.532	1.011	1.589
23.81	0.796	0.565	1.004	1.664
20.80	0.906	0.710	0.968	2.103
20.43	0.914	0.732	0.971	2.086
19.49	0.939	0.784	0.966	2.262
17.03	1	1	1.011	

Margules equations

Valid experimental data must satisfy several thermodynamic constraints that are obtained solely by application of first principles. These constraints can be used to set up fitting equations that allow us to obtain a quantity representing the non-ideality of a liquid mixture, such as the excess Gibbs function. For a binary mixture, Equation (8.55) turns into

$$\frac{\hat{g}^E}{\hat{R}T} = x_1 \ln \gamma_1 + x_2 \ln \gamma_2. \tag{8.62}$$

Plotting the values of $\frac{\hat{g}^E}{\hat{R}T}$ that can be obtained from Table 8.2 as a function of x_1 provides the points indicated by green circles in the chart of Figure 8.11. We are looking for a continuous and differentiable function that can interpolate these data, as the activity coefficient can be obtained from it using (8.54). This function must also go to zero as $x_1 \to 0$ and $x_1 \to 1$, as we know that the excess Gibbs function of a pure fluid is zero by definition.

Furthermore the sought function must satisfy the *Gibbs–Duhem* equation for the partial molar property γ (8.56), which for a binary mixture reads

$$x_1 d \ln \gamma_1 + x_2 d \ln \gamma_2 = 0.$$

Let us divide this equation by dx_1 and observe that $\frac{d \ln \gamma_1}{dx_1}$ and $\frac{d \ln \gamma_2}{dx_1}$ are the slopes of the curves $\ln \gamma_1$ and $\ln \gamma_2$ plotted on a $\ln \gamma - x_1$ chart as in Figure 8.11. The *Gibbs–Duhem* equation imposes a constraint on

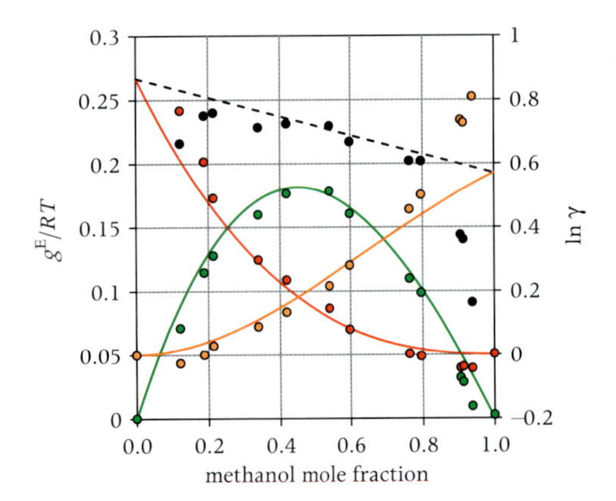

Figure 8.10 The *P–xy* diagram of methanol/acetone mixtures at $T = 298.15$ K. The circles are experimental data from A. Tamir, A. Apelblat, and M. Wagner, "An evaluation of thermodynamic analyses of the vapor–liquid equilibria in the ternary system acetone + chloroform + methanol and its binaries," *Fluid Phase Equilib.*, 6, (1–2), 113–139, 1981. The continuous lines (blue: bubble, red: dew line) are calculated with the help of the Margules equations, the dashed lines with Raoult's law.

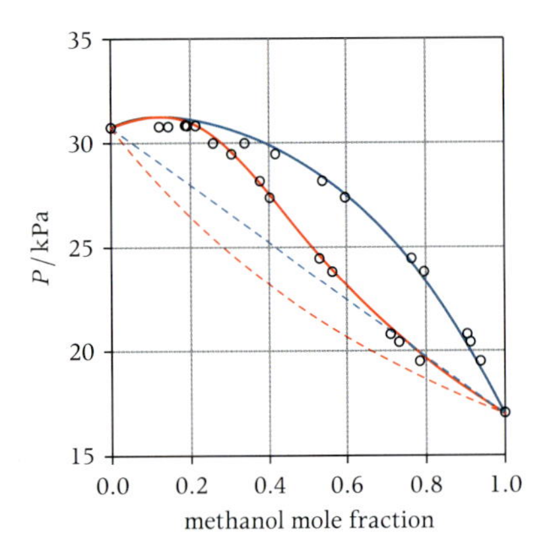

Figure 8.11 Values of $\hat{g}^E/\hat{R}T$ (green line), $\ln \gamma_{\text{methanol}}$, $\ln \gamma_{\text{acetone}}$ (purple and orange line, respectively), and $g^E/x_1x_2\hat{R}T$ (dashed line) as a function of x_1 for methanol/acetone mixture calculated with the Margules equation. The circles represent values calculated from the experimental data of Table 8.2.

the $\dfrac{\hat{g}^E}{\hat{R}T}$ function that the slopes of the $\ln \gamma_{1,2}$ functions obey the relation

$$\frac{d \ln \gamma_1}{dx_1} = -\frac{x_2}{x_1}\frac{d \ln \gamma_2}{dx_2}. \tag{8.63}$$

This relation leads to the following constraints that can be visually verified on the chart of Figure 8.11:

- both $\ln \gamma_i(x_1)$ curves must become horizontal as $x_i \to 1$; and
- the slopes of the $\ln \gamma_1(x_1)$ and $\ln \gamma_2(x_1)$ lines must have opposite sign for any value of the composition.

We also notice by examining the data in Figure 8.11 that $\ln \gamma_1$ tend to a finite value for $x_1 \to 0$ and $\ln \gamma_2$ does the same for $x_2 \to 0$. This means that both of them tend to a certain value at *infinite dilution*. We have therefore that

$$\lim_{x_1 \to 0} \frac{\hat{g}^E}{\hat{R}T} = 0 \times \ln \gamma_{1,\infty} + 1 \times 0 = 0.$$

Let us demonstrate that

$$\ln \gamma_{1,\infty} = \lim_{x_1 \to 0} \frac{\hat{g}^E}{x_1 x_2 \hat{R}T}.$$

The function $\dfrac{\hat{g}^E}{x_1x_2\hat{R}T}$ is not determined for $x_1 = 0$ and $x_1 = 1$ because $\hat{g}^E = 0$ in both cases, as is the product x_1x_2. We can though calculate its limiting value for $x_1 \to 0$,

$$\lim_{x_1 \to 0} \frac{\hat{g}^E}{x_1x_2\hat{R}T} = \lim_{x_1 \to 0} \frac{1}{x_1}\left(\frac{\hat{g}^E}{\hat{R}T}\right) = \lim_{x_1 \to 0} \frac{d\left(\hat{g}^E/\hat{R}T\right)}{dx_1},$$

where the last term is obtained by means of l'Hôpital's rule. From (8.62) we have that

$$\frac{d\left(\hat{g}^E/\hat{R}T\right)}{dx_1} = x_1\frac{d\ln\gamma_1}{dx_1} + \ln\gamma_1 + x_2\frac{d\ln\gamma_2}{dx_1} - \ln\gamma_2.$$

Because of (8.63) we therefore obtain

$$\frac{d\left(\hat{g}^E/\hat{R}T\right)}{dx_1} = \ln\frac{\gamma_1}{\gamma_2},$$

which allows us to demonstrate what we wanted, as we have that

$$\lim_{x_1 \to 0} \frac{\hat{g}^E}{x_1x_2\hat{R}T} = \lim_{x_1 \to 0} \frac{d\left(\hat{g}^E/\hat{R}T\right)}{dx_1} = \lim_{x_1 \to 0} \ln\frac{\gamma_1}{\gamma_2} = \ln\gamma_{1,\infty}.$$

Similarly, we have that

$$\ln \gamma_{2,\infty} = \lim_{x_1 \to 1} \frac{\hat{g}^E}{x_1 x_2 \hat{R}T}.$$

The thermodynamic function $\frac{\hat{g}^{\mathrm{E}}}{x_1 x_2 \hat{R} T}$ must therefore have these properties, and it looks like a good idea to also plot the values obtained from experimental data as a function of x_1, which we can see as black circles on Figure 8.11. We note that these points can be interpolated reasonably well with a linear function of the mole fractions x_1 and x_2, especially for values of the composition that are not too close to $x_1 = 0$ and $x_1 = 1$. Note also that the infinite dilution values of the activity coefficients serve as the coefficients of such a function. This linear function is

$$\frac{\hat{g}^{\mathrm{E}}}{x_1 x_2 \hat{R} T} = \ln \gamma_{2,\infty} x_1 + \ln \gamma_{1,\infty} x_2.$$

This relation, *by construction*, provides us with an excess Gibbs function for binary mixtures that satisfies the necessary thermodynamic constraints. The *Margules equation* for the correlation of low pressure non-ideal VLE data is therefore

$$\frac{\hat{g}^{\mathrm{E}}}{\hat{R} T} = x_1 x_2 \left(A_{21} x_1 + A_{12} x_2 \right), \qquad (8.64)$$

with

$$A_{21} = \ln \gamma_{2,\infty}, \qquad A_{12} = \ln \gamma_{1,\infty}.$$

The equations for the activity coefficients are readily obtained from (8.54), and they are

$$\ln \gamma_1 = x_2^2 \left[A_{12} + 2 \left(A_{21} - A_{12} \right) x_1 \right], \qquad (8.65a)$$

$$\ln \gamma_2 = x_1^2 \left[A_{21} + 2 \left(A_{12} - A_{21} \right) x_2 \right]. \qquad (8.65b)$$

The Margules equation, its corresponding $\frac{g^{\mathrm{E}}}{x_1 x_2 \hat{R} T}$ function, together with its activity coefficients are reported as continuous lines on the chart of Figure 8.11. VLE states at various temperatures and pressures may be calculated with

$$y_i P = x_i \gamma_i P_i^{\mathrm{sat}} \qquad i = 1, 2.$$

Results of calculations employing this equation are shown in Figure 8.10, where it can be seen that the model allows for sufficient accuracy as far as engineering use is concerned. On the contrary, it can be seen that the ideal solution approximation does not hold for this mixture of polar compounds.

Inspection of Figure 8.11 reveals that, even though the Margules equations provide a fairly good match for the data obtained from experiments, there exists a deviation for a range of compositions for which measurements are available. In general this means that either the functional form is not optimal, or that the experimental data are affected by errors. Experimental data of other more non-ideal mixtures are poorly correlated by the Margules equation, as shown for example in Figure 8.12. The reason is that this simple model does not incorporate physical information on molecular interaction, but only correlates experimental data and complies with basic thermodynamic contraints. This activity coefficient model is affected therefore by some limitations, namely

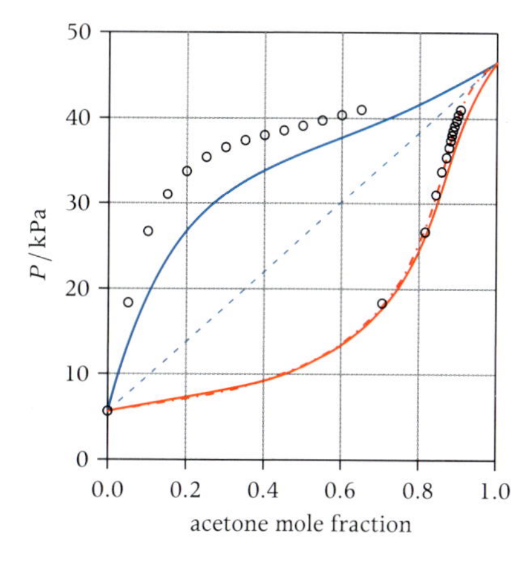

Figure 8.12 The *P*–*xy* chart of acetone/water mixtures at $T = 308$ K. The symbols represent experimental data of I. Lieberwirth and H. Schuberth, "Das isotherme dampf-flüssigkeits-phasengliechgewichtsverhalten des systems aceton/wasser bei 35 °C," *Z. Phys. Chem. (Leipzig)*, 260, 669–672, 1979. The continuous lines (blue: bubble, red: dew line) are calculated with the help of the Margules equations, the dashed lines with Raoult's law.

- it is only valid for binary mixtures,
- it is applicable only to mildly non-ideal mixtures.

A large body of literature documents a variety of activity coefficient models of various degree of complexity that are applicable to multicomponent mixtures and also if the degree of non-ideality is high.[2] This treatment is outside the scope of this book. Section 8.10 describes how the information on molecular interaction coming from liquid activity coefficient models that are valid only at low pressure can be incorporated into a complete and consistent thermodynamic model that is therefore valid also at high pressure and temperature. This model for the estimation of fluid-mixture thermodynamic properties is implemented in STANMIX and can calculate the needed excess Gibbs function at low pressure with the non-random two-liquid (NRTL) activity coefficient model. The NRTL activity coefficient model is suitable for the modeling of multicomponent mixtures of a fairly high degree of non-ideality, and even partially miscible systems. For example, the NRTL model correlates quite well with the measurements of Figure 8.12.

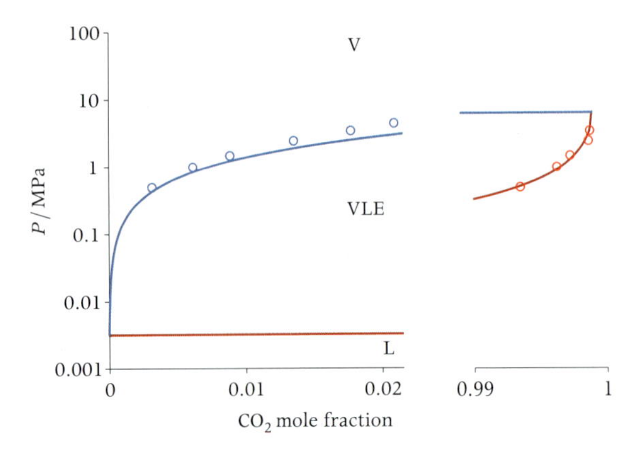

Figure 8.13 Vapor–liquid equilibrium of H_2O/CO_2 mixtures at 298 K and for $x_{CO_2} \to 0$ and $x_{CO_2} \to 1$. Circles are experimental data taken from A. Valtz, A. Chapoy, C. Coquelet, *et al.*, "Vapour–liquid equilibria in the carbon dioxide–water system, measurement and modelling from 278.2 to 318.2 K," *Fluid Phase Equilib.*, 226, (1–2), 333–344, 2004, while the bubble line (blue) and the dew line (red) are calculated with STANMIX .

8.9 Dilute Solution

Mixtures formed by a liquid in which a small amount of gas is dissolved are relevant in engineering applications. Let us consider an even more familiar example: a bottle of sparkling water. Water and a little bit of carbon dioxide are mixed at ambient temperature, say 25 °C, and at a pressure which is a little higher than atmospheric. At the given conditions, water is liquid, while carbon dioxide is a gas. When they are mixed, a small amount of carbon dioxide (solute) is dissolved into the water (solvent), and they form a solution. Figure 8.13 shows parts of the phase diagram of this mixture. The vapor–liquid equilibrium for compositions other than $x_{CO_2} \to 0$ and $x_{CO_2} \to 1$ is very complex, and can entail chemical reactions and more phases. It is therefore very challenging to model, thus only values for compositions in which there is almost solely water or almost solely carbon dioxide are shown. Certainly the ideal mixture assumption does not hold, nor is the Margules model able to describe this mixture correctly due to the highly non-ideal behavior. For the concentrations of CO_2 we are interested in, the phase equilibrium is still simple and we notice that the liquid phase contains mostly water and the gaseous phase mostly carbon dioxide. We can exploit these observations to obtain a simple model that is applicable only to dilute solutions.

The vapor phase of this mixture can be treated as a perfect gas, and the liquid as a non-ideal solution, therefore the vapor–liquid equilibrium conditions are

$$y_{H_2O}P = \gamma_{H_2O}x_{H_2O}P^{sat}_{H_2O},$$
$$y_{CO_2}P = x_{CO_2}\tilde{f}^{L}_{CO_2}.$$

Figure 8.14 displays the fugacity of carbon dioxide in water at the temperature of interest using a complex thermodynamic model that is suitable for this highly non-ideal mixture. The fugacity curve can be approximated with a straight line for values of x_{CO_2} that are close to zero. This observation can be generalized to many similar mixtures of what are commonly called gases and liquids, and is the foundation of the

[2] See, e.g., G. M. Kontogeorgis and G. K. Folas, *Thermodynamic Models for Industrial Applications: From Classical and Advanced Mixing Rules to Association Theories*. New York: John Wiley & Sons, 2010, and B. E. Poling, J. M. Prausnitz, and J. P. O'Connell, *The Properties of Gases and Liquids*, 5th ed. McGraw-Hill, 2001.

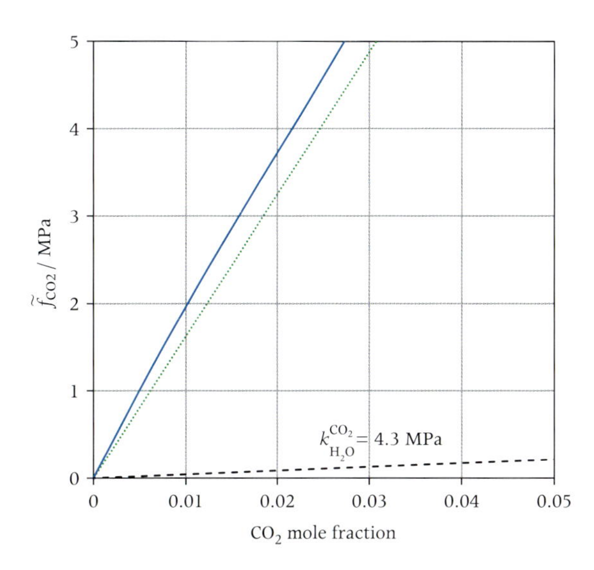

Figure 8.14 Fugacity of carbon dioxide in water as a function of x_{CO_2} at $T = 298.15$ K. Values of the dashed line are calculated with StanMix, the continuous line with Henry's law.

approximation that goes under the name of *Henry's law*, that is

$$\tilde{f}_{solute}(x_{solute} \to 0) = k_{solute} x_{solute}. \quad (8.66)$$

In the case of the CO_2/H_2O mixture, for example, the linear interpolation of $\tilde{f}^L_{CO_2}$ at 25 °C gives a value of 4.3 MPa for the Henry's constant. If data are available at other temperatures, several other values of the Henry's constant can be obtained and they can be fitted with a function of the temperature.

Henry's constants as a function of temperature are tabulated for many mixtures in databases, providing also their limits of applicability. In the chosen example, the fugacity of carbon dioxide progressively departs from the linear approximation for increasing x_{CO_2} as shown in Figure 8.14.

Databases collecting values of the Henry's Law constants are, *e.g.*,

- The DECHEMA DETHERM database (www.dechema.de/en/detherm.html)
- Dortmund Data Bank (www.ddbst.com)
- Infotherm Database of Thermophysical Properties (www.infotherm.com),
- IUPAC-NIST Solubility Database (https://srdata.nist.gov/solubility/)
- NIST Chemistry Webbook (http://webbook.nist.gov/chemistry/)

Furthermore, the solubility of the gas in the liquid is often very low, therefore the liquid phase is formed almost entirely of solvent molecules, therefore $\gamma_{solvent} \approx 1$. In this case solving the system of equations of the VLE problem is simple, and in our example the equilibrium conditions reduce to

$$y_{H_2O} P = x_{H_2O} P^{sat}_{H_2O}, \quad (8.67a)$$

$$y_{CO_2} P = k^{CO_2}_{H_2O} x_{CO_2}. \quad (8.67b)$$

Example: concentrations with Henry's constant. The bubble pressure of the H_2O/CO_2 mixture at hand can be easily calculated by summing y_{H_2O} and y_{CO_2} as obtained from (8.67) and noting that it must be equal to one, resulting in

$$P^{bubble} = x_{H_2O} P^{sat}_{H_2O} + x_{CO_2} k^{CO_2}_{H_2O}. \quad (8.68)$$

The corresponding concentration of the solute can be found by inserting $x_{H_2O} = 1 - x_{CO_2}$ into (8.68), which gives

$$x_{CO_2} = \frac{P^{bubble} - P^{sat}_{H_2O}}{k^{CO_2}_{H_2O} - P^{sat}_{H_2O}}.$$

The other concentrations can be calculated by substitution. The dependency of the bubble pressure on the CO_2 concentration is shown in Figure 8.15. Henry's law (continuous line) provides a good approximation of the experimental data up to $x_{CO_2} \approx 0.01$ as expected. The values calculated with the complex equation of state model described in Section 8.10 (dashed line)

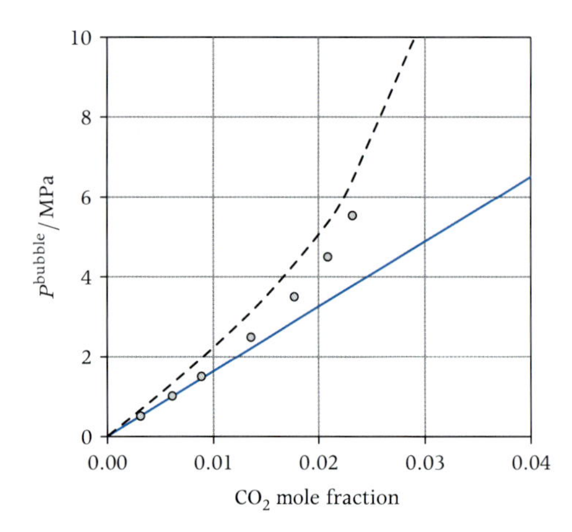

Figure 8.15 Bubble pressure curve of a carbon dioxide/water solution at $T = 298$ K. The continuous line is computed with Henry's law ($k_{H_2O}^{CO_2} = 162.5$ MPa) and the dashed line with STANMIX, the circles are experimental data from A. Valtz, A. Chapoy, C. Coquelet, *et al.*, "Vapour–liquid equilibria in the carbon dioxide–water system, measurement and modelling from 278.2 to 318.2 K," *Fluid Phase Equilib.*, 226, (1–2), 333–344, 2004.

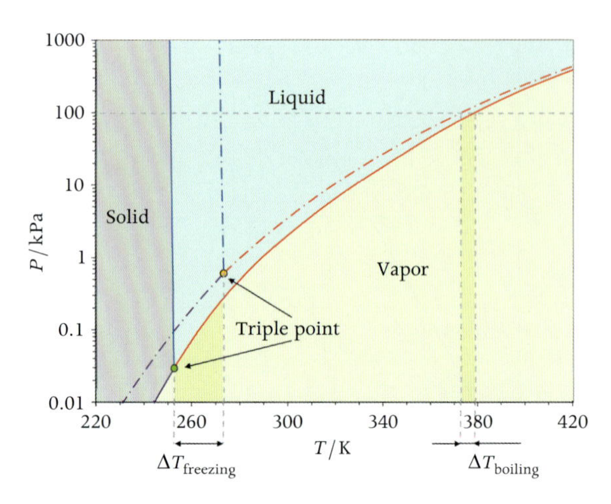

Figure 8.16 Freezing point depression and boiling point elevation: comparison between equilibrium properties of water (dashed-dotted lines, orange circle) and an 80/20 mole % mixture of water and ethylene glycol (continuous lines, green circle). The properties for water are calculated with the IF97 model (red line for the saturation line and orange circle for the triple point) and with equations available at www.iapws.org, and published in W. Wagner, A. Saul, and A. Pruß, "International equations for the pressure along the melting and along the sublimation curve of ordinary water substance," *J. Phys. Chem. Ref. Data*, 23, (3), 515–527, 1994 (purple for the sublimation line and blue for the melting line of water). The freezing point depression and boiling point elevation are calculated with (8.80) and (8.81), respectively.

are not extremely accurate, but correctly reproduce the trend of the experimental points also for $x_{CO_2} > 0.02$.

Freezing point depression and boiling point elevation in binary solutions

Addition of a certain quantity of solute to a solvent decreases the freezing temperature and increases the boiling temperature of the homogeneous mixture (solution). "Antifreeze" additives are mixed with a suitable fluid in order to lower the temperature at which heat transfer fluids solidify and increase the temperature at which the fluid evaporates. A well-known example is the coolant of automotive engines, in which a certain quantity of an organic compound, for example ethylene glycol, is added to water.

The freezing temperature decrease and the boiling temperature increase with respect to those of the fluid without additive can be quite easily calculated, if it can be assumed that the binary mixture is an ideal solution.

If the low volatility fluid is B, with $P_A^{sat} \gg P_B$, Raoult's law applied in the case of a dilute solution tells us that for the boiling point

$$P^{bubble} \approx P_A^{sat}(T)(1 - x_B). \tag{8.69}$$

Therefore, at given T, P^{bubble} decreases with increasing x_B. Conversely, at given pressure the boiling temperature increases as the amount of the less volatile compound increases. This holds down to the freezing point of the mixture, as shown in Figure 8.16, thus this effect is called *freezing point depression*.

At the triple point, one of the equilibrium conditions is

$$\bar{g}_A^L(T, x_A) = \hat{g}_A^S(T), \tag{8.70}$$

where it is assumed that: i) the solid phase is pure A, ii) the change in the composition of the liquid phase is negligible as a consequence of the very small amount of liquid that turns into a solid, and iii) the phases are incompressible, thus the pressure dependence of $\bar{g}$ can be neglected.

Equation (8.70) can also be written as

$$d\bar{g}_A^L = d\hat{g}_A^S. \tag{8.71}$$

Given that we are considering an ideal solution, we have

$$\bar{g}_A^L(T, P, x_A) = \hat{g}_A^L(T, P) + \hat{R}T \ln x_A. \tag{8.72}$$

At fixed P, and with $x_A \gg x_B$ we have that

$$d\bar{g}_A^L = \left(\frac{\partial \hat{g}_A^L}{\partial T}\right)_{P,x_A} dT + \left(\frac{\partial \hat{g}_A^L}{\partial x_A}\right)_{P,T} dx_A, \tag{8.73}$$

and

$$d\hat{g}_A^S = \left(\frac{\partial \hat{g}_A^S}{\partial T}\right)_P dT. \tag{8.74}$$

By combining the definition of specific Gibbs function (6.33) with the Gibbs equation (6.34), it can be shown that

$$\left(\frac{\partial \hat{g}}{\partial T}\right)_P = -\hat{s}. \tag{8.75}$$

Substituting (8.73) and (8.74) into (8.71), and making use of (8.75) in (8.73) and (8.74), provides

$$\left(\frac{\partial \hat{g}_A^L}{\partial x_A}\right)_{P,T} dx_A = \left(\bar{s}_A^L - \hat{s}_A^S\right) dT. \tag{8.76}$$

For an ideal solution, we obtain from (8.72) that, at constant T and P,

$$d\bar{g}_A^L(T, P, x_A) = \hat{R}Td \ln x_A = \left(\frac{\partial \hat{g}_A^L}{\partial x_A}\right)_{P,T} dx_A. \tag{8.77}$$

Equation (8.76) can thus be written as

$$\frac{\hat{R}T}{x_A} dx_A = \left(\bar{s}_A^L - \hat{s}_A^S\right) dT. \tag{8.78}$$

Since $\hat{g} \equiv \hat{h} - T\hat{s}$ and $\bar{g}_A^L = \hat{g}_A^S$, we have that

$$\left(\bar{s}_A^L - \hat{s}_A^S\right) = \frac{\bar{h}_A^L - \hat{h}_A^S}{T},$$

which, substituted in (8.78), yields

$$\frac{\hat{R}T}{x_A} dx_A = \frac{\bar{h}_A^L - \hat{h}_A^S}{T} dT.$$

With $\bar{h}_A^L - \hat{h}_A^S \approx \hat{h}_A^{LS}$, that is, the molar heat of fusion of A, and noting that $\ln x_A^L = \ln\left(1 - x_B^L\right) \approx -x_B^L$ because $x_B^L \ll 1$, we can finally write that

$$\frac{dT}{dx_B^L} = -\frac{\hat{R}T^2}{\hat{h}_A^{LS}}. \tag{8.79}$$

This simple equation, approximately valid if the mixture is a dilute solution of B into A, allows us to compute the so called freezing point depression as

$$\Delta T^{FPD} = T_A^{freeze} - T_{AB}^{freeze} = \frac{\hat{R}\left(T_A^{freeze}\right)^2}{\hat{h}_A^{LS}} x_B. \tag{8.80}$$

It can be analogously demonstrated that the boiling point elevation can be calculated with

$$\Delta T^{BPE} = T_{AB}^{boil} - T_A^{boil} = \frac{\hat{R}\left(T_A^{boil}\right)^2}{\hat{h}_A^{LV}} x_B. \tag{8.81}$$

Note that, because of the simplifications, (8.80) and (8.81) depend only on the mole fraction of the solute, and are thus independent of the selected solute. However, if the amount of solute in the mixture is specified by mass fraction instead of by mole fraction, the equation for the approximate estimation of the freezing point depression and boiling point elevation depend on the specified solute because the calculation involves its molecular weight.

Example: ethylene glycol and methanol as antifreeze. The enthalpy of fusion of water is $\hat{h}_w^{LS} = 6002$ J/mol, and the freezing temperature is $T_w^{freeze} = 273.16$ K. Let us consider ethylene glycol (EG) and methanol (M) as antifreeze. We want to compare how effective they are if the goal is to obtain a solution whose freezing temperature is 10 K lower than that of water. The most effective antifreeze is the one that requires the least amount of mass. The needed mole fraction of antifreeze can be obtained with (8.80), and is

$$x_{solute} = \frac{\Delta T^{FPD} \hat{h}_w^{LS}}{\hat{R}\left(T_w^{freeze}\right)^2} = \frac{10 \times 6002}{8.3143 \times 273.16^2} = 0.10.$$

The mole fraction is small enough that the use of the approximate Equation (8.80) is justified. The needed mass fractions can be computed once the molar masses of water, ethylene glycol, and methanol are specified. These are 18.0, 62.1, and 32.0 kg/kmol, respectively. The needed mass fraction of ethylene glycol is therefore

$$\xi_{EG} = \frac{x_{solute}\hat{M}_{EG}}{x_{solute}\hat{M}_{EG} + (1 - x_{solute})\,\hat{M}_{w}} = \frac{0.10 \times 62.1}{0.10 \times 62.1 + (1 - 0.10) \times 18.0} = 0.27\ \mathrm{kg_{EG}/kg_{solution}},$$

while that of methanol is

$$\xi_{M} = \frac{x_{solute}\hat{M}_{M}}{x_{solute}\hat{M}_{M} + (1 - x_{solute})\,\hat{M}_{w}} = \frac{0.10 \times 32.0}{0.10 \times 32.0 + (1 - 0.10) \times 18.0} = 0.16\ \mathrm{kg_{M}/kg_{solution}}.$$

A smaller amount of methanol is needed in order to obtain a solution that freezes at −10 °C.

The calculated values compare well with experimental values reported in the literature, which are $\xi_{EG} = 0.25$[3] and $\xi_{M} = 015$.[4] The small differences can be attributed to the approximations introduced in the derivation of (8.80).

8.10 A Complete and Consistent Thermodynamic Model

Engineering analysis and design often demand mixture thermodynamic models that are computationally efficient, valid over a large range of properties, reasonably accurate, and consistent. *Consistent* means that the model satisfies all the main constraints of thermodynamics coming from the application of its laws. Therefore, models based on fundamental equations of state or on an equation of state in the $P = P(T, \hat{v})$ form complemented by $\hat{c}_P^{IG}(T)$ are an attractive option, though others are available. The advantages compared to the statistical mechanics approach are that computer calculations can be fast, and that properties can be obtained for the entire range of fluid states of interest, though they require a fair amount of experimental information.

The difficulty is to obtain thermodynamically correct relations between the equation-of-state parameters of the mixture and the corresponding parameters of the constituents. In addition, the amount of experimental information that is needed in order to model molecular interaction must be kept to a minimum.

If we take as example the iPRSV equation of state (6.94), then this means that we want to obtain a mixture equation of state in the form

$$P = \frac{\hat{R}T}{\left(\hat{v} - b_{mix}\right)} - \frac{a_{mix}(T)}{\hat{v}(\hat{v} + b_{mix}) + b_{mix}(\hat{v} - b_{mix})}, \tag{8.82}$$

where

$$a_{mix} = a_{mix}(T, a_i, b_i, z_i), \qquad b_{mix} = b_{mix}(T, a_i, b_i, z_i).$$

The a_{mix} and b_{mix} equations therefore also depend on the mixture composition and are commonly called *mixing rules*. From a physical point of view, mixture parameters depend on the interaction among the molecules, exactly as we have seen for pure fluids, but with the added difficulty of the interaction between different molecules.

Having a $P = P(T, \hat{v})$ equation of state, mixing rules, and also data for the isobaric (or isochoric) specific heat in the ideal gas state $\hat{c}_P = \sum_{i=1}^{nc} \chi_i \hat{c}_{P_i}$, allows us to compute all thermodynamic properties in the same way that we illustrated for a pure fluid. This model is complete and consistent, if mixing rules also comply with thermodynamic constraints.

No general theory exists that will allow us to obtain a mixture thermodynamic model that is suitable and accurate for any conceivable mixture of fluids. However, years of intense research have led to models that are suitable for complex engineering

[3] D. R. Cordray, L. R. Kaplan, P. M. Woyciesjes, *et al.*, "Solid–liquid phase diagram for ethylene glycol + water," *Fluid Phase Equilib.*, 117, (1), 146–152, 1996.

[4] "Physical constants of organic compounds," in *CRC Handbook of Chemistry and Physics*, W. M. Haynes, Ed., 97th ed., Internet version, Boca Raton, FL.: Taylor and Francis, 2017.

calculations and that are applicable to many classes of fluid mixtures. Here we illustrate one of these models, which is based on mixing rules for cubic equations of state. This model therefore can estimate thermodynamic properties in a wide range of temperatures and pressures and for highly non-ideal simple-fluid mixtures. However, it suffers from the limitations of cubic equations of state, namely that they are relatively inaccurate close to the critical point, and for some classes of fluids liquid density estimations are also affected by larger errors.

Thermodynamically correct mixing rules

One way of obtaining mixing rules comes from the idea that they must obey thermodynamic constraints and that the appropriate formulation of the constraints can be exploited in order to obtain a suitable functional form. The mixing rules of *Wong and Sandler*[5] are of this type, and their derivation is illustrated in the following, as a paradigmatic example. The Wong–Sandler mixing rules are widely adopted in process and energy systems engineering, and are implemented in STANMIX .

The first fundamental observation is that the vapor phase at low and moderate pressure should obey the virial equation of state for mixtures, as the virial equation can be rigorously derived with statistical mechanics from the mathematical representation of the interaction between different molecules. In the case of mixtures, this treatment at molecular level leads to the second virial coefficient of a mixture being quadratically dependent on the composition, thus

$$B_{\text{mix}} = \sum_{i=1}^{nc}\sum_{j=1}^{nc} y_i y_j B_{ij}. \tag{8.83}$$

Here the $B_{ij,i\neq j}$ parameters contain information about the interaction between two molecules (binary interaction parameters), while the B_is are the parameters of the constituents if unmixed.

[5] H. Wong, D. Shan, and S. I. Sandler, "Theoretically correct mixing rule for cubic equations of state," *AICHE J.*, **38**, (5), 671–680, 1992.

This physically based relation can be used in order to obtain a relation between the parameters of a cubic equation of state. Here we carry out the derivation of the mixing rule for the van der Waals (vdW) cubic equation of state, as the resulting algebraic expressions are simpler, though the procedure is applicable to any cubic equation of state. If we equate the compressibility factor obtained from the vdW equation of state (6.71) to the compressibility factor expressed as a virial expansion (6.83) we have that

$$Z = \frac{P\hat{v}}{\hat{R}T} = \frac{\hat{v}}{\hat{R}T}P_{\text{vdW}} = \frac{\hat{v}}{\hat{R}T}\left[\frac{\hat{R}T}{(\hat{v}-\mathsf{b})} - \frac{\mathsf{a}}{\hat{v}^2}\right]$$
$$= 1 + \frac{B_{\text{vdW}}(T)}{\hat{v}} + \cdots .$$

Algebraic manipulation leads to

$$B_{\text{vdW}} = \mathsf{b} - \frac{\mathsf{a}}{\hat{R}T}.$$

Imposing the condition that the second virial coefficient derived from the mixture equation of state complies with Equation (8.83) results in the relation

$$\mathsf{b}_{\text{mix}} - \frac{\mathsf{a}_{\text{mix}}}{\hat{R}T} = \sum_{i=1}^{nc}\sum_{j=1}^{nc} y_i y_j \left(\mathsf{b}_{ij} - \frac{\mathsf{a}_{ij}}{\hat{R}T}\right). \tag{8.84}$$

This equation has been obtained by exploiting a condition based on physics valid at low pressure.

In order to fully determine the mixture equation of state parameters a_{mix} and b_{mix} we need another relation involving pure-fluid parameters and composition. Conversely, this condition can be searched at higher pressure, that is, a condition valid for the liquid phase, where mixing non-ideality can be accounted for using activity coefficient models. Analogously with the excess Gibbs function (8.49), the mixture molar excess Helmholtz function is

$$\hat{a}^{\text{E}}_{\text{mix}} = \hat{a}_{\text{mix}} - \hat{a}^{\text{IS}}_{\text{mix}} = \hat{a}_{\text{mix}} - \left(\sum_{i=1}^{nc} x_i \hat{a}_i + \hat{R}T\sum_{i=1}^{nc} x_i \ln x_i\right). \tag{8.85}$$

Since for a perfect gas, see (8.7),

$$\hat{a}^{\text{IG}}_{\text{mix}} = \sum_{i=1}^{nc} x_i \hat{a}^{\text{IG}}_i + \hat{R}T\sum_{i=1}^{nc} x_i \ln x_i,$$

substituting $\hat{R}T\sum_{i=1}^{nc} x_i \ln x_i = \hat{a}_{\mathrm{mix}}^{\mathrm{IG}} - \sum_{i=1}^{nc} x_i \hat{a}_i^{\mathrm{IG}}$ into (8.85) gives

$$\begin{aligned}\hat{a}_{\mathrm{mix}}^{\mathrm{E}} &= \hat{a}_{\mathrm{mix}} - \sum_{i=1}^{nc} x_i \hat{a}_i - \hat{a}_{\mathrm{mix}}^{\mathrm{IG}} + \sum_{i=1}^{nc} x_i \hat{a}_i^{\mathrm{IG}} \\ &= \hat{a}_{\mathrm{mix}}^{\mathrm{R}} - \sum_{i=1}^{nc} x_i \hat{a}_i^{\mathrm{R}}. \end{aligned} \tag{8.86}$$

From its definition, we have that $da = -Pdv - sdT$, therefore the residual molar Helmholtz function at constant P and T is

$$\begin{aligned}\hat{a}^{\mathrm{R}} = \hat{a} - \hat{a}^{\mathrm{IG}} &= -\int_{\hat{a}^{\mathrm{IG}}}^{\hat{a}} d\hat{a} \\ &= -\int_{\hat{v}=\infty}^{\hat{v}} P_{\mathrm{vdW}}(T,\hat{v})d\hat{v} - \left(-\int_{\hat{v}=\infty}^{\hat{v}=\frac{\hat{R}T}{P}} \frac{\hat{R}T}{v} d\hat{v} \right),\end{aligned}$$

which allows us to obtain an expression for both $\hat{a}_i^{\mathrm{R}}$ and $\hat{a}_{\mathrm{mix}}^{\mathrm{R}}$, once the vdW equation for P is substituted and the integrations are carried out. We have therefore that

$$\hat{a}_i^{\mathrm{R}} = -\hat{R}T \ln\left[\frac{P\left(\hat{v}_i - b_i\right)}{\hat{R}T}\right] - \frac{a_i}{\hat{v}_i},$$

and

$$\hat{a}_{\mathrm{mix}}^{\mathrm{R}} = -\hat{R}T \ln\left[\frac{P\left(\hat{v}_{\mathrm{mix}} - b_{\mathrm{mix}}\right)}{\hat{R}T}\right] - \frac{a_{\mathrm{mix}}}{\hat{v}_{\mathrm{mix}}},$$

can be substituted into (8.86), thus obtaining an expression for the mixture excess Helmholtz function that contains the mixture equation of state parameters and those of the pure components and the composition, which reads

$$\begin{aligned}\hat{a}_{\mathrm{mix}}^{\mathrm{E}} = &-\frac{a_{\mathrm{mix}}}{\hat{v}_{\mathrm{mix}}} + \sum_{i=1}^{nc} x_i \frac{a_i}{\hat{v}_i} - \hat{R}T \ln\left[\frac{P\left(\hat{v}_{\mathrm{mix}} - b_{\mathrm{mix}}\right)}{\hat{R}T}\right] \\ &+ \hat{R}T \sum_{i=1}^{nc} x_i \ln\left[\frac{P\left(\hat{v}_i - b_i\right)}{\hat{R}T}\right].\end{aligned} \tag{8.87}$$

At high pressure, or, more rigorously, in the limit of infinite pressure, the molecules are so packed together that there is no free volume between them, which means that the volume $\hat{v}$ and the covolume b tend to coincide. Under the assumption of infinite pressure we therefore have that

$$\lim_{P\to\infty} \hat{v}_i = b_i,$$

and

$$\lim_{P\to\infty} \hat{v}_{\mathrm{mix}} = b_{\mathrm{mix}}.$$

Equation (8.87) thus becomes

$$\hat{a}_{\mathrm{mix},P\to\infty}^{\mathrm{E}} = -\frac{a_{\mathrm{mix}}}{b_{\mathrm{mix}}} + \sum_{i=1}^{nc} x_i \frac{a_i}{b_i}. \tag{8.88}$$

Equations (8.84) and (8.88) form a system of two equations in the two unknowns a_{mix} and b_{mix}, whose solution is

$$b_{\mathrm{mix}} = \frac{\sum_{i=1}^{nc}\sum_{j=1}^{nc} z_i z_j \left(b_{ij} - \frac{a_{ij}}{\hat{R}T}\right)}{1 + \frac{\hat{a}_{\mathrm{mix},P\to\infty}^{\mathrm{E}}}{\hat{R}T} - \sum_{i=1}^{nc} z_i \frac{a_i}{b_i \hat{R}T}}, \tag{8.89a}$$

and

$$a_{\mathrm{mix}} = b_{\mathrm{mix}} \left(\sum_{i=1}^{nc} z_i \frac{a_i}{b_i} - \hat{a}_{\mathrm{mix},P\to\infty}^{\mathrm{E}} \right). \tag{8.89b}$$

There is still a problem: the molar excess Helmholtz function at infinite (or large) pressure of mixtures is not something that has been measured or is available in any other way. Luckily it can be related to the mixture excess Gibbs function at low pressure.

The molar excess Helmholtz free energy is related to the molar excess Gibbs function according to

$$\hat{g}^{\mathrm{E}} = \hat{a}^{\mathrm{E}} + P\hat{v}^{\mathrm{E}}.$$

We note that at low pressure (large volume) the excess molar volume $\hat{v}^{\mathrm{E}}$ is small, therefore the difference between $\hat{g}^{\mathrm{E}}$ and $\hat{a}^{\mathrm{E}}$ is small, that is

$$\hat{g}^{\mathrm{E}}(T, x_{i=1,\ldots,nc}, P\downarrow) \approx \hat{a}^{\mathrm{E}}(T, x_{i=1,\ldots,nc}, P\downarrow).$$

Furthermore, it can be verified that at constant temperature, the molar excess Helmholtz function depends very weakly on the pressure, to the point that

$$\hat{a}^{\mathrm{E}}(T, x_{i=1,\ldots,nc}, P\downarrow) \approx \hat{a}^{\mathrm{E}}(T, x_{i=1,\ldots,nc}, P\to\infty).$$

is a good approximation.

We can conclude therefore that

$$\hat{a}_{\mathrm{mix}}^{\mathrm{E}}(T, x_{i=1,\ldots,nc}, P\to\infty) \approx \hat{g}_{\mathrm{mix}}^{\mathrm{E}}(T, x_{i=1,\ldots,nc}, P\downarrow),$$

whereby $\hat{g}^{\mathrm{E}}_{\mathrm{mix}}(T, x_{i=1,\ldots,nc}, P\downarrow)$ is a value that can be obtained by correlating vapor–liquid measurements at low pressure with activity coefficient models, like the Margules equations of Section 8.8. For simplicity we now indicate $\hat{g}^{\mathrm{E}}_{\mathrm{mix}}(T, x_{i=1,\ldots,nc}, \text{low } P)$ simply by $\hat{g}^{\mathrm{E}}_{\mathrm{mix}}$. We can therefore rewrite (8.90) as

$$b_{\mathrm{mix}} = \frac{\sum_{i=1}^{nc}\sum_{j=1}^{nc} z_i z_j \left(b_{ij} - \frac{a_{ij}}{\hat{R}T}\right)}{1 + \frac{\hat{g}^{\mathrm{E}}_{\mathrm{mix}}}{\hat{R}T} - \sum_{i=1}^{nc} z_i \frac{a_i}{b_i \hat{R}T}}, \tag{8.90a}$$

and

$$a_{\mathrm{mix}} = b_{\mathrm{mix}} \left(\sum_{i=1}^{nc} z_i \frac{a_i}{b_i} - \hat{g}^{\mathrm{E}}_{\mathrm{mix}}\right). \tag{8.90b}$$

It can finally be worked out that the Wong–Sandler mixing rules for the van der Waals equation of state (8.89) can be generalized to any cubic equation of state, resulting in

$$a_{\mathrm{mix}} = \hat{R}T b_{\mathrm{mix}} D, \tag{8.91a}$$

$$b_{\mathrm{mix}} = \frac{Q}{1 - D}, \tag{8.91b}$$

with

$$Q = \sum_{i=1}^{nc}\sum_{j=1}^{nc} z_i z_j \left(b_{ij} - \frac{a_{ij}}{\hat{R}T}\right),$$

$$D = \frac{\hat{g}^{\mathrm{E}}_{\mathrm{mix}}}{\hat{R}T\sigma} + \sum_{i=1}^{nc} z_i \frac{a_i}{\hat{R}T b_i},$$

and where parameter σ depends on the equation of state (e.g., $\sigma = \ln\left(\sqrt{2}-1\right)/\sqrt{2}$ for the Peng–Robinson equation of state).

In Equations (8.91), information on how the Q term can be calculated is still missing. Observe that for $i \neq j$ we must be able to compute the terms $b_{ij} - \frac{a_{ij}}{\hat{R}T}$ from the values of the pure-fluid equations of state a_i and b_i. This is done with a so-called empirical *combining rule* in the form

$$b_{ij} - \frac{a_{ij}}{\hat{R}T} = \frac{1}{2}\left[\left(b_i - \frac{a_i}{\hat{R}T}\right) + \left(b_j - \frac{a_j}{\hat{R}T}\right)\right]\left(1 - k_{ij}\right), \tag{8.92}$$

where, though, we need to be able to calculate the *binary interaction parameter* k_{ij}. It can be done by assuming that its dependence on temperature is weak, therefore it becomes a datum of the model. k_{12} $(= k_{21})$ is calculated at the temperature and pressure at which the data for the activity coefficient model are available by equating $\frac{\hat{g}^{\mathrm{E}}_{\mathrm{mix}}}{\hat{R}T}$ as obtained from the selected activity coefficient model for the equimolar composition ($x_1 = x_2 = 0.5$) to the same value calculated by means of the equation of state model. This means, in the case of a binary mixture, solving the equation

$$x_1 \ln\gamma_1 + x_2 \ln\gamma_2 = x_1\left(\ln\tilde{\phi}_1(k_{12}) - \ln\phi_1\right) + x_2\left(\ln\tilde{\phi}_2(k_{12}) - \ln\phi_2\right)$$

for k_{12} at the T and P at which $\ln\gamma_1$ and $\ln\gamma_2$ were correlated to VLE measurements.

Property calculations

A mixture thermodynamic model like the one described in this section allows us to compute all properties using the same relations illustrated in Section 6.3. For example, the entropy at a certain temperature and pressure is obtained from

$$\hat{s} = \hat{s}^{\mathrm{IG}} + \hat{R}\ln Z + \int_{\infty}^{\hat{v}} \left[\left(\frac{\partial P}{\partial T}\right)_{\hat{v}} - \frac{\hat{R}}{\hat{v}}\right] d\hat{v},$$

whereby the ideal-gas isobaric specific heat is used to compute the $\hat{s}^{\mathrm{IG}}$ term, while the entropy departure can be computed once the equation for the selected model is obtained by substituting the appropriate $P = P(T, \hat{v})$ equation into the residual terms and solving the integral.

The calculation of VLE states requires solution of the equilibrium conditions, which are normally expressed in terms of fugacity coefficients because this function is numerically well behaved with respect to the calculation of the non-linear system of equations. The equations to be solved are thus

$$y_i \tilde{\phi}^{\mathrm{V}}_i(T, P, y_i) = x_i \tilde{\phi}^{\mathrm{L}}_i(T, P, x_i) \qquad i = 1, \ldots, nc, \tag{8.93}$$

with the fugacity coefficient given by

$$\ln\tilde{\phi}_i = \frac{1}{\hat{R}T}\int_{\infty}^{\hat{v}} \left[\frac{\hat{R}T}{\hat{v}} - N\left(\frac{\partial P}{\partial N_i}\right)_{T,\hat{v},N_{j\neq i}}\right] d\hat{v} - \ln Z. \tag{8.94}$$

The equation for the fugacity coefficient for species i in the mixture is obtained by substituting the equation of state in the form $P = P(T, \hat{v})$ into (8.94), and solving the integral.

VLE calculations at given T and/or P

Three types of VLE calculation are of most common engineering interest, and they are defined based on what is specified when solving the system of Equations (8.93):

- the *bubble point* calculation occurs when either the pressure (bubble temperature calculation, T^{bubble}) or temperature (bubble pressure calculation, P^{bubble}) are specified together with the composition of the liquid ($x_i = z_i$). The output of the calculation is either the temperature or the pressure and the composition of the vapor phase (y_i). The blue line in the T-s diagram of Figure 8.1 is formed by all the bubble points of the equimolar mixture of propane and n-pentane from the freezing point up to the critical point;
- the *dew point* calculation occurs when either the pressure (dew temperature calculation, T^{dew}) or temperature (dew pressure calculation, P^{dew}) are specified together with the composition of the vapor ($y_i = z_i$). The output of the calculation is either the temperature or the pressure and the composition of the liquid phase (x_i). The red line in the T-s diagram of Figure 8.1 is formed by all the dew points of the equimolar mixture of propane and n-pentane; and
- the *(isothermal) flash point* calculation occurs when both a temperature and a pressure corresponding to a saturated state are specified, together with the overall mixture composition z_i. The output of the calculation is the composition of the liquid and vapor phase (x_i, y_i). The name of this type of calculation derives from the fact that it calculates the pressure and the phase compositions of a mixture of composition z_i and at temperature T when it is suddenly brought to pressure P. The line connecting the bubble point to the dew point on an isobar in Figure 8.1 is calculated with a flash point calculation at given P and varying T. The P–xy chart of Figure 8.17 can be obtained by performing the flash point calculation at the given temperature and for pressures varying between the saturation pressures of the two components, for $0 \leq z_1 \leq 1$, and by plotting the corresponding x_1 and y_1; and

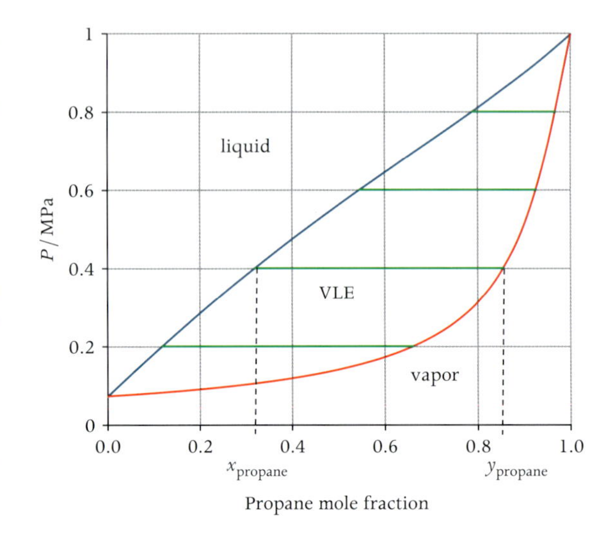

Figure 8.17 The P–xy chart of mixtures of propane and n-pentane for $T = 300$ K with the bubble line (blue) and the dew (red) connected by isobars (green) indicating the composition of the liquid and vapor phases.

Simple numerical schemes for the calculation of bubble, dew, and flash points are illustrated in Appendix C. The system of algebraic equations that must be solved in order to perform these types of calculations is highly nonlinear. In addition, its degree of nonlinearity depends also on the type of mixture and on how close to the critical point the calculation is performed. For these reasons, very complex numerical schemes must sometimes be employed. The solution schemes of Appendix C are suitable only for mildly nonideal mixtures and fail in the critical-point region.

The critical point of a mixture can be calculated directly, by exploiting thermodynamic constraints that define the critical point of a mixture or by interpolation from the calculation of the bubble and dew lines.

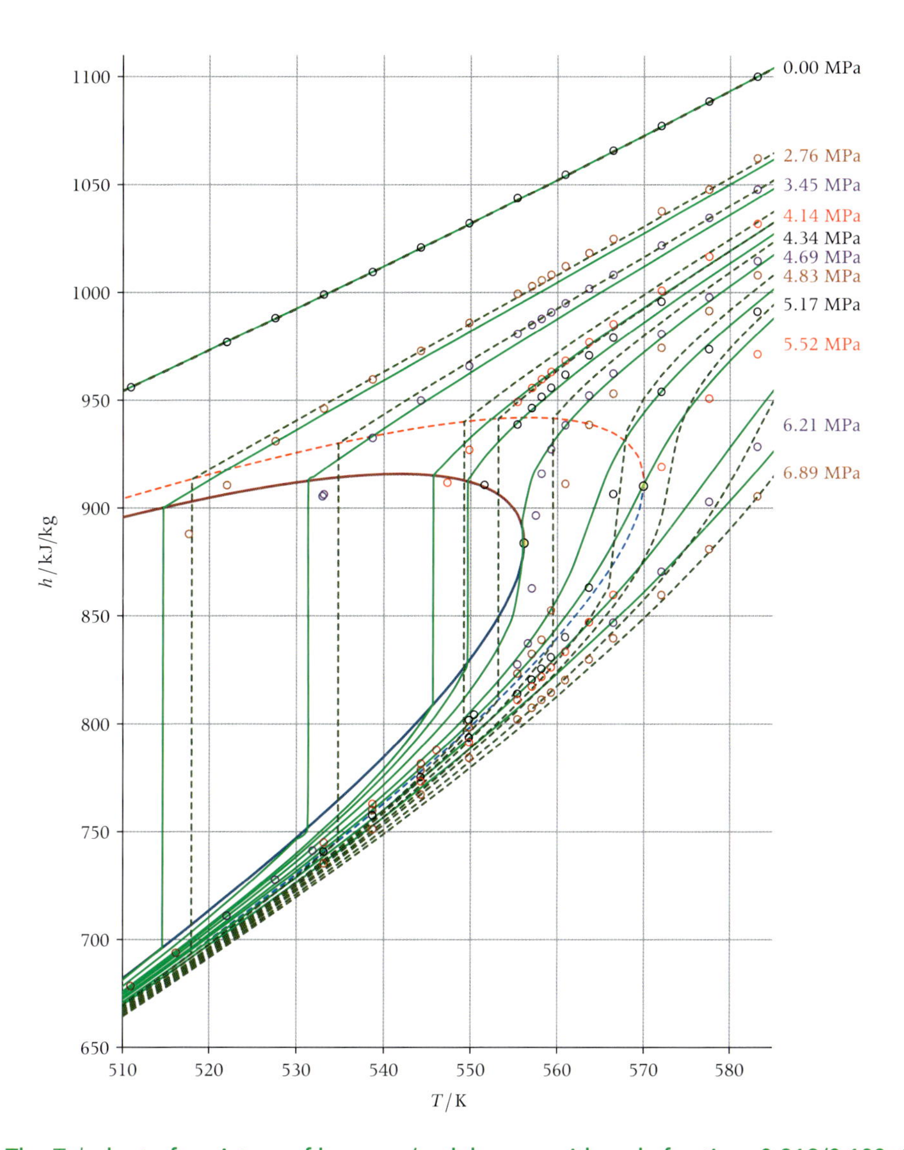

Figure 8.18 The T–h chart of a mixture of benzene/cyclohexane with mole fractions 0.812/0.188. Circles represent experimental values for several isobars from J. Lenoir and K. Hayworth, "Enthalpies of mixtures of benzene and cyclohexane," *J. Chem. Eng. Data*, **16**, (3), 285–288, 1971, continuous lines are the bubble line (blue), dew line (red), and isobars (green) calculated with StanMix, while the corresponding dashed lines were calculated with PCP-SAFT.

The derivation of the complete thermodynamic model formed by the iPRSV cubic equation of state and the Wong–Sandler mixing rules is reported in Appendix A.12, and is the one implemented in StanMix.

Examples of thermodynamic diagrams for several mixtures are given in Figures 8.18 and 8.19. These charts illustrate the capabilities and limitations of complete models for calculation of mixture properties. The cubic equation of state model implemented in StanMix computes states in the dense vapor region that are close to the measured ones. The same holds for saturated vapor states and for the vapor branch of the critical isobar. The difference from

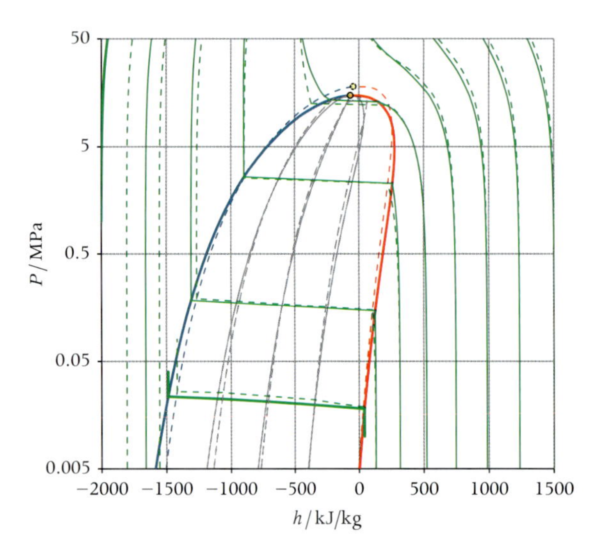

Figure 8.19 The *P*–*h* chart of a mixture of ethanol/water with mole fractions 0.35/0.65. Continuous lines are the bubble line (blue), dew line (red), and isotherms (green) of T = −100, 0, 50, 100, 200, . . . , 700 °C, respectively, calculated by StanMix. The corresponding dashed lines are calculated by PCP-SAFT.

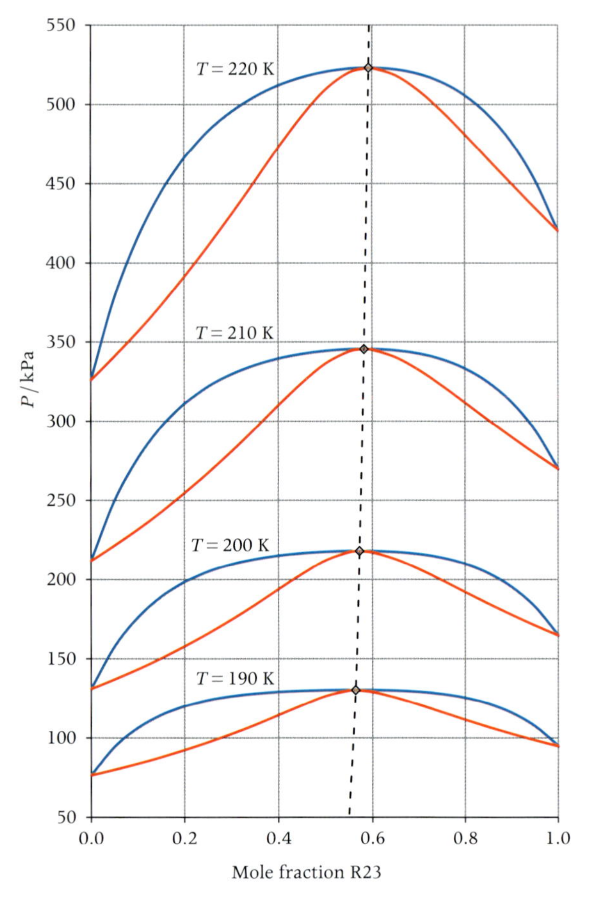

Figure 8.20 The *P*–*xy* chart of mixtures of the refrigerants R23 (CHF_3) and R116 (C_2F_6) for various temperatures. The dashed line connects the azeotropes which are formed at different compositions at different temperatures.

experimental data becomes larger for liquid states at high reduced temperature. The PCP-SAFT model, a model based on molecular parameters, is able to predict values that are closer to experimental data for both vapor and liquid states, provided they are not close to the critical point. The critical temperature and pressure are inputs for the StanMix model, while they are calculated for the PCP-SAFT model. The PCP-SAFT model is not able to accurately predict the critical temperature and pressure nor therefore the critical thermodynamic region: a general model for engineering use that is also able to predict mixture properties in the vicinity of the critical point is still missing.

8.11 Azeotropes

A particular case of mixture vapor–liquid equilibrium occurs when, for a certain composition, the vapor and the liquid phase feature the same composition as the overall composition ($x_i = y_i = z_i$). This depends on the particular combination of interacting forces between the molecules forming the mixture. Mixtures characterized by this behavior are called *azeotropes*. Examples of compositions forming azeotropes, in this case refrigerants R23 (trifluoromethane, CHF_3) and R116 (perfluoroethane, C_2F_6) are the constituents, are shown in Figure 8.20.

Azeotropes are sometimes wanted in technical applications. This is the case for several refrigerant fluids whose azeoptropic composition has been selected in such a way that the properties are very similar to the pure fluids that have been banned because they cause depletion of the atmospheric

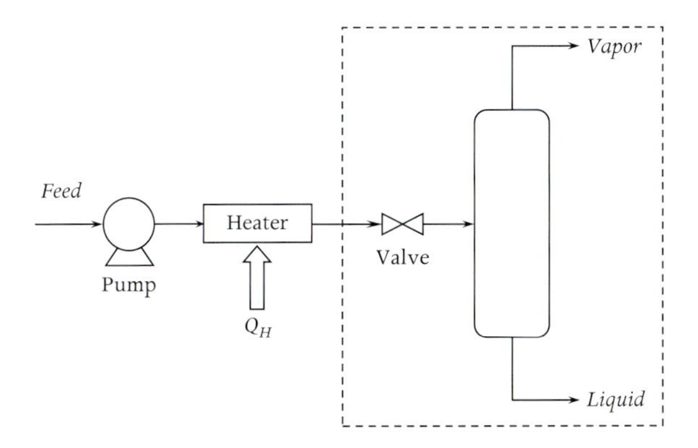

Figure 8.21 Simple scheme of a flash distillation stage.

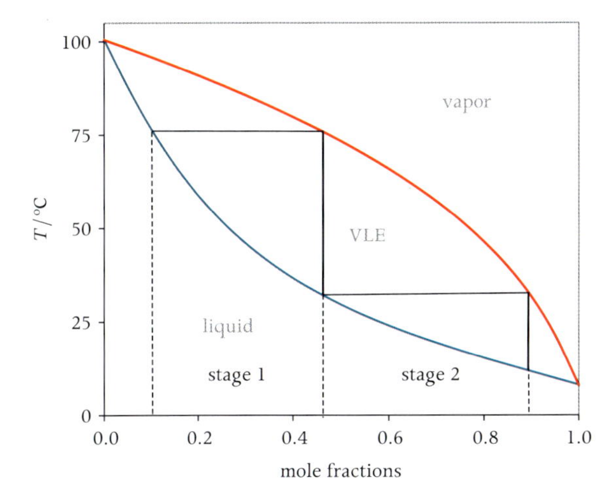

Figure 8.22 The states of a multi-stage distillation process applied to a simple mixture (black lines) on a *T–xy* chart.

ozone layer of the Earth. In other cases azeotropes are not wanted, like, for example, when distillation would be needed in order to separate mixture constituents. In the simplest case, flash distillation (Figure 8.21) is performed when a binary mixture is fed to a vessel with a certain composition and then it is flashed at constant pressure. The vapor and the liquid are continuously extracted: the vapor phase is richer in the more volatile compound while the liquid mostly contains the heavier compound. The *T-xy* diagram on which the process can be visualized is given in Figure 8.22. If the mixture forms an azeoptrope distillation is not possible.

8.12 Concluding Remarks

This chapter treats only simple fluid mixtures and their vapor–liquid equilibrium. Many more complex mixtures and mixture equilibria exist. You might ask yourself, what can be done in order to compute the equilibrium properties of water with a gas in case the dilute solution approximation cannot be applied. One can try to use a thermodynamic model like the one described in Section 8.10, but in many cases the physical phenomena, ultimately the interactions between molecules, are too complex for this type of model. Other examples of very complex phase equilibria are those in which several liquid phases are present, or equilibria in which constituent molecules dissociate, or even react. All models to compute equilibrium deal with the fundamental condition that the Gibbs function must be at a minimum if the pressure and the temperature (the most common independent variables in engineering) are fixed. If we think of a model that is capable of predicting the equilibrium of a mixture that forms a vapor phase, multiple liquid phases, and in addition constituents undergo chemical reactions, the starting point is always the possibility of formulating the Gibbs function of this complex mixture in mathematical terms. This abstract function describes the thermodynamic system at macroscopic level, but often the best path to model development is to start from a description of the molecular (microscopic) level of the phenomena. A large body of specialized literature treats all these complex cases.

EXERCISES

8.1 Obtain the relation between the isochoric specific heat of an ideal gas mixture and those of its constituents.

8.2 A gas having an ideal gas specific-heat ratio γ of 1.5 is required for a certain gas dynamics experiment. Specify a mixture that could be used.

8.3 The analysis of a gas mixture shows that the composition by weight is 15 % He, 60 % N_2, and 25 % O_2. Calculate the mole fractions and molecular weight of the mixture.

8.4 A mixture of gases used for a special heat transfer application consists of 25 % Ar, 50 % He, 25 % H_2 by weight. The operating temperature is such that the mixture behaves like an ideal gas at the operating pressure of 10 bar. Determine the partial pressure of the constituents, the c_P of the mixture in kJ/(kg · K) and the mixture molecular weight.

8.5 A mixture of 40 % argon and 60 % nitrogen by weight is compressed from 1 atm, 20 °C, to 4 atm. Assuming that the fluid complies with the ideal gas model, calculate the mixture temperature at the final pressure, the work required and the entropy change of the mixture and of the constituents.

8.6 The flue gas of a gas turbine contains water vapor and $y_{H_2O} = 0.05$. The temperature after the gas turbine outlet, downstream in the process, is 350 K. The flue gas contains minute quantities of sulfur which is present in the fuel. Sulfur can combine with liquid water forming sulfuric acid which is highly corrosive, thus water condensation is unwanted. The flue gas can be modeled as an ideal gas mixture. Calculate the maximum pressure in the stack that will avoid water condensation.

8.7 The following table reports the measurements of excess molar volume for the water–ethanol mixture at $T = 298$ K of Figure 8.4.

x_{water}	$\Delta\hat{v}_{\mathrm{mixing}}$ /cm^3/mol
0.063	−0.222
0.134	−0.454
0.220	−0.678
0.426	−0.999
0.538	−1.067
0.650	−1.060
0.737	−0.986
0.796	−0.884
0.844	−0.750
0.887	−0.575
0.919	−0.409
0.950	−0.234
0.987	−0.050

Data on excess molar properties are often correlated with a polynomial of the so-called Redlich–Kister form

$$\Delta\hat{v}_{\mathrm{mixing}} = x_1 x_2 \sum_{i=0}^{n} \alpha_i \left(x_1 - x_2\right)^i .$$

Correlate the given data with a Redlich–Kister polynomial of third degree, and thus calculate the coefficients of the polynomial.

8.8 Obtain the relation between the molar entropy change of an ideal gas vapor mixture and the molar entropy of each constituent.

8.9 The excess enthalpy of a non-ideal liquid mixture is also often called *heat of mixing* or *enthalpy of mixing*, as it is the energy that gets transferred as heat to or from a mixture at constant temperature and pressure when the constituents are mixed. The result depends on the interaction of dissimilar molecules, which is different from that of molecules of the same type. The excess enthalpy can thus be measured with a calorimeter. Consider a liquid binary mixture for which the heat of mixing has been measured and correlated with

$$\hat{h}^{\mathrm{E}} = x_1 x_2 \left(50 x_1 - 15 x_2\right) .$$

Plot the $\hat{h}^{\mathrm{E}}(x_1)$ function and derive the equations for $\bar{h}^{\mathrm{E}}_{i=1,2}(x_1)$. What is the physical meaning of $\bar{h}^{\mathrm{E}}_i$?

8.10 A generic partial molar property is defined as

$$\bar{\phi}_i \equiv \left(\frac{\partial \Phi}{\partial N_i}\right)_{T,P,N_{j\neq i}} = \left(\frac{\partial \left(N\hat{\phi}\right)}{\partial N_i}\right)_{T,P,N_{j\neq i}},$$

where ϕ can be $v, h, s, u, g, \ldots$; if you are hasty, you might wrongly conclude that $\bar{\phi}_i = \left(\frac{\partial \hat{\phi}}{\partial z_i}\right)_{T,P,z_{j\neq i}}$. Show that the correct relation is

$$\bar{\phi}_i = \hat{\phi} + \left(\frac{\partial \hat{\phi}}{\partial z_i}\right)_{T,P,z_{j\neq i}} - \sum z_k \left(\frac{\partial \hat{\phi}}{\partial z_k}\right)_{T,P,z_{j\neq k}}.$$

8.11 How many variables can be fixed independently in order to fix the state of a binary mixture in vapor–liquid equilibrium if, as is most often the case, the two phases contain both components in different amounts? Which variables would you decide to control in an actual experiment?

How many degrees of freedom characterize a mixture in vapor–liquid equilibrium in the case where the mixture forms an azeotrope, that is a mixture in vapor–liquid equilibrium featuring vapor and liquid with the same composition?

8.12 A fluid formed by pentane and hexane obeys Raoult's law quite well. Such a mixture can be used for example as the working fluid of an organic Rankine cycle power plant. The vapor pressure of pure pentane can be computed rather accurately with the Antoine equation

$$\log_{10} P^{\mathrm{sat}} = A - \frac{B}{T + C - 273.15},$$

with P in bar and T in K, $A = 3.978$ bar, $B = 1065$ bar·K, and $C = 232.01$ K. The vapor pressure of hexane can be computed similarly, whereby the coefficients of the Antoine equation are $A = 4.001$ bar, $B = 1171$ bar · K, and $C = 224.32$ K.

(a) Make a *P–xy* chart, or *phase diagram*, by applying the ideal mixture model for a temperature of 30 °C (a value which would be suitable for the condenser of an ORC power plant).

(b) Similarly, make a *T–xy* chart for a pressure of 0.5 bar.

(c) Indicate on the *T–xy* chart the isobaric condensation process undergone by the working fluid of an ORC power plant in the case where it is an equimolar mixture of pentane and hexane, starting from a superheated state and ending at a subcooled state. Indicate also the loci of the points giving the vapor and the liquid composition during condensation.

(d) If the overall mixture composition is equimolar, and for $T = 32$ °C and $P = 0.5$ bar, determine the mass fraction of liquid, and the composition of the two phases.

(e) Compare the charts you have made with those you can obtain by using StanMix and plot the deviations in temperature/pressure versus the composition. What are the causes of the difference? What can be the advantage of the ideal mixture model?

8.13 Assume that a binary mixture and its components can be modeled using the van der Waals equation of state, both for the liquid and the vapor phase. Assume also that the mixture equation of state parameters can be calculated using so-called van der Waals one-fluid mixture rules,

$$a_{\mathrm{mix}} = \sum_i \sum_j z_i z_j a_{ij}, \quad b_{\mathrm{mix}} = \sum z_i b_i.$$

(a) Derive the fugacity coefficient for component i in the mixture and show that

$$\ln \tilde{\phi}_i \equiv \ln \frac{\tilde{f}_i}{y_i P} = \frac{K_i}{(Z-K)} - \ln (Z-K) - \frac{2\sum_j z_j a_{ij}}{\hat{R}T\hat{v}},$$

where $K = \dfrac{Pb}{\hat{R}T}$.

(b) Similarly, obtain the equation for the activity coefficient of component i, γ_i.

8.14 Given the set of VLE experimental data for the non-ideal mixture propane/n-pentane at 344.26 K from B. Sage and W. Lacey, "Phase equilibria in hydrocarbon systems, propane-n-pentane system," *Industrial and Engineering Chemistry*, **32**, (7), 992–996, 1940,

P /kPa	x_1	y_1
413.7	0.058	0.295
551.6	0.124	0.482
689.5	0.189	0.600
861.8	0.270	0.701
1034	0.350	0.770
1379	0.504	0.863
1724	0.657	0.920
2068	0.799	0.958
2413	0.925	0.986

correlate them with the Margules equations and make the T–xy chart at P = 400 kPa under the assumption that the vapor phase obeys the ideal gas law, while the liquid is a non-ideal liquid mixture.

8.15 Given the set of experimental VLE data for a CO_2/toluene mixture at T = 80 °C of W. Morris and M. Donohue, "Vapor–liquid equilibria in mixtures containing carbon dioxide, toluene, and i-methylnaphthalene," *J. Chem. Eng. Data*, **30**, (3), 259–263, 1985,

P /kPa	x_1	y_1
259	0.0120	0.856
579	0.0287	0.926
905	0.0463	0.952
1950	0.105	0.968
3125	0.170	0.978
4290	0.239	0.980
5350	0.300	0.981
6475	0.366	0.980
8040	0.468	0.975
9495	0.546	0.974
10335	0.613	0.971
11930	0.783	0.939

estimate the Henry's constant. Use the result to calculate the bubble pressure at 400 K for a molar concentration of CO_2 in the liquid phase of 0.015.

8.16 Use STANMIX to make a P–xy chart similar to that in Figure 8.20 for the equimolar methanol/water mixture. Plot the bubble and dew lines for T = 350 K, 375 K, and 400 K.

8.17 Use STANMIX to make a T–s chart similar to that in Figure 8.1 for the equimolar methanol/water mixture. Plot two iso-lines for each type, that is, isobar, isochore, isenthalp, iso-quality.

8.18 Use STANMIX to find the azeotropic composition for the mixture 2-propanol/water at P = 60 kPa. Explain why distillation of 2-propanol is not possible.

9 Exergy Analysis

CONTENTS

Application of the First and Second Laws of Thermodynamics, together with entropy analysis, has already showed us that a useful effect like work, cooling, or heating can be obtained if a thermodynamic system is not in equilibrium with its surroundings. The Carnot efficiency tells us which is the maximum amount of work (or power) that we can extract from a thermal source with a device operating between the temperature of the thermal source and a lower temperature. We also know that all the processes in a Carnot cycle are reversible, and this is an unattainable limit for real processes. Similar reasoning can be applied to refrigeration systems, whereby mechanical work is used to obtain the cooling effect. Entropy production always occurs in real processes and we generally speak of thermodynamic losses associated with entropy production.

Now we look closer at this concept, which becomes a powerful analysis and optimization tool in energy conversion engineering.

9.1 What for?

We are about to define another thermodynamic function, called *exergy*. This function is useful to evaluate thermodynamic losses in quantitative terms and to answer questions like:

- *Given a certain thermodynamic system, made of several components each realizing a certain set of processes of different nature (think for example of a Rankine cycle power plant, with heat exchangers performing heat transfer, turbines expanding vapor to obtain mechanical power, and pumps used to compress liquid water), which of the processes is most responsible for thermodynamic losses?*

By using exergy analysis you can compare a heat exchanger with a turbine and decide which of the two components is more dissipative with respect to the desired energy conversion goal of the entire system. If you know which components account for more irreversibility within a system, thus depleting the efficiency of the entire process with respect to the ideal limit of the Carnot cycle, you can decide how to best invest resources to improve your system design.

- *Given two thermodynamic processes or energy systems operating between different temperature levels, which one of the two performs best (has the best thermodynamic quality)?*

  We must take into account their different potential for converting energy in order to realize a useful effect, due to the different temperature levels at which they operate. The potential for useful energy conversion is higher for larger temperature differences, but is such a potential exploited to its maximum extent? Is the quality of the thermodynamic processes involved good enough? Exergy system analysis can answer these questions. For example, consider a geothermal ORC power plant operating between 150 °C (423 K) and 30 °C (303 K) with a thermal efficiency of 15 % and a gas turbine operating between the same lower temperature and 1200 °C (1473 K) with a thermal efficiency of 33 %. The Carnot efficiency associated with the geothermal ORC power plant is 28 %, while that associated with the gas turbine is 42 %. The Carnot cycle efficiency prescribes the maximum efficiency for a given difference between the maximum and minimum temperatures of the cycle, therefore the ratio between the thermal and the Carnot efficiency is an indicator of the thermodynamic quality of the system, as it compares its efficiency with the maximum attainable. We see in this case that the geothermal ORC power plant makes much better use of the available potential, as its thermal efficiency is 53 % of the associated Carnot efficiency, while that of the gas turbine is only 41 %.

Based on these observations note that, if it is possible to define a unit cost for the primary energy source (*e.g.*, fuel) based on exergy, it is also possible to evaluate in correct economic terms the cost of the diminished capacity of obtaining power due to sources of inefficiency which are very different in nature.

9.2 Available Energy

Consider the steady-flow device of Figure 9.1: a fluid enters the control volume at state 1 and exits at state 2. The boundary of the control volume separates it from a large environment at temperature T_0. The device rejects energy as heat to the environment at T_0 through an infinitesimal temperature difference, thus reversibly. Any sort of processes can occur within the control volume, including thermodynamic cycles involving other fluids. What is the maximum power output that this device can convert? Therefore, given a certain amount of energy, how much of this energy is actually available for conversion into useful work? The general approach here is to apply the first and the second laws of thermodynamics, thereby obtaining an answer to these questions.

Neglecting kinetic and potential energy of the flow streams, the energy balance yields

$$\underbrace{\dot{M}h_1}_{\text{Energy inflow rate}} = \underbrace{\dot{M}h_2 + \dot{W} + \dot{Q}_0}_{\text{Energy outflow rate}}.$$

The entropy bookkeeping for the control volume is

$$\underbrace{\dot{\mathcal{P}}_S}_{\text{Rate of entropy production}} = \underbrace{\left(\dot{M}s_2 + \frac{\dot{Q}_0}{T_0}\right)}_{\text{Entropy–outflow rate}} - \underbrace{\dot{M}s_1}_{\text{Entropy inflow rate}}.$$

Combining these two equations and solving for $\dot{W}$ gives

$$\dot{W} = \dot{M}\left[(h_1 - T_0 s_1) - (h_2 - T_0 s_2)\right] - T_0 \dot{\mathcal{P}}_S.$$

The second law requires that $\dot{\mathcal{P}}_S \geq 0$, therefore

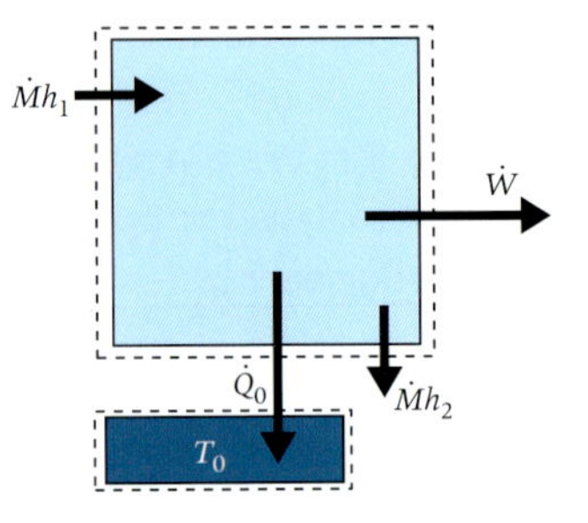

Figure 9.1 Control volume for the available-energy analysis of a steady-flow device.

$$\dot{W} \leq \dot{M}\left[(h_1 - T_0 s_1) - (h_2 - T_0 s_2)\right],$$

where $h - T_0 s$ is a function of both the thermodynamic state of the fluid and of the state of the environment.

The specific power output cannot exceed the decrease in the function $h - T_0 s$. If the equality holds, the processes within the control volume must be reversible, and also the energy-transfer process between the device and the environment at T_0.

The difference between the maximum possible useful work output and the actual work output is called *irreversibility*, or *thermodynamic loss*. The rate of thermodynamic loss $T_0 \dot{\mathcal{P}}_S$ represents useful mechanical power that could be obtained and instead has been dissipated. As we will see, in the analysis of complex energy systems or processes, one can locate the primary sources of irreversibility, which all have environmental and financial consequences, by calculating the amount of irreversibility associated with each component of the system or each step or even phenomenon of the process. Efforts to improve the performance of a system can then be concentrated in areas where the greatest gains can be made.

In this example we see that one can vary the discharge state 2 and the power output also varies. You might correctly guess that more power can be obtained if the discharge is at atmospheric temperature T_0 and pressure P_0. If the discharge temperature is higher than the temperature of the environment, a thermal engine can be added to generate more power. If the discharge pressure is greater than atmospheric, an expander can be used to obtain additional power. If the discharge pressure is lower than atmospheric the fluid cannot flow out of the system.

Example: Improving the use of energy in a chemical process. Suppose that a certain chemical process needs 2 kg/s of steam at 1 MPa and 320 °C, and the steam generator of a nearby power plant makes steam available at 2 MPa and 600 °C. The needed steam could simply be diverted from the boiler and passed through a valve in order to reduce the pressure to the desired value, and then through a heat exchanger in order to cool it down to the desired temperature ($1 - 1' - 2$ in Figure 9.2). In the interest of an efficient use of energy we explore the possibility of generating power as an additional benefit of this steamstate change. How much power could be obtained? From FLUIDPROP, by setting the thermodynamic model for water to IF97 and the input specification to PT we obtain:

State 1

$$P_1 = 2 \text{ MPa},$$
$$T_1 = 873.15 \text{ K},$$
$$h_1 = 3691 \text{ kJ/kg},$$
$$s_1 = 7.704 \text{ kJ/(kg} \cdot \text{K), and}$$

State 2

$$P_2 = 1 \text{ MPa},$$
$$T_2 = 693.15 \text{ K},$$
$$h_2 = 3094 \text{ kJ/kg},$$
$$s_2 = 7.198 \text{ kJ/(kg} \cdot \text{K).}$$

Assuming that all processes within the control volume encompassing the device realizing the steam state change are reversible, including the transfer of energy as heat to the environment, and taking $T_0 = 20$ °C $= 293.15$ K, we have that

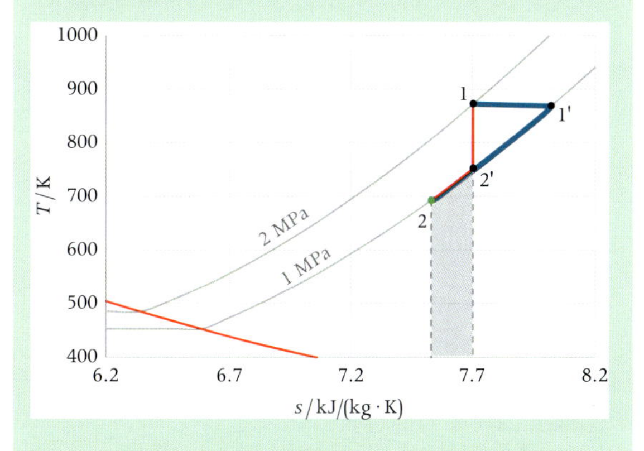

Figure 9.2 Representation in the T–s diagram of water of possible processes undergone by steam bringing the fluid from state 1 at higher temperature, pressure, and entropy to state 2 at lower temperature, pressure, and entropy.

$$\dot{W}_{max} = \dot{M}\,[(h_1 - T_0 s_1) - (h_2 - T_0 s_2)] = 896 \text{ kW}.$$

How can such power be obtained, given that all processes leading to it must be reversible? Observe that the entropy of *state 2* is lower than that of *state 1* ($s_2 < s_1$), therefore the Second Law of Thermodynamics tells us that the fluid cannot be expanded adiabatically from *state 1* to *state 2*, but only to *state 2′*, with $s_{2'} = s_1$, and $P_{2'} = P_2$, see Figure 9.2. Relevant thermodynamic properties of *state 2′* are

$$s_{2'} = s_1 = 7.704 \text{ kJ/(kg} \cdot \text{K)},$$
$$P_{2'} = 1 \text{ MPa},$$
$$T_{2'} = 752.1 \text{ K},$$
$$h_{2'} = 3433 \text{ kJ/kg}.$$

The maximum power associated with the $1 - 2'$ process is therefore

$$\dot{W}_{1-2',max} = \dot{M}\,[(h_1 - T_0 s_1) - (h_{2'} - T_0 s_{2'})]$$
$$= \dot{M}\,(h1 - h_{2'}) = 2 \text{ kg/s}$$
$$\times (3691 - 3433) \text{ kJ/kg} = 516 \text{ kW}.$$

Furthermore, the thermal power that must be transferred from the steam to the environment in order to cool it down to the desired temperature ($2' - 2''$ in Figure 9.2), can then be further converted into mechanical power by an engine implementing a suitable thermodynamic cycle. The power is maximum if the processes within such a thermal engine are reversible. We can imagine that a large number of Carnot engines (infinite in the limit) in parallel can be operated between a temperature $\bar{T}$, with $T_2 < \bar{T} < T'_2$, and T_0. Both the energy transfer as heat from the steam to the Carnot engine and from the Carnot engine to the environment must be reversible. The summation of the power converted by each Carnot engine must be

$$\dot{W}_{2'-2,max} = \dot{M}\,[(h_{2'} - T_0 s_{2'}) - (h_2 - T_0 s_2)] = 380 \text{ kW},$$

because this upper limit of power conversion can be achieved only if all processes within the pertinent control volume are reversible.

In reality all processes are irreversible and cost is associated with any system or component. We could think of expanding the steam in a turbine and obtain additional power by transferring energy as heat to a suitable engine (for example, an Organic Rankine Cycle turbogenerator), but such a solution might be too expensive.

A good engineer could realize (or buy) a steam turbine in order to expand the steam to pressure $P = P_2$, and a heat exchanger in order to cool it to temperature $T = T_2$. This turbine could be approximately 75 % efficient for this low power capacity, thus its power output would be $\dot{W} = 0.75 \cdot 516 \text{ kW} = 387 \text{ kW}$. At 0.1 €/(kW · hr), or 0.07 $ /(kW · hr), the value of the electricity that can be produced annually amounts to approximately €230 000 or $260 000 for a utilization factor of 70 % (the amount depends on the current and local value of electricity). The investment and maintenance cost of the system employing a steam turbine would have to be weighed against the utility bill saving in deciding whether or not to implement this solution. In the future, regulation in some countries might make it impossible to dissipate available energy to a certain extent.

Example: Geothermally driven cooling system. Let us now assume that in a remote desert area a small geothermal well produces 50 kg/hr of saturated steam at 150 °C. The environment temperature is 45 °C on average during the day and it is thought that a clever engineer might be able to devise a system to use the geothermal steam to produce cooling for homes at 23 °C (Figure 9.3a). The steam will emerge from this system as condensate at 1 atm. Figure 9.3b shows the process undergone by steam in the T–s diagram; the dotted line indicates that we are not committed to any particular process connecting the inlet and discharge states. What is the maximum cooling rate that could be provided by this system?

The control volume we analyze is shown in Figure 9.3a. We assume steady-flow, steady-state, one-dimensional flows at the inlet and outlet, and neglect the kinetic and potential energy changes of the flow. The energy input is provided by the cooling of steam. Any thermodynamic refrigeration system must reject energy as heat to the environment. An idealized example of a reversible cooling system is that operating according to the reversed Carnot cycle. We thus assume that $\dot{Q}_0$ is transferred reversibly to the environment.

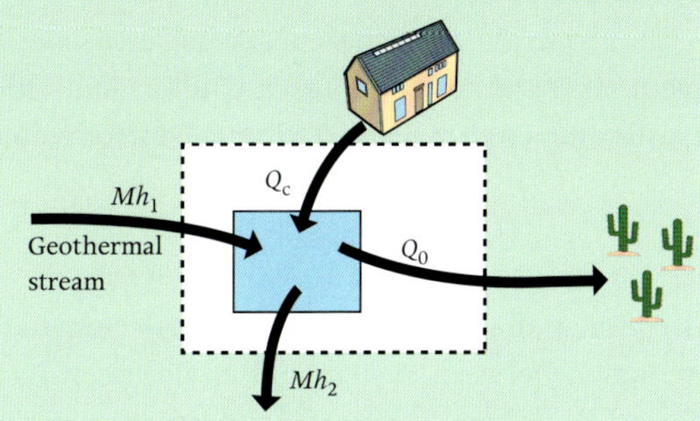

(a) The system and its environment

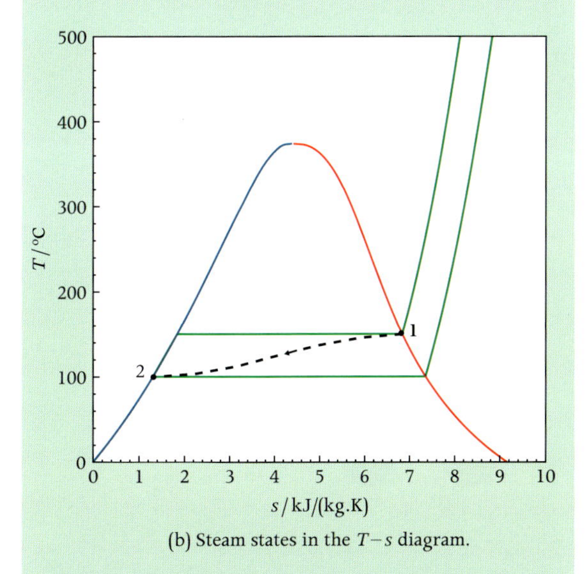

(b) Steam states in the $T-s$ diagram.

Figure 9.3 A geothermally driven cooling system.

The energy balance gives

$$\underbrace{\dot{Q}_{\text{cooling}} + \dot{M}h_1}_{\text{energy input rate}} = \underbrace{\dot{Q}_0 + \dot{M}h_2}_{\text{energy output rate}} .$$

The entropy book-keeping is

$$\dot{\mathcal{P}}_S = \left(\frac{\dot{Q}_0}{T_0} + \dot{M}s_2\right) - \left(\frac{\dot{Q}_{\text{cooling}}}{T_{\text{cooling}}} + \dot{M}s_1\right).$$

Combining and solving for $\dot{Q}_{\text{cooling}}$ gives

$$\dot{Q}_{\text{cooling}} = \frac{\dot{M}\left[(h_1 - T_0 s_1) - (h_2 - T_0 s_2)\right] - T_0 \dot{\mathcal{P}}_S}{T_0/T_{\text{cooling}} - 1}.$$

The second law requires that $\dot{\mathcal{P}}_S \geq 0$. Therefore, for a given discharge state 2, the maximum that $\dot{Q}_{\text{cooling}}$ can be is

$$\dot{Q}_{\text{cooling,max}} = \frac{\dot{M}\left[(h_1 - T_0 s_1) - (h_2 - T_0 s_2)\right]}{T_0/T_{\text{cooling}} - 1}.$$

Note that any irreversibility in the device makes $\dot{\mathcal{P}}_S > 0$ thus reducing $\dot{Q}_{\text{cooling}}$ with respect to its maximum value.

Using the IF97 model for water we obtain:

State 1

$$T_1 = 150\ ^\circ\text{C}\ (423.15\ \text{K}),$$
$$q_1 = 1\ \text{(saturated vapor)},$$
$$h_1 = 2745.9\ \text{kJ/kg},$$
$$s_1 = 6.837\ \text{kJ/(kg} \cdot \text{K), and}$$

State 2

$$T_2 = 100\ ^\circ\text{C}\ (373.15\ \text{K}),$$
$$q_2 = 0\ \text{(saturated liquid)},$$
$$h_2 = 419.1\ \text{kJ/kg},$$
$$s_2 = 1.307\ \text{kJ/(kg} \cdot \text{K)}.$$

Therefore, since $T_0 = 318.15$ K (45 °C), we have that

$$h_1 - T_0 s_1 = 2745.9 - 318.15 \times 6.837 = 570.7\ \text{kJ/kg},$$
$$h_2 - T_0 s_2 = 419.1 - 318.15 \times 1.307 = 3.3\ \text{kJ/kg}.$$

Finally, with $T_{\text{cooling}} = 296.15$ K, we obtain

$$\dot{Q}_{\text{cooling,max}} = \frac{50\ \text{kg/hr} \times (570.7 - 3.3)\text{kJ/kg}}{318.15/296.15 - 1} = 3.82 \cdot 10^5\ \text{kJ/hr} = 106\ \text{kW}.$$

Any actual device would have a lower cooling capacity, because of irreversibilities. A good engineer could easily come up with a design that could provide about 50 kW of cooling. We again note that the maximum performance was determined without any particular choice of process; this can be done because of the great power of fundamental thermodynamics.

9.3 Exergy

Starting from the concept of *available energy*, we can define a function called *exergy* as the

> **maximum theoretical work that can be obtained from a thermodynamic system as it interacts with the environment and reaches equilibrium with it.**

It is therefore important to define what the *environment* is, because exergy is inherently linked to it. The environment is characterized by the fact that, if the system is in equilibrium with it, no more work can be extracted. In *practical* terms therefore the state of the environment is defined by a temperature T_0 and a pressure P_0 which are typical of the surroundings of a system, at a certain distance. It is also assumed that these values cannot change as a result of the interaction with the system. If the system and the environment are in equilibrium, they both possess energy, but the exergy of the system is zero because there is no possibility that the system spontaneously changes or interacts with the environment.

Here, a *practical* approach is adopted in the sense that the results of an analysis conducted under the stated assumptions can be precise up to a certain level of accuracy for engineering purposes. The environment is practically at equilibrium if any disequilibrium within the environment is of negligible significance compared to the thermodynamic system under scrutiny. Wind holds exergy because it is not in equilibrium with adjacent quiescent air. Similarly, geysers can be used for energy conversion because the temperature of the hot water is not in equilibrium with the environment. Ground water at 10 °C can be used as a source of exergy in order to cool a home if its temperature is 30 °C.

Exergy of a system

In order to evaluate exergy, we apply its definition to a well defined system. We want to calculate the maximum amount of energy that can be converted into work within a control volume formed by a system combined with its surrounding environment. This situation corresponds, for example, to a piston generator fed by a pressurized air tank, which is operating until it reaches thermal equilibrium with the surrounding environment, as shown in Figure 9.4. The control volume of the system includes the source and sink tanks so that no air is exchanged with the environment.

We formulate first the energy balance over the time it takes the system to reach equilibrium with its surrounding environment. The total control volume

$$\mathrm{CV}_{\mathrm{tot}} = \mathrm{CV}_{\mathrm{sys}} + \mathrm{CV}_{\mathrm{env}}$$

is large enough such that the work that occurs due the interaction between system and environment is not affected by the energy transfer as heat between them,

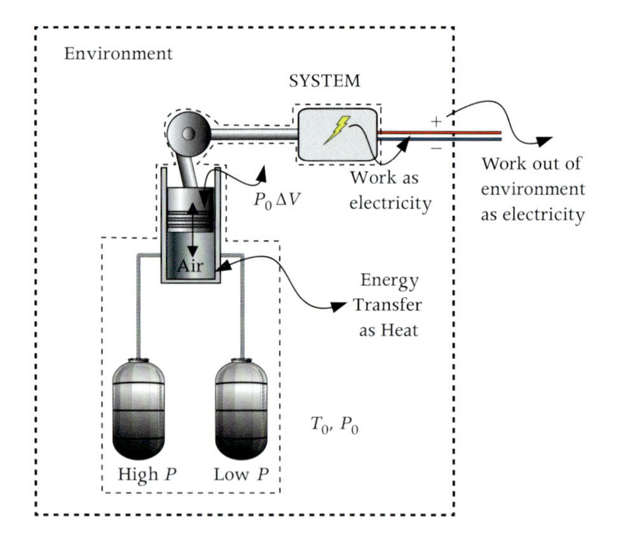

Figure 9.4 A system and its control mass (dotted line) enclosed in a larger control volume (dashed line) representing its surrounding environment. The system exchanges energy as work and heat with the environment, and changes its volume due to the movement of the piston. The environment and what is outside the environment exchange energy only as work.

thus the pressure and temperature of the environment, P_0 and T_0, remain the same during the process. In addition, the volume of the system, and consequently of the environment, can vary but not their sum.

Recalling that *accumulation* = *final* − *initial*, the energy balance over the stated period of time applied to $\mathrm{CV_{tot}}$ is

$$\Delta E_{\mathrm{tot}} = \cancelto{0}{Q} - W.$$

Where ΔE_{tot} is the variation of total energy, therefore it is equal to the sum of the change of energy of the system and that of the environment, hence

$$\Delta E_{\mathrm{tot}} = \Delta E_{\mathrm{env}} + \Delta E_{\mathrm{sys}}.$$

The total energy of the system (control mass) in the general case is given by

$$E_{\mathrm{sys}} = U_{\mathrm{sys}} + E_{\mathrm{k,sys}} + E_{\mathrm{p,sys}}.$$

The variation of the total energy of the environment is due only to its variation of internal energy, as its kinetic and potential energy do not vary, thus

$$\Delta E_{\mathrm{env}} = \Delta U_{\mathrm{env}}.$$

The variation of the total energy of the system is given by its energy in the final state minus its energy in the initial state E_{sys}. Once it reaches equilibrium, the system holds only internal energy. By denoting with U_0 its energy in the final state, the energy balance for the system is

$$\Delta E_{\mathrm{sys}} = U_0 - E_{\mathrm{sys}}.$$

The variation of total energy of $\mathrm{CV_{tot}}$ is therefore

$$\Delta E_{\mathrm{tot}} = \Delta U_{\mathrm{env}} + \left(U_0 - E_{\mathrm{sys}}\right).$$

Recalling that, from the Gibbs equation,

$$du = Tds - Pdv,$$

and given that the pressure and temperature of the environment do not change and remain equal to P_0 and T_0, we obtain

$$\Delta U_{\mathrm{env}} = T_0 \Delta S_{\mathrm{env}} - P_0 \Delta V_{\mathrm{env}}.$$

The expression for the variation of total energy therefore becomes

$$\Delta E_{\mathrm{tot}} = \left(U_0 - E_{\mathrm{sys}}\right) + T_0 \Delta S_{\mathrm{env}} - P_0 \Delta V_{\mathrm{env}}.$$

Substituting for ΔE_{tot} in the energy balance for $\mathrm{CV_{tot}}$, $\Delta E_{\mathrm{tot}} = -W$, leads to

$$W = \left(E_{\mathrm{sys}} - U_0\right) - \left(T_0 \Delta S_{\mathrm{env}} - P_0 \Delta V_{\mathrm{env}}\right).$$

The volume change of the environment is equal, but opposite in sign, to the volume change of the system, therefore

$$\Delta V_{\mathrm{env}} = -\left(V_0 - V_{\mathrm{sys}}\right),$$

thus

$$W = \left(E_{\mathrm{sys}} - U_0\right) + P_0\left(V_{\mathrm{sys}} - V_0\right) - T_0 \Delta S_{\mathrm{env}}. \quad (9.1)$$

The entropy balance for $\mathrm{CV_{tot}}$ is

$$\Delta S_{\mathrm{tot}} = \mathcal{P}_S.$$

The entropy production $\mathcal{P}_S$ is due to all the irreversibilities occurring in the system as it reaches equilibrium with the environment. It is therefore given by the summation of the entropy change of the system and that of the environment resulting from their interaction, thus

$$\mathcal{P}_S = \Delta S_{\mathrm{sys}} + \Delta S_{\mathrm{env}}.$$

The entropy change of the system is given by the difference between its entropy in the final state and its entropy in the initial state

$$\Delta S_{\mathrm{sys}} = \left(S_0 - S_{\mathrm{sys}}\right).$$

The entropy production over the process can therefore be written as

$$\mathcal{P}_S = \left(S_0 - S_{\mathrm{sys}}\right) + \Delta S_{\mathrm{env}}.$$

Finally, the entropy change of the environment is therefore

$$\Delta S_{\mathrm{env}} = \mathcal{P}_S - \left(S_0 - S_{\mathrm{sys}}\right),$$

which can be substituted in (9.1) and gives

$$W = \left(E_{\mathrm{sys}} - U_0\right) + P_0\left(V_{\mathrm{sys}} - V_0\right) - T_0\left(S_{\mathrm{sys}} - S_0\right) - T_0 \mathcal{P}_S. \quad (9.2)$$

The term $T_0 \mathcal{P}_S$ is always positive, and it depends on how an actual process is realized.

We now apply the definition of exergy to (9.2) in order to obtain the exergy of the system. Exergy is the maximum amount of work that can be obtained from the interaction of the system with the environment, once the system has reached equilibrium with the environment. The maximum amount of work is obtained if the process is reversible, that is if $\mathcal{P}_S = 0$. The exergy of the system Ξ_{sys} is therefore

$$\Xi_{sys} = \left(E_{sys} - U_0\right) + P_0 \left(V_{sys} - V_0\right) - T_0 \left(S_{sys} - S_0\right), \tag{9.3a}$$

or

$$\Xi_{sys} = E_{sys} + P_0 V_{sys} - T_0 S_{sys} - G_0. \tag{9.3b}$$

The value of the exergy of the system depends only on its initial and final state, which is set by the conditions of the environment.

We chose the Greek letter Ξ (pronounced *Xi*) as the symbol for exergy because of its assonance with the word exergy, and in order to avoid abusing x, which is already used for the vapor fraction of a pure fluid (quality) and for the mole fraction of a mixture component in the liquid phase.

Specific, physical, kinetic, potential, and flow exergy

As we have seen, it is often convenient to perform analysis and calculations with intensive properties. The specific exergy function is given by

$$\xi = e + P_0 v - T_0 s - g_0, \tag{9.4a}$$

or

$$\xi = (e - u_0) + P_0 (v - v_0) - T_0 (s - s_0), \tag{9.4b}$$

where

$$e = u + \mathcal{V}^2/2 + gz.$$

The term $(u - u_0) + P_0(v - v_0) - T_0(s - s_0)$ is often called *specific physical exergy*, while the exergies resulting from kinetic and potential energy are called kinetic and potential exergy respectively. In summary, these forms of exergy are defined as

$$\xi_{phys} \equiv (u - u_0) + P_0(v - v_0) - T_0(s - s_0),$$
$$\xi_k \equiv \mathcal{V}^2/2,$$
$$\xi_p \equiv gz.$$

Such subdivision allows us to evaluate their relative contribution to the total specific exergy. Both kinetic and potential energy are forms of mechanical energy and can be converted completely into work, therefore in this case there is no difference between energy and exergy.

Finally, the *specific flow exergy* is the maximum potential of performing work by a fluid flow of 1 kg/s, that is, according to such definition,

$$\xi_f = (h - h_0) - T_0 (s - s_0) + \frac{\mathcal{V}^2}{2} + gz. \tag{9.5}$$

These forms of exergy are not sufficient to evaluate the total specific exergy if work can be obtained from the combination of the system and the environment as a result of chemical disequilibrium or difference in composition. In this case *chemical exergy* (ξ_{chem}) must be accounted for. A more comprehensive treatment, including the exergy of reacting systems and of systems involving multicomponent fluids can be found in more specialized textbooks.

9.4 Control Volume Exergy Analysis

In order to perform exergy analysis, it is useful to obtain one more balance equation in addition to the mass and energy balance. Let us derive the exergy balance on a rate basis for a control volume CV exchanging mass and energy as work and heat with the environment. The procedure is similar to the one related to control volume energy analysis in Section 4.3.

First we write the mass, energy, and entropy balances for the control volume of Figure 9.5. Here we consider an unsteady situation for broader applicability, unlike the examples of Section 9.2. Note also that energy is transferred as heat at a boundary b, which is at a constant temperature T_b higher than the temperature of the environment.

Thus, the mass balance is

$$\frac{dM_{CV}}{dt} = \dot{M}_1 - \dot{M}_2, \tag{9.6}$$

the energy balance is

$$\frac{dE_{CV}}{dt} = (e + Pv)_1 \dot{M}_1 - (e + Pv)_2 \dot{M}_2 - \dot{Q} - \dot{W}, \tag{9.7}$$

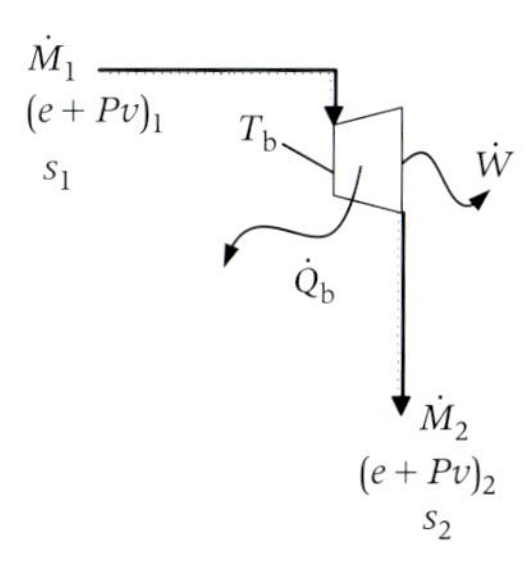

Figure 9.5 Control volume for the exergy analysis.

and the entropy balance is

$$\frac{dS_{CV}}{dt} = \dot{M}_1 s_1 - \dot{M}_2 s_2 - \frac{\dot{Q}}{T_b} + \dot{\mathcal{P}}_S. \quad (9.8)$$

Subtracting (9.8) multiplied by T_0 from (9.7) gives

$$\frac{dE_{CV}}{dt} - T_0 \frac{dS_{CV}}{dt} = (e+Pv)_1 \dot{M}_1 - (e+Pv)_2 \dot{M}_2 - \dot{Q} - \dot{W} - T_0 \left(\dot{M}_1 s_1 - \dot{M}_2 s_2\right) + \frac{T_0}{T_b}\dot{Q} - T_0 \dot{\mathcal{P}}_S.$$

We can rearrange the terms on the right-hand side and obtain

$$\frac{dE_{CV}}{dt} - T_0 \frac{dS}{dt} = +\dot{M}_1 \left[(e+Pv)_1 - T_0 s_1\right] - \dot{M}_2 \left[(e+Pv)_2 - T_0 s_2\right] - \left(1 - \frac{T_0}{T_b}\right)\dot{Q} - \dot{W} - T_0 \dot{\mathcal{P}}_S. \quad (9.9)$$

We now want to introduce the exergy function. We have seen that the exergy of the system is given by (9.3b). Hence we have that

$$\frac{d\Xi}{dt} = \frac{dE}{dt} + P_0 \frac{dV}{dt} - T_0 \frac{dS}{dt} - g_0 \frac{dM}{dt},$$

therefore, making use of the mass balance (9.6), we can write

$$\frac{dE}{dt} - T_0 \frac{dS}{dt} = \frac{d\Xi}{dt} - P_0 \frac{dV}{dt} + (h_0 - T_0 s_0)\left(\dot{M}_1 - \dot{M}_2\right). \quad (9.10)$$

The term $P_0 \frac{dV}{dt}$ must be accounted for only if the boundary of the CV moves in time (in the example illustrated in Figure 9.5 the boundary of the CV is fixed, but the boundary of the CV would vary in time if it coincided with the inner surface of the cylinder of a piston engine).

By applying (9.10) to the left-hand side of (9.9) we get

$$\frac{d\Xi}{dt} + (h_0 - T_0 s_0)\left(\dot{M}_1 - \dot{M}_2\right) = \dot{M}_1 \left[(e+Pv)_1 - T_0 s_1\right] - \dot{M}_2 \left[(e+Pv)_2 - T_0 s_2\right] - \left(1 - \frac{T_0}{T_b}\right)\dot{Q} - \dot{W} - T_0 \dot{\mathcal{P}}_S. \quad (9.11)$$

With $h \equiv u + Pv$ and $e \equiv u + \frac{v^2}{2} + gz$, and the definition of flow exergy (9.5), the terms associated with mass flow on the right-hand side of (9.11) can be written as

$$\begin{aligned}(e+Pv) - T_0 s &= u + \frac{v^2}{2} + gz + Pv - T_0 s = h + \frac{v^2}{2} + gz - T_0 s \\ &= (h - h_0) + \frac{v^2}{2} + gz - T_0 (s - s_0) + h_0 - T_0 s_0 \\ &= \xi_f + (h_0 - T_0 s_0).\end{aligned}$$

The introduction of the specific flow exergy in (9.11) and some rearrangement of terms therefore gives

$$\frac{d\Xi}{dt} + \cancel{(h_0 - T_0 s_0)\left(\dot{M}_1 - \dot{M}_2\right)} = \dot{M}_1 \xi_{f,1} - \dot{M}_2 \xi_{f,2} + \cancel{(h_0 - T_0 s_0)\left(\dot{M}_1 - \dot{M}_2\right)} - \left(1 - \frac{T_0}{T_b}\right)\dot{Q} - \dot{W} - T_0 \dot{\mathcal{P}}_S.$$

The rate of flow exergy is $\dot{\Xi}_f = \dot{M}\xi_f$. Furthermore, the term $T_0 \dot{\mathcal{P}}_S$ is called *rate of exergy destruction* $\dot{\Xi}_d$.

We see therefore that the *exergy balance on a rate-basis* for a control volume exchanging with the environment i) thermal power at a boundary temperature T_b, ii) mechanical power, and iii) mass flow of a simple compressible substance at the system inlet and outlet, can be written as

$$\frac{d\Xi}{dt} = \dot{\Xi}_{f,1} - \dot{\Xi}_{f,2} - \left(1 - \frac{T_0}{T_b}\right)\dot{Q} - \dot{W} - \dot{\Xi}_d. \quad (9.12)$$

The terms of the exergy balance Equation (9.12) can all be interpreted in relation to the exergy concept. $\left(1 - \frac{T_0}{T_b}\right)\dot{Q}$ is the maximum mechanical power that can be converted from the heat transfer rate $\dot{Q}$ occurring at the boundary where the temperature is T_b. It is therefore an *exergy transfer associated with energy transfer as heat*. We see that if the system is closed ($\dot{M}_1 = \dot{M}_2 = 0$) and in steady-state, that maximum amount of mechanical power is obtained with a reversible process ($\dot{\mathcal{P}}_S = 0$) promoted by that amount of energy transfer as heat, namely $\dot{W} = \left(1 - \frac{T_0}{T_b}\right)\dot{Q}$. If the temperature of the boundary is lower than the temperature of the environment, $\left(1 - \frac{T_0}{T_b}\right)$ is negative, therefore the exergy flow associated with the energy transfer as heat is in an opposite direction with respect to the energy transfer.

The term $\dot{W}$ represents the *exergy transfer associated with mechanical work*. The rate of exergy transfer is equal to the reversible mechanical power if the CV does not exchange energy as heat nor mass with the environment.

The terms $\dot{\Xi}_{f,1}$ and $\dot{\Xi}_{f,2}$ are rates of flow exergy. In order to understand what they stand for, let us assume that the system is at steady state ($\dot{M}_1 = \dot{M}_2 = \dot{M}$), and that there is no energy transfer as heat from or to the environment. The system exchanges mass with the environment at the inlet and outlet ports. We have that, for a reversible process ($\dot{\mathcal{P}}_S = 0$), the maximum amount of mechanical power that can be obtained is given by $\dot{W} = \dot{\Xi}_{f,1} - \dot{\Xi}_{f,2}$. We recall, see (4.13), that energy associated with mass transfer at the boundary is the sum of its total energy and flow work Pv. We conclude that the specific flow exergy is *exergy transfer associated with mass transfer and flow work*.

Finally, note that the rate of exergy destruction, or exergy loss, is closely related to entropy production within the control volume. These are two conceptual quantities used to describe the same phenomenon. Exergy destruction represents the power (energy per unit time) that is dissipated due to the unavoidable irreversibilities within the system. The microscopic phenomenon that we call dissipation, which is responsible for the decreased capability of converting energy into a useful effect, is due to the fact that the transfer of energy always entails an increase in the disorder of molecules. As a consequence of the second law of thermodynamics, the exergy of a closed system always decreases, or remains constant in the limit of a reversible process. Exergy analysis is performed in energy engineering and whenever energy efficiency is a concern, in order to minimize exergy destruction.

To clarify the relation between energy, entropy and exergy, it is useful to consider the example of the heat transfer through a metallic wall from one fluid to another, as it occurs, for instance in a heat exchanger (Figure 9.6). A certain amount of energy is transferred unaltered as heat. In this process, some entropy is generated (molecules are shuffled around and become more unorganized), therefore the entropy of the colder fluid increases. Conversely, this means that some exergy is destroyed, i.e., that some potential to perform work is lost.

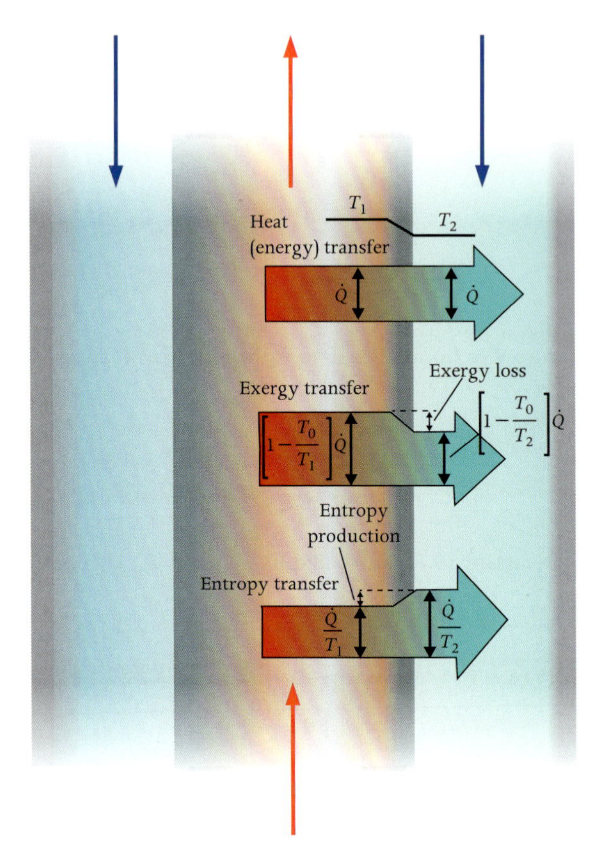

Figure 9.6 Heat transfer from a fluid to another fluid through the wall of a heat exchanger, and representation of its associated energy, entropy, and exergy transfer.

Steady-state exergy balance

Many engineering problems can be studied under the steady-state assumption, therefore it is worth writing out the time-invariant form of the exergy balance for a control volume like that in Figure 9.5. From (9.12) we obtain

$$\underbrace{\dot{\Xi}_{f,1}}_{\substack{\text{rate of exergy transfer}\\ \text{associated with } f,1}} = \underbrace{\dot{\Xi}_{f,2}}_{\substack{\text{rate of exergy transfer}\\ \text{associated with } f,2}} + \underbrace{\left(1-\frac{T_0}{T_b}\right)\dot{Q}}_{\substack{\text{rate of exergy transfer}\\ \text{associated with } \dot{Q}}} + \underbrace{\dot{W}}_{\substack{\text{rate of exergy transfer}\\ \text{associated with } \dot{W}}} + \underbrace{\dot{\Xi}_d}_{\text{rate of exergy loss}} .$$

In a steady state $\dot{M}_1 = \dot{M}_2 = \dot{M}$, therefore we can also write

$$\dot{M}\left(\xi_{f,1}-\xi_{f,2}\right) = \left(1-\frac{T_0}{T_b}\right)\dot{Q} + \dot{W} + \dot{\Xi}_d. \quad (9.13)$$

The change of flow exergy in (9.13) is given by

$$\xi_{f,1}-\xi_{f,2} = (h_1-h_2) - T_0(s_1-s_2) + \left(\frac{v_1^2}{2}-\frac{v_2^2}{2}\right) + g(z_1-z_2). \quad (9.14)$$

Temperature of the boundary lower than the temperature of the environment

In general the temperature of the environment T_0 can also be higher than the temperature of the boundary T_b. For example, in cryogenic processes, fluid flows can be at temperatures far lower than the ambient temperature, and it might be interesting to evaluate the possibility of using this temperature difference in order to convert energy into mechanical work. In this case, we see from (9.13) that thermal power entering the system depletes its exergy, that is its ability of performing work. This is opposite to what happens for example in a power plant discharging energy as heat to the environment, whereby the energy that is transferred as heat to the system at its boundary increases its exergy.

Let us now learn how we can derive the steady-state exergy balance for a few typical components, as a first step, which will enable us to perform the exergy analysis of entire energy systems.

Example: Valve. The exergy balance for the throttling valve of Figure 9.7 operating in steady-state and adiabatic conditions is

$$\dot{M}\left(\xi_{f,1}-\xi_{f,2}\right) = \dot{\Xi}_d.$$

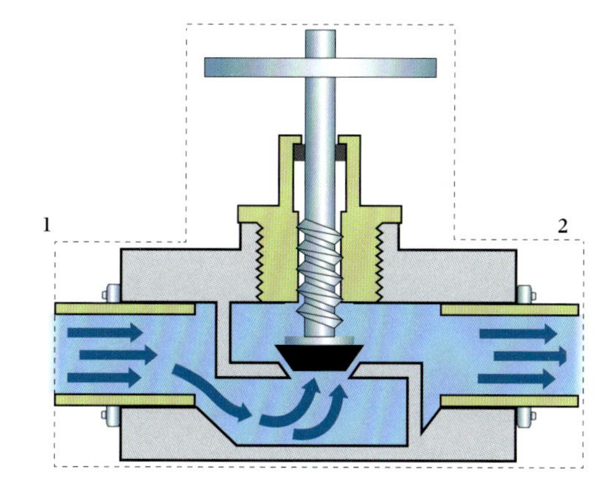

Figure 9.7 A throttling valve and the control volume for its exergy analysis.

The exergy balance equation shows that all the potential of the fluid flow of performing mechanical work is wasted because all the exergy associated with the flow gets destroyed. A valve promotes a chaotic (turbulent) expansion of the molecules of the fluid, as opposed to a turbine, whereby the orderly motion of the fluid through the blades allows for the extraction of useful power.

We can now calculate the value of the exergy loss for a valve laminating a mass flow of 0.5 kg/s of supercritical CO_2 at $T_1 = 100$ °C and $P_1 = 90$ bar down to $P_2 = 50$ bar. The environment is at $T_0 = 25$ °C and $P_0 = 1$ atm $= 1.013$ bar, therefore $h_0 = h_0(T_0, P_0) = 505.8$ kJ/kg and $s_0 = s_0(T_0, P_0) = 2.737$ kJ/(kg · K).

The combination of the energy and mass balances on the control volume of Figure 9.7 gives

$$h_2 = h_1 = h(T_1 = 100\ ^\circ\text{C}, P_1 = 90\ \text{bar}) = 512.1\ \text{kJ/kg},$$

and

$$s_1 = s(T_1, P_1) = 1.967\ \text{kJ/}\left(\text{kg}\cdot\text{K}\right),$$
$$s_2 = s(P_2, h_2) = 2.057\ \text{kJ/}\left(\text{kg}\cdot\text{K}\right).$$

With the exergy associated with kinetic energy being negligible with respect to the other terms and the elevation of the flanges of the valve being the same, the inlet and outlet flow exergy is given by

$$\xi_{f,1} = h_1 - h_0 - T_0(s_1 - s_0) = 235.9 \text{ kJ/kg},$$

and

$$\xi_{f,2} = 209.1 \text{ kJ/kg}.$$

Therefore the specific exergy loss is

$$\frac{\dot{\Xi}_d}{\dot{M}} = \xi_{f,1} - \xi_{f,2} = 26.8 \text{ kJ/kg}.$$

You might have been quicker and noticed that, in this case, because $h_1 = h_2$, the difference in the specific flow exergy is just $\xi_{f,1} - \xi_{f,2} = -T_0(s_1 - s_2)$.

Example: Heat exchanger. We now apply the steady-state exergy balance equation (9.13) to the regenerator of Figure 7.17, appropriately modified in order to take into account the double mass flow inlets and outlets. We want to compute the value of the rate of exergy destruction within that heat exchanger. In addition we want to compute the change in flow exergy from inlet to outlet for both streams and discuss the results.

The working fluid is siloxane MDM. Furthermore, as is very often the case in power systems analysis, we assume that the heat exchanger is perfectly insulated: if the heat exchanger is designed correctly, the amount of thermal energy that is transferred within the device is orders of magnitude larger than the energy that is transferred to the environment as heat. This is easy to achieve with proper insulating material. The environment is assumed at $T_0 = 10$ °C and $P_0 = 1.013$ bar.

The thermodynamic states at the inlets and outlets are identified by the following values of pressure and temperature:

$$P_2 = 28.0 \text{ bar},$$
$$T_2 = 146.2 \text{ °C};$$
$$P_3 = 27.0 \text{ bar},$$
$$T_3 = 161.0 \text{ °C};$$
$$P_5 = 0.8 \text{ bar},$$
$$T_5 = 183.2 \text{ °C; and}$$
$$P_6 = 0.8 \text{ bar}.$$

The mass flow circulating in the system at steady state is $\dot{M} = 13.5$ kg/s.

The state of the superheated vapor exiting the hot side of the regenerator can be obtained from the steady-state energy balance on a rate basis applied to the control volume enclosing the heat exchanger,

$$\underbrace{\dot{M}\left[(h_2 + h_5) + \frac{v_2^2 + v_5^2}{2} + g(z_2 + z_5)\right]}_{\text{rate of energy input}} =$$
$$\underbrace{\dot{M}\left[(h_3 + h_6) + \frac{v_3^2 + v_6^2}{2} + g(z_3 + z_6)\right] + \dot{Q}}_{\text{rate of energy output}}.$$

The assumption that the heat exchanger is perfectly insulated, and the other very common assumption that the kinetic and potential energy differences are negligible, lead to the equation for the energy balance

$$h_2 - h_3 = h_6 - h_5,$$

which gives

$$h_6 = h_2 + h_5 - h_3.$$

StanMix provides the following values for the enthalpies:

$$h_2 = h_2(T_2, P_2) = 31.8 \text{ kJ/kg},$$
$$h_5 = h_2(T_5, P_5) = 253.1 \text{ kJ/kg},$$
$$h_3 = h_3(T_3, P_3) = 62.4 \text{ kJ/kg},$$
$$h_6 = 31.8 + 253.1 - 62.4 = 222.5 \text{ kJ/kg}.$$

Now state 6 is also identified, because, according to the phase rule for a simple compressible fluid, the thermodynamic state is fixed by the two properties P_6 and h_6.

In order to calculate the rate of exergy loss within the heat exchanger we have to write the exergy balance. In this case, it reduces to

$$\dot{\Xi}_{f,2} + \dot{\Xi}_{f,5} = \dot{\Xi}_{f,3} + \dot{\Xi}_{f,6} + \left(1 - \frac{T_0}{T_b}\right)\cancelto{0}{\dot{Q}} + \dot{\Xi}_d,$$

because of the assumption of perfect isolation. The rate of exergy destruction is therefore given by

$$\dot{\Xi}_d = \dot{M}\left(\xi_{f,2} - \xi_{f,3}\right) + \dot{M}\left(\xi_{f,5} - \xi_{f,6}\right).$$

The change in the rate of flow exergy between the inlet and the outlet of the liquid stream is

$$\begin{aligned}\dot{M}\left(\xi_{f,2} - \xi_{f,3}\right) &= \dot{M}\left[(h_2 - h_3) - T_0\left(s_2 - s_3\right)\right]\\ &= -137.9 \text{ kJ/kg},\end{aligned}$$

and for the vapor stream

$$\begin{aligned}\dot{M}\left(\xi_{f,5} - \xi_{f,6}\right) &= \dot{M}\left[(h_5 - h_6) - T_0\left(s_5 - s_6\right)\right]\\ &= 152.1 \text{ kJ/kg},\end{aligned}$$

where the values for s have been obtained with StanMix as $s = s(T, P)$ for states 2, 3, 5 and $s = s(P, h)$ for state 6. By comparing the values of the change in the rate of flow exergy we see that the exergy increase of the liquid stream is lower than the exergy decrease of the vapor stream. The difference is the exergy that gets lost (the potential for performing useful work) in the process of transferring energy as heat and because of the flow friction in the pipes of the heat exchanger. Flow friction losses are proportional to the pressure drop between the inlet and the outlet, while the average temperature difference between the hot and the cold stream is an indicator of the thermodynamic loss due to heat transfer.

Example: Compressor. We consider the compressor of a gas turbine and calculate its associated rate of exergy destruction. With reference to Figure 7.23 the inlet and outlet conditions are:

$$P_1 = 1.013 \text{ bar},$$

$$T_1 = 25\ °\text{C},$$

$$v_1 = 200 \text{ m/s; and}$$

$$P_2 = 4 \text{ bar},$$

$$T_2 = 234.5\ °\text{C},$$

$$v_2 = 100 \text{ m/s}.$$

The mass flow is $\dot{M} = 20$ kg/s. Furthermore it can be assumed that the heat transfer from the casing to the surroundings occurs at the uniform temperature of $T_b = 135$ °C, and the power transferred as heat to the environment is $\dot{Q} = 200$ kW. We take $T_0 = 25$ °C and $P_0 = 1.013$ bar.

Analogously with (9.13), the rate-basis exergy balance for the control volume enclosing the compressor is

$$\dot{M}\left(\xi_{f,2} - \xi_{f,1}\right) = \left(1 - \frac{T_0}{T_b}\right)\dot{Q} - \dot{W} + \dot{\Xi}_d. \quad (9.15)$$

The specific flow exergy associated with the air stream being compressed is given by

$$\xi_{f,2} - \xi_{f,1} = (h_2 - h_1) - T_0\left(s_2 - s_1\right) + \frac{v_2^2 - v_1^2}{2}.$$

We can use the GasMix library to calculate air properties, as we can assume that it behaves as an ideal gas through the process, and we obtain

$$\begin{aligned}\xi_{f,2} - \xi_{f,1} &= 212.6 - 0.0 - 298.15\,(7.00 - 6.86)\\ &\quad + \frac{200^2 - 100^2}{2} \cdot 10^{-3}\\ &= 154.3 \text{ kJ/kg}.\end{aligned}$$

At steady state, the exergy transfer accompanying mechanical power is the power itself. Therefore the rate of exergy loss is computed as

$$\begin{aligned}\dot{\Xi}_d &= \dot{M}\left(\xi_{f,1} - \xi_{f,2}\right) - \left(1 - \frac{T_0}{T_b}\right)\dot{Q} + \dot{W}\\ &= 1.1 \text{ MW}.\end{aligned}$$

Let us use the concept of exergy and look at the magnitude of the various exergy transfers involved in the process. The distribution of the various exergy rates for the compression process is given in Table 9.1.

Table 9.1 Exergy rate bookkeeping for a compressor.

Net rate of exergy out	3.1 MW	72.6%
Mechanical power in (exergy rate with mech. power)	4.3 MW	100 %
Thermal power out (exergy rate with heat transfer)	0.05 MW	1.3 %
Rate of exergy destruction	1.1 MW	26.1 %

As we see, most of the exergy associated with the input power goes into the flow exergy, some is dissipated within the flow, and the portion accompanying heat transfer is small.

Causes of exergy losses

These examples have shown that various phenomena within the realm of thermomechanical systems can cause exergy losses. Exergy loss is proportional to entropy production. An example is provided by flow friction, whereby some of the mechanical energy of a fluid in motion is dissipated by being transferred as heat to the fluid because of viscosity. Another example is the dissipation that is always associated with energy transfer as heat under a finite temperature difference. Other common causes of exergy destruction are mixing and chemical reactions.

9.5 A Useful Thermodynamic Efficiency Based on Exergy

We can now conclude that the correct assessment of thermodynamic systems can be performed by comparing the actual process to an ideal process whereby all irreversibilities are zero. The definition of ideal process is also important and must be done according to the final useful effect that is pursued. For a thermal power system the ideal term for comparison is the Carnot cycle, that is a set of processes which provide the maximum efficiency for the conversion of thermal into mechanical power. For a cooling system, the term of comparison becomes the efficiency of the inverse Carnot cycle.

We define therefore as the *II-law efficiency*, or *exergetic efficiency* for thermal power systems

$$\eta_{\mathrm{II}} = \frac{\eta_{\mathrm{I}}}{\eta_{\max}},$$

and for a cooling system

$$\eta_{\mathrm{II}} = \frac{\mathrm{COP_I}}{\mathrm{COP_{max}}}.$$

The II-law efficiency for a heat pump is analogously defined.

In many ways the II-law efficiency is most suitable to evaluate the quality of a system, as it compares it with the maximum potential for a useful effect. The I-law efficiency, the efficiency based on energy conservation, is a performance index by which one could assume that all the available thermal energy could be converted into work, which is not going to happen in this universe.

For a process whose final objective is not the conversion of thermal power into mechanical power, or vice versa, the exergetic efficiency is defined by comparison with a reversible process ($\dot{\mathcal{P}}_S = 0$, or $\dot{\Xi}_d = 0$). This is the case for example of a reactor, e.g., a gasifier (Figure 9.8). A gasifier is a reactor where a solid fuel undergoes chemical reactions at high temperature and with a deficit of oxygen in order to produce a gas with

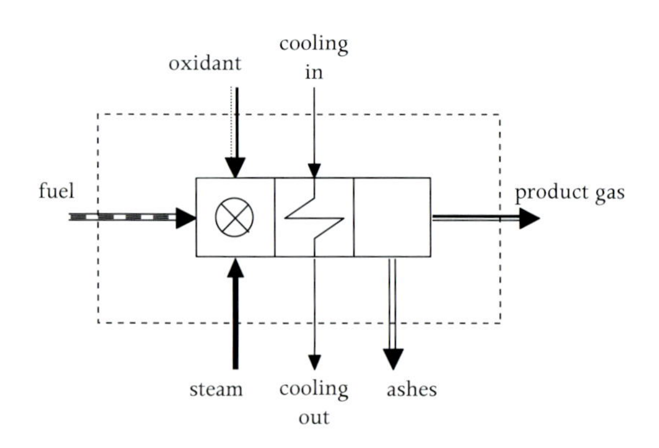

Figure 9.8 A gasifier with a representation of the interaction involving exergy transfer that needs to be accounted for when computing its exergetic efficiency.

a high calorific value, which can also be used as a fuel, for instance in a gas turbine.

In general terms the II-law efficiency is

$$\eta_{II} \equiv \frac{\text{Rate of exergy obtained}}{\text{Rate of exergy supplied}}.$$

In the case of a gasifier, the obtained exergy rate is the exergy associated with the products of gasification, which depends on the chemical species that are obtained and their temperature, while the rate of exergy that is supplied is the exergy of the fuel inflow (its potential of performing work, reversibly) and that of the oxidant. Moreover, the exergy of the cooling applied to keep the reactor's wall from melting, that is the flow exergy of the steam used to this purpose, and the exergy associated with the ashes that are extracted from the gasifier, must also be taken into account.

This more general definition of exergetic efficiency covers also the efficiency of cyclic processes. For example, for a thermal power cycle, the rate of exergy that is obtained is the net power output, while the rate of exergy that is supplied is the difference between the rate of exergy associated with the energy that is transferred as heat to the system, and that associated with the energy rejected as heat by the system to its surroundings.

Example: Domestic heating. To get an idea of the difference between the second and the first-law efficiency and of its significance, let us compare two solutions to the engineering problem of providing a household with thermal power, namely an electrical heater and a heat pump. Let us assume the same operating conditions as in Section 4.7, where the outside temperature is 7 °C and the thermostated temperature inside the house is 20 °C. The maximum COP for any heating device is

$$\text{COP}_{\text{max}} = \frac{1}{1 - \dfrac{T_0}{T_{\text{set}}}} = \frac{1}{1 - \dfrac{278.15\ \text{K}}{293.15\ \text{K}}} = 22.5.$$

The COP of the electrical heater, neglecting minor sources of exergy losses, is 1. All the electrical power $\dot{W}_{el}$ is dissipated into heat, which is the desired useful effect. The exergetic efficiency of the electrical heater is therefore

$$\eta_{\text{II,el.heater}} = \frac{\text{COP}_{\text{I,el.heater}}}{\text{COP}_{\text{max}}} = \frac{1}{22.5} = 0.044,$$

that is 4.4 %. As we have calculated in the example of Section 4.7, the COP of a heat pump suitable for these operating conditions is 4.67. The II-law efficiency of the heat pump is therefore

$$\eta_{\text{II,heat pump}} = \frac{\text{COP}_{\text{I,heat pump}}}{\text{COP}_{\text{max}}} = \frac{4.7}{22.5} = 0.207,$$

or 20.7 %. A comparison could also include a home-boiler operated with natural gas. In this case one should take into account that the primary source of energy is different. Electricity is generated in a power station, possibly using natural gas as fuel. A fair comparison therefore should include the exergy that is lost at the power station to convert the fuel into electricity and the exergy that is lost because of the transportation of the electricity over the grid. From the exergetic point of view, the best solution would be a system whereby a small and efficient internal combustion engine operated on natural gas provides the power for the compressor of the heat pump and all the heat loss from the engine (flue gas and engine and lubricant cooling) is also recovered to provide the heating. However, the large difference in the initial investment between such a system and an electrical heater, together with the maintenance needs and reliability issues, must also be factored in when evaluating the best solution for a given energy engineering problem.

Example: Heat exchanger. We can now use the result of the control volume energy analysis applied to a simple heat exchanger, the regenerator of Figure 7.17, in order to define its *exergetic* efficiency. Recall that we made the following assumptions: kinetic and potential energy associated with the two streams are negligible and the heat exchanger is perfectly insulated. The exergetic efficiency of the heat exchanger is given by the ratio between the rate of exergy provided and the rate of exergy obtained, therefore

$$\eta_{\mathrm{II,HX}} \equiv \frac{\text{Rate of exergy obtained}}{\text{Rate of exergy provided}} = \frac{\dot{M}_2\left(\xi_{\mathrm{f},2}-\xi_{\mathrm{f},3}\right)}{\dot{M}_5\left(\xi_{\mathrm{f},5}-\xi_{\mathrm{f},6}\right)} = \frac{\left[(h_2-h_3)-T_0\left(s_2-s_3\right)\right]}{\left[(h_5-h_6)-T_0\left(s_5-s_6\right)\right]}.$$

Note that in this example the useful effect is the increase in temperature of the liquid, therefore the rate of exergy that is obtained is the one of the liquid stream.

Note also that heat transfer in the heat exchanger occurs at variable temperatures along the heat exchanger. Let us consider only one of the streams of the heat exchanger and apply the steady-state exergy balance,

$$\dot{M}\left(\xi_{\mathrm{f,in}}-\xi_{\mathrm{f,out}}\right) = \int_{x_{\mathrm{in}}}^{x_{\mathrm{out}}} \left(1-\frac{T_0}{T}\right)\dot{q}\left(x\right)dx,$$

where $\dot{q}\left(x\right)$ is the heat transfer to the fluid per unit length. The only source of exergy destruction is the flow friction and this is negligible with respect to the overall exergy transfer in the vast majority of technical applications. We have therefore that

$$\dot{M}\left(\xi_{\mathrm{f,in}}-\xi_{\mathrm{f,out}}\right) = \left(1-\frac{T_0}{\overline{T}}\right)\dot{Q}.$$

Hence we see that it is possible to define an *equivalent thermodynamic temperature* $\overline{T}$ as the constant temperature at which the heat transfer should occur in order to generate the same amount of exergy transfer. Noting that $\dot{Q} = \dot{M}\left(h_{\mathrm{in}}-h_{\mathrm{out}}\right)$, we have

$$\dot{M}\left[(h_{\mathrm{out}}-h_{\mathrm{in}})-T_0\left(s_{\mathrm{out}}-s_{\mathrm{in}}\right)\right] = \left(1-\frac{T_0}{\overline{T}}\right)\dot{M}\left(h_{\mathrm{out}}-h_{\mathrm{in}}\right).$$

The *equivalent thermodynamic temperature* for the heat transfer to a fluid flow is therefore defined as

$$\overline{T} \equiv \frac{h_{\mathrm{out}}-h_{\mathrm{in}}}{s_{\mathrm{out}}-s_{\mathrm{in}}}.$$

The exergetic efficiency of a heat exchanger can therefore also be written as

$$\eta_{\mathrm{II,HX}} = \frac{1-T_0/\overline{T}_{\mathrm{hot}}}{1-T_0/\overline{T}_{\mathrm{cold}}}.$$

This result is very important as it shows that the thermodynamic quality of heat transfer increases if the average temperature under which such heat transfer occurs is minimal.

Often the transfer of energy as heat in a heat exchanger is represented in a $\dot{Q}-T$ diagram or a *value diagram* such as those shown in Figure 9.9. The visual evaluation of the shaded area in the $\dot{Q}-T$ diagram depicting the temperature difference along the heat exchanger helps in identifying by comparison which of the two heat exchangers performs best. Value diagrams, having $(1-T_0/T)$ instead of T on the y-axis, are used to easily compare amounts of exergy and exergy losses, because the shaded area in such a diagram is the exergy that is lost due to the temperature transfer. The graph in Figure 9.10 shows the exergetic efficiency as a function of the mean temperature difference for values of the temperature difference that are meaningful in energy applications.

Example: Turbine. Also in this case we assume steady-state and that there is no energy transfer as heat from the turbine casing to the surroundings. The steady-state form of the exergy balance equation (9.13) becomes

$$\dot{M}\left(\xi_{f,1}-\xi_{f,2}\right) = -\left(1-\frac{T_0}{T_\mathrm{b}}\right)\cancelto{0}{\dot{Q}} + \dot{W} + \dot{\Xi}_\mathrm{d},$$

which shows that the rate of exergy provided to the turbine $\dot{M}\left(\xi_{f,1}-\xi_{f,2}\right)$ becomes partly mechanical power ($\dot{W}$, rate of exergy obtained) and partly is destroyed ($\dot{\Xi}_\mathrm{d}$). The exergetic efficiency of a turbine is therefore

$$\eta_{\mathrm{II,turb}} = \frac{\dot{W}}{\dot{M}\left(\xi_{f,1}-\xi_{f,2}\right)}.$$

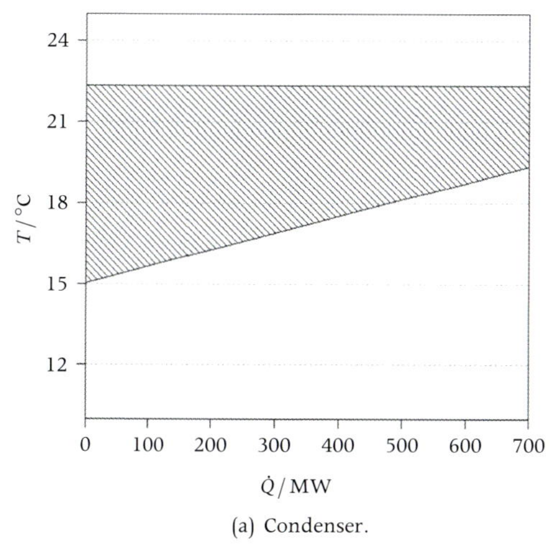

(a) Condenser.

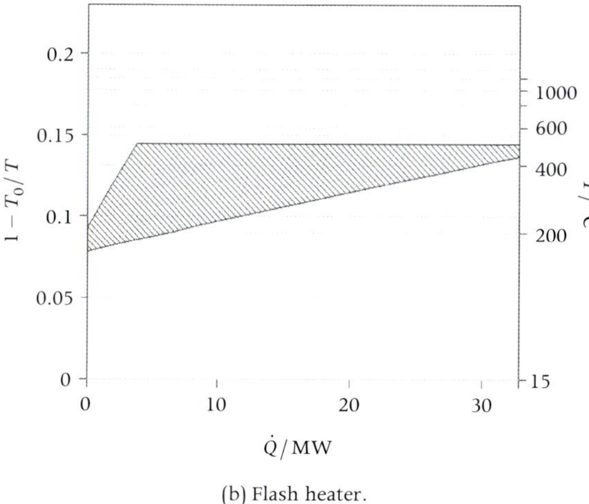

(b) Flash heater.

Figure 9.9 Heat transfer diagrams for two heat exchangers of a steam power plant.

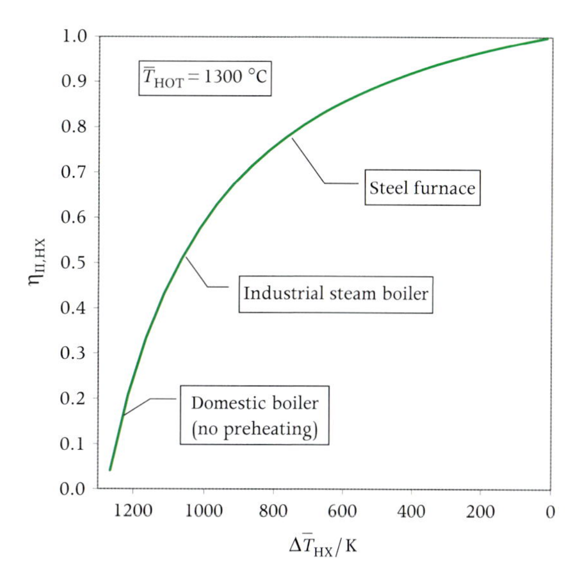

Figure 9.10 Exergetic efficiency of heat exchangers as a function of the difference between a fixed-temperature heat source (1300 °C) and the equivalent thermodynamic temperature of the cold stream.

As expected, for a given rate of exergy provided to the turbine, the smaller the rate of the exergy destruction, the larger its exergetic efficiency. This exergetic efficiency allows for the evaluation of the thermodynamic quality of two turbines operating under different conditions (inlet and outlet pressures and temperatures), while a comparison based on the isentropic efficiency does not take into account the different values of the energy, depending on the temperature at which energy is made available.

Example: Mixer or contact heat exchanger. Mixing of two streams (gaseous or liquid) is common in many process plants. One of the typical components of steam power plants is the deaerator, whereby air is extracted from the condensate by mixing it with steam extracted from the turbine. Such a type of heat exchanger is also called a contact heater (Figure 7.9a). We again assume that the device is perfectly insulated and that the exergy associated with kinetic and potential energy of the streams is negligible. Hence, the exergy balance on a rate basis applied to a contact heater is

$$\dot{M}_1\xi_{f,1} + \dot{M}_2\xi_{f,2} = \dot{M}_3\xi_{f,3} + \dot{\Xi}_d.$$

The mass balance is

$$\dot{M}_1 + \dot{M}_2 = \dot{M}_3.$$

Combining the balances gives

$$\dot{M}_2\left(\xi_{f,2} - \xi_{f,3}\right) = \dot{M}_1\left(\xi_{f,3} - \xi_{f,1}\right).$$

Table 9.2 Fluid properties for the thermodynamic states of the ORC plant of Figure 7.16.

State no.	Model	Fluid	$\dot{M}$ / kg/s	P /bar	T / °C	h / kJ/kg	s / kJ/ (kg · K)
1	REFPROP	R245fa	9.9	1.4	22.7	229.4	1.103
2	REFPROP	R245fa	9.9	8.9	23.7	231.0	1.107
3	REFPROP	R245fa	9.9	8.9	85.6	465.9	1.785
4	REFPROP	R245fa	9.9	1.4	40.9	438.3	1.808
5	IF97	water	70.7	1.0	10.0	42.1	0.151
6	IF97	water	70.7	1.5	10.0	42.2	0.151
7	IF97	water	70.7	1.0	17.0	71.4	0.253
8	IF97	water	22.5	6.0	104.0	436.3	1.351
9	IF97	water	22.5	6.0	79.4	332.8	1.068

Consistently with our definition of exergetic efficiency, we can write

$$\eta_{\mathrm{II,deaerator}} = \frac{\dot{M}_1 \left(\xi_{f,3} - \xi_{f,1}\right)}{\dot{M}_2 \left(\xi_{f,2} - \xi_{f,3}\right)},$$

which shows how much of the exergy provided by the hot stream (2) is transferred to the cold stream (1).

Table 9.3 Power transferred by the components of the ORC plant of Figure 7.16.

Component	Power / kW
Boiler	2328
Turbine	273.5
Pump	18.7
Condenser	2070

9.6 Example: Exergy Analysis of a Simple Rankine Cycle Power Plant

In order to summarize and use the knowledge we have gained so far in this chapter, we now perform the exergy analysis for the geothermal organic Rankine cycle power plant of Figure 7.16. The energy source is pressurized geothermal water at 6 bar and 104 °C. Cooling water is available at 10 °C and atmospheric pressure. The configuration of the plant is based on a simple saturated cycle using R245fa as the working fluid. Pressure losses in the ducts and heat exchangers are negligible and all components are assumed perfectly insulated.

The cycle design data are:

Boiler pressure $P_{\mathrm{boiler}} = 8.89$ bar
Turbine inlet temperature $T_{\mathrm{in,turb}} = 85.6$ °C
Condensation pressure $P_{\mathrm{cond}} = 1.36$ bar
Isentropic efficiency of $\eta_{\mathrm{turb},s} = 0.80$
Isentropic efficiency of the feeding pump $\eta_{\mathrm{pump},s} = 0.35$
Minimum temperature difference in the boiler $\Delta T_{\mathrm{boiler}} = 4$ K
Minimum temperature difference in the condenser $\Delta T_{\mathrm{cond}} = 4$ K.

The thermodynamic properties of the working fluid at the states identifying the cycle design conditions are given in Table 9.2.

The mechanical and thermal power transferred by the components in the system are reported in Table 9.3.

The exergy balances for the four components are as follows:

$$\text{Boiler}: \dot{M}\left(\xi_{\mathrm{f},8} - \xi_{\mathrm{f},9}\right) + \dot{M}\left(\xi_{\mathrm{f},2} - \xi_{\mathrm{f},3}\right) = \dot{\Xi}_{\mathrm{d}};$$

$$\text{Turbine}: \dot{M}\left(\xi_{f,3} - \xi_{f,4}\right) = \dot{W}_{\mathrm{turb}} + \dot{\Xi}_{\mathrm{d}};$$

Table 9.4 Exergy losses for the components of the ORC plant of Figure 7.16.

Component	Exergy loss / kW	Exergy efficiency / %
Boiler	97.7	81.3
Turbine	62.6	81.4
Pump	12.7	32.2
Condenser	70.8	24.4

Condenser : $\dot{M}\left(\xi_{f,4}-\xi_{f,1}\right)+\dot{M}\left(\xi_{f,6}-\xi_{f,7}\right)=\dot{\Xi}_d$; and

Pump : $\dot{M}\left(\xi_{f,2}-\xi_{f,1}\right)=-\dot{W}_{pump}+\dot{\Xi}_d$.

Let us assume that the environment is at $T = 10$ °C and $P = 1$ bar. The data to perform such an analysis are all available from Tables 9.2 and 9.3. By plugging the correct values into the exergy balances and solving for $\dot{\Xi}_d$, the results listed in Table 9.4 are obtained.

Exergy flows and losses are also commonly presented in so-called Grassman diagrams such as the one in Figure 9.11. From the exergy analysis we see that the component that is responsible for the largest thermodynamic loss is the boiler, while the condenser and the turbine cause a similar amount of exergy destruction. The designer then knows that the major potential for improvement comes from reducing the losses in the primary heat exchanger. This can be achieved usually by adding heat exchanging surface, which translates into an additional cost for each unit that is manufactured. A better investment might be performing research on increasing the fluid dynamic performance of the turbine. If the research results are positive, this is a one-time investment, because modifications to the shape of the turbine blades do not arguably involve any additional manufacturing or material cost, while the benefit applies to each and every unit produced.

Exergy analysis can of course become much more complicated than in the case of the ORC plant, though its significance for the design of energy conversion systems is apparent also from this simple example. Imagine for instance that the task is the optimization of a large combined-cycle power plant like the one of Figure 9.12. Among the many choices to

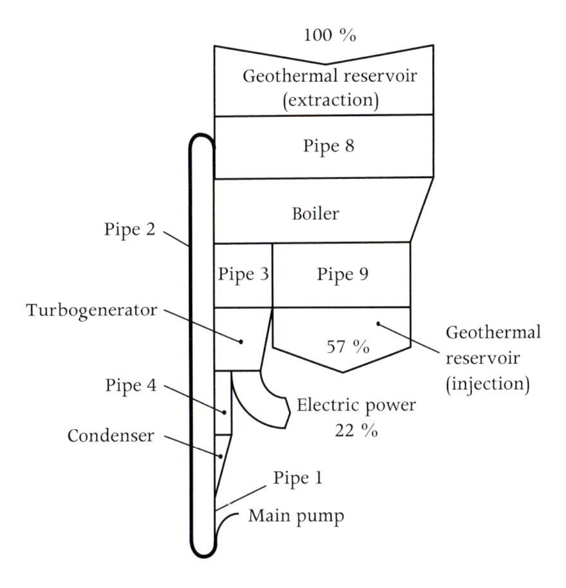

Figure 9.11 Grassman diagram representing the specific exergy flows of the ORC plant of Figure 7.16.

be made, one of the most important aspects that influence the overall performance is the selection of the configuration of the heat recovery steam generator (HRSG). Figure 9.13 presents the results of the exergy analysis applied to a single-pressure and a triple-pressure HRSG configuration. These calculations are performed with the Cycle-Tempo program, T. P. van der Stelt, N. Woudstra, and P. Colonna, *Cycle-Tempo (version 5.0): A program for thermodynamic modeling and optimization of energy conversion systems*, A computer program since 1980, Asimptote, 2013. [Online]. Available: www.asimptote.nl/software/cycle-tempo/.

The diagram shows that more than 35 % of the fuel exergy is dissipated in the combustion process, and this is the major cause of thermodynamic loss. The exergy losses in the HRSG, stack and steam cycle system together are only 11% in the case of the single-pressure HRSG, and are 9% for the triple-pressure configuration. The diagrams convey much useful information about the exergetic efficiency of all the different parts and components of the plant and we leave to the student the exercise of commenting further on the results of this exergy analysis. Note that combined-cycle technology is reaching its peak today, and the only area where large improvements

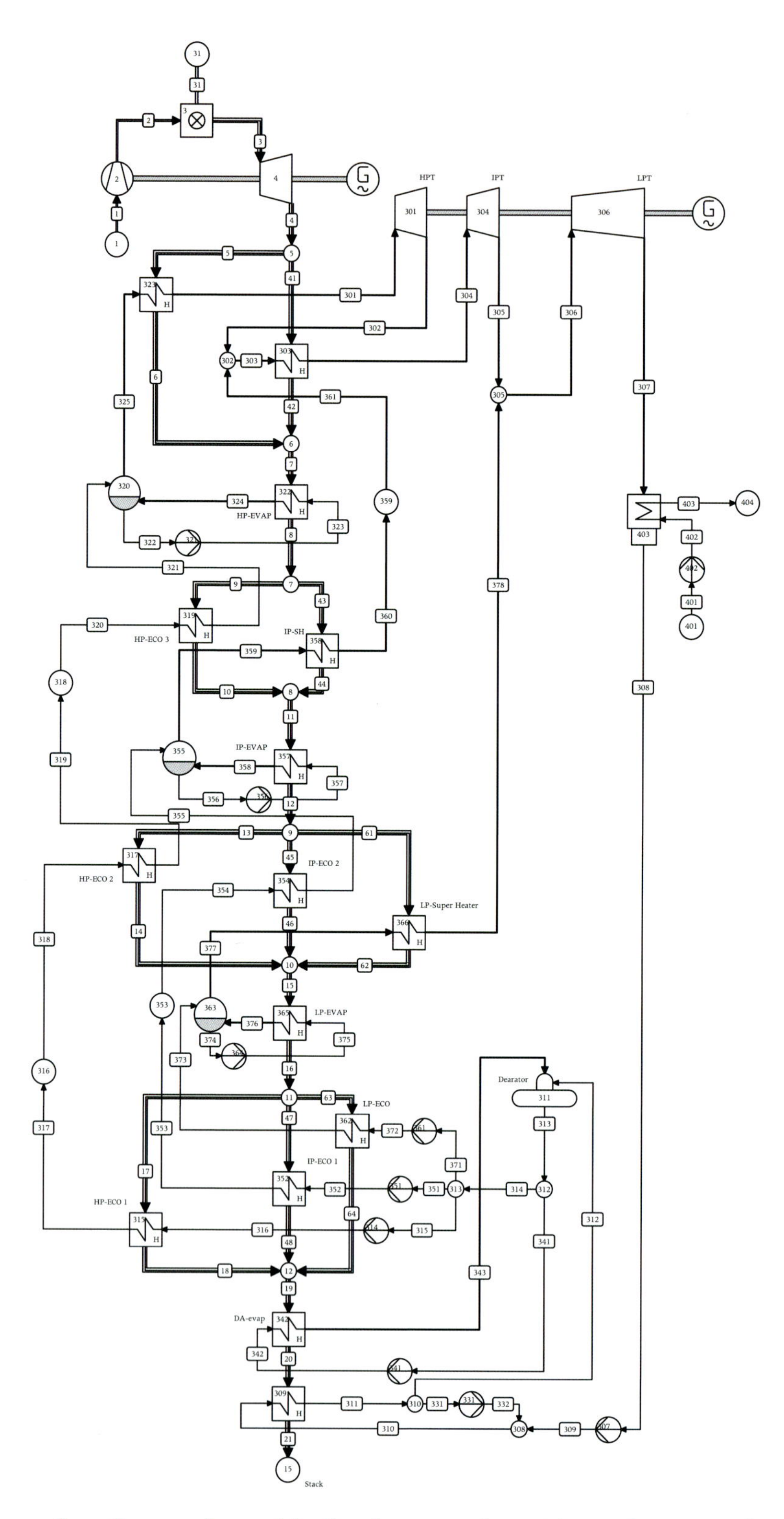

Figure 9.12 Process flow diagram of a combined cycle power plant with a triple-pressure heat recovery steam generator; Reprinted from N. Woudstra, T. Woudstra, A. Pirone, *et al.*, "Thermodynamic evaluation of combined cycles," *Energ. Convers. Manage.*, **51**, (5), 1099–1110, 2010, with permission of Elsevier.

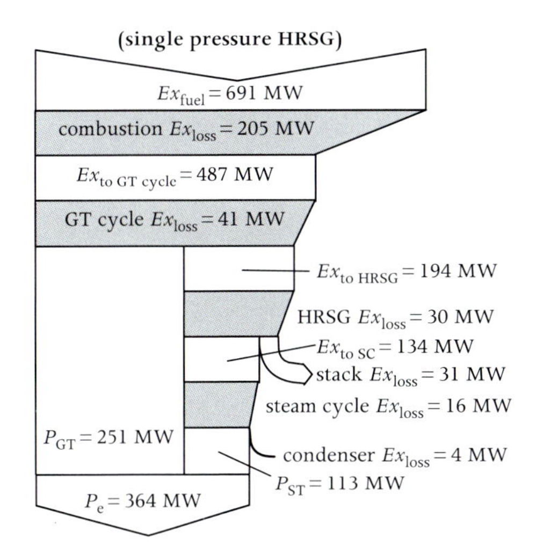

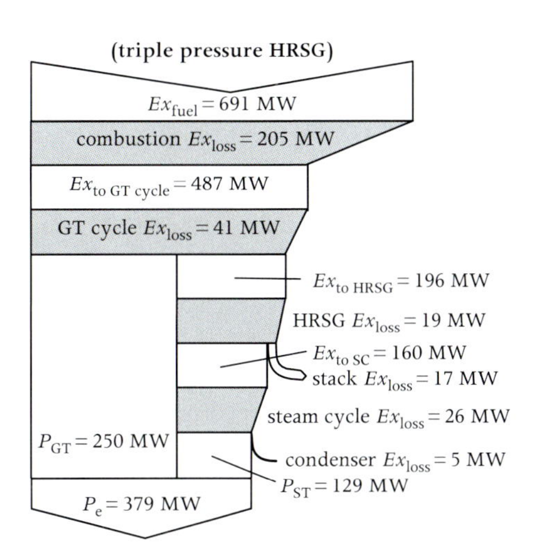

Figure 9.13 Grassman diagrams reporting the results of the exergy analysis of combined-cycle power plants. Comparison between single and a triple-pressure HRSG configuration; Reprinted from N. Woudstra, T. Woudstra, A. Pirone, *et al.*, "Thermodynamic evaluation of combined cycles," *Energ. Convers. Manage.*, 51, (5), 1099–1110, 2010, with permission of Elsevier.

are still possible is indeed with respect to the conversion of the chemical energy of the fuel. This is where high-temperature fuel cells may play a major role in the future, as the electrochemical conversion of the fuel taking place in a fuel cell is considerably more exergetically efficient than the old and *moleculary chaotic* process of burning fuels.

9.7 Concluding Remarks

Now that you know about exergy analysis you can answer correctly if someone tells you that a combined cycle power plant is way more efficient than a geothermal power plant, because its thermal or I-law efficiency is way higher than that of the geothermal power plant. From the point of view of thermodynamic quality and a fair comparison, the level at which energy is available is of the utmost importance, therefore the two technologies might well display a similar level of exergetic efficiency.

Exergy analysis can also be combined with costing, and such a combination takes the name of *exergoeconomics*. If the values of exergy are given an economic value, such a tool can be used in order to make important design decisions related to the realization of an energy system based on sound thermodynamic reasoning. Exergy analysis can even be used to take into account the impact on the environment, if the scope of the analysis is enlarged. Note, though, that the correct selection of economic data and data related to the so-called life-cycle of industrial installations are far more arbitrary than what has been illustrated in this chapter. Nonetheless the results of these broad-scope analyses are important tools for the correct evaluation of technologies and industrial installations with large economic and environmental impact. Last but not least, observe that also here we made several choices with regard to the definitions of quantities related to the exergy concept, and these choices are not unique. For example, the calculation of the exergetic efficiency can lead to different results depending on what is considered as provided exergy in

$$\eta_{II} \equiv \frac{\text{Rate of exergy obtained}}{\text{Rate of exergy provided}}.$$

Exergy can be supplied to a certain system in various forms (chemical, thermal, mechanical, kinetic, potential energy), and also the definition of the system is somewhat arbitrary. In general, the definition of the conditions under which the best thermodynamic performance can be achieved present a level of arbitrariness.

EXERCISES

9.1 A converging nozzle used in the stator of a high pressure steam turbine expands a steam mass flow rate of 1.0 kg/s. The nozzle inlet conditions of the steam are $T = 500$ °C and $P = 150$ bar. The steam at the outlet of the nozzle is at a pressure of 100 bar. Note that for this expansion ratio the nozzle is not chocked. A testing campaign is performed in order to characterize the nozzle performance by measuring the velocity ratio – i.e, the ratio of the actual nozzle discharge velocity over the ideal discharge velocity, the one corresponding to an isentropic process. The measured velocity ratio is 0.95. The nozzle can be considered adiabatic since the exergy wasted as a consequence of thermal losses is negligible if compared to the exergy associated with the steam flowing through the nozzle, also because the turbine casing is properly insulated in order to limit heat transfer to the environment. Being a static component, the nozzle does not convert energy into work, and its only purpose is to accelerate the flow of steam. Calculate the rate of exergy destruction in the nozzle during steady operation. Calculate both the isoentropic and the exergy efficiency of the nozzle, compare them and comment on the results.

9.2 Calculate the maximum power output of an adiabatic steam turbine in kW, whereby the mass flow rate of steam is 10 kg/s, the inlet is at 20 MPa and 400 °C and the discharge pressure is 1 atm.

9.3 Potassium enters a turbine as saturated vapor at 1400 K and is discharged at 90 kPa. Calculate the minimum possible quality at the turbine outlet.

9.4 Compute the maximum percent liquefaction (by mass) of O_2 that can be achieved by expanding it adiabatically in a piston–cylinder system from the saturated vapor state at 10 atm to twice the initial volume.

9.5 A solar-powered heat pump receives energy as heat from a solar collector at T_H, rejects energy as heat to the atmosphere at T_A, and *pumps energy as heat* from a cold space at T_C. Derive an expression for the minimum ratio $\dot{Q}_H/\dot{Q}_C$, in terms of the three temperatures. If $T_H = 350$ K, $T_A = 290$ K, $T_C = 200$ K, and $\dot{Q}_C = 10$ kW, what is the minimum area of the collector that is required (in m^2)?

9.6 Derive the reference I-law cycle efficiency for the evaluation of the II-law efficiency of refrigeration systems.

9.7 Obtain the expression for the specific flow exergy (9.5) from its definition.

9.8 Nitrogen is contained in a closed pressurized vessel. The vessel and its contents are in thermal equilibrium with the environment at temperature $T_0 = 300$ K. The environment pressure is $P_0 = 1.013$ bar. The vessel does not move and is at zero elevation. Assume that the nitrogen in the vessel obeys the ideal gas law.

(a) Derive an expression for the exergy content of the vessel in the form $\xi = \xi(P, P_0, T_0)$.
(b) Evaluate ξ by using one of the multiparameter equations of state for nitrogen implemented in FLUIDPROP.
(c) Make a P-ξ chart by considering pressures up to 500 bar, and compare the results obtained with the two models.

9.9 Calculate the exergy associated with the drum of an industrial steam boiler that is shutting down. Assume all the mass flow rates of water in and out of the drum are negligible. The environment is at $T_0 = 293$ K and 1 atm. The drum is at 15 m height, and its water at

12.4 MPa in saturated conditions, and half of its volume is occupied by liquid. The drum diameter is 1.5 m and its length 5 m. Which is its largest form of exergy? What should be the height of the drum for its potential exergy to equal its physical exergy?

9.10 In a chemical process butane vapor is needed at some point. At that time 3 kg of butane are stored in a closed well-insulated vessel in saturated conditions at 15.4 bar (100 °C) and an electric heater is turned on in order to completely vaporize the fluid. Assuming that the environment is at 293 K and 1 atm, how much exergy is destroyed in the process?

9.11 Derive the exergy balance for the exhaust stroke valid for the internal combustion engine model of Section 4.9, and comment on the outcome.

9.12 1kg/s of nitrogen enters the turboexpander of a nitrogen liquefaction plant at 8 bar and 310 K with a velocity of 1.5 m/s and expands with pressure ratio of 6, exiting from the turbine with a velocity of 30 m/s. The environment is at 10 °C and 1 atm. The turbine is poorly insulated and energy is transferred as heat from the environment to the turbine at a rate of 8 % of its power output. The overall thermodynamic transformation undergone by the working fluid can be modeled as a polytropic process with $n = 1.2$. Nitrogen in the thermodynamic region of interest can be treated as a perfect gas with a constant isobaric specific heat $c_P = 1.05$ kJ/(kg · K). Assuming that the heat transfer from the environment to the turbine casing occurs at the uniform temperature of −10 °C, calculate the rate of exergy destruction and the exergy efficiency of the turbine.

9.13 The ammonia mass flow rate of the compressor of a refrigeration plant is 1 kg/s, and its inlet is at 266 K and 3 bar. The compressor outlet pressure is 16 bar while the outlet temperature is kept at 333 K thanks to a water cooling jacket. 5000 kg/hr of liquid water enters the cooling jacket at 288 K and 1.5 bar and leaves it with a temperature increase of 45 K. Pressure losses in the cooling jacket can be considered negligible and the body of the cooled compressor can be considered perfectly insulated. The environment is at $T_0 = 298$ K and $P_0 = 1$ atm. Derive the exergy balance for the cooled compressor and calculate the exergy destruction rate in the component assuming that kinetic energy effects are negligible.

9.14 Consider the recuperator of the supercritical CO_2 cycle power plant of Exercise 7.19 Assume that the energy rejected to the environment from the surface of the insulating material covering the recuperator is negligible, and that the environment is at $T_0 = 308$ K and $P_0 = 1$ atm. Calculate the change in flow exergy rate of each CO_2 stream, the rate of exergy loss in the heat exchanger, and its exergetic efficiency.

9.15 Calculate the rate of exergy loss of each component of the example steam power plant treated in Section 7.2. Cooling water is available at environmental conditions, which are $T_0 = 15$ °C and $P_0 = 1$ bar. The outlet pressure of the cooling water pump is 2 bar and the isentropic efficiency of this pump is 75 %. The temperature rise of the cooling water in the condenser equals 10 K. The exergy flow of the fuel to the boiler is 80 MW and assume that the fuel is available at environmental conditions. Determine also the exergetic efficiencies of the components and comment on the result.

9.16 Calculate the rate of exergy loss of each component of the example gas turbine illustrated in Section 7.5, and its exergetic efficiency. Comment on the results and compare them to results of the exergy analysis of Exercise 9.15.

9.17 Perform the exergy analysis of the recuperated gas turbine of Exercise 7.15 knowing that i) the exergy input of the system is 14.26 MW, ii) the chemical exergy of the flue gas is 11 kJ/kg, and iii) the environment is at 15 °C and 1 bar. Draw the Grassman diagram of the system. If you were the engineer charged with deciding which component could be improved first, which one would you choose and why?

10 Thermodynamics of Reacting Mixtures

CONTENTS

This chapter deals with the study of mixtures of substances undergoing chemical reactions. Combustion is an example of a thermodynamic process governed by many complex chemical reactions. It is at the core of the majority of the power and propulsion systems in use today. The First Law of Thermodynamics is the basis for quantitative analysis of chemical reactions. The Second Law of Thermodynamics tells us if a certain reaction is possible, and it is at the basis of the calculation of the state of thermodynamic equilibrium reached by a reacting mixture. Here we will introduce the Third Law of Thermodynamics in order to correctly compute the entropy of different species involved in the chemical reaction. The prediction of reaction rates, and the investigation of reactions mechanisms, that is chemical kinetics, is beyond the scope of this book.

10.1 Some Concepts and Terms

Associated with every chemical reaction is a chemical equation, derived by applying the law of conservation of atoms to each of the atomic species involved in some sort of *units reaction*. The reaction begins with the collection of certain chemical constituents, called *reactants*. The chemical reaction causes a rearrangement of atoms and electrons to form different constituents, called the *products*.

The reaction between the reactants hydrogen and oxygen to form the product water can be expressed as

$$2\,\mathrm{H_2} + \mathrm{O_2} \rightleftharpoons 2\,\mathrm{H_2O}. \tag{10.1}$$

This expression indicates that two molecules of hydrogen and one molecule of oxygen can be combined to form two molecules of water. The arrows in both directions indicate that the opposite reaction may also occur. The coefficients in the chemical equation are called *stoichiometric coefficients* (2, 1, and 2 in this example). The number of hydrogen atoms is conserved, as is the number of oxygen atoms. Chemical equations can also be interpreted as equations relating the molar masses of the species involved in the reaction; here two moles of hydrogen and one mole of oxygen combine to form two moles of water. Note that the number of moles of the reactants may differ from the number of moles of the products.

A *stoichiometric mixture of reactants* is one in which the molar proportions of the reactants are exactly as given by the stoichiometric coefficients so that no excess of any constituent is present. A *stoichiometric combustion* is one in which all the oxygen atoms in the oxidizer react chemically to appear in the products.

The most common oxidizer is air, which for many purposes can be considered a mixtures of 21 % oxygen and 79 % nitrogen (mole or volume fractions, they are the same if the gas behaves as ideal). The chemical equation for stoichiometric combustion of methane (CH_4) is then

$$\underbrace{CH_4 + 2\left(O_2 + \tfrac{79}{21} N_2\right)}_{\text{Reactants}} \rightleftharpoons \underbrace{CO_2 + 2\,H_2O + 7.52\,N_2}_{\text{Products}}. \tag{10.2}$$

If more air is supplied, not all the reactants will participate in the reaction, and the composition of the products will differ from that of a stoichiometric combustion. The additional air supplied is called *excess air*. For example, if methane is burned with 25 % excess air, the chemical equation becomes

$$\underbrace{CH_4 + 1.25 \times 2\left(O_2 + 3.76\,N_2\right)}_{\text{Reactants}} \rightleftharpoons \underbrace{CO_2 + 2\,H_2O + 0.5\,O_2 + 9.4\,N_2}_{\text{Products}}. \tag{10.3}$$

and the products mixture would not be stoichiometric. In order to complete a combustion reaction, excess air is normally supplied. Note that (10.3) is balanced and that the stoichiometric coefficients are not necessarily integers. One can balance the equation (find the products' stoichiometric coefficients) by first considering the C atoms, then the H atoms, and finally the O atoms.

Power plant boilers normally run with about 10 to 20 % excess air. The combustor of a gas turbine runs very *lean*, with 300 % excess air. Incomplete combustion, typical of *rich* mixtures, results in the production of CO instead of CO_2 and the production of unburned or partially burned hydrocarbons.

Another important combustion parameter is the air/fuel ratio (AFR) in a reaction. For the stoichiometric combustion of CH_4 we have, on a molar basis,

$$\mathrm{AFR} = \frac{N_{\mathrm{air}}}{N_{CH_4}} = \frac{2 + 7.52}{1} = 9.52\,\frac{\text{moles air}}{\text{mole fuel}}. \tag{10.4}$$

Since $\hat{M}_{\mathrm{air}} = 28.97$ kg/kmol and $\hat{M}_{CH_4} = 16$ kg/kmol, the AFR can also be expressed on a mass basis as

$$\mathrm{AFR} = 9.52 \times \frac{28.97}{16} = 17.2\,\frac{\text{kg air}}{\text{kg fuel}}. \tag{10.5}$$

If a reaction occurs in an isolated vessel, the internal energy of the products and reactants will be the same, and the entropy of the products will be greater than that of the reactants. Reactions can be made to occur at constant pressure and temperature by allowing the volume change and transferring energy as heat from the reacting mixture. The internal energy of the products will then be different from that of the reactants. If the products have less internal energy (at constant P and T), the reaction is said to be *exothermic*, and energy must be transferred as heat from the mixture in order to keep the temperature constant. A reaction at constant P and T for which the opposite is true is called *endothermic*. Exothermic reactions are particularly useful in engineering as a means for supplying energy as heat to thermal power systems.

10.2 Fuel Analysis and Product Composition

In an actual combustion process it may be necessary to determine experimentally both the products of combustion and their respective amounts. Often the fuel itself is unknown.

The analysis of gaseous products of combustion was once performed with an *Orsat gas analyzer*. This piece of laboratory equipment is replaced nowadays by more modern instruments, like gas-chromatographs. The Orsat is, however, an easy-to-use reliable instrument, and understanding its working principle can be conducive to the study of current techniques, which are treated in more advanced courses.

A sample of the gas to be analyzed is introduced into the Orsat analyzer at a known temperature and atmospheric pressure. The gas is then placed in contact with a liquid capable of absorbing CO_2, such as a potassium hydroxide solution (KOH). The temperature and the total pressure are maintained constant during the measurements so that the volume occupied by the gas decreases as CO_2 is absorbed, and this decrease is noted. The gas is successively placed in contact with a liquid that absorbs O_2, such as pyrogallol ($C_6H_6O_3$), and then with a liquid that absorbs CO, such as cuprous chloride (CuCl). The further decrease of volume of the gas sample is noted. The change in the volume of the gas sample during each separate absorption test is a measure of the volumetric fraction of that particular gas in the sample. Since the measurements are on a volumetric basis at a constant temperature and pressure, and we assume the gases to obey the ideal gas law, the values obtained are also a measure of the mole fraction of each of the measured constituents. If additional gases (other than N_2 and H_2O) are expected to be present in significant amounts in the sample, additional absorbers will be required. At each stage of absorption, the mixture is assumed to be saturated with water vapor. Since the temperature and pressure of the water vapor remain constant, the partial pressure of water vapor is unchanged by the absorptions, and consequently an appropriate amount of H_2O condenses out with each absorption. The net effect is that the indicated mole fractions are exactly those that would be obtained in a mixture without any water vapor (a *dehydrated* or *dry mixture*); the proof is left as a problem for the student. The gas remaining in the analyzer after these steps of the procedure have been completed determines the nitrogen mole fraction of the dry mixture. The actual H_2O content is then calculated from a mass balance.

Example: Fuel, AFR, and chemical equation from flue gas composition. The result of the analysis of dry-mixture combustion products is listed in Table 10.1. We want to estimate:

- the fuel composition,
- the chemical equation, and
- the AFR.

Table 10.1 Dry mixture composition from gas analysis.

Species	Mole fraction
CO_2	0.11
O_2	0.03
CO	0.01
Subtotal	0.15
N_2	0.85
Total	1.00

Let us assume that the fuel is a hydrocarbon whose chemical formula can be written as C_nH_m. We also assume that the oxidant is air, and that the chemical equation is of the form

$$C_nH_m + a\,(O_2 + 3.76\,N_2) \rightleftharpoons b\,CO_2 + c\,H_2O + d\,CO + e\,O_2 + f\,N_2.$$

Since the dry mixture product mole fractions are known, it is convenient to write the chemical equation on the basis of 100 moles of dry mixture, hence

$$b = 11, \quad d = 1, \quad e = 3, \text{and} \quad f = 85.$$

The conservation of nitrogen atoms (see the chemical equation) yields

$$a = \frac{85}{3.76} = 22.6.$$

Conservation of oxygen atoms requires

$$2a = 2b + c + d + 2e,$$

therefore

$$c = 2 \times 22.6 - 2 \times 11 - 1 - 2 \times 3 = 16.2.$$

Conservation of carbon atoms requires

$$n = b + d = 11 + 1 = 12,$$

and conservation of hydrogen atoms requires

$$m = 2c = 32.4.$$

Accordingly, the fuel can be considered as an equivalent hydrocarbon with representative formula $C_{12}H_{32.4}$. The representative empirical chemical formula gives the simplest ratio of atoms of each element present in the mixture, but it is clearly not representative of the chemical structure of each of the compounds present in the fuel. The chemical equation is then

$$C_{12}H_{32.4} + 22.6\left(O_2 + 3.76\,N_2\right) \rightleftharpoons 11\,CO_2 + 16.2\,H_2O + CO + 3\,O_2 + 85\,N_2. \qquad (10.6)$$

We could calculate the AFR on a molar basis but, as the fuel molecular structure is not known, the result would be meaningless. However, we can calculate the AFR on a mass basis. The molar mass of the fuel having the representative formula $C_{12}H_{32.4}$ is

$$\hat{M} = 12 \times 12.0 + 32.4 \times 1.0 = 176.4\ \text{kg/kmol}.$$

The AFR is then calculated as

$$\text{AFR} = \frac{N_{\text{air}}\hat{M}_{\text{air}}}{N_{\text{fuel}}\hat{M}_{\text{fuel}}} = \frac{(22.6 \times 4.76) \times 28.97}{1 \times 176.4} = 17.7\ \text{kg}_{\text{air}}/\text{kg}_{\text{fuel}}.$$

A useful parameter in technical applications of combustion is the percent excess air. In order to compute it we start from the stoichiometric chemical equation

$$C_{12}H_{32.4} + 20.1\left(O_2 + 3.76\,N_2\right) \rightleftharpoons 12\,CO_2 + 16.2\,H_2O + 75.5\,N_2.$$

The excess air fraction is

$$\frac{22.6 - 20.1}{20.1} = 0.124,$$

corresponding to 12.4 % excess air, or 112.4 % theoretical air.

Because of the high water content in typical combustion products, and the very corrosive effects of high-temperature condensate, knowledge of the dew point of products of combustion is often important. If the product composition is known, the partial pressure of water is readily determined, as this corresponds to the saturation pressure of water at the dew point to a very good approximation, see Section 8.3. For instance, for the products of the previous example,

$$\chi_{H_2O} = \frac{16.2}{11 + 16.2 + 1 + 3 + 85} = 0.139.$$

Then, given that the mixture is assumed to be an ideal gas mixture, the pressure fraction is equal to the mole fraction,

$$P_{H_2O} = \chi_{H_2O} \times P_{\text{atm}} = 0.139 \times 1.013\ \text{bar} = 0.141\ \text{bar}.$$

By using one of the suitable thermodynamic models in FLUIDPROP we can therefore obtain a saturation (dew-point) temperature of about 326 K.

10.3 Standardized Energy and Enthalpy

In order to apply the principle of energy conservation to processes involving chemical reactions we need to know the equations of state for all participating species. As we have seen in Sections 3.6 and 6.3, formulation of these thermodynamic models of the fluid requires the definition of some arbitrary datum state at which the internal energy (or the enthalpy) and the entropy are taken to be zero. These equations of state may be used in any thermodynamic analysis involving one single substance or mixture of non-reacting substances, since the balance equations of this type of analysis require only differences in u (or h) and s. However, with chemically reacting systems it is necessary to use a common basis for the evaluation of u, h, and s of all substances included in any particular analysis. Suppose for example that we arbitrarily select some state T_0 and P_0 as the reference state for H_2, O_2, and H_2O, then any reaction carried out at this temperature and pressure would appear to result in no change in the internal energy or entropy of the mixture, which is obviously an incorrect outcome.

One might think that, since the internal energy is a monotonically increasing function of temperature, the energy of all substances at the absolute zero of temperature is zero. This is certainly not the case, because a considerable amount of energy may be associated with molecular and nuclear binding forces, and with other energy modes. It is therefore not practical to define the "absolute energy" of matter. A reference state must be selected in order to properly account for the internal energy of different substances in energy analyses involving chemical reactions.

The procedure we follow is to select some arbitrary reference state (defined by an appropriate temperature and pressure) at which we set to zero the values of the enthalpy of the basic elements. It is customary to give the enthalpy, rather than the internal energy, a zero value at the reference state; this then fixes the value of the internal energy at the reference state, since absolute values for P and v are measurable. One option could be to choose the absolute-zero temperature and, say, 1 atmosphere pressure, but this would require accurate low-pressure data not generally available at present. A more practical alternative conventionally adopted for chemical thermodynamic calculations is 298.15 K (exactly 25 °C) and 1 bar (100 kPa). This is called the *standard reference state*. By convention, the enthalpy of every *elemental* substance is zero at the standard reference state. Elemental substance is a substance composed of only one kind of atom in the form in which it exists in equilibrium at the standard reference state. For example, the enthalpy of mercury (Hg) – a liquid – is zero at the standard reference state, as is the enthalpy of oxygen (O_2) – a gas. Different isotopes of an element have the same chemical properties. If no nuclear reactions are involved, the enthalpy of each isotope can be considered zero at this standard state.

Note that the *standard* reference state is most often different from the arbitrary reference state chosen for the equation of state of a substance. The standard reference state is primarily used in chemical thermodynamics, as for instance in the NIST-JANAF tables, the database from which STANJAN gets thermodynamic species data. The user of equations-of-state information should always check the reference state of the equation of state, if chemical reactions are also part of the problem, and make the calculation of properties consistent.

Enthalpy of formation

The enthalpy of formation of a compound is defined as the difference between the enthalpy of the compound $h^{\mathrm{o}}_{\mathrm{comp}}$ and the enthalpies of the elemental substances from which it is formed, all evaluated at the standard reference state. It is conventionally denoted by $\Delta h^{\mathrm{o}}_{\mathrm{f}}$, and on a molar basis, is given by

$$\Delta \hat{h}^{\mathrm{o}}_{\mathrm{f}} \equiv \hat{h}^{\mathrm{o}}_{\mathrm{comp}} - \sum \nu_i \hat{h}^{\mathrm{o}}_i, \tag{10.7}$$

where $\hat{h}^{\mathrm{o}}_i$ is the molar enthalpy of the ith elemental substance participating in the formation reaction, and ν_i is the number of moles of the ith elemental substance involved in forming a single mole of the compound. The enthalpy of formation can be evaluated by appropriate measurements of energy transfers as heat and work, and it is one of the properties collected in thermodynamic databases. Since by convention the elemental substances have zero enthalpy at the standard reference state, the enthalpy of a compound at the standard reference state is merely its enthalpy of formation. The enthalpy at other states may be calculated using the standard reference state enthalpy. We call the enthalpy calculated in this way the *standardized enthalpy*, meaning simply that it is properly related to the enthalpy of other elements and compounds. Values of $\Delta h^{\mathrm{o}}_{\mathrm{f}}$ can be determined by laboratory measurements or by methods of statistical thermochemistry. They can be obtained from GASMIX by calculating the enthalpy of a compound at 1 atm and 25 °C. If $\Delta h^{\mathrm{o}}_{\mathrm{f}}$ is known, the standardized enthalpy can be obtained by simple adjustment of enthalpy data obtained for instance from an equation-of-state model. Note that the enthalpy of formation and standardized enthalpy often have negative values, as a result of the choice of reference state. The term "heat of formation" is used sometimes to refer to the enthalpy of formation and sometimes to mean the negative of the enthalpy of formation; the user of such a database must be sure which, if either, the database compiler had in mind. The term arises because it may be shown that the energy which must be transferred as heat from the mixture to keep the temperature and pressure constant is equal to the enthalpy of formation.

10.4 Heat of Reaction, Heating Values

We now have all the information to perform the energy analysis of chemical reactions. The combustion of fuels is a reaction of particular technical interest. It is common practice to write the energy balance on a per-mole-basis. Consider the combustion of a hydrocarbon fuel C_nH_m in air in a steady-flow burner, Figure 10.1. It is commonly assumed that the products contain only H_2O, CO_2, and N_2, with additional O_2, if excess air is supplied. For a stoichiometric reaction the chemical equation is

$$C_nH_m + \left(n + \tfrac{m}{4}\right)\left(O_2 + 3.76\,N_2\right) \rightleftharpoons nCO_2 + \frac{m}{2}H_2O + \left(n + \tfrac{m}{4}\right)3.76\,N_2.$$

Note that in this form the stoichiometric coefficients give the amounts of species in the products and reactants per mole of fuel. The energy balance on a per-mole-of-fuel basis is

$$\hat{H}_R = \hat{H}_P + \hat{Q},$$

where for brevity we denote the *total* enthalpies of the products and reactants, *per mole of fuel*, by $\hat{H}_P$ and $\hat{H}_R$, respectively. $\hat{Q}$ is then the energy transfer as heat from the combustor per mole of fuel burned. It is common practice to evaluate $\hat{H}_P$, $\hat{H}_R$, and $\hat{Q}$ considering the reaction to occur *at the standard reference state* (1 atm, 25 °C = 298.15 K), for this provides a sensible way to compare various fuels. Under these conditions we have

$$\hat{H}_P^o = \left(\sum_i N_i \hat{h}_i\right)_{\text{Products}}, \tag{10.8a}$$

$$\hat{H}_R^o = \left(\sum_i N_i \hat{h}_i\right)_{\text{Reactants}}, \tag{10.8b}$$

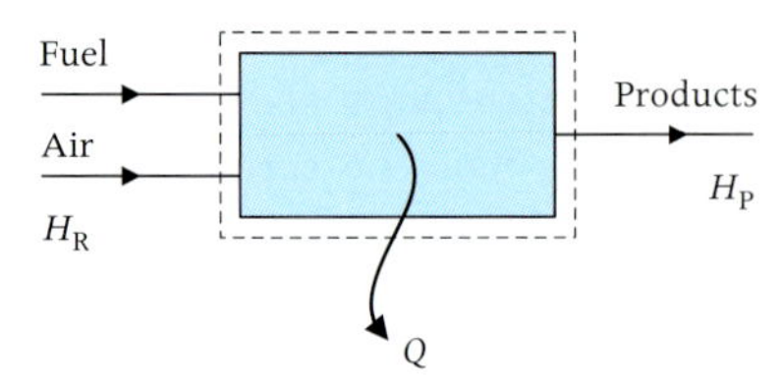

Figure 10.1 Control volume for the energy balance of combustion in a burner.

where the mole numbers N_i are simply stoichiometric coefficients in the chemical equation. It follows also that

$$\hat{Q}^o = \hat{H}_R^o - \hat{H}_P^o. \tag{10.9}$$

Where $\hat{Q}^o$ is termed the *heating value* or *heat of reaction* of the fuel. Note that $\hat{Q}^o$ represents the energy that must be transferred as heat from the system, per mole of fuel, in order to maintain the system at constant temperature. The term *enthalpy of combustion* is sometimes used for the *negative* of $\hat{Q}^o$, which is then denoted by $\hat{H}_{RP}^o$, therefore

$$\hat{H}_{RP}^o = -\hat{Q}^o = \hat{H}_P^o - \hat{H}_R^o. \tag{10.10}$$

The superscript ^{o} is often omitted in other treatments, but we included it here to emphasize that the reaction is assumed to take place at the standard reference state.

The H_2O in the products may be in either the liquid or vapor phase. If water is in the liquid phase, Q^o is called the *higher heating value* (HHV), while Q^o is termed the *lower heating value* (LHV) if water is in the vapor phase. Noting that condensation of water releases the latent heat of vaporization Δh^{LV}, which adds to Q^o, it follows that HHV > LHV.

To exemplify, let us evaluate the quantities discussed above for the combustion of ethane, C_2H_6. The chemical equation is

$$C_2H_6 + 3.5\left(O_2 + 3.76\,N_2\right) \rightleftharpoons 2\,CO_2 + 3\,H_2O + 13.16\,N_2.$$

Using enthalpy data from Table A.1, or data retrieved from STANJAN, which are the same,

$$\hat{H}_R^o = 1 \times -83.82 = -83.82\ \text{kJ/mol},$$
$$\hat{H}_P^o = 2 \times -393.52 + 3 \times -285.83 = -1644.53\ \text{kJ/mol}.$$

Note that we consider the H_2O to be liquid. Hence,

$$\hat{Q}^o = -83.82 - (-1644.53) = 1560.70\ \text{kJ/mol},$$
$$\hat{H}_{RP}^o = -\hat{Q}^o = -1560.70\ \text{kJ/mol}.$$

Since we assumed the H_2O to be liquid, we have that

$$\text{HHV} = 1560.70\ \text{kJ/mol}.$$

In order to calculate the LHV, we need the molar enthalpy of vaporization of water at 298.15 K. With IF97 we can compute

$$\Delta h^{LV}(T = 298.15\ \mathrm{K}) = 2441.71\ \mathrm{kJ/kg},$$

so

$$\Delta \hat{h}^{LV} = 2441.71 \times 18.015 \times 10^{-3} = 43.998\ \mathrm{kJ/mol}.$$

Hence, for water vapor at the standard reference state

$$\begin{aligned}\hat{h}^{o} &= -285.83\ \mathrm{kJ/mol} + 43.998\ \mathrm{kJ/mol}\\ &= -241.84\ \mathrm{kJ/mol}.\end{aligned}$$

Alternatively, we could have obtained from Table A.1 or from STANJAN

$$\hat{h}^{o} = -241.99\ \mathrm{kJ/mol},$$

which is less accurate since the perfect-gas approximation is employed. Then, with water vapor in the products,

$$\hat{H}_{P}^{o} = 2(-393.52) + 3(-241.84) = -1512.57\ \mathrm{kJ/mol},$$

hence, the lower heating value is

$$\mathrm{LHV} = -83.82 - (-1512.57) = 1428.75\ \mathrm{kJ/mol}.$$

Often the heating values are reported on a mass basis. Since the molar mass of ethane is 30.07 kg/kmol, the lower heating value can be expressed as

$$\mathrm{LHV} = \frac{1428.75}{30.07} \times 10^{3} = 47\,514\ \mathrm{kJ/kg_{fuel}}.$$

Table 10.2 presents lower and higher heating values for several molecules.

Table 10.2 Heating values at 298.15 K for several gaseous fuels.

Compound	HHV / MJ/kg fuel	LHV / MJ/kg fuel
Methane, CH_4	55.5	50.0
Ethane, C_2H_6	51.9	47.5
Hexane, C_6H_{14}	48.7	45.1
Octane, C_8H_{18}	48.3	44.8
Hydrogen, H_2	141.8	120.0

Example: LHV calculation. Let us first calculate the LHV for methane (CH_4) burned with air in a stoichiometric reaction at 25 °C and atmospheric pressure. The chemical equation is

$$CH_4 + 2\left(O_2 + 3.76\,N_2\right) \rightleftharpoons CO_2 + 2\,H_2O + 7.52\,N_2.$$

An energy balance on the control volume (see Figure 10.1) on a per-mole-of-fuel basis, gives

$$\hat{H}_{R} = \hat{H}_{P} + \hat{Q},$$

and

$$\mathrm{LHV} = \hat{Q} = \hat{H}_{R} - \hat{H}_{P}.$$

Using the chemical equation, we obtain

$$\mathrm{LHV} = \hat{h}^{o}_{CH_4} + 2\hat{h}^{o}_{O_2} + 7.52\hat{h}^{o}_{N_2} - \hat{h}^{o}_{CO_2} - 2\hat{h}^{o}_{H_2O} - 7.52\hat{h}^{o}_{N_2},$$

where H_2O is in the vapor phase. Using data from Table A.1 or from GASMIX, we have that

$$\mathrm{LHV} = 802.30\ \mathrm{kJ/mol}.$$

Note the difference from the molar LHV of ethane computed in the previous section. However, the molar mass of methane is only 16.04 kg/kmol, hence on a mass basis

$$\mathrm{LHV} = \frac{802.30}{16.04} \times 10^{-3} = 50\,019\ \mathrm{kJ/kg} \approx 50\ \mathrm{MJ/kg},$$

which is quite similar to that of ethane. Most hydrocarbon fuels have lower heating values in the range from 45 to 55 MJ/kg.

The enthalpy of combustion of methane is

$$\hat{H}^{o}_{RP} = -\mathrm{LHV} = -802.30\ \mathrm{kJ/mol}.$$

Example: Temperature effects. In order to study the effect of temperature on the LHV, or $\hat{H}^{o}_{RP}$, which is the same, we consider now the same reaction at an elevated temperature, say 1100 K. In this case

$$\begin{aligned}\hat{H}_{RP} = \hat{H}_{P} - \hat{H}_{R} &= \hat{h}_{CO_2} + 2\hat{h}_{H_2O} + 7.52\hat{h}_{N_2} - \hat{h}_{CH_4}\\ &\quad - 2\hat{h}_{O_2} - 7.52\hat{h}_{N_2},\end{aligned}$$

but now all the enthalpies must be evaluated at 1100 K, under the assumption that all species are in the gaseous form and that thermodynamic properties can be calculated with the ideal-gas equation of state, $h = h(T)$. The nitrogen contributions cancel out, since the products and reactant temperature are identical. GASMIX provides all the enthalpies of the chemicals at 1100 K, starting from their standardized enthalpies, and by integrating the temperature-dependent $\hat{c}_P$ from the reference-state temperature up to 1100 K. Over this wide temperature range the variation of $\hat{c}_P$ cannot be neglected. Therefore, combining the enthalpy data provided by GASMIX gives

$$\hat{H}_{RP} = -49\,941 \text{ kJ/kg}.$$

At 25 °C we find $\hat{H}^{o}_{RP} = -50\,018.7$ kJ/kg. We see that the enthalpy of combustion changes only by a small amount with temperature, which is true for many hydrocarbon fuels. This fact is often useful in making simplified combustion calculations.

Example: Adiabatic flame temperature.

We now want to further extend the calculations. Suppose the reactants enter at 25 °C and 1 atm. We wish to determine the exit temperature. Whereas in the previous calculations we considered exothermic reactions at constant temperature, we now assume that the burner is adiabatic. Our procedure will be to make an energy balance and solve for the enthalpies of the products in terms of the enthalpies of the reactants. The enthalpy, composition, and pressure of the products will then suffice to fix the exhaust temperature. We assume the following idealizations hold:

- Products contain only CO_2, H_2O, N_2, and O_2;
- O_2, N_2, CH_4, and CO_2 are perfect gases;
- The products of combustion form a mixture of independent perfect gases;
- Steady flow, steady state;
- Adiabatic control volume;
- Thermodynamic equilibrium at inlets and exhaust; and
- Kinetic and potential energies negligible.

With these idealizations, an energy balance on the control volume (Figure 10.2) yields

$$\left(\dot{N}\hat{h}\right)_{CH_4} + \left(\dot{N}\hat{h}\right)_{O_2} + \left(\dot{N}\hat{h}\right)_{N_2} = \left(\dot{N}\hat{h}\right)_{\text{products}}, \tag{10.11}$$

where $\dot{N}$ denotes the molar flow rate. The chemical equation is again

$$CH_4 + 2\,(O_2 + 3.76\,N_2) \rightleftharpoons CO_2 + 2\,H_2O + 7.52\,N_2. \tag{10.12}$$

Note that 10.52 moles of products are formed by combustion of 1 mole of fuel (methane). Hence,

$$\frac{\dot{N}_{CH_4}}{\dot{N}_{\text{products}}} = \frac{1}{10.52} = 0.095,$$

$$\frac{\dot{N}_{O_2}}{\dot{N}_{\text{products}}} = \frac{2}{10.52} = 0.190,$$

$$\frac{\dot{N}_{N_2}}{\dot{N}_{\text{products}}} = \frac{7.52}{10.52} = 0.715.$$

So, the energy balance can be written as

$$\hat{h}_{\text{products}} = 0.095\hat{h}_{CH_4} + 0.19\hat{h}_{O_2} + 0.715\hat{h}_{N_2},$$

which is equivalent to an energy balance written on a per-mole-of-products basis. Since the state of the reactants is known, the energy balance becomes an equation for $\hat{h}_{\text{products}}$. Again, using the data from GASMIX we obtain

$$\hat{h}_{\text{products}} = -7.118 \text{ kJ/kmol}.$$

We now need to determine what temperature will give this value of $\hat{h}_{\text{products}}$. The product enthalpy is related to the individual species enthalpies via the mole fractions as $\hat{h} = \sum_i \chi_i \hat{h}_i(T, P)$, therefore

$$\hat{h}_{\text{products}} = \left(\chi\hat{h}\right)_{N_2} + \left(\chi\hat{h}\right)_{CO_2} + \left(\chi\hat{h}\right)_{H_2O}. \tag{10.13}$$

The mole fractions are easily obtained from

$$\chi_{N_2} = \frac{7.52}{10.52} = 0.715,$$

Table 10.3 Some adiabatic flame temperatures with oxygen and air as oxidizers. The reactants are at 298.15 K and 1 atm. The mixture composition is stoichiometric.

Fuel	Oxygen T/K	Air T/K
Hydrogen, H_2	3079	2384
Methane, CH_4	3054	2227
Propane, C_3H_8	3095	2268
Octane, C_8H_{18}	3108	2277

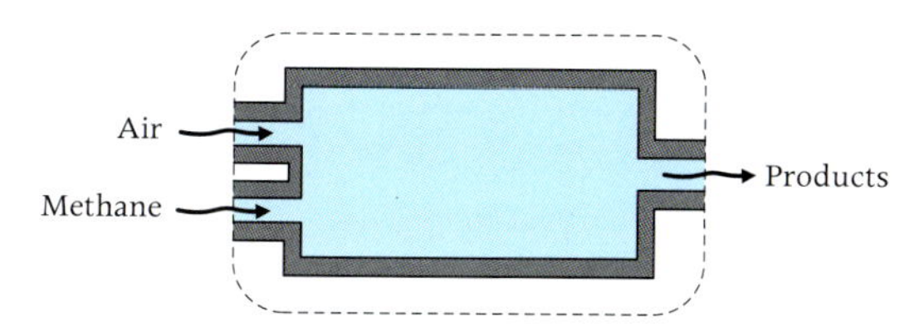

Figure 10.2 Control volume for the energy balance of an adiabatic burner.

$$\chi_{CO_2} = \frac{1}{10.52} = 0.095,$$

$$\chi_{H_2O} = \frac{2}{10.52} = 0.190,$$

which provides us with

$$\hat{h}_{products}(T) = 0.715\,\hat{h}_{N_2}(T) + 0.095\,\hat{h}_{CO_2}(T) + 0.190\,\hat{h}_{H_2O}(T). \quad (10.14)$$

The enthalpies of the mixture components can be obtained with GasMix. The implicit equation (10.14) can be solved with an iterative method, and the result is $T = 2328$ K. The product temperature is called *adiabatic flame temperature*, and is the temperature reached in an adiabatic steady-flow combustion process. Table 10.3 presents some results for the adiabatic flame temperature for the oxidizers air and O_2. Note that the value of the adiabatic flame temperature calculated for CH_4 differs from the value of 2328 K we have just calculated because the tabular results also include as products of combustion the proper amounts of CO, O, H, OH, and NO, which were neglected in our calculation.

Example: Nonstandard reactant states. Typically, the reactants of a combustion process would not be exactly at the standard state. To illustrate the effects of the deviations, let us modify the previous example. Suppose the air enters the combustor at 303 K and 2.7 bar and the CH_4 enters at 288 K and 1.4 bar. First we evaluate the standardized enthalpies of the reactants. Since the inlet temperatures are not far from 25 °C, the constant-specific-heat assumption provides a result that is accurate enough. Therefore, by applying

$$\hat{h} = \hat{h}_0 + \hat{c}_P\,(T - T_0),$$

and values of c_P at the standard reference temperature (see Table A.1), or using GasMix we obtain

$$\hat{h}_{O_2} = 0 + \frac{29.4}{1000}(303 - 298) = 0.147 \text{ kJ/kmol},$$

$$\hat{h}_{N_2} = 0 + \frac{29.1}{1000}(303 - 298) = 0.146 \text{ kJ/kmol},$$

$$\hat{h}_{CH_4} = -74.6 + \frac{35.7}{1000}(288 - 298) = -74.957 \text{ kJ/kmol}.$$

As already shown, $\hat{h}_{products}$ is calculated from the energy balance, and the result is $\hat{h}_{products} = -6.993$ kJ/kmol. The temperature can be determined, as in the previous example, by iterating on the energy balance equation to find its root, and results in 2330 K. Note that a constant-specific-heat assumption can only be used for the calculation of $\hat{h}_{products}$ and cannot be used for calculating the temperature by iterating over the energy balance. Note also that the modest change in inlet conditions has only slightly affected the burner discharge temperature.

Example: Excess air. A typical combustion process utilizes excess air to ensure that complete combustion takes place. Excess air also causes a decrease in the adiabatic flame temperature, since energy is required to increase the temperature of the nonreacting air. In fact, excess air is often introduced to control the adiabatic flame temperature and maintain it within the limits set by the materials in the system.

Again, let us consider the combustion of methane in a steady-flow burner. The oxidizer is air, and we suppose that twice as much air is supplied as is necessary for the combustion (200 % theoretical air). The air enters at 298 K and 2.7 bar, the CH_4 at 288 K and 1.4 bar. The pressure of the combustion products at the burner outlet is atmospheric. The chemical equation is, cf. (10.12),

$$CH_4 + 4\left(O_2 + 3.76\,N_2\right) \rightleftharpoons CO_2 + 2H_2O + 2\,O_2 + 15.04\,N_2. \qquad (10.15)$$

The energy balance again produces equation (10.13), which we use as the basis for the discharge-temperature calculation. The reactant enthalpies are calculated for these inlet states in the previous example, and they are

$$\hat{h}_{CH_4} = -74.957 \text{ kJ/kmol},$$
$$\hat{h}_{O_2} = 0.147 \text{ kJ/kmol},$$
$$\hat{h}_{N_2} = 0.146 \text{ kJ/kmol}.$$

Now, the mole fractions of the reactant mixture are obtained from the chemical equation as

$$\frac{\dot{N}_{CH_4}}{\dot{N}_{products}} = \frac{1}{20.04} = 0.0499,$$
$$\frac{\dot{N}_{O_2}}{\dot{N}_{products}} = \frac{4}{20.04} = 0.1996,$$
$$\frac{\dot{N}_{N_2}}{\dot{N}_{products}} = \frac{4 \times 3.76}{20.04} = 0.7505.$$

The standardized enthalpy of the product mixture is then

$$\hat{h}_{products} = 0.0499 \times (-74.957) + 0.1996 \times 0.147 + 0.7505 \times 0.146 = -3.616 \text{ kJ/kmol}.$$

Similarly, the mole fractions of the species in the product mixture are

$$\chi_{CO_2} = \frac{1}{20.04} = 0.0499,$$
$$\chi_{H_2O} = \frac{2}{20.04} = 0.0998,$$
$$\chi_{O2} = \frac{2}{20.04} = 0.0998,$$
$$\chi_{N_2} = \frac{15.04}{20.04} = 0.7505.$$

The enthalpy of the products may therefore be written as

$$\hat{h}_{products}(T) = \chi_{CO_2}\,\hat{h}_{CO_2}(T) + \chi_{H_2O}\,\hat{h}_{H_2O}(T) + \chi_{O_2}\,\hat{h}_{O_2}(T) + \chi_{N_2}\,\hat{h}_{N_2}(T).$$

Data for the standardized enthalpies of the constituents in the product mixture may be obtained from GASMIX. Calculation of the product temperature again requires iteration, and we find $T = 1485$ K. Note the dramatic decrease in flame temperature caused by excess air.

Figure 10.3 indicates the temperature at which the products of combustion are released when methane is burned in air. If no energy is removed from the combustor, the temperature reaches the adiabatic flame value. Suppose we are burning methane in the combustor of a small gas turbine, and the maximum admissible continuous temperature for the turbine blades of the first stage is 1150 K. We see that a mixture containing about 200 percent excess air is required to reduce the exit temperature to this value. The other curves on the figure indicate the exhaust temperature for the indicated thermal energy extraction from the system. If we treat the internal combustion engine as a steady-flow system of the type shown in Figure 10.3a, we can estimated the exhaust temperature from a car fueled with natural gas (largely methane). For the internal combustion engine of a modern car the work output is about one-third of the fuel energy input. The thermal energy removed from the engine via the cooling water is also about one-third of the input. This leaves about one-third of the energy to go out with the exhaust. Interpolating in the figure for $r = 0.67$ and 0 % excess air, the exhaust temperature should be about 850 K.

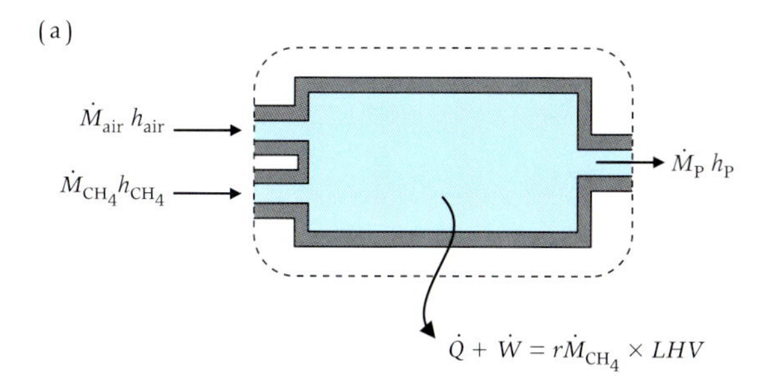

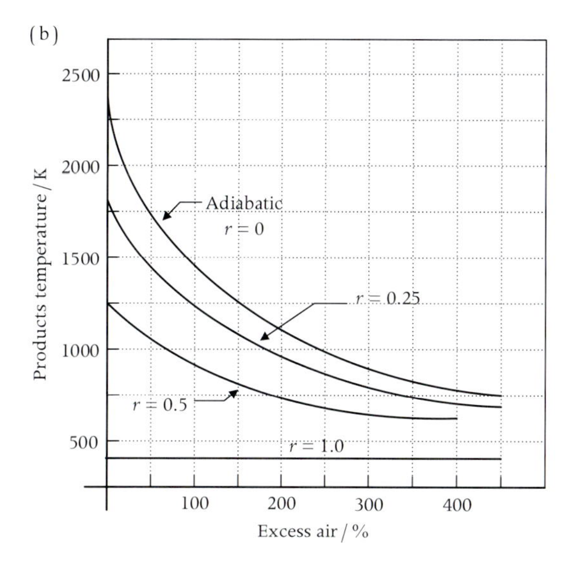

Figure 10.3 (a) A combustor control volume. (b) Combustion of methane with air at an inlet condition of 25 °C.

10.5 Absolute Entropy and the Third Law of Thermodynamics

Thus far we have considered only the first-law aspects of chemical reactions, with particular emphasis on the combustion of hydrocarbons in air. In these examples we always assumed complete combustion and made certain assumptions about the products of reaction. The Second Law of Thermodynamics provides the basis for the theoretical prediction of the composition of a product mixture, hence it plays a fundamental role also in chemical thermodynamics. Analogously to the need for a common reference enthalpy encountered for the energy balance of reacting systems, in order to apply the second law we need to properly relate the entropies of all elements and compounds.

Let us first recall the physical meaning of entropy. As discussed in Section 5.8, the quantity entropy is defined in terms of quantum state probability $p_i(t)$ as

$$S = -k \sum_i p_i \ln p_i.$$

Hence, a state of zero entropy is one for which a single system quantum state is always observed. In this case, at zero entropy, we would know the microscopic state of a system precisely, with zero uncertainty. Imagine now that we add one quantum of energy to the system, and that this permits it suddenly to take any one of a very large number of quantum states. Assuming each quantum state is equally likely, the associated entropy changes would be

$$\delta S = k \ln \Omega - 0,$$

where Ω is the number of quantum states available to the system with the single quantum of energy. The internal energy change δU would be simply ε, that is the very small amount of energy added. Since Ω is likely to be very large, a tiny increase in energy would give a huge increase in entropy. Recalling the thermodynamic definition of temperature (5.20), we are led to suspect that T must be very small when the entropy is zero.

We can illustrate this fact with a numerical example. Consider 1 cm^3 of matter having a mass of the order of 1 g and a molar mass of the order of 20 g/mol. This system contains roughly 3×10^{22} atoms. The quantum of energy added might be possessed by any one of the atoms in the system, so the entropy in the slightly energized state is approximately

$$S \approx k \ln N_{atoms} \approx 1.38 \times 10^{-23}\,\text{J/K} \times \ln\left(3 \times 10^{22}\right)$$
$$\approx 70 \times 10^{-23}\,\text{J/K}.$$

We imagine adding a quantum of energy by means of a photon having a wavelength equal to 1 cm (this is a reasonable estimate for the least energetic photon that might be captured by 1 cm^3 of matter). Its energy is

$$\varepsilon = \frac{hc}{\lambda} = \frac{6.62 \times 10^{-34}\,\text{J/s} \times 3 \times 10^{8}\,\text{m/s}}{0.01\text{m}} \approx 2 \times 10^{-23}\,\text{J}.$$

The temperature is therefore estimated to be

$$T \approx \frac{1}{\delta S/\varepsilon} \approx \frac{1}{70 \times 10^{-23}/2 \times 10^{-23}} \approx \frac{1}{35}\,\text{K}.$$

Such temperature should be interpreted as the average temperature over the range $0 < S <$

70×10^{-23} J/K. Since the temperature is a monotonic function of the entropy, the temperature at the zero entropy state may be expected to be even smaller.

On the basis of these considerations the *Third Law of Thermodynamics* states as a principle that the temperature of any pure substance in thermodynamic equilibrium approaches zero as the entropy approaches zero. Conversely, the entropy of any pure substance in thermodynamic equilibrium approaches zero as the temperature approaches zero. In mathematical terms, for a pure substance in thermodynamic equilibrium

$$\lim_{T \to 0} S = 0. \tag{10.16}$$

A considerable body of experimental data supporting the third law exists.

Example: A second-law application. Starting from the last example provided for the calculation of the adiabatic flame temperature, we now examine again the combustion of methane in air, as described by the chemical Equation (10.12). There we assumed, without justification, that the reaction could in fact occur. The assumption can be tested with the aid of the Second Law of Thermodynamics. Applying the second law to the control volume of Figure 10.2, we find

$$\dot{\mathcal{P}}_S = \left(\dot{N}\hat{s}\right)_{\text{products}} - \left(\dot{N}\hat{s}\right)_{CH_4} - \left(\dot{N}\hat{s}\right)_{O_2} - \left(\dot{N}\hat{s}\right)_{N_2} \geq 0, \tag{10.17}$$

where $\hat{s}$ is the absolute molar entropy. If the entropy production is positive, the reaction is possible; if it is zero, the reaction is reversible, that is, could go in either direction; if a negative value is calculated, the assumed reaction cannot occur.

The absolute molar entropies of the reactant gases can be calculated from the entropies at standard reference state: if we treat the gases as ideal, with constant specific heats over the small temperature range, we can write

$$\hat{s} = \hat{s}^o + \hat{c}_P \ln \frac{T}{T^o} - \hat{R} \ln \frac{P}{P^o},$$

where $\hat{s}^o$ is the entropy in the standard reference state (T^o, P^o). The partial pressures of the O_2 and N_2 in the air are proportional to their mole fractions, thus

$$P_{O_2} = 0.21 \times 2.7 = 0.57 \text{ bar},$$
$$P_{N_2} = 0.79 \times 2.7 = 2.13 \text{ bar}.$$

The absolute molar entropies of the reactants at the inlet are therefore, for CH_4,

$$\hat{s} = 186.37 + 35.70 \times \ln \frac{288}{298} - 8.3143 \times \ln \frac{1.4}{1.01325}$$
$$= 182.45 \text{ J/(mol} \cdot \text{K)},$$

for O_2,

$$\hat{s} = 205.15 + 29.38 \times \ln \frac{303}{298} - 8.3143 \times \ln \frac{0.57}{1.01325}$$
$$= 210.45 \text{ J/(mol} \cdot \text{K)},$$

and for N_2,

$$\hat{s} = 191.61 + 29.12 \times \ln \frac{303}{298} - 8.3143 \times \ln \frac{2.13}{1.01325}$$
$$= 185.89 \text{ J/(mol} \cdot \text{K)}.$$

Assuming atmospheric pressure for the products, we obtain the molar entropy of the products with GasMix, i.e.,

$$\hat{s}_{\text{products}} = 225.60 \text{ J/(mol} \cdot \text{K)}.$$

Substituting the appropriate values into (10.17) yields

$$\frac{\dot{\mathcal{P}}_S}{\dot{N}_{\text{products}}} = 225.60 - 0.0499 \times 182.45$$
$$- 0.1995 \times 210.45 - 0.7505 \times 185.89$$
$$= 34.98 \text{ J/(mol} \cdot \text{K)} \geq 0. \tag{10.18}$$

Since a positive entropy-production rate is obtained, the assumed process does not violate the second law. In addition, note that the reaction is not reversible because entropy is produced. We assumed that the products of combustion do not contain any CO, NO, or other compound, and that the reaction is complete. The second law provides also a means for determining the equilibrium composition of a reacting mixture, as we shall see in the following.

10.6 Chemical Equilibrium

A fundamental aspect of the theory of chemical reactions is the determination of the equilibrium composition of a mixture of chemically reactive constituents. Also in this case we start from first principles, in order to derive some general conditions for chemical equilibrium. These conditions are expressed in terms of thermodynamic properties of the mixture; note that it is not possible to define a thermodynamic state for a system that is not in equilibrium, therefore it is difficult to associate properties like temperature and entropy with such system. Our approach is therefore similar to the one we adopted in Section 5.4, whereby, in order to define the conditions for thermal equilibrium of a system, we initially subdivide it in two parts, each one in internal equilibrium, but not necessarily in equilibrium with one another. In order to discuss the properties of a mixture not in chemical equilibrium, i.e., not having the composition of the equilibrium mixture, we imagine shutting off the reactions, and evaluating the properties of the mixture as if it were a mixture of nonreacting gases in a thermodynamic equilibrium state. In this way we can establish the conditions for chemical equilibrium, and the composition of the equilibrium mixtures by application of the First and Second Laws of Thermodynamics.

Let us consider a mixture of simple compressible substances at temperature T and pressure P. The energy balance gives

$$dU = đQ + dW,$$

and the entropy production is

$$đ\dot{\mathcal{P}}_S = dS - \frac{đQ}{T} \geq 0.$$

Since the pressure is uniform,

$$dW = -PdV.$$

Combining these equations results in

$$Tđ\dot{\mathcal{P}}_S = TdS - dU - PdV \geq 0. \qquad (10.19)$$

Equation (10.19) tells us that any reactions that take place must produce entropy. In particular, a reaction taking place in an insulated constant-volume vessel, where $dU = 0$ and $dV = 0$, must be such that

$$dS \geq 0. \qquad (10.20)$$

If we want to determine the equilibrium composition of a mixture reacting under these conditions, we need only determine the mixture entropy as a function of composition; the equilibrium composition for an isolated reaction is then the one which maximizes the entropy (Figure 10.4a).

We are more often interested in calculating the equilibrium compositions under other conditions. The condition for equilibrium in a reaction taking place at constant volume and temperature can be expressed in terms of the Helmholtz function $A \equiv U - TS$. The differential of the Helmholtz function is

$$dA = dU - TdS - SdT.$$

Upon combination with (10.19) we obtain

$$dA + SdT + PdV \leq 0, \qquad (10.21)$$

hence a reaction occurring at constant temperature and volume reaches equilibrium when the composition of the mixture is such that the Helmholtz function is at its minimum (Figure 10.4b).

From the point of view of engineering applications, the most common situation is a reaction at constant pressure and temperature. Analogously with the calculation condition valid for a vapor–liquid system at equilibrium (Section 6.11), here the equilibrium condition can be conveniently expressed in terms of the Gibbs function $G \equiv U + PV - TS$. By differentiating its definition we obtain

$$dG = dU + PdV + VdP - Tds - SdT,$$

which, combined with (10.19), yields

$$dG - VdP + SdT \leq 0. \qquad (10.22)$$

This condition tells us that any reaction occurring at fixed temperature and pressure will be such that the Gibbs function of the mixture at equilibrium will reach its minimum value for a certain composition, that is therefore called the equilibrium composition.

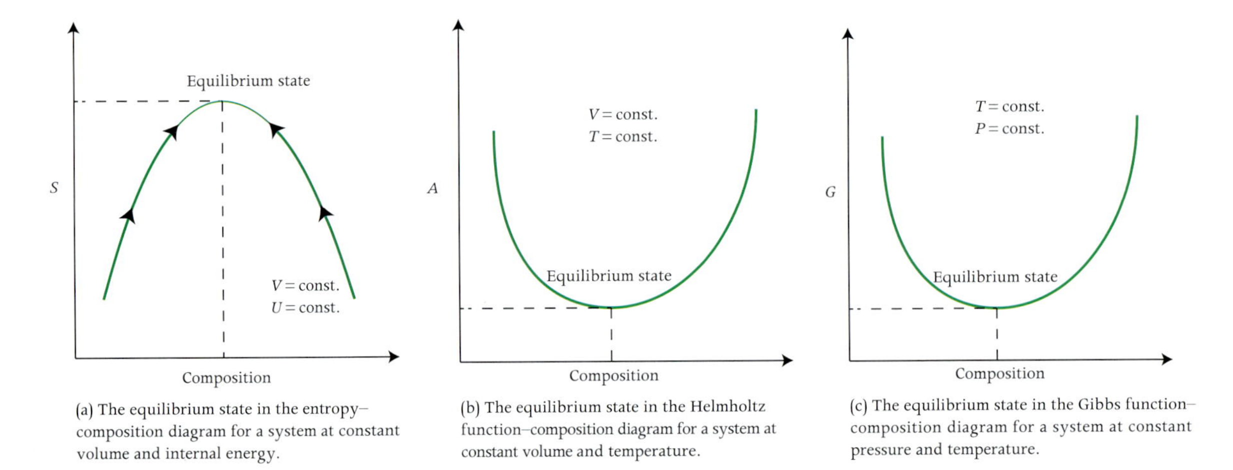

(a) The equilibrium state in the entropy–composition diagram for a system at constant volume and internal energy.

(b) The equilibrium state in the Helmholtz function–composition diagram for a system at constant volume and temperature.

(c) The equilibrium state in the Gibbs function–composition diagram for a system at constant pressure and temperature.

Figure 10.4 Equilibrium state of a reacting mixture depending on the system's constraints.

Simple reactive mixture

Any reaction occurring in a reactive mixture is bound to have the number of atoms conserved. Consider for example a mixture of CH_4, O_2, H_2O, CO, and CO_2. If no material is added to the mixture, any change in the composition must result from chemical reactions. Let us denote a change in mole number by dN. As in calculus, dN always denotes an infinitesimal *increase* in N. Each mole of CH_4 contains the same number of carbon atoms as 1 mole of CO, and twice as many hydrogen atoms as 1 mole of H_2O. The constraining conditions are therefore,

for the conservation of C

$$dN_{CH_4} + dN_{CO} + dN_{CO_2} = 0,$$

for the conservation of O

$$2dN_{O_2} + dN_{H_2O} + dN_{CO} + 2dN_{CO_2} = 0,$$

for the conservation of H

$$4dN_{CH_4} + 2dN_{H_2O} = 0.$$

We thus have three constraining equations and five constituents, thus only two *degrees of reaction freedom* exist for the mixture. For instance, we could consider CH_4 and CO as the two independent components; then from the three constraining equations we find

$$dN_{H_2O} = -2dN_{CH_4}, \tag{10.23a}$$

$$dN_{CO_2} = -dN_{CH_4} - dN_{CO}, \tag{10.23b}$$

$$dN_{O_2} = \frac{1}{2}dN_{CO} + 2dN_{CH_4}. \tag{10.23c}$$

It is evident that the number of degrees of reaction freedom of a mixture is equal to the number of possible compounds minus the number of kinds of atoms represented. We shall call a mixture having one degree of reaction freedom a *simple reactive mixture*. For example, the mixture of CH_4, O_2, H_2O, and CO_2 is a simple reactive mixture.

Equations of reaction equilibrium

Consider a simple reactive mixture having a chemical equation in the form

$$\nu_1 C_1 + \nu_2 C_2 \rightleftharpoons \nu_3 C_3 + \nu_4 C_4,$$

where $C_1, \ldots, C_4$ denote chemical species, and $\nu_1, \ldots, \nu_4$ are the stoichiometric coefficients. From the definition of partial molar Gibbs function $\bar{g}$, see (8.21), it follows that, if we consider the mixture as non-reacting, the difference in the Gibbs function of the mixture between any two states having the same temperature and pressure, but infinitesimally different composition, is

$$dG_{T,P} = \bar{g}_1\, dN_1 + \bar{g}_2\, dN_2 + \bar{g}_3\, dN_3 + \bar{g}_4\, dN_4. \tag{10.24}$$

However, the changes in mole numbers are related through the chemical equation, and

$$dN_2 = \frac{\nu_2}{\nu_1} dN_1, \tag{10.25a}$$

$$dN_3 = -\frac{\nu_3}{\nu_1} dN_1, \tag{10.25b}$$

$$dN_4 = -\frac{\nu_4}{\nu_1} dN_1. \tag{10.25c}$$

From (10.24) and (10.25), we see that for any reaction at constant temperature and pressure, since G must decrease,

$$\left(\bar{g}_1\, \nu_1 + \bar{g}_2\, \nu_2 - \bar{g}_3\, \nu_3 - \bar{g}_4\, \nu_4\right)\, dN_1 \leq 0.$$

As a consequence, we have that

if

$$\bar{g}_1\, \nu_1 + \bar{g}_2\, \nu_2 > \bar{g}_3\, \nu_3 + \bar{g}_4\, \nu_4$$

then $dN_1 < 0$ *and the reaction proceeds to the right, and if*

$$\bar{g}_1\, \nu_1 + \bar{g}_2\, \nu_2 < \bar{g}_3\, \nu_3 + \bar{g}_4\, \nu_4$$

then $dN_1 > 0$ *and the reaction proceeds to the left.*

The condition for chemical equilibrium for a simple reactive mixture is therefore

$$\bar{g}_1\, \nu_1 + \bar{g}_2\, \nu_2 = \bar{g}_3\, \nu_3 + \bar{g}_4\, \nu_4. \tag{10.26}$$

Equation (10.26) is called the *equation of reaction equilibrium*; it is a relation between intensive properties of the products and the reactants. We see therefore that in order to determine the composition of a mixture after chemical equilibrium has been reached, we must know the partial molar Gibbs functions depending on temperature, pressure, and the mole fractions.

The conditions for equilibrium of more complex mixtures are obtained in a similar way. For example for the mixture of CH_4, O_2, H_2O, CO, and CO_2, the change in the Gibbs function for any infinitesimal reaction at constant temperature and pressure is

$$dG_{T,P} = \bar{g}_{\mathrm{CH_4}} dN_{\mathrm{CH_4}} + \bar{g}_{\mathrm{O_2}} dN_{\mathrm{O_2}} + \bar{g}_{\mathrm{H_2O}} dN_{\mathrm{H_2O}} + \bar{g}_{\mathrm{CO}} dN_{\mathrm{CO}} + \bar{g}_{\mathrm{CO_2}} dN_{\mathrm{CO_2}}.$$

Using the constraining Equations (10.23), we find

$$dG_{T,P} = \left(\bar{g}_{\mathrm{CH_4}} + 2\bar{g}_{\mathrm{O_2}} - 2\bar{g}_{\mathrm{H_2O}} - \bar{g}_{\mathrm{CO_2}}\right) dN_{\mathrm{CH_4}} + \left(\frac{1}{2}\bar{g}_{\mathrm{O_2}} + \bar{g}_{\mathrm{CO}} - \bar{g}_{\mathrm{CO_2}}\right) dN_{\mathrm{CO}}.$$

At equilibrium the Gibbs function must be at a minimum with respect to each and every independent variation of the mixture's composition. In other words, dG must be zero for any $dN_{\mathrm{CH_4}}$ and any dN_{CO}. This results in the *two* equations of reaction equilibrium

$$\bar{g}_{\mathrm{CH_4}} + 2\bar{g}_{\mathrm{O_2}} = 2\bar{g}_{\mathrm{H_2O}} + \bar{g}_{\mathrm{CO_2}}, \tag{10.27a}$$

$$\bar{g}_{\mathrm{CO_2}} = \frac{1}{2}\bar{g}_{\mathrm{O_2}} + \bar{g}_{\mathrm{CO}}. \tag{10.27b}$$

These equations can be directly associated with the two chemical equations

$$\mathrm{CH_4} + 2\,\mathrm{O_2} \rightleftharpoons 2\,\mathrm{H_2O} + \mathrm{CO_2}, \tag{10.28a}$$

$$\mathrm{CO_2} \rightleftharpoons \frac{1}{2}\mathrm{O_2} + \mathrm{CO}. \tag{10.28b}$$

In general, one equation of reaction equilibrium will be obtained for each degree of reaction freedom. The coefficients of the partial molar Gibbs functions for each equation of reaction equilibrium will be identical with the coefficients of the constituents in the associated simple chemical equation. It must be mentioned that the task of determining what constituents to put in the mixtures list is not a simple one, and generally requires considerable experience. This can be overcome with the method of element potentials illustrated in Section 10.7, but we prefer to first illustrate the conventional *equilibrium constant* method, which can be used for simple problems. Knowing the two methods allows one to appreciate the advantages of the element potentials method.

Equilibrium reactions in a perfect-gas mixture. The definition of the equilibrium constant

The equation of reaction equilibrium provides the means for determining the equilibrium composition of reacting gas mixtures. Previously we found that the partial molar Gibbs function of one constituent of a perfect-gas mixture is given by

$$\bar{g}_i = \hat{g}_i(T, P) + \hat{R}T \ln x_i.$$

Since the equation of reaction equilibrium involves the Gibbs functions of the constituents, it is convenient to define the Gibbs-function change for a complete unit reaction ΔG_{react} as

$$\Delta G_{\text{react}} = \sum_{\text{products}} \nu_i \hat{g}_i - \sum_{\text{reactant}} \nu_i \hat{g}_i. \tag{10.29}$$

For the special reaction of Equation (10.26),

$$\Delta G_{\text{react}} = \nu_3 \hat{g}_3(T, P) + \nu_4 \hat{g}_4(T, P) - \nu_1 \hat{g}_1(T, P) - \nu_2 \hat{g}_2(T, P), \tag{10.30}$$

thus

$$\Delta G_{\text{react}} = \nu_3 \left(\bar{g}_3 - \hat{R}T \ln x_3\right) + \nu_4 \left(\bar{g}_4 - \hat{R}T \ln x_4\right) - \nu_1 \left(\bar{g}_1 - \hat{R}T \ln x_1\right) - \nu_2 \left(\bar{g}_2 - \hat{R}T \ln x_2\right). \tag{10.31}$$

Thanks to (10.26), the $\bar{g}$ terms cancel out, therefore the ln terms can be combined to give

$$\Delta G_{\text{react}} = \hat{R}T \ln \frac{\chi_1^{\nu_1} \chi_2^{\nu_2}}{\chi_3^{\nu_3} \chi_4^{\nu_4}}. \tag{10.32}$$

If ΔG_{react} is known, (10.32) together with the equations indicating the relative quantity of the elements in the mixture may be solved simultaneously to obtain the constituent fractions at equilibrium.

The value of ΔG_{react} depends on both pressure and temperature. However, the pressure effect is easily separated. Denoting with P_0 an arbitrarily chosen reference pressure, it follows from (6.11) and the definition of the Gibbs function that

$$\bar{g}(T, P) = \bar{g}(T, P_0) + \hat{R}T \ln \frac{P}{P_0}. \tag{10.33}$$

Hence, substituting it into (10.31), gives

$$\Delta G_{\text{react}} = \Delta G_{\text{react}}(T, P_0) + \hat{R}T \ln \left(\frac{P}{P_0}\right)^{\nu_3+\nu_4-\nu_1-\nu_2}.$$

Note that ΔG_{react} is a function only of T for a selected reference pressure. Equation (10.32) may then be written as

$$\frac{\chi_3^{\nu_3} \chi_4^{\nu_4}}{\chi_1^{\nu_1} \chi_2^{\nu_2}} \left(\frac{P}{P_0}\right)^{\nu_3+\nu_4-\nu_1-\nu_2} = \exp\left[\frac{-\Delta G_{\text{react}}(T, P_0)}{\hat{R}T}\right].$$

Since the right-hand side is a function of temperature, the left-hand side is independent of pressure. We can therefore define the *equilibrium constant* for the reaction as

$$K(T) \equiv \frac{\chi_3^{\nu_3} \chi_4^{\nu_4}}{\chi_1^{\nu_1} \chi_2^{\nu_2}} \left(\frac{P}{P_0}\right)^{\nu_3+\nu_4-\nu_1-\nu_2}. \tag{10.34}$$

We see also that the equilibrium constant is a dimensionless property of the equilibrium mixture and is a function only of the mixture temperature.

The term ΔG_{react} is sometimes called the *free-energy change* of the reaction. The reference pressure P_0 is normally taken as 1 atm; values of $\Delta G_{\text{react}}(298\ \text{K}, 1\ \text{atm})$ can be found tabulated in handbooks as *standard free-energy change* for a given reaction. Given that

$$K(T) = \exp\left[\frac{-\Delta G_{\text{react}}(T, P_0)}{\hat{R}T}\right], \tag{10.35}$$

the equilibrium constant can be calculated if ΔG_{react} is known. Values of K for several reactions are given in Table D.1 in the $\log_{10} K$ form.

The equilibrium composition of a given simple reactive mixture of perfect gases can be determined from the equilibrium constant. Perfect-gas mixtures with more degrees of reaction freedom require one additional equilibrium constant for each additional degree of freedom.

Example: Equilibrium constant calculation.
Consider the reaction

$$CO_2 \rightleftharpoons CO + \frac{1}{2} O_2$$

at 1000 K. Let us calculate the equilibrium constant for this reaction. We must first calculate ΔG_{react}, using (10.29), and then we calculate $K(1000\ \text{K})$ from (10.35). Now, by definition

$$\bar{g} \equiv \bar{h} - T\bar{s}.$$

With molar enthalpy and entropy data calculated with GASMIX at 1000 K and 1 atm, we obtain

$$\bar{g}_{O_2} = 42.26 - 1000 \times 0.25905 = -216.79\ \text{kJ/mol},$$

$$\bar{g}_{CO} = -70.08 - 1000 \times 0.24935 = -319.42\ \text{kJ/mol},$$

$$\bar{g}_{CO_2} = -329.16 - 1000 \times 0.29382 = -622.98\ \text{kJ/mol},$$

therefore

$$\ln K = \frac{-\left(0.5(-216.79) + (-319.42) - (-622.98)\right) \times 1000}{8.3143 \times 1000} = -23.5,$$

$$\log_{10} K = \frac{1}{2.30} \ln K = -10.2.$$

Note that this value agrees well with the value that can be interpolated from Table D.1.

Example: Equilibrium composition. As a further example, let us determine the equilibrium composition of a mixture of CO, CO_2, and O_2 as a function of temperature at a pressure of 1 atm. We also suppose that the mixture contains 1 C atom per 3.125 O atoms. This happens to be the stoichiometric C/O ratio for the combustion of iso-octane (C_8H_{18}) with air. While the equilibrium resulting from that combustion process is not the same as would occur from the single reaction we are considering, it is instructive to use the corresponding C/O ratio. We then have for this single reaction

$$CO_2 \rightleftharpoons CO + \frac{1}{2}O_2. \tag{10.36}$$

The equilibrium condition is therefore

$$K(T) = \frac{\chi_{CO}^{\nu_{CO}} \cdot \chi_{O_2}^{\nu_{O_2}}}{\chi_{CO_2}^{\nu_{CO_2}}} \left(\frac{P}{P_0}\right)^{\nu_{CO}+\nu_{O_2}-\nu_{CO_2}},$$

where $K(T)$ is the equilibrium constant, x is the mole fraction, ν is the stoichiometric coefficient and P_0 is the reference pressure, 1 atm. For this reaction we have

$$K(T) = \frac{\chi_{CO} \cdot \chi_{O_2}^{1/2}}{\chi_{CO_2}} \left(\frac{P}{P_0}\right)^{1+1/2-1},$$

or, since $P = P_0$

$$K(T) = \frac{\chi_{CO} \cdot \chi_{O_2}^{1/2}}{\chi_{CO_2}}.$$

An additional condition is that the mole fractions must total unity, therefore

$$\chi_{O_2} + \chi_{CO_2} + \chi_{O} = 1.$$

A third relation follows from the given initial atomic composition, thus

$$\frac{\chi_{CO_2} + \chi_{CO}}{\chi_{CO_2} + \chi_{CO} + \chi_{O_2}} = \frac{1}{3.125}.$$

In order to solve these three simultaneous equations let us set $\chi_{CO_2} = a$, which shall then be our unknown. We obtain

$$\chi_{CO} + \chi_{O_2} = 1 - a,$$

or

$$\chi_{O_2} = 1 - a - \chi_{CO}.$$

We can now substitute it in the former equations and obtain

$$\frac{\chi_{CO} + a}{2a + \chi_{CO} + 2\chi_{CO_2}} = \frac{1}{3.125},$$

$$\frac{\chi_{CO} + a}{\chi_{CO} + 2 - 2\chi_{CO}} = \frac{1}{3.125},$$

or

$$\chi_{CO} = 0.484 - 0.758a,$$

$$\chi_{O_2} = 0.516 - 0.242a.$$

From the remaining relation we thus get

$$K(T) = \frac{(0.484 - 0.758a)\,(0.516 - 0.242a)^{\frac{1}{2}}}{a}. \tag{10.37}$$

We can now solve this equation for a, and for several different temperatures, at a pressure of 1 atm. The equilibrium constant is tabulated usually as $\log_{10}$, and several values for this example are supplied in Table 10.4.

Solving the system of equations iteratively allows us to obtain the concentration *at equilibrium*. Table 10.5 lists the results for several temperatures, but also for different C/O ratios and pressures.

What can we conclude from this exercise and these results? First, at low temperature only CO_2 and O_2 are present. The dissociation of CO_2 does not begin until about 2000 K. The reaction $CO_2 \rightleftharpoons CO + \frac{1}{2}O_2$ is endothermic when going from left to right. The fact that endothermic reactions are more complete at higher temperature is a general conclusion, demonstrated also by this example. In addition, we see that the equilibrium state is also affected by pressure. For clarity, the effect of the temperature, pressure, and the C/O ratio has also been calculated with STANJAN and the result is shown in Figures 10.5a–10.5c. Note that STANJAN also takes into account the dissociation into oxygen atoms. As a consequence the concentration of O_2 as a function of temperature decreases strongly, the concentration of CO decreases slightly instead of increasing, and the concentration of CO_2 decreases a little more compared to the case where dissociation into oxygen atoms is not taken into account. The calculations with STANJAN also show that at high temperatures ($T > 3000$ K) traces of carbon are formed.

Table 10.4 Equilibrium constants for the reaction $CO_2 \rightleftharpoons CO + \frac{1}{2}O_2$.

$\log_{10} K$	T / K	K
−45.059	298	8.73×10^{-46}
−10.195	1000	6.39×10^{-11}
−2.875	2000	1.33×10^{-3}
−0.481	3000	0.331
0.695	4000	4.956

Example: Effect of pressure. Let us rework the previous example for $T = 3000$ K, but for pressure $P = 2$ bar. Since $K = K(T)$, the equilibrium constant has the same value. Application of (10.34) yields

$$\frac{\chi_{CO} \cdot \chi_{O_2}^{1/2}}{\chi_{CO_2}} = 0.331 \left(\frac{2}{1.01325}\right)^{-1/2} = 0.235.$$

Solving with this value in (10.37), we find

$$\chi_{O_2} = 0.430,$$
$$\chi_{CO} = 0.159,$$
$$\chi_{O_2} = 0.411.$$

The equilibrium composition is indeed affected by the pressure. A higher pressure tends to reduce the number of moles (hence the number of molecules) of CO in the equilibrium mixture. Consequently, the effect here of higher pressure is to increase the CO_2 concentration. The C/O ratio must of course also affect the equilibrium composition. We see that an initial state with an excess of O atoms, that is if the mixture is lean, results in lower concentrations of CO.

Conversely, excess fuel, or, as here, an increased ratio of C to O atoms, promotes the formation of CO. At 3000 K we find for C/O ratios of 1:2, 1:3.12, and 1:5, CO mole fractions of 0.363, 0.198, and 0.113, respectively. A hasty

Table 10.5 Equilibrium concentrations for the single reaction $CO_2 \rightleftharpoons CO + \frac{1}{2}O_2$.

	T / K	x_{CO}	x_{CO_2}	x_{O_2}
$P = 1.01325$ bar				
C/O = 1 : 3.12	298	negligible	0.640	0.360
	1000	< 0.001	0.640	0.360
	2000	0.001	0.638	0.361
	3000	0.195	0.383	0.422
	4000	0.438	0.062	0.500
C/O = 1 : 2	3000	0.359	0.461	0.180
C/O = 1 : 5	3000	0.112	0.266	0.622
$P = 0.7$ bar				
C/O = 1 : 5	3000	0.126	0.249	0.625
$P = 2$ bar				
C/O = 1 : 3.12	3000	0.158	0.431	0.411
C/O = 1 : 5	3000	0.088	0.294	0.618
$P = 0.7$ bar				
C/O = 1 : 5	5000	0.326	0.009	0.665

conclusion might be simply to run engines with excess air in order to reduce CO concentrations. Fuel combustion is more complicated though, and excess air causes the formation of other pollutants, therefore additional precautions are taken in order to run engines with a lean fuel/air mixture.

Example: Effect of temperature. Again consider the reaction between CO_2, CO, and O_2. This time we shall calculate the equilibrium for $T = 5000$ K and $P = 0.7$ bar. For variety, let us write the reaction the other way around as

$$CO + \frac{1}{2}O_2 \rightleftharpoons CO_2, \qquad (10.38)$$

which interchanges the role of products and reactants from our previous approach. Note, however, that the equilibrium composition is independent of the direction of the reaction. Table D.1 does not explicitly provide $K(T)$ for this reaction. However, ΔG_{react} for reaction (10.38) is simply the negative of that for reaction (10.36). Hence the $K(T)$ for one reaction is simply the *reciprocal* of the $K(T)$ for the other, see (10.35), and the logarithms of K are simply the *negatives* of one another. Hence, for the reaction (10.38) Table D.1 implies

$$\log_{10} K = -1.468,$$
$$K = 0.0340.$$

For an initial ratio now of 1 C atom to 5 O atoms we find that $\chi_{CO} = 0.326$, $\chi_{CO_2} = 0.009$, and $\chi_{O_2} = 0.665$. Reaction (10.38) is exothermic when proceeding from left to right. Note that in our higher-temperature example less CO_2 is formed and the reaction is thus less complete. As mentioned previously, it is a general rule that exothermic reactions are less complete at higher temperature and endothermic reactions are more complete at higher temperatures.

The above example is a particularly simple one involving only three species. The more general combustion equation for gasoline, treated here as iso-octane, is

$$C_8H_{18} + X\cdot O_2 + Y\cdot N_2 \rightleftharpoons a\cdot H_2O + c\cdot O_2 + d\cdot H_2 + e\cdot CO + f\cdot H + g\cdot OH + h\cdot O + i\cdot N_2 + j\cdot N + k\cdot NO + \cdots.$$

Working out all the coefficients would require first of all to know all the possible reactions, such as

$$H_2O + CO_2 \rightleftharpoons CO + H_2O,$$
$$N_2 \rightleftharpoons 2\,N,$$
$$O_2 \rightleftharpoons 2\,O,$$
$$NO \rightleftharpoons \frac{1}{2}O_2 + \frac{1}{2}N_2,$$
$$H_2O \rightleftharpoons H + OH,$$
$$H_2 \rightleftharpoons 2\,H,$$
$$2\,H_2O \rightleftharpoons 2\,H_2 + O_2.$$

In addition, finding the equilibrium composition with this large number of possible reactions clearly requires the use of a computer.

A convenient method to solve this type of complex equilibrium reaction problems is presented in Section 10.7. In Section 10.2 we assumed that all the reactants are used up and no CO or NO is produced in the combustion process. In reality, often combustion is incomplete, and therefore noxious products can result. The unburned hydrocarbons, or partially burned ones, along with CO and the oxides of nitrogen, are the major pollutants from automotive internal combustion engines.

The understanding of why reactions do not go to completion requires knowledge of both the equilibrium conditions for reactions and the kinetics of reactions. Reaction rates are strongly dependent upon temperature. If the temperature drops before the reactions have had time to reach equilibrium, the mixture composition may be *frozen* at a non-equilibrium composition. This occurs for example for some of the

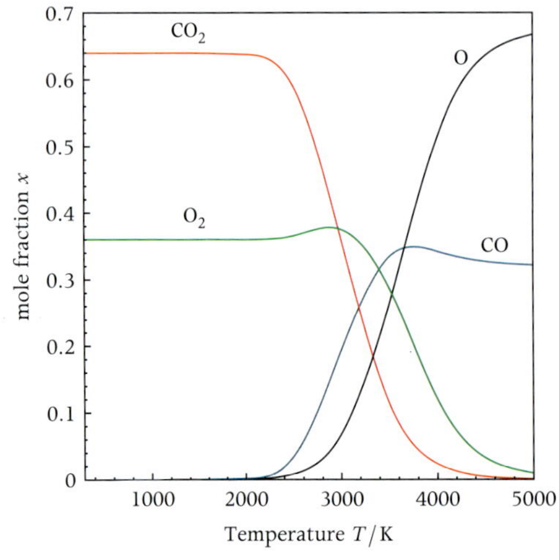

(a) As function of the temperature. $P = 1.01325$ bar, $C/O = 1:3.125$.

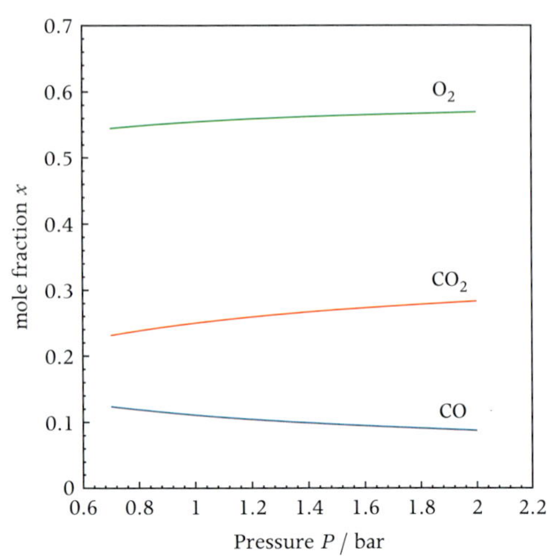

(b) As function of the pressure. $T = 3000$ K, $C/O = 1:5$.

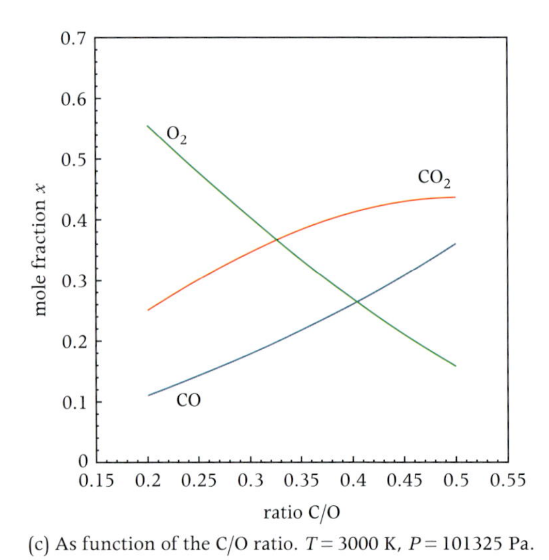

(c) As function of the C/O ratio. $T = 3000$ K, $P = 101325$ Pa.

Figure 10.5 Equilibrium compositions calculated with STANJAN for the single reaction $CO_2 \rightleftharpoons CO + \frac{1}{2}O_2$.

critical pollutant-formation reactions in internal combustion engines.

Van't Hoff equation

The dependence of the equilibrium constant on temperature can be related to the enthalpy change for a complete unit reaction. Differentiating (10.35) with respect to temperature gives

$$\frac{dK}{dT} = \frac{1}{\hat{R}T^2}\left[\Delta G_{\text{react}}(T, P_0) - T\frac{d\Delta G_{\text{react}}(T, P_0)}{dT}\right]K.$$

From the definition of specific Gibbs function we have that $g = h - Ts$, therefore

$$dg = -sdT + vdP,$$

cf. for example (6.111). By applying this relation to the changes for a complete unit reaction at fixed pressure we find

$$\frac{\Delta G_{\text{react}}(T, P_0)}{dT} = -\Delta S_{\text{react}}.$$

Also from the definition of Gibbs function it follows that

$$\Delta G_{react} = \Delta H_{react} - T\Delta S_{react}.$$

Hence the derivative of the equilibrium constant may be expressed as

$$\frac{1}{K}\frac{dK}{dT} = \frac{\Delta H_{react}(T)}{\hat{R}T^2},$$

or

$$\frac{d(\ln K)}{dT} = \frac{\Delta H_{react}(T)}{\hat{R}T^2}. \qquad (10.39)$$

This is called the van't Hoff equation, and it is important as it allows us to determine the enthalpy of change for a complete reaction solely from equilibrium-composition data. Conversely, it permits us to evaluate changes in the equilibrium constant (a second-law parameter) solely from energy balance (first-law) data.

Equation (10.39) shows that for endothermic reactions ($\Delta H_{react} > 0$), K increases with increasing temperature, while for exothermic reactions ($\Delta H_{react} < 0$), K decreases with increasing temperature. Since it can be shown that ΔH_{react} is only weakly dependent on temperature, we can integrate (10.39) from reference state 1, assuming that ΔH_{react} is constant, and obtain

$$\frac{(\ln K)}{K_1} = -\frac{\Delta H_{react}}{\hat{R}}\left(\frac{1}{T} - \frac{1}{T_1}\right). \qquad (10.40)$$

Note that $\ln K$ is linear in $\frac{1}{T}$ in this approximation; consequently, plots of $\ln K$ vs $\frac{1}{T}$ can be used to determine ΔH_{react} from experimental data.

As an example let us consider a reaction of interest in magnetohydrodynamics: the determination of ΔH_{react} for the ionization reaction of cesium. We have that

$$Cs \rightleftharpoons Cs^+ + e^-.$$

From the data of Table D.1 we obtain

$$T_1 = 4000\,\text{K}, \quad \log_{10} K_1 = 4.07, \quad K_1 = 11\,749,$$
$$T = 4500\,\text{K}, \quad \log_{10} K = 4.43, \quad K = 26\,915.$$

Assuming that ΔH_{react} is approximately constant, we have from (10.40)

$$\Delta H_{react} = \frac{-\hat{R}\ln(26\,915/11\,749)}{1/4\,500 - 1/4\,000} = 248 \times 10^3\ \text{J/mol}.$$

If this result is converted to eV/molecule we have

$$\Delta H_{react} = 248 \times 10^3\ \frac{\text{J}}{\text{mol}} \times \frac{\text{mol}}{6.023 \times 10^{23}\ \text{molecules}} \times \frac{\text{eV}}{1.06 \times 10^{-18}\ \text{J}} = 0.39\ \text{eV/molecule}.$$

Since ΔH_{react} is positive, the reaction is endothermic and requires 0.39 eV to ionize each molecule. This is the *ionization potential* of cesium at these high temperatures.

Example: Fuel cell. Consider a fuel cell processing hydrogen and oxygen, as shown in Figure 10.6. The maximum electric energy output per mole of water formed and the maximum cell voltage can be determined with the theory developed in this chapter.

We assume steady flow, steady state, and negligible kinetic- and potential-energy changes for the fluid, and consider only the electrostatic potential energy of the electrons. The cell temperature and the fluid temperature are assumed to be equal. The energy balance made over an infinitesimal period gives

$$\left(\bar{h}dN\right)_{H_2O} - \left(\bar{h}dN\right)_{H_2} - \left(\bar{h}dN\right)_{O_2} + đQ + e\left(\mathcal{E}_b - \mathcal{E}_a\right)dN_e = 0,$$

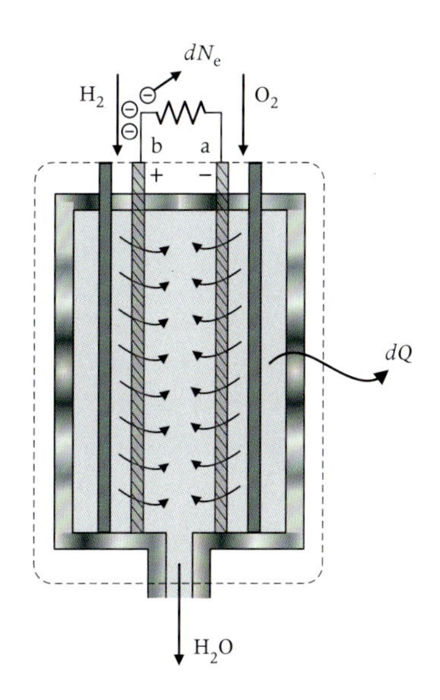

Figure 10.6 Representation of a fuel cell stack.

where e denotes the charge of an electron, and dN_e is the number of electrons passing through the control volume. Application of the second law gives

$$\bar{d}\mathcal{P}_S = (\bar{s}dN)_{H_2O} - (\bar{s}dN)_{H_2} - (\bar{s}dN)_{O_2} + \frac{\bar{d}Q}{T} \geq 0.$$

The equality holds if the processes within the cell can be approximated as reversible, otherwise the inequality sign must be used. The combination of the first- and second-law equations gives

$$\mathrm{e}\,(\mathcal{E}_b - \mathcal{E}_a) \leq -\left[(\bar{g}dN)_{H_2O} - (\bar{g}dN)_{H_2} - (\bar{g}dN)_{O_2}\right]. \tag{10.41}$$

The chemical equation is

$$H_2 + \frac{1}{2}O_2 \rightleftharpoons H_2O.$$

Since stoichiometric reaction is assumed, (10.41) may be written as

$$dE_{elect} = (\mathcal{E}_b - \mathcal{E}_a)\,\mathrm{e} \cdot dN_e \leq -\Delta G_{react} \cdot dN_{H_2O}, \tag{10.42}$$

where dE_{elect} is the useful electric energy output. The value of ΔG_{react} for this reaction is equivalent to $-238\,000$ J/mol of H_2O. This means that the maximum possible electric energy output of such a cell is 238 000 J for each mole of water produced.

The maximum cell voltage can be obtained by considering the reaction at the electrodes. At the anode we have that

$$H_2 + 2\,OH^- \longrightarrow 2\,H_2O + 2\,e^-,$$

while at the cathode

$$\frac{1}{2}O_2 + H_2O + 2\,e^- \longrightarrow 2\,OH^-.$$

The net amount of water production associated with the transfer of two electrons through the load is one molecule. The number of electrons passed for each mole of water formation is then $2N_0$, where N_0 is Avogadro's number. If we divide (10.42) by $-2\mathrm{e}N_0$ (a positive quantity), we obtain

$$\mathcal{E}_b - \mathcal{E}_a \leq -\frac{\Delta G_{react}}{2\mathrm{e}N_0}$$

$$= \frac{-238\,000}{2 \times 6.023 \times 10^{23} \times \left(-1.602 \times 10^{-19}\right)}$$

$$= 1.23 \text{ Volt}.$$

The efficiency of the fuel cell can be defined as the ratio between the actual energy output per mole of water formation and the maximum possible output, as we have calculated. Like the isentropic efficiency of a compressor, it compares the actual output to the output of an idealized device.

10.7 The Element Potential Method

The solution of complex chemical equilibrium problems is challenging. The problem may be formulated in several ways. If one adopts the equilibrium constants approach described in Section 10.6, it is then necessary to identify the set of reactions that take place and to determine the associated equilibrium constants. A set of nonlinear algebraic equations for the mole numbers of each species must be solved, a difficult task if the system is large. Other methods based on minimization of the Gibbs function adjust the moles of each species, consistently with the atomic constraints, until the minimum Gibbs function state is found. Again, there are many variables involved, and great care must be taken to be sure that all moles are non-negative. If there are important species in the system, but they are present in very small quantities, the solution might be hard to achieve.

The method of element potentials for the calculation of chemical equilibrium also targets the minimization of the Gibbs function, but via a theory relating the mole fractions of each species to quantities traditionally called *element potentials*. There is one element potential for each independent atom in the system, and these element potentials, plus the total number of moles in each phase, are the only variables that must be adjusted to find the solution. In large problems, this is a much smaller number of variables than the number of species, and hence far fewer variables need to be adjusted. There are many advantages to the element-potential method that quickly become obvious when one starts to use it. We believe that the method of element potentials should be part of modern instruction in thermodynamics.

The analysis illustrated here, and the corresponding program STANJAN, assumes that the gas phase is a mixture of ideal gases and that condensed phases are ideal solutions. These are good approximations for many practical problems of interest.

The basic theory of element potentials

The Gibbs function of a system (mixture) is, cf. (8.16c),

$$G = \sum_{j=1}^{nc} N_j \bar{g}_j.$$

If we treat each phase as either a mixture of ideal gases or as an ideal solution, the partial molar Gibbs functions of each species are

$$\bar{g}_j = \hat{g}_j(T, P) + \hat{R}T \ln z_j, \qquad (10.43)$$

where $g_j(T, P)$ is the Gibbs function of pure j evaluated at the system temperature and pressure, z_j is the mole fraction of j in its phase, and R is the universal gas constant.

In this book, see also Chapter 8, we use z_i for the mole fraction of a compound i in an unspecified phase, χ_i if the substance is in the ideal gas phase, y_i if it is in the vapor phase, x_i if it is in the liquid phase, and w_i if it is in the solid phase.

The atomic population constraints are given by

$$\sum_{j=1}^{nc} n_{ij} N_j = p_i, \qquad i = 1, \ldots, na, \qquad (10.44)$$

where n_{ij} is the number of i atoms in a j molecule, p_i is the population (moles) of i atoms in the system, and na is the number of different elements (atom types) present in the system.

The equilibrium mixture at the given T and P is the distribution of N_j that minimizes G, subject to the atomic constraints (10.43), *for non-negative N_j.*

Minimization problems with constraints can be conveniently handled by the methods of Lagrange multipliers. The following development does not assume knowledge of the method, but in essence describes the method for this particular problem. However, since the method of Lagrangian multipliers is very important in the numerical solution, a brief review of the method is presented in Appendix E.

For convenience, we define $\tilde{g}_j \equiv \hat{g}_j(T, P)/\hat{R}T$, and seek the minimum of $G/\hat{R}T$. Using (10.43) we find that, for arbitrary variations in the mole numbers,

$$d\left(\frac{G}{\hat{R}T}\right) = \sum_{j=1}^{nc} \left(\tilde{g}_j + \ln z_j\right) dN_j + \sum_{j=1}^{nc} N_j \frac{1}{z_j} dz_j. \qquad (10.45)$$

If we denote the number of moles in phase m by $\bar{N}_m$, then

$$z_j = \frac{N_j}{\bar{N}_{(j)}}, \qquad (10.46)$$

where $\bar{N}_{(j)}$ is the number of moles in the phase containing species j (a species present in more than one phase is assigned distinct j indices for each phase). The second sum in (10.45) can therefore be replaced by

$$\sum_{m=1}^{ph} \bar{N}_m \sum_{\substack{j=1 \\ \text{in } m}}^{nc} dz_j,$$

where ph is the total number of phases which might be present. This term vanishes, because the mole fractions in each phase always sum to unity.

Now, the dN_j are not all independent, because of the atomic constraints. Relationships between the dN_j are obtained by differentiating (10.44),

$$\sum_{j=1}^{nc} n_{ij} dN_j = 0, \qquad i = 1, \ldots, na. \qquad (10.47)$$

We must solve for the na restricted dN_j in terms of the $nc - na$ free dN_j, and then substitute these relationships into (10.45) in order to express the G variation in terms of the freely variable N_j. This process is equivalent to subtracting multiples of (10.47) from (10.45), that is

$$d\left(\frac{G}{\hat{R}T}\right) = \sum_{j=1}^{nc} \left(\tilde{g}_j + \ln z_j\right) dN_j - \sum_{i=1}^{na} \lambda_i \sum_{j=1}^{na} n_{ij} dN_j. \qquad (10.48)$$

The λ_i are the multipliers that are required to drop out of the set of restricted dN_j from (10.48). Thus, by setting the coefficient of these dN_j to zero, one has

$$\tilde{g}_j + \ln z_j - \sum_{i=1}^{na} \lambda_i n_{ij} = 0 \qquad (10.49)$$

for the restricted j. With these dN_j absent from (10.48), the remaining dn_j may be freely varied, and at the minimum G point there must be no variations that change G (to first order). This will be true only if the coefficient of each free dN_j is zero; hence, (10.49) also applies to the free j. So, for every species,

$$z_j = \exp\left(-\tilde{g}_j + \sum_{i=1}^{na} \lambda_i n_{ij}\right). \qquad (10.50)$$

Equation (10.50) is the main result of the element potentials method for the solution of the chemical equilibrium problem, valid for mixtures of ideal gases or for ideal solutions. It relates the phase mole fraction of each species to its value of $G(T, P)/\hat{R}T$, to the atomic makeup of its molecule, and to a set of undetermined multipliers (the "Lagrange multipliers") to be determined from the atomic constraints. The multiplier λ_i is called the *element potential* of the i atoms. Using (10.43) and (10.49), we see that

$$\frac{\bar{g}_j}{\hat{R}T} = \sum_{i=1}^{na} \lambda_i n_{ij}, \qquad (10.51)$$

and hence λ_i represents the Gibbs function divided by $\hat{R}T$ per mole of i atoms. What is even more amazing is that each atom of an element contributes the same amount to the Gibbs function of the system, irrespective of which molecule or phase it is in! The λ_i are properties of the system, however, they cannot be tabulated as functions of the atom or molecule, as can the g_j. It is perhaps for this reason that the method of element potentials was not widely used.

The values of the element potentials are determined by the atomic constraints of Equation (10.44), which we rewrite as

$$\sum_{j=1}^{na} n_{ij} \bar{N}_{(j)} z_j = p_i, \qquad i = 1, \ldots, na. \qquad (10.52)$$

Using (10.50), this becomes a set of na equations for the na unknown λ_i and the p unknown $\bar{N}_m$. To this we add the ph equations

$$\sum_{\substack{j=1 \\ \text{in } m}}^{nc} z_j = 1, \qquad m = 1, \ldots, ph. \qquad (10.53)$$

Equations (10.52) and (10.53) must be solved simultaneously to determine the element potentials and phase moles. This might appear to be a difficult task, but it is possible to do it accurately and quickly.

In many problems there will be a set of dominant species, the mole fractions of which can be estimated from simple balances. These can be used to estimate the element potentials, which can in turn be used to calculate the mole fractions of the minor species. As we shall see in the examples to follow, this involves only the solutions of linear algebraic equations. Thus there are advantages to the method of element potentials, even in "small" problems.

In problems with many species, the method has many significant advantages. There is no need to identify a set of reactions or to make use of the associated equilibrium constants. One has to deal only with $(na + np)$ variables, whereas other methods work with the nc unknown moles as variables. In a gas-phase problem with 100 species containing C, H, O, and N, the element-potential method has only five unknowns, whereas mole-interaction methods must work with 100 unknowns. Mole-interaction methods must guard against negative mole fractions, which can never occur with mole fractions generated by Equation (10.50). Furthermore, they can suffer from serious problems when some species feature very small mole fractions, but this is not a problem in a well-designed implementation of the method of element potentials. The power of the element potential method is exceptional in dealing with systems containing multiple phases. This is illustrated by examples available with the STANJAN program. The underlying so-called *dual problem* and its numerical solution method implemented in STANJAN are described in Appendix D.2. Here we go through some simpler examples that illustrate the basic concepts.

Table 10.6 Values of the nondimensional partial molar Gibbs function for the mixture constituents at 3000 K and 1 atm.

Species	$\tilde{g} = \hat{g}(T, P)/\hat{R}T$
CO	−33.578
CO_2	−49.830
O_2	−30.273
C(S)	−3.686

Element potentials in hand calculations

Two simple examples illustrate the use of element potentials in hand calculations. Both involve a mixture consisting of CO, CO_2, O_2, and C(S) (solid carbon) at 3000 K and 1 atm. The molar Gibbs functions of these species are listed in Table 10.6.

a) Same number of dominant species as elements
Suppose that the system contains 2 moles of C atoms and 1 mole of O atoms. Some solid carbon must therefore be present, and its mole fraction in the solid phase must be 1. Using Equation (10.50),

$$w_{C(S)} = 1 = \exp\left(-\tilde{g}_{C(S)} + \lambda_C\right). \tag{10.54}$$

The element potential for carbon can be found immediately from Equation (10.54):

$$\lambda_C = \tilde{g}_{C(S)} + \ln(1). \tag{10.55}$$

If we can estimate one other mole fraction, we can calculate the element potential for oxygen. At first glance, it looks as though the dominant gas species should be CO_2, which has the lowest Gibbs function. Now, the dissociation of CO_2 gives 1 mol of CO and 0.5 mole of O_2, for which

$$\begin{aligned} G/\hat{R}T &= -33.578 + \ln(2/3) + 0.5 \times \left[-30.273 + \ln(1/3)\right] \\ &= -49.668. \end{aligned}$$

This is just slightly greater than the Gibbs function of the mole of CO_2, so there still appears to be a slight preference for CO_2. However, a half-mole of CO_2 can combine with a half-mole of C(S) to make a mole of CO. For the CO_2 and C(S),

$$\begin{aligned} G/\hat{R}T &= 0.5 \times \left[-49.830 + \ln(1/2)\right] + 0.5 \times (-3.686) \\ &= -27.104, \end{aligned}$$

which is much greater than that of the mole of CO. Hence, any free carbon will tend to react with the CO_2 to form CO, and consequently we expect the dominant gas species to be CO. Assuming that the CO mole fraction is 1, Equation (10.50) gives

$$\chi_{CO} = \exp\left(-\tilde{g}_{CO} + \lambda_C + \lambda_O\right) = 1, \tag{10.56}$$

from which we obtain our second linear equation in the element potentials,

$$\lambda_C + \lambda_O = \tilde{g}_{CO} + \ln(1). \tag{10.57}$$

Solving (10.55) and (10.57) simultaneously for the potentials, we have

$$\lambda_C = -3.686 \quad \text{and} \quad \lambda_O = -29.892.$$

From this we use (10.56) to estimate the mole fraction of CO_2 with

$$\chi_{CO_2} = \exp\left(-\tilde{g}_{CO_2} + \lambda_C + 2\lambda_O\right) = 0.1193 \times 10^{-5}.$$

The assumption that CO was the dominant gas species was clearly correct. If we wished, we could correct our estimates by lowering the mole fraction of CO, but in this case we are so close to the exact solution that the iteration is not worthwhile. Indeed, the solution is exact to four decimal places!

Suppose at this point we wished to estimate the concentration of a species that we have not thus far included in the system, for example O. We can do this easily using the element potentials. At 3000 K and 1 atm, $\tilde{g}_O = -12.951$, so

$$\chi_O = \exp(12.951 - 29.892) = 2.38 \times 10^{-8}.$$

This is a very accurate estimate, since the inclusion of O in the system with this mole fraction will not significantly influence the element potentials.

In summary, whenever we have a system in which one dominant species can be identified for each element in the system, the mole fractions of these species can be used to estimate the element potentials. With these element potentials, estimates of all of the other mole fractions can be made, and corrections can be made to the element potentials by iteration, if necessary. The element potentials can then be used to estimate the concentrations of minor species.

b) Fewer dominant species than elements

Suppose instead that the system contains 1 mole of C atoms and 2 moles of O atoms. Here the condensed phase will be almost absent and the gas phase will consist almost entirely of CO_2. Therefore, we have only one dominant mole fraction from which we want to estimate two element potentials. We can still do this by using a concept called "balancing." The atomic constraints can be written as

$$N_{CO} + N_{CO_2} + N_{C(S)} = 1, \quad (10.58a)$$

$$N_{CO} + 2N_{CO_2} + 2N_{O_2} = 2. \quad (10.58b)$$

The idea of balancing is to select a set of *base species*, and then to recast the constraints so that each equation contains only one of these base species. The base set should include the dominant species. We select the dominant species CO_2 as one base, and O_2 as the other. Equation (10.58a) contains only the base CO_2 and tells us that there is approximately 1 mole of CO_2 in the system. By combining the equations to eliminate CO_2 from (10.58b), we obtain

$$2N_{O_2} - N_{CO} - 2N_{C(S)} = 0.$$

This tells us that the second base species O_2 must be "balanced" by CO and/or C(S). Since CO has a much smaller Gibbs function than C(S) the balance will be primarily with CO, and so approximately

$$2N_{O_2} = N_{CO}. \quad (10.59)$$

Since both O_2 and CO are in the same phase, this translates into a requirement that the mole fraction of CO must be twice that of CO_2. Then, using (10.50) in (10.59) and taking the log of both sides, a linear equation relating the element potentials is obtained, namely

$$\ln(2) - \tilde{g}_{O_2} + 2\lambda_O = -\tilde{g}_{CO} + \lambda_C + \lambda_O. \quad (10.60)$$

A second linear equation relating the element potential is obtained from the estimate that the mole fraction of CO_2 is unity,

$$-\tilde{g}_{CO_2} + \lambda_C + 2\lambda_O = \ln(1). \quad (10.61)$$

We can solve these two equations and obtain

$$\lambda_C = -18.351 \quad \text{and} \quad \lambda_O = -15.739.$$

Using these potentials, the mole fractions of the species in the mixture are estimated as

$$\chi_{CO} = 0.599, \quad \chi_{O_2} = 0.299, \quad \chi_{CO_2} = 1.$$

Clearly, the assumption that CO and O_2 are rare species was not very good. However, we can correct our estimate by rescaling the χ_j so that they sum to unity, therefore

$$\chi_{CO} = 0.316, \quad \chi_{O_2} = 0.158, \quad \chi_{CO_2} = 0.526.$$

These estimates are within 10 % of the exact values. An improvement can be obtained by iterating, using our revised estimate for χ_{CO_2} in (10.61). The result is

$$\lambda_C = -18.565 \quad \text{and} \quad \lambda_O = -15.953,$$

and these produce

$$\chi_{CO} = 0.390, \quad \chi_{O_2} = 0.195, \quad \chi_{CO_2} = 0.526.$$

These normalize to give

$$\chi_{CO} = 0.352, \quad \chi_{O_2} = 0.175, \quad \chi_{CO_2} = 0.473.$$

The exact solution obtained with STANJAN is

$$\chi_{CO} = 0.3582, \quad \chi_{O_2} = 0.1791, \quad \chi_{CO_2} = 0.4627,$$

and so we see that with only two iterations we are very close.

The iterative process used here might be used as the basis for a numerical method for general problems. However, a general method must work irrespective of the structure of any particular problem, and thus the multiphase, many-species problem presents a greater challenge. A suitable numerical method was developed by the late Professor W. C. Reynolds to meet this challenge and it is based on the so-called *dual problem*. It is illustrated in Appendix D.2, where other details about the STANJAN program are also given.

An even more general and complex problem arises when one wants to solve the thermodynamic equilibrium problem whereby species are present in multiple phases, do not obey the ideal gas law nor the ideal solution approximation and undergo many chemical reactions. In this case the element potential method as such cannot be applied. The problem of predicting the equilibrium of non-ideal multiphase reacting mixtures is treated in specialized literature,[1] and is actively researched.

[1] See, for example, A. V. Phoenix and R. A. Heidemann, "A non-ideal multiphase chemical equilibrium algorithm," *Fluid Phase Equilib.*, 150–151, 255–265, 1998.

EXERCISES

10.1 Write out the chemical equation for the stoichiometric reaction of octane (C_8H_{18}) with air. Determine the theoretical air/fuel ratio for this reaction on both a volumetric (mole) and mass basis.

10.2 Determine the chemical equation for the reaction of octane (C_8H_{18}) with: a) 100 % excess air, and b) 250 % theoretical air.

10.3 An oil-based fuel contains 84 % C and 16 % H_2 by mass. Calculate the stoichiometric amount of air for complete combustion of 0.5 kg of fuel. Determine the air/fuel ratio.

10.4 Compute the air/fuel ratio by mass for the combustion of methane, if the exhaust gas dry analysis in percent by volume is: $\chi_{CO_2} = 0.01537$, $\chi_{O_2} = 0.4917$, $\chi_{H_2O} = 0.03073$. The presence of minor species like CO and H_2 can be neglected.

10.5 Determine the dew point of the products of combustion of octane with 400 % theoretical dry air if the pressure is 0.1013 MPa.

10.6 Compute the standardized enthalpy and entropy of a mixture formed by an equal mass amount of CO_2 and H_2O at 2 atm and 311 K. If liquid is present for these conditions, assume that the vapor phase can be approximated as an ideal gas mixture, and the liquid phase is pure water. Use: a) values obtained from Table A.1, and b) values obtained with GasMix. If there is a difference, explain it.

10.7 Consider n-butane (C_4H_{10}) reacting stoichiometrically with dry air at a pressure of 125 kPa: determine the dew point. What is the effect of excess air if this reaction occurs with 100 % excess air?

10.8 A hydrocarbon whose composition is unknown undergoes combustion such that an Orsat apparatus measures the following composition: $\chi_{CO_2} = 0.09$, $\chi_{O_2} = 0.088$, $\chi_{CO} = 0.011$, $\chi_{N_2} = 0.811$. Determine the reaction equation, air/fuel ratio, and the percent of theoretical air on a mass basis.

10.9 How much energy is released when 1 mole of propane (C_3H_8) reacts with 80 % theoretical air and both the inlet and the outlet of the combustor are kept at the standard reference state? How much energy is released if this reaction is stoichiometric?

10.10 Make a table of the enthalpy of combustion at the standard reference state for hydrogen (H_2), carbon monoxide (CO), methane (CH_4), ethane(C_2H_6), and propane (C_3H_8), assuming that the water in the products is liquid.

10.11 Ammonia (NH_3) and hydrogen (H_2) have been proposed as automotive fuels as a way of reducing related pollution. Calculate a) the stoichiometric air/fuel ratio for these two fuels and compare it to that for hydrocarbons, b) for ammonia, the HHV and LHV at standard conditions. Obtain $\hat{h}^o$ using StanMix .

10.12 If methane is burned in air and the excess air is 15 %, calculate the volume percent of oxygen in the flue gas, both on a wet basis (including water vapor) and on a dry basis (neglecting water vapor).

10.13 Compare the energy released as heat by burning stoichiometrically 1 mole of CH_4 with air and that released by burning 1 mole of C and 2 moles of H_2. Assume that combustion occurs at the standard reference state. Why is the energy change different?

10.14 Calculate the enthalpy of combustion for the so called water-gas shift reaction $CO(g) + H_2O(g) \rightleftharpoons CO_2(g) + H_2(g)$, at standard reference conditions and at 300 °C.

10.15 A gas turbine uses liquid n-butane (C_4H_{10}) as fuel. Calculate the required excess air such that the turbine inlet temperature does not exceed 1000 K. Assume for simplicity that the combustor inlet is at the standard reference condition.

10.16 Determine the adiabatic flame temperature for the combustion of propane in air if: (a) the air/fuel ratio is stoichiometric, (b) the air is twice the stoichiometric amount, and (c) the air is four times the stoichiometric amount. Assume complete combustion in each case.

10.17 Propane (C_3H_8) and oxygen in stoichiometric proportions react in a steady flow water-cooled burner. The reactants enter the burner at 305 K and leave the burner at 2 atm and emerge at 730 K. The flow rate of the products is 115 kg/hr. What is the rate of energy transfer as heat to the cooling water? Assume that the products contain only CO_2 and H_2O.

10.18 A turboprop engine has the following characteristics:

Compressor pressure ratio	6
Turbine inlet temperature	1000 °C
Compressor isentropic efficiency, $\eta_{\circ,s}$	0.85
Turbine isentropic efficiency, $\eta_{\text{turb},s}$	0.90
Fuel	n-octane (C_8H_{18})

The engine propels an airplane at 640 km/hr at an altitude of 7600 m (−35 °C, 38 kPa). Analyze the thermodynamic cycle on a per kg basis, making suitable assumptions. Calculate the state points of the cycle, and the engine specific consumption in $\text{kg}_{\text{fuel}}/\text{kWh}$.

10.19 The burner of a power station must provide thermal power to the boiler at a rate of 88 MW. The maximum flame temperature is to be 1100 K, and the products of combustion can leave the boiler at no less than 920 K. Design a burner system, specifying the fuel, flow rates, and any other datum deemed necessary.

10.20 A laboratory analysis shows that CO_2 becomes 10 % dissociated into CO and O_2 at 2390 K if the total pressure is 1 atm. Obtain the equilibrium constant from this information and compare it to the value given in Table D.1.

10.21 Hydrogen is to be burned with oxygen to produce a 1922 K flame. Neglecting dissociation, determine the composition of the product gases and specify the ratios of oxygen- and hydrogen-flow rates. The burner is adiabatic and atmospheric.

10.22 Find the mole fractions of O present in equilibrium O_2 at 1000 K and at 5000 K. Assume $P = 1$ atm. Repeat the calculation for N_2 and the same conditions.

10.23 Assuming that air is composed of O_2, O, N_2, N, and NO, and that only O_2 and N_2 are present in significant amounts at room temperature in the ratio of 3.76 moles of N_2 per mole of O_2, determine the composition of equilibrium air at 1000 K and 1 atm.

10.24 Using data for standardized enthalpy and absolute entropy obtained from GASMIX calculate the equilibrium constant for the gaseous reaction $CO_2 + H_2 \rightleftharpoons CO + H_2O$ at 298 K. Derive an expression which would let you obtain this, as an alternative to using Table D.1 in terms of the equilibrium constants for simpler reactions.

10.25 The water-gas shift reaction is $CO + H_2O \rightleftharpoons CO_2 + H_2$. Processes based on this reaction have been studied as a means of providing H_2 as an automotive fuel. Assume that 1 mole of CO reacts with one mole of H_2O at 1800 K and 1 atm. Determine the equilibrium composition. Will lowering the temperature increase or decrease the yield of H_2? Neglect species other than CO, H_2O, CO_2, H_2. Values of $K = K(T)$ for the water-gas shift reaction are

T/K	$\log_{10} K$
298	5.018
1000	0.159
1500	−0.409
1800	−0.577
2000	−0.656

What effect does an increase in CO have on the H_2 yield at 1800 K?

10.26 Consider the reaction $N_2 \rightleftharpoons 2\,N$. (a) At the peak combustion temperature in a car (2500 to 2800 K), and neglecting the effect of pressure, will this reaction affect the production of NO? (b) If the peak pressure is about 35 bar in the cylinder of a car engine, does this high pressure increase or decrease the production of N via this reaction?

10.27 For the reaction $H_2O \rightleftharpoons H_2 + \frac{1}{2}O_2$ the equilibrium constant is $\log_{10} K = -3.531$ at 2000 K. Estimate the constant at $T = 2225$ K using the van t'Hoff equation.

10.28 One mole of CO reacts with one mole of O_2 in a steady flow process. Both reactants at the inlet are at 298 K and 1 atm. The final products of the reaction are a mixture of CO, CO_2, and O_2 at 1 atm. Determine the equilibrium composition at 3000 K and 2800 K. Determine also the energy transferred as heat from the reactor for the two temperatures. At what exit temperature is the reactor adiabatic?

10.29 Determine the maximum energy output per gmole of CO_2 for a fuel cell operating on CO and O_2 at 298 K and 1 atm. Assume that the products at the outlet of the fuel cell contain only CO_2.

10.30 Consider the reaction of air to form a gaseous mixture containing N_2, O_2, and NO and assume that the perfect gas law is applicable. Calculate the mole fractions of the species at equilibrium at $T = 2000$ K and $P = 5$ atm using the element potentials method.

The nondimensional Gibbs function values of the pertinent species at these conditions are

Species	$\tilde{g} = \hat{g}\{T, P)/\hat{R}T$
N_2	−25.32
O_2	−27.14
NO	−22.31

10.31 One of the production processes of gallium-arsenide used in the semiconductor industry involves the equilibrium reaction of a gas mixture of As_2, As_4, and Ga at 800 K and 0.1 atm, thus the perfect gas approximation is valid. If there are the same number of atoms of As and Ga, and 1 mole of As_2, use STANJAN to compute the equilibrium composition.

10.32 Calculate the equilibrium composition of the water-gas shift reaction of Exercise 10.25 with the element potential method. Compare the accuracy of the results with those obtained with the equilibrium constant method and with STANJAN, commenting on the difference and on the differences between the methods. Make a chart of the equilibrium composition as a function of temperature similar to the one of Figure 10.5a. Comment on the trends visible in the chart.

APPENDICES

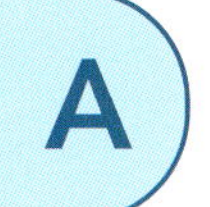

Thermodynamic Properties of Fluids

A.1 Values of Several Molar Properties for Some Common Fluids

Table A.1 Values of molecular mass, gas constant, isobaric heat capacity, reference enthalpy, and entropy for several fluids

Fluid	Chemical formula	$\hat{M}$ / kg/kmol	R / J/(kg · K)	$\hat{c}_P$ / J/mol K	$\hat{h}^0$ / kJ/mol	$\hat{s}^0$ / J/(mol · K)
Argon	Ar	39.948	208.1	20.786	0	154.845
Hydrogen	H_2	2.016	4124.7	28.836	0	130.680
Nitrogen	N_2	28.013	296.8	29.124	0	191.608
Oxygen	O_2	31.999	259.8	29.378	0	205.148
Carbon	C	12.011	692.2	8.540	0	5.694
Water (liquid)	H_2O	18.015	461.5	75.300	−285.830	69.950
Carbon monoxide	CO	28.010	296.8	29.141	−110.530	197.658
Carbon dioxide	CO_2	44.010	188.9	37.135	−393.510	213.783
Ammonia	NH_3	17.031	488.2	35.630	−45.940	192.768
Methane	CH_4	16.043	518.3	35.695	−74.600	186.369
Ethane	C_2H_6	30.069	276.5	52.487	−84.000	229.161
Propane	C_3H_8	44.096	188.6	73.597	−103.847	270.313
n-Butane	C_4H_{10}	58.123	143.0	99.167	−124.727	310.030
Acetylene	C_2H_2	26.038	319.3	44.036	227.400	200.927
Benzene (liquid)	C_6H_6	78.113	106.4	135.950	49.080	173.450

A.2 Low-density Thermodynamic Properties of Air

Properties in the tables are calculated with TPSI, which implements the thermodynamic model for air documented in W. C. Reynolds, *Thermodynamic Properties in SI*. Dept. of Mechanical Engineering, Stanford University, 1979:

$$\begin{aligned} s_2 - s_1 &= \phi(T_2) - \phi(T_1) - R \ln \frac{P_2}{P_1} \\ &= \psi(T_2) - \psi(T_1) + R \ln \frac{v_2}{v_1}, \end{aligned}$$

$$\ln P_r = \frac{\phi(T)}{R}, \quad \ln v_r = -\frac{\psi(T)}{R}.$$

Table A.2 Properties of air as an ideal gas with temperature-dependent specific heat.

T / K	h / kJ/kg	u / kJ/kg	ϕ / kJ/(kg · K)	ψ / kJ/(kg · K)	P_r	v_r	c_P / kJ/(kg · K)	c_v / kJ/(kg · K)	γ
200	359.8	302.4	7.4367	4.2910	17.81	3224	1.002	0.715	1.402
205	364.8	306.0	7.4614	4.3086	19.41	3032	1.002	0.715	1.402
210	369.8	309.5	7.4855	4.3258	21.11	2855	1.002	0.715	1.402
215	374.8	313.1	7.5091	4.3427	22.92	2693	1.002	0.715	1.401
220	379.8	316.7	7.5322	4.3591	24.84	2543	1.002	0.715	1.401
225	384.9	320.3	7.5547	4.3752	26.86	2404	1.002	0.715	1.401
230	389.9	323.8	7.5767	4.3909	29.01	2276	1.002	0.715	1.401
235	394.9	327.4	7.5983	4.4063	31.27	2157	1.002	0.715	1.401
240	399.9	331.0	7.6194	4.4213	33.65	2047	1.002	0.715	1.401
245	404.9	334.6	7.6400	4.4361	36.17	1945	1.002	0.715	1.401
250	409.9	338.1	7.6603	4.4505	38.81	1849	1.003	0.715	1.401
255	414.9	341.7	7.6801	4.4647	41.59	1760	1.003	0.716	1.401
260	419.9	345.3	7.6996	4.4786	44.51	1677	1.003	0.716	1.401
265	425.0	348.9	7.7187	4.4922	47.57	1599	1.003	0.716	1.401
270	430.0	352.5	7.7375	4.5056	50.78	1526	1.003	0.716	1.401
275	435.0	356.0	7.7559	4.5187	54.14	1458	1.003	0.716	1.401
280	440.0	359.6	7.7739	4.5317	57.66	1394	1.003	0.716	1.401
285	445.0	363.2	7.7917	4.5443	61.34	1334	1.004	0.716	1.401
290	450.0	366.8	7.8092	4.5568	65.19	1277	1.004	0.717	1.401
295	455.1	370.4	7.8263	4.5690	69.20	1224	1.004	0.717	1.400
300	460.1	374.0	7.8432	4.5811	73.39	1173	1.004	0.717	1.400
305	465.1	377.5	7.8598	4.5930	77.76	1126	1.004	0.717	1.400
310	470.1	381.1	7.8761	4.6046	82.32	1081	1.005	0.718	1.400
315	475.1	384.7	7.8922	4.6161	87.06	1039	1.005	0.718	1.400
320	480.2	388.3	7.9080	4.6274	91.99	998.6	1.005	0.718	1.400
325	485.2	391.9	7.9236	4.6385	97.12	960.6	1.006	0.718	1.400

Table A.2 (cont.)

T / K	h / kJ/kg	u / kJ/kg	ϕ / kJ/(kg · K)	ψ / kJ/(kg · K)	P_r	v_r	c_P / kJ/(kg · K)	c_v / kJ/(kg · K)	γ
330	490.2	395.5	7.9390	4.6495	102.5	924.6	1.006	0.719	1.399
335	495.3	399.1	7.9541	4.6603	108.0	890.4	1.006	0.719	1.399
340	500.3	402.7	7.9690	4.6710	113.8	857.9	1.007	0.720	1.399
345	505.3	406.3	7.9837	4.6815	119.7	827.1	1.007	0.720	1.399
350	510.4	409.9	7.9982	4.6919	125.9	797.8	1.007	0.720	1.399
355	515.4	413.5	8.0125	4.7021	132.4	769.9	1.008	0.721	1.398
360	520.4	417.1	8.0266	4.7122	139.0	743.3	1.008	0.721	1.398
365	525.5	420.7	8.0405	4.7221	145.9	718.0	1.009	0.722	1.398
370	530.5	424.3	8.0542	4.7319	153.1	693.8	1.009	0.722	1.398
375	535.6	427.9	8.0678	4.7416	160.5	670.8	1.010	0.723	1.397
380	540.6	431.5	8.0812	4.7512	168.1	648.8	1.010	0.723	1.397
385	545.7	435.2	8.0944	4.7607	176.1	627.8	1.011	0.724	1.397
390	550.7	438.8	8.1074	4.7700	184.2	607.7	1.011	0.724	1.396
395	555.8	442.4	8.1203	4.7792	192.7	588.4	1.012	0.725	1.396
400	560.8	446.0	8.1330	4.7884	201.4	570.0	1.013	0.725	1.396
405	565.9	449.6	8.1456	4.7974	210.5	552.4	1.013	0.726	1.395
410	571.0	453.3	8.1581	4.8063	219.8	535.5	1.014	0.727	1.395
415	576.0	456.9	8.1703	4.8151	229.4	519.3	1.014	0.727	1.395
420	581.1	460.6	8.1825	4.8238	239.3	503.8	1.015	0.728	1.394
425	586.2	464.2	8.1945	4.8324	249.5	488.9	1.016	0.729	1.394
430	591.3	467.8	8.2064	4.8410	260.1	474.6	1.017	0.729	1.394
435	596.4	471.5	8.2182	4.8494	271.0	460.8	1.017	0.730	1.393
440	601.5	475.1	8.2298	4.8577	282.2	447.6	1.018	0.731	1.393
445	606.5	478.8	8.2413	4.8660	293.7	434.9	1.019	0.732	1.392
450	611.6	482.5	8.2527	4.8742	305.6	422.7	1.020	0.733	1.392
455	616.7	486.1	8.2640	4.8823	317.8	410.9	1.020	0.733	1.391
460	621.8	489.8	8.2751	4.8903	330.4	399.6	1.021	0.734	1.391
465	627.0	493.5	8.2862	4.8982	343.4	388.7	1.022	0.735	1.391
470	632.1	497.1	8.2971	4.9061	356.7	378.2	1.023	0.736	1.390
475	637.2	500.8	8.3079	4.9139	370.4	368.1	1.024	0.737	1.390
480	642.3	504.5	8.3187	4.9216	384.5	358.3	1.025	0.738	1.389
485	647.4	508.2	8.3293	4.9293	399.0	348.9	1.026	0.739	1.389
490	652.6	511.9	8.3398	4.9369	413.9	339.8	1.027	0.739	1.388
495	657.7	515.6	8.3502	4.9444	429.3	331.0	1.027	0.740	1.388
500	662.8	519.3	8.3606	4.9518	445.0	322.6	1.028	0.741	1.387

Table A.2 (cont.)

T / K	h / kJ/kg	u / kJ/kg	ϕ / kJ/(kg · K)	ψ / kJ/(kg · K)	P_r	v_r	c_P / kJ/(kg · K)	c_v / kJ/(kg · K)	γ
505	668.0	523.0	8.3708	4.9592	461.1	314.4	1.029	0.742	1.387
510	673.1	526.7	8.3809	4.9665	477.7	306.5	1.030	0.743	1.386
515	678.3	530.4	8.3910	4.9738	494.8	298.8	1.031	0.744	1.386
520	683.4	534.2	8.4010	4.9810	512.3	291.4	1.032	0.745	1.385
525	688.6	537.9	8.4109	4.9881	530.2	284.3	1.033	0.746	1.385
530	693.8	541.6	8.4207	4.9952	548.6	277.3	1.034	0.747	1.384
535	699.0	545.4	8.4304	5.0022	567.5	270.6	1.035	0.748	1.384
540	704.1	549.1	8.4400	5.0092	586.9	264.1	1.036	0.749	1.383
545	709.3	552.9	8.4496	5.0161	606.7	257.9	1.038	0.750	1.383
550	714.5	556.6	8.4590	5.0229	627.1	251.8	1.039	0.752	1.382
555	719.7	560.4	8.4684	5.0297	648.0	245.9	1.040	0.753	1.381
560	724.9	564.1	8.4778	5.0365	669.4	240.2	1.041	0.754	1.381
565	730.1	567.9	8.4870	5.0432	691.3	234.6	1.042	0.755	1.380
570	735.3	571.7	8.4962	5.0499	713.8	229.2	1.043	0.756	1.380
575	740.5	575.5	8.5053	5.0565	736.8	224.0	1.044	0.757	1.379
580	745.8	579.3	8.5144	5.0630	760.4	219.0	1.045	0.758	1.379
585	751.0	583.1	8.5234	5.0695	784.6	214.0	1.046	0.759	1.378
590	756.2	586.9	8.5323	5.0760	809.3	209.3	1.047	0.760	1.378
595	761.5	590.7	8.5411	5.0824	834.6	204.6	1.049	0.762	1.377
600	766.7	594.5	8.5499	5.0888	860.6	200.1	1.050	0.763	1.376
605	772.0	598.3	8.5586	5.0951	887.1	195.8	1.051	0.764	1.376
610	777.2	602.1	8.5673	5.1014	914.2	191.5	1.052	0.765	1.375
615	782.5	605.9	8.5758	5.1077	942.0	187.4	1.053	0.766	1.375
620	787.8	609.8	8.5844	5.1139	970.4	183.4	1.054	0.767	1.374
625	793.0	613.6	8.5929	5.1201	999.5	179.5	1.056	0.768	1.374
630	798.3	617.5	8.6013	5.1262	1029	175.7	1.057	0.770	1.373
635	803.6	621.3	8.6096	5.1323	1060	172.0	1.058	0.771	1.372
640	808.9	625.2	8.6179	5.1383	1091	168.4	1.059	0.772	1.372
645	814.2	629.0	8.6262	5.1443	1123	164.9	1.060	0.773	1.371
650	819.5	632.9	8.6344	5.1503	1155	161.5	1.061	0.774	1.371
655	824.8	636.8	8.6425	5.1563	1188	158.2	1.063	0.776	1.370
660	830.1	640.7	8.6506	5.1622	1222	155.0	1.064	0.777	1.370
665	835.4	644.5	8.6586	5.1680	1257	151.9	1.065	0.778	1.369
670	840.8	648.4	8.6666	5.1739	1292	148.8	1.066	0.779	1.368

Table A.2 (cont.)

T / K	h / kJ/kg	u / kJ/kg	ϕ / kJ/(kg · K)	ψ / kJ/(kg · K)	P_r	v_r	c_P / kJ/(kg · K)	c_v / kJ/(kg · K)	γ
675	846.1	652.3	8.6745	5.1796	1329	145.9	1.067	0.780	1.368
680	851.4	656.2	8.6824	5.1854	1365	143.0	1.069	0.782	1.367
685	856.8	660.1	8.6903	5.1911	1403	140.1	1.070	0.783	1.367
690	862.1	664.1	8.6980	5.1968	1442	137.4	1.071	0.784	1.366
695	867.5	668.0	8.7058	5.2025	1481	134.7	1.072	0.785	1.366
700	872.9	671.9	8.7135	5.2081	1521	132.1	1.073	0.786	1.365
705	878.2	675.9	8.7211	5.2137	1563	129.5	1.075	0.788	1.365
710	883.6	679.8	8.7287	5.2193	1604	127.0	1.076	0.789	1.364
715	889.0	683.7	8.7363	5.2248	1647	124.6	1.077	0.790	1.363
720	894.4	687.7	8.7438	5.2304	1691	122.2	1.078	0.791	1.363
725	899.8	691.7	8.7512	5.2358	1735	119.9	1.079	0.792	1.362
730	905.2	695.6	8.7587	5.2413	1781	117.7	1.081	0.794	1.362
735	910.6	699.6	8.7660	5.2467	1827	115.5	1.082	0.795	1.361
740	916.0	703.6	8.7734	5.2521	1875	113.3	1.083	0.796	1.361
745	921.4	707.5	8.7807	5.2575	1923	111.2	1.084	0.797	1.360
750	926.8	711.5	8.7879	5.2628	1972	109.2	1.085	0.798	1.360
755	932.3	715.5	8.7952	5.2681	2022	107.2	1.087	0.799	1.359
760	937.7	719.5	8.8023	5.2734	2073	105.2	1.088	0.801	1.359
765	943.1	723.5	8.8095	5.2786	2126	103.3	1.089	0.802	1.358
770	948.6	727.5	8.8166	5.2839	2179	101.4	1.090	0.803	1.358
775	954.0	731.6	8.8236	5.2891	2233	99.63	1.091	0.804	1.357
780	959.5	735.6	8.8306	5.2942	2288	97.85	1.092	0.805	1.356
785	965.0	739.6	8.8376	5.2994	2345	96.11	1.094	0.806	1.356
790	970.4	743.7	8.8446	5.3045	2402	94.41	1.095	0.808	1.355
795	975.9	747.7	8.8515	5.3096	2461	92.75	1.096	0.809	1.355
800	981.4	751.7	8.8584	5.3147	2520	91.12	1.097	0.810	1.354
805	986.9	755.8	8.8652	5.3197	2581	89.53	1.098	0.811	1.354
810	992.4	759.8	8.8720	5.3248	2643	87.98	1.099	0.812	1.353
815	997.9	763.9	8.8788	5.3298	2706	86.46	1.100	0.813	1.353
820	1003.4	768.0	8.8855	5.3347	2770	84.97	1.102	0.814	1.352
825	1008.9	772.1	8.8922	5.3397	2836	83.52	1.103	0.816	1.352
830	1014.4	776.1	8.8989	5.3446	2902	82.10	1.104	0.817	1.351
835	1019.9	780.2	8.9055	5.3495	2970	80.70	1.105	0.818	1.351
840	1025.5	784.3	8.9121	5.3544	3039	79.34	1.106	0.819	1.351
845	1031.0	788.4	8.9187	5.3593	3109	78.01	1.107	0.820	1.350
850	1036.5	792.5	8.9252	5.3641	3181	76.71	1.108	0.821	1.350

Table A.2 (cont.)

T / K	h / kJ/kg	u / kJ/kg	ϕ / kJ/(kg · K)	ψ / kJ/(kg · K)	P_r	v_r	c_P / kJ/(kg · K)	c_v / kJ/(kg · K)	γ
855	1042.1	796.6	8.9317	5.3689	3254	75.43	1.109	0.822	1.349
860	1047.6	800.7	8.9382	5.3737	3328	74.18	1.110	0.823	1.349
865	1053.2	804.9	8.9446	5.3785	3404	72.95	1.112	0.824	1.348
870	1058.7	809.0	8.9510	5.3833	3481	71.76	1.113	0.826	1.348
875	1064.3	813.1	8.9574	5.3880	3559	70.58	1.114	0.827	1.347
880	1069.9	817.3	8.9638	5.3927	3638	69.43	1.115	0.828	1.347
885	1075.4	821.4	8.9701	5.3974	3719	68.31	1.116	0.829	1.346
890	1081.0	825.5	8.9764	5.4021	3802	67.20	1.117	0.830	1.346
895	1086.6	829.7	8.9826	5.4067	3885	66.12	1.118	0.831	1.345
900	1092.2	833.8	8.9889	5.4114	3971	65.07	1.119	0.832	1.345
905	1097.8	838.0	8.9951	5.4160	4057	64.03	1.120	0.833	1.345
910	1103.4	842.2	9.0012	5.4206	4146	63.01	1.121	0.834	1.344
915	1109.0	846.4	9.0074	5.4252	4235	62.02	1.122	0.835	1.344
920	1114.6	850.5	9.0135	5.4297	4327	61.04	1.123	0.836	1.343
925	1120.2	854.7	9.0196	5.4342	4419	60.08	1.124	0.837	1.343
930	1125.9	858.9	9.0256	5.4388	4514	59.15	1.125	0.838	1.342
935	1131.5	863.1	9.0317	5.4433	4610	58.23	1.126	0.839	1.342
940	1137.1	867.3	9.0377	5.4477	4707	57.33	1.127	0.840	1.342
945	1142.8	871.5	9.0437	5.4522	4806	56.44	1.128	0.841	1.341
950	1148.4	875.7	9.0496	5.4566	4907	55.58	1.129	0.842	1.341
955	1154.1	879.9	9.0556	5.4611	5009	54.73	1.130	0.843	1.340
960	1159.7	884.1	9.0615	5.4655	5114	53.89	1.131	0.844	1.340
965	1165.4	888.4	9.0673	5.4698	5219	53.07	1.132	0.845	1.340
970	1171.0	892.6	9.0732	5.4742	5327	52.27	1.133	0.846	1.339
975	1176.7	896.8	9.0790	5.4786	5436	51.49	1.134	0.847	1.339
980	1182.4	901.1	9.0848	5.4829	5547	50.71	1.135	0.848	1.338
985	1188.1	905.3	9.0906	5.4872	5660	49.96	1.136	0.849	1.338
990	1193.7	909.5	9.0964	5.4915	5775	49.21	1.137	0.850	1.338
995	1199.4	913.8	9.1021	5.4958	5891	48.49	1.138	0.851	1.337
1000	1205.1	918.1	9.1078	5.5001	6009	47.77	1.139	0.852	1.337
1005	1210.8	922.3	9.1135	5.5043	6129	47.07	1.140	0.853	1.337
1010	1216.5	926.6	9.1191	5.5086	6252	46.38	1.141	0.854	1.336
1015	1222.2	930.9	9.1248	5.5128	6375	45.70	1.142	0.855	1.336
1020	1227.9	935.1	9.1304	5.5170	6501	45.04	1.143	0.856	1.336

Table A.2 (cont.)

T / K	h / kJ/kg	u / kJ/kg	ϕ / kJ/(kg · K)	ψ / kJ/(kg · K)	P_r	v_r	c_P / kJ/(kg · K)	c_v / kJ/(kg · K)	γ
1025	1233.7	939.4	9.1360	5.5212	6629	44.39	1.144	0.856	1.335
1030	1239.4	943.7	9.1416	5.5253	6759	43.75	1.144	0.857	1.335
1035	1245.1	948.0	9.1471	5.5295	6891	43.12	1.145	0.858	1.334
1040	1250.8	952.3	9.1526	5.5336	7025	42.50	1.146	0.859	1.334
1045	1256.6	956.6	9.1581	5.5378	7161	41.89	1.147	0.860	1.334
1050	1262.3	960.9	9.1636	5.5419	7298	41.30	1.148	0.861	1.333
1055	1268.0	965.2	9.1691	5.5460	7439	40.71	1.149	0.862	1.333
1060	1273.8	969.5	9.1745	5.5500	7581	40.14	1.150	0.863	1.333
1065	1279.5	973.8	9.1799	5.5541	7725	39.58	1.151	0.864	1.332
1070	1285.3	978.1	9.1853	5.5581	7871	39.02	1.152	0.864	1.332
1075	1291.1	982.5	9.1907	5.5622	8020	38.48	1.152	0.865	1.332
1080	1296.8	986.8	9.1960	5.5662	8171	37.94	1.153	0.866	1.331
1085	1302.6	991.1	9.2013	5.5702	8324	37.42	1.154	0.867	1.331
1090	1308.4	995.5	9.2066	5.5742	8479	36.90	1.155	0.868	1.331
1095	1314.1	999.8	9.2119	5.5782	8637	36.39	1.156	0.869	1.330
1100	1319.9	1004.1	9.2172	5.5821	8797	35.90	1.157	0.870	1.330
1105	1325.7	1008.5	9.2224	5.5861	8959	35.41	1.157	0.870	1.330
1110	1331.5	1012.8	9.2277	5.5900	9124	34.92	1.158	0.871	1.330
1115	1337.3	1017.2	9.2329	5.5939	9291	34.45	1.159	0.872	1.329
1120	1343.1	1021.6	9.2381	5.5978	9460	33.99	1.160	0.873	1.329
1125	1348.9	1025.9	9.2432	5.6017	9632	33.53	1.161	0.874	1.329
1130	1354.7	1030.3	9.2484	5.6056	9806	33.08	1.161	0.874	1.328
1135	1360.5	1034.7	9.2535	5.6094	9983	32.64	1.162	0.875	1.328
1140	1366.3	1039.1	9.2586	5.6133	10163	32.20	1.163	0.876	1.328
1145	1372.1	1043.4	9.2637	5.6171	10344	31.77	1.164	0.877	1.327
1150	1378.0	1047.8	9.2688	5.6209	10529	31.35	1.165	0.878	1.327
1155	1383.8	1052.2	9.2738	5.6248	10716	30.94	1.165	0.878	1.327
1160	1389.6	1056.6	9.2789	5.6286	10905	30.53	1.166	0.879	1.327
1165	1395.4	1061.0	9.2839	5.6323	11098	30.14	1.167	0.880	1.326
1170	1401.3	1065.4	9.2889	5.6361	11293	29.74	1.168	0.881	1.326
1175	1407.1	1069.8	9.2939	5.6399	11490	29.36	1.168	0.881	1.326
1180	1413.0	1074.2	9.2988	5.6436	11691	28.97	1.169	0.882	1.325
1185	1418.8	1078.6	9.3038	5.6473	11894	28.60	1.170	0.883	1.325
1190	1424.7	1083.0	9.3087	5.6511	12100	28.23	1.171	0.884	1.325
1195	1430.5	1087.5	9.3136	5.6548	12308	27.87	1.171	0.884	1.325
1200	1436.4	1091.9	9.3185	5.6585	12520	27.51	1.172	0.885	1.324

Table A.2 (cont.)

T / K	h / kJ/kg	u / kJ/kg	ϕ / kJ/(kg · K)	ψ / kJ/(kg · K)	P_r	v_r	c_P / kJ/(kg · K)	c_v / kJ/(kg · K)	γ
1205	1442.2	1096.3	9.3234	5.6621	12734	27.16	1.173	0.886	1.324
1210	1448.1	1100.7	9.3282	5.6658	12952	26.82	1.174	0.886	1.324
1215	1454.0	1105.2	9.3331	5.6695	13172	26.48	1.174	0.887	1.324
1220	1459.8	1109.6	9.3379	5.6731	13395	26.14	1.175	0.888	1.323
1225	1465.7	1114.1	9.3427	5.6767	13621	25.82	1.176	0.889	1.323
1230	1471.6	1118.5	9.3475	5.6804	13851	25.49	1.176	0.889	1.323
1235	1477.5	1123.0	9.3523	5.6840	14083	25.17	1.177	0.890	1.323
1240	1483.4	1127.4	9.3570	5.6876	14318	24.86	1.178	0.891	1.322
1245	1489.3	1131.9	9.3618	5.6912	14557	24.55	1.178	0.891	1.322
1250	1495.2	1136.3	9.3665	5.6947	14798	24.25	1.179	0.892	1.322
1255	1501.0	1140.8	9.3712	5.6983	15043	23.95	1.180	0.893	1.322
1260	1507.0	1145.2	9.3759	5.7018	15291	23.65	1.180	0.893	1.321
1265	1512.9	1149.7	9.3806	5.7054	15542	23.36	1.181	0.894	1.321
1270	1518.8	1154.2	9.3852	5.7089	15796	23.08	1.182	0.895	1.321
1275	1524.7	1158.7	9.3899	5.7124	16054	22.80	1.182	0.895	1.321
1280	1530.6	1163.1	9.3945	5.7159	16315	22.52	1.183	0.896	1.320
1285	1536.5	1167.6	9.3991	5.7194	16579	22.25	1.184	0.897	1.320
1290	1542.4	1172.1	9.4037	5.7229	16847	21.98	1.184	0.897	1.320
1295	1548.3	1176.6	9.4083	5.7264	17118	21.72	1.185	0.898	1.320
1300	1554.3	1181.1	9.4129	5.7298	17393	21.46	1.186	0.899	1.319
1305	1560.2	1185.6	9.4174	5.7333	17671	21.20	1.186	0.899	1.319
1310	1566.1	1190.1	9.4220	5.7367	17952	20.95	1.187	0.900	1.319
1315	1572.1	1194.6	9.4265	5.7402	18237	20.70	1.188	0.901	1.319
1320	1578.0	1199.1	9.4310	5.7436	18526	20.45	1.188	0.901	1.319
1325	1584.0	1203.6	9.4355	5.7470	18818	20.21	1.189	0.902	1.318
1330	1589.9	1208.1	9.4400	5.7504	19114	19.97	1.189	0.902	1.318
1335	1595.8	1212.6	9.4444	5.7538	19414	19.74	1.190	0.903	1.318
1340	1601.8	1217.1	9.4489	5.7571	19717	19.51	1.191	0.904	1.318
1345	1607.8	1221.6	9.4533	5.7605	20024	19.28	1.191	0.904	1.317
1350	1613.7	1226.2	9.4577	5.7639	20335	19.06	1.192	0.905	1.317
1355	1619.7	1230.7	9.4622	5.7672	20650	18.84	1.192	0.905	1.317
1360	1625.6	1235.2	9.4665	5.7706	20968	18.62	1.193	0.906	1.317
1365	1631.6	1239.8	9.4709	5.7739	21290	18.40	1.194	0.907	1.317
1370	1637.6	1244.3	9.4753	5.7772	21616	18.19	1.194	0.907	1.316

Table A.2 (cont.)

T / K	h / kJ/kg	u / kJ/kg	ϕ / kJ/(kg · K)	ψ / kJ/(kg · K)	P_r	v_r	c_P / kJ/(kg · K)	c_v / kJ/(kg · K)	γ
1375	1643.5	1248.8	9.4796	5.7805	21947	17.99	1.195	0.908	1.316
1380	1649.5	1253.4	9.4840	5.7838	22281	17.78	1.195	0.908	1.316
1385	1655.5	1257.9	9.4883	5.7871	22619	17.58	1.196	0.909	1.316
1390	1661.5	1262.5	9.4926	5.7904	22961	17.38	1.196	0.909	1.316
1395	1667.5	1267.0	9.4969	5.7936	23307	17.18	1.197	0.910	1.315
1400	1673.5	1271.6	9.5012	5.7969	23658	16.99	1.198	0.911	1.315
1405	1679.4	1276.1	9.5055	5.8001	24012	16.80	1.198	0.911	1.315
1410	1685.4	1280.7	9.5097	5.8034	24371	16.61	1.199	0.912	1.315
1415	1691.4	1285.2	9.5140	5.8066	24734	16.42	1.199	0.912	1.315
1420	1697.4	1289.8	9.5182	5.8098	25102	16.24	1.200	0.913	1.315
1425	1703.4	1294.4	9.5224	5.8130	25473	16.06	1.200	0.913	1.314
1430	1709.4	1298.9	9.5266	5.8162	25849	15.88	1.201	0.914	1.314
1435	1715.4	1303.5	9.5308	5.8194	26229	15.71	1.201	0.914	1.314
1440	1721.4	1308.1	9.5350	5.8226	26614	15.53	1.202	0.915	1.314
1445	1727.5	1312.6	9.5392	5.8258	27003	15.36	1.203	0.915	1.314
1450	1733.5	1317.2	9.5433	5.8289	27397	15.19	1.203	0.916	1.313
1455	1739.5	1321.8	9.5475	5.8321	27795	15.03	1.204	0.916	1.313
1460	1745.5	1326.4	9.5516	5.8352	28198	14.86	1.204	0.917	1.313
1465	1751.5	1331.0	9.5557	5.8384	28605	14.70	1.205	0.918	1.313
1470	1757.6	1335.6	9.5598	5.8415	29017	14.54	1.205	0.918	1.313
1475	1763.6	1340.2	9.5639	5.8446	29434	14.39	1.206	0.919	1.313
1480	1769.6	1344.7	9.5680	5.8477	29855	14.23	1.206	0.919	1.312
1485	1775.6	1349.3	9.5721	5.8508	30281	14.08	1.207	0.920	1.312
1490	1781.7	1353.9	9.5761	5.8539	30712	13.93	1.207	0.920	1.312
1495	1787.7	1358.5	9.5802	5.8570	31148	13.78	1.208	0.921	1.312
1500	1793.7	1363.1	9.5842	5.8601	31589	13.63	1.208	0.921	1.312

A.3 Water

Properties in tables and charts are calculated with IF97, which implements the thermodynamic model for water documented in W. Wagner and A. Kruse, *Properties of Water and Steam: The Industrial Standard IAPWS-IF97 for the Thermodynamic Properties and Supplementary Equations for Other Properties*. Springer-Verlag, 1998. Reference state: $u_0 = u^{\mathrm{sat,L}}(T_{\mathrm{triple}} = 273.16\ \mathrm{K}) = 0\ \mathrm{kJ/kg}$, $s_0 = s^{\mathrm{sat,L}}(T_{\mathrm{triple}} = 273.16\ \mathrm{K}) = 0\ \mathrm{kJ/(kg \cdot K)}$.

Table A.3 Properties of saturated water as function of temperature.

T / K	p^{sat} / MPa	v^{L} / m^3/kg	v^{V} / m^3/kg	h^{L} / kJ/kg	h^{LV} / kJ/kg	h^{V} / kJ/kg	s^{L} / kJ/(kg · K)	s^{LV} / kJ/(kg · K)	s^{V} / kJ/(kg · K)
273.16	0.0006117	0.001000	206.0	0.0	2500.9	2500.9	0.0	9.1555	9.1555
275	0.0006985	0.001000	181.6	7.8	2496.5	2504.3	0.0283	9.0783	9.1066
280	0.0009918	0.001000	130.2	28.8	2484.7	2513.5	0.1041	8.8738	8.9779
285	0.001389	0.001001	94.61	49.8	2472.8	2522.6	0.1784	8.6766	8.8550
290	0.001920	0.001001	69.63	70.7	2461.0	2531.7	0.2513	8.4862	8.7375
295	0.002621	0.001002	51.87	91.7	2449.2	2540.8	0.3228	8.3023	8.6251
300	0.003537	0.001003	39.08	112.6	2437.3	2549.9	0.3931	8.1244	8.5175
305	0.004719	0.001005	29.77	133.5	2425.4	2558.9	0.4622	7.9523	8.4145
310	0.006231	0.001007	22.91	154.4	2413.5	2567.9	0.5302	7.7856	8.3158
315	0.008145	0.001009	17.80	175.3	2401.6	2576.8	0.5970	7.6241	8.2211
320	0.01055	0.001011	13.96	196.2	2389.6	2585.7	0.6629	7.4674	8.1303
325	0.01353	0.001013	11.04	217.1	2377.5	2594.6	0.7277	7.3154	8.0431
330	0.01721	0.001015	8.806	238.0	2365.4	2603.3	0.7915	7.1678	7.9593
335	0.02172	0.001018	7.079	258.9	2353.2	2612.1	0.8544	7.0244	7.8787
340	0.02719	0.001021	5.734	279.8	2340.9	2620.7	0.9164	6.8849	7.8013
345	0.03378	0.001024	4.678	300.8	2328.5	2629.3	0.9775	6.7492	7.7267
350	0.04168	0.001027	3.842	321.7	2316.0	2637.7	1.0378	6.6171	7.6549
355	0.05108	0.001030	3.176	342.7	2303.4	2646.1	1.0973	6.4884	7.5857
360	0.06219	0.001034	2.642	363.7	2290.7	2654.4	1.1560	6.3629	7.5189
365	0.07526	0.001037	2.210	384.8	2277.8	2662.5	1.2140	6.2405	7.4545
370	0.09054	0.001041	1.859	405.8	2264.8	2670.6	1.2713	6.1209	7.3922
375	0.1083	0.001045	1.572	426.9	2251.6	2678.5	1.3279	6.0042	7.3320
380	0.1289	0.001049	1.336	448.0	2238.2	2686.3	1.3838	5.8900	7.2738
385	0.1525	0.001053	1.141	469.2	2224.7	2693.9	1.4390	5.7783	7.2174
390	0.1796	0.001058	0.9794	490.4	2210.9	2701.3	1.4937	5.6690	7.1627
395	0.2106	0.001062	0.8440	511.7	2197.0	2708.6	1.5477	5.5619	7.1096
400	0.2458	0.001067	0.7303	532.9	2182.8	2715.7	1.6012	5.4569	7.0581
405	0.2856	0.001072	0.6345	554.3	2168.3	2722.6	1.6542	5.3539	7.0080

Table A.3 (cont.)

T / K	P^{sat} / MPa	v^L / m^3/kg	v^V / m^3/kg	h^L / kJ/kg	h^{LV} / kJ/kg	h^V / kJ/kg	s^L / kJ/(kg · K)	s^{LV} / kJ/(kg · K)	s^V / kJ/(kg · K)
410	0.3304	0.001077	0.5533	575.7	2153.6	2729.3	1.7065	5.2528	6.9593
415	0.3808	0.001082	0.4842	597.1	2138.7	2735.8	1.7584	5.1534	6.9119
420	0.4372	0.001087	0.4253	618.7	2123.4	2742.1	1.8098	5.0558	6.8656
425	0.5002	0.001093	0.3747	640.2	2107.9	2748.1	1.8607	4.9597	6.8205
430	0.5702	0.001098	0.3311	661.9	2092.0	2753.9	1.9112	4.8652	6.7764
435	0.6478	0.001104	0.2935	683.6	2075.8	2759.4	1.9612	4.7720	6.7333
440	0.7335	0.001110	0.2609	705.4	2059.3	2764.7	2.0109	4.6802	6.6911
445	0.8281	0.001117	0.2326	727.3	2042.4	2769.7	2.0601	4.5897	6.6498
450	0.9320	0.001123	0.2078	749.3	2025.1	2774.4	2.1089	4.5003	6.6092
455	1.046	0.001130	0.1862	771.4	2007.4	2778.8	2.1574	4.4120	6.5694
460	1.171	0.001137	0.1672	793.5	1989.4	2782.9	2.2056	4.3247	6.5303
465	1.307	0.001144	0.1504	815.8	1970.8	2786.7	2.2534	4.2384	6.4918
470	1.455	0.001152	0.1356	838.2	1951.9	2790.1	2.3010	4.1529	6.4539
475	1.616	0.001159	0.1226	860.7	1932.4	2793.2	2.3483	4.0683	6.4165
480	1.790	0.001167	0.1110	883.4	1912.5	2795.9	2.3953	3.9843	6.3796
490	2.183	0.001185	0.09140	929.1	1871.0	2800.1	2.4886	3.8183	6.3069
500	2.639	0.001203	0.07577	975.5	1827.1	2802.6	2.5811	3.6543	6.2354
510	3.165	0.001223	0.06317	1022.5	1780.7	2803.3	2.6731	3.4916	6.1647
520	3.769	0.001245	0.05291	1070.4	1731.5	2801.9	2.7646	3.3298	6.0944
530	4.457	0.001268	0.04451	1119.2	1679.1	2798.3	2.8559	3.1680	6.0240
540	5.237	0.001294	0.03756	1169.1	1623.1	2792.2	2.9473	3.0056	5.9530
550	6.117	0.001323	0.03177	1220.3	1563.0	2783.3	3.0391	2.8418	5.8809
560	7.106	0.001355	0.02692	1272.9	1498.3	2771.2	3.1315	2.6755	5.8070
570	8.213	0.001392	0.02282	1327.2	1428.2	2755.4	3.2250	2.5056	5.7305
580	9.448	0.001433	0.01933	1383.7	1351.6	2735.3	3.3201	2.3303	5.6505
590	10.82	0.001482	0.01633	1442.8	1267.1	2709.9	3.4177	2.1477	5.5654
600	12.34	0.001540	0.01373	1505.2	1172.8	2678.0	3.5188	1.9546	5.4734
610	14.03	0.001611	0.01145	1572.1	1065.1	2637.3	3.6250	1.7461	5.3711
620	15.90	0.001704	0.009407	1645.7	938.2	2583.9	3.7396	1.5133	5.2528
630	17.97	0.001837	0.007525	1730.7	780.1	2510.8	3.8697	1.2382	5.1079
640	20.27	0.002076	0.005637	1842.0	552.4	2394.4	4.0378	0.8632	4.9010
647.10	22.064	0.003106	0.003106	2087.5	0.0	2087.5	4.4120	0.0	4.4120

Table A.4 Properties of saturated water as function of pressure.

P / MPa	T^{sat} / K	v^L / m^3/kg	v^V / m^3/kg	h^L / kJ/kg	h^{LV} / kJ/kg	h^V / kJ/kg	s^L / kJ/(kg · K)	s^{LV} / kJ/(kg · K)	s^V / kJ/(kg · K)
0.00080	276.91	0.001000	159.6	15.8	2492.0	2507.8	0.0575	8.9992	9.0567
0.0010	280.12	0.001000	129.2	29.3	2484.4	2513.7	0.1059	8.8690	8.9749
0.0012	282.80	0.001000	108.7	40.6	2478.0	2518.6	0.1460	8.7624	8.9083
0.0014	285.12	0.001001	93.90	50.3	2472.5	2522.8	0.1802	8.6720	8.8521
0.0016	287.16	0.001001	82.75	58.8	2467.7	2526.6	0.2101	8.5935	8.8036
0.0018	288.99	0.001001	74.01	66.5	2463.4	2529.9	0.2366	8.5242	8.7609
0.0020	290.65	0.001001	66.99	73.4	2459.5	2532.9	0.2606	8.4621	8.7227
0.0025	294.23	0.001002	54.24	88.4	2451.0	2539.4	0.3119	8.3303	8.6422
0.0030	297.23	0.001003	45.66	101.0	2443.9	2544.9	0.3543	8.2222	8.5766
0.0040	302.11	0.001004	34.79	121.4	2432.3	2553.7	0.4224	8.0510	8.4735
0.0050	306.03	0.001005	28.19	137.8	2423.0	2560.8	0.4763	7.9177	8.3939
0.0060	309.31	0.001006	23.73	151.5	2415.2	2566.7	0.5209	7.8083	8.3291
0.0080	314.66	0.001008	18.10	173.9	2402.4	2576.2	0.5925	7.6349	8.2274
0.010	318.96	0.001010	14.67	191.8	2392.1	2583.9	0.6492	7.4997	8.1489
0.012	322.57	0.001012	12.36	206.9	2383.4	2590.3	0.6963	7.3887	8.0850
0.014	325.70	0.001013	10.69	220.0	2375.8	2595.8	0.7366	7.2945	8.0312
0.016	328.46	0.001015	9.431	231.6	2369.1	2600.7	0.7720	7.2127	7.9847
0.018	330.95	0.001016	8.443	241.9	2363.1	2605.0	0.8035	7.1403	7.9437
0.020	333.21	0.001017	7.648	251.4	2357.5	2608.9	0.8320	7.0753	7.9072
0.025	338.11	0.001020	6.203	271.9	2345.5	2617.4	0.8931	6.9371	7.8302
0.030	342.25	0.001022	5.229	289.2	2335.3	2624.6	0.9439	6.8235	7.7675
0.040	349.01	0.001026	3.993	317.6	2318.5	2636.1	1.0259	6.6431	7.6690
0.050	354.47	0.001030	3.240	340.5	2304.7	2645.2	1.0910	6.5020	7.5930
0.060	359.08	0.001033	2.732	359.8	2293.0	2652.9	1.1452	6.3859	7.5311
0.080	366.64	0.001038	2.087	391.6	2273.5	2665.2	1.2328	6.2011	7.4339
0.10	372.76	0.001043	1.694	417.4	2257.5	2674.9	1.3026	6.0562	7.3588
0.101325	373.12	0.001043	1.673	419.0	2256.5	2675.5	1.3067	6.0477	7.3544
0.12	377.93	0.001047	1.428	439.3	2243.8	2683.1	1.3608	5.9369	7.2976
0.14	382.44	0.001051	1.237	458.4	2231.6	2690.0	1.4109	5.8352	7.2460
0.16	386.45	0.001054	1.091	475.3	2220.7	2696.0	1.4549	5.7464	7.2014
0.18	390.06	0.001058	0.9775	490.7	2210.7	2701.4	1.4944	5.6677	7.1620
0.20	393.36	0.001061	0.8857	504.7	2201.6	2706.2	1.5301	5.5968	7.1269
0.25	400.56	0.001067	0.7187	535.4	2181.2	2716.5	1.6072	5.4452	7.0524
0.30	406.68	0.001073	0.6058	561.5	2163.4	2724.9	1.6718	5.3198	6.9916

Table A.4 (cont.)

P / MPa	T^{sat} / K	v^L / m^3/kg	v^V / m^3/kg	h^L / kJ/kg	h^{LV} / kJ/kg	h^V / kJ/kg	s^L / kJ/(kg · K)	s^{LV} / kJ/(kg · K)	s^V / kJ/(kg · K)
0.40	416.76	0.001084	0.4624	604.7	2133.3	2738.1	1.7766	5.1188	6.8954
0.50	424.99	0.001093	0.3748	640.2	2107.9	2748.1	1.8606	4.9600	6.8206
0.60	431.98	0.001101	0.3156	670.5	2085.6	2756.1	1.9311	4.8281	6.7592
0.80	443.56	0.001115	0.2403	721.0	2047.3	2768.3	2.0460	4.6156	6.6615
1.0	453.04	0.001127	0.1943	762.7	2014.4	2777.1	2.1384	4.4465	6.5850
1.2	461.11	0.001139	0.1632	798.5	1985.3	2783.8	2.2163	4.3054	6.5217
1.4	468.20	0.001149	0.1408	830.1	1958.8	2788.9	2.2839	4.1836	6.4675
1.6	474.53	0.001159	0.1237	858.6	1934.3	2792.9	2.3438	4.0762	6.4200
1.8	480.27	0.001168	0.1104	884.6	1911.4	2796.0	2.3978	3.9798	6.3776
2.0	485.53	0.001177	0.09958	908.6	1889.8	2798.4	2.4470	3.8921	6.3392
2.5	497.11	0.001197	0.07995	962.0	1840.1	2802.0	2.5544	3.7015	6.2560
3.0	507.01	0.001217	0.06666	1008.4	1794.9	2803.3	2.6456	3.5402	6.1858
4.0	523.51	0.001253	0.04978	1087.4	1713.5	2800.9	2.7967	3.2731	6.0697
5.0	537.09	0.001286	0.03945	1154.5	1639.7	2794.2	2.9207	3.0530	5.9737
6.0	548.74	0.001319	0.03245	1213.7	1570.8	2784.6	3.0274	2.8626	5.8901
7.0	558.98	0.001352	0.02738	1267.4	1505.1	2772.6	3.1220	2.6926	5.8146
8.0	568.16	0.001385	0.02353	1317.1	1441.5	2758.6	3.2077	2.5372	5.7448
10	584.15	0.001453	0.01803	1407.9	1317.6	2725.5	3.3603	2.2556	5.6159
11	591.23	0.001489	0.01599	1450.3	1256.1	2706.4	3.4300	2.1246	5.5545
12	597.83	0.001526	0.01427	1491.3	1194.3	2685.6	3.4965	1.9977	5.4941
13	604.01	0.001566	0.01279	1531.4	1131.5	2662.9	3.5606	1.8733	5.4339
14	609.82	0.001610	0.01149	1570.9	1067.2	2638.1	3.6230	1.7500	5.3730
16	620.51	0.001710	0.009308	1649.7	931.1	2580.8	3.7457	1.5006	5.2463
18	630.14	0.001839	0.007499	1732.0	777.5	2509.5	3.8717	1.2339	5.1055
20	638.90	0.002039	0.005858	1827.1	584.3	2411.4	4.0154	0.9145	4.9299
22.064	647.10	0.003106	0.003106	2087.5	0.0	2087.5	4.4120	0.0000	4.4120

Table A.5 Properties of gaseous water.

P/MPa (T^{sat}/K)		T/K sat	350	400	450	500	550	600	650	700
0.0010	v / m^3/kg	129.2	161.5	184.6	207.7	230.8	253.8	276.9	300.0	323.1
280.1	h / kJ/kg	2513.7	2644.9	2739.5	2835.2	2932.3	3030.8	3130.9	3232.5	3335.7
	s / kJ/(kg · K)	8.9749	9.3930	9.6456	9.8712	10.0757	10.2635	10.4376	10.6002	10.7532
	u / kJ/kg	2384.5	2483.4	2554.9	2627.5	2701.5	2777.0	2854.0	2932.5	3012.7
0.0020	v / m^3/kg	66.99	80.73	92.29	103.8	115.4	126.9	138.5	150.0	161.5
290.6	h / kJ/kg	2532.9	2644.7	2739.4	2835.2	2932.3	3030.8	3130.8	3232.5	3335.7
	s / kJ/(kg · K)	8.7227	9.0727	9.3256	9.5512	9.7557	9.9436	10.1176	10.2803	10.4333
	u / kJ/kg	2398.9	2483.2	2554.8	2627.5	2701.5	2777.0	2853.9	2932.5	3012.6
0.0040	v / m^3/kg	34.79	40.35	46.13	51.91	57.68	63.45	69.22	74.99	80.76
302.1	h / kJ/kg	2553.7	2644.3	2739.2	2835.1	2932.2	3030.7	3130.8	3232.4	3335.7
	s / kJ/(kg · K)	8.4735	8.7520	9.0053	9.2311	9.4357	9.6236	9.7977	9.9604	10.1134
	u / kJ/kg	2414.5	2482.9	2554.7	2627.4	2701.5	2776.9	2853.9	2932.5	3012.6
0.0070	v / m^3/kg	20.53	23.04	26.35	29.66	32.96	36.26	39.55	42.85	46.15
312.2	h / kJ/kg	2571.8	2643.8	2738.9	2834.9	2932.1	3030.7	3130.7	3232.4	3335.6
	s / kJ/(kg · K)	8.2746	8.4926	8.7465	8.9725	9.1773	9.3652	9.5393	9.7020	9.8550
	u / kJ/kg	2428.1	2482.5	2554.5	2627.3	2701.4	2776.9	2853.9	2932.4	3012.6
0.010	v / m^3/kg	14.67	16.12	18.44	20.76	23.07	25.38	27.69	29.99	32.30
319.0	h / kJ/kg	2583.9	2643.3	2738.7	2834.7	2932.0	3030.6	3130.7	3232.3	3335.6
	s / kJ/(kg · K)	8.1489	8.3268	8.5814	8.8076	9.0125	9.2005	9.3746	9.5373	9.6904
	u / kJ/kg	2437.2	2482.1	2554.3	2627.2	2701.3	2776.8	2853.8	2932.4	3012.6
0.020	v / m^3/kg	7.648	8.044	9.211	10.37	11.53	12.68	13.84	15.00	16.15
333.2	h / kJ/kg	2608.9	2641.6	2737.8	2834.2	2931.6	3030.3	3130.5	3232.2	3335.5
	s / kJ/(kg · K)	7.9072	8.0029	8.2597	8.4868	8.6920	8.8802	9.0545	9.2172	9.3703
	u / kJ/kg	2456.0	2480.7	2553.6	2626.8	2701.0	2776.6	2853.7	2932.3	3012.5
0.040	v / m^3/kg	3.993	4.005	4.596	5.179	5.760	6.339	6.917	7.495	8.073
349.0	h / kJ/kg	2636.1	2638.0	2736.0	2833.1	2930.8	3029.8	3130.0	3231.8	3335.2
	s / kJ/(kg · K)	7.6690	7.6747	7.9363	8.1650	8.3710	8.5595	8.7340	8.8970	9.0501
	u / kJ/kg	2476.3	2477.8	2552.1	2625.9	2700.5	2776.2	2853.3	2932.0	3012.3

Table A.5 (cont.)

P/MPa (T^{sat}/K)		T/K 750	800	850	900	950	1000	1050	1100	1150
0.0010	v / m³/kg	346.1	369.2	392.3	415.4	438.4	461.5	484.6	507.7	530.8
280.1	h / kJ/kg	3440.6	3547.2	3655.6	3765.7	3877.6	3991.2	4106.6	4223.9	4342.8
	s / kJ/(kg · K)	10.8980	11.0356	11.1669	11.2928	11.4137	11.5303	11.6429	11.7520	11.8577
	u / kJ/kg	3094.5	3178.0	3263.3	3350.3	3439.1	3529.7	3622.0	3716.2	3812.0
0.0020	v / m³/kg	173.1	184.6	196.1	207.7	219.2	230.8	242.3	253.8	265.4
290.6	h / kJ/kg	3440.6	3547.2	3655.6	3765.7	3877.6	3991.2	4106.6	4223.8	4342.8
	s / kJ/(kg · K)	10.5780	10.7156	10.8470	10.9729	11.0938	11.2104	11.3230	11.4321	11.5378
	u / kJ/kg	3094.5	3178.0	3263.3	3350.3	3439.1	3529.7	3622.0	3716.2	3812.0
0.0040	v / m³/kg	86.53	92.30	98.07	103.84	109.61	115.38	121.15	126.92	132.69
302.1	h / kJ/kg	3440.6	3547.2	3655.6	3765.7	3877.5	3991.2	4106.6	4223.8	4342.8
	s / kJ/(kg · K)	10.2581	10.3957	10.5271	10.6529	10.7739	10.8905	11.0031	11.1122	11.2179
	u / kJ/kg	3094.5	3178.0	3263.3	3350.3	3439.1	3529.7	3622.0	3716.2	3812.0
0.0070	v / m³/kg	49.45	52.74	56.04	59.34	62.63	65.93	69.23	72.52	75.82
312.2	h / kJ/kg	3440.6	3547.2	3655.5	3765.6	3877.5	3991.2	4106.6	4223.8	4342.8
	s / kJ/(kg · K)	9.9998	10.1374	10.2688	10.3947	10.5156	10.6322	10.7448	10.8539	10.9596
	u / kJ/kg	3094.4	3178.0	3263.2	3350.3	3439.1	3529.7	3622.0	3716.2	3812.0
0.010	v / m³/kg	34.61	36.92	39.23	41.54	43.84	46.15	48.46	50.77	53.07
319.0	h / kJ/kg	3440.5	3547.1	3655.5	3765.6	3877.5	3991.2	4106.6	4223.8	4342.7
	s / kJ/(kg · K)	9.8352	9.9728	10.1041	10.2300	10.3510	10.4676	10.5802	10.6893	10.7950
	u / kJ/kg	3094.4	3177.9	3263.2	3350.3	3439.1	3529.6	3622.0	3716.1	3812.0
0.020	v / m³/kg	17.30	18.46	19.61	20.77	21.92	23.08	24.23	25.38	26.54
333.2	h / kJ/kg	3440.4	3547.0	3655.4	3765.6	3877.4	3991.1	4106.5	4223.8	4342.7
	s / kJ/(kg · K)	9.5151	9.6528	9.7842	9.9101	10.0310	10.1476	10.2603	10.3693	10.4751
	u / kJ/kg	3094.3	3177.9	3263.2	3350.2	3439.0	3529.6	3622.0	3716.1	3812.0
0.040	v / m³/kg	8.651	9.228	9.805	10.38	10.96	11.54	12.11	12.69	13.27
349.0	h / kJ/kg	3440.2	3546.9	3655.3	3765.4	3877.3	3991.0	4106.4	4223.7	4342.6
	s / kJ/(kg · K)	9.1950	9.3327	9.4641	9.5900	9.7110	9.8276	9.9403	10.0494	10.1551
	u / kJ/kg	3094.1	3177.7	3263.0	3350.1	3438.9	3529.5	3621.9	3716.0	3811.9

Table A.5 (cont.)

P / MPa (T^{sat} / K)		sat	400	450	500	550	600	650	700	750
0.070	v / m^3/kg	2.365	2.618	2.954	3.287	3.619	3.950	4.281	4.612	4.942
363.1	h / kJ/kg	2659.4	2733.2	2831.4	2929.7	3028.9	3129.4	3231.3	3334.8	3439.8
	s / kJ/(kg · K)	7.4790	7.6726	7.9039	8.1111	8.3001	8.4750	8.6381	8.7914	8.9364
	u / kJ/kg	2493.9	2550.0	2624.7	2699.6	2775.6	2852.9	2931.6	3011.9	3093.9
0.101325	v / m^3/kg	1.673	1.802	2.037	2.268	2.498	2.727	2.956	3.185	3.413
373.1	h / kJ/kg	2675.5	2730.3	2829.7	2928.5	3028.1	3128.7	3230.8	3334.3	3439.5
	s / kJ/(kg · K)	7.3544	7.4961	7.7303	7.9386	8.1283	8.3035	8.4668	8.6203	8.7653
	u / kJ/kg	2506.0	2547.7	2623.3	2698.7	2774.9	2852.4	2931.2	3011.6	3093.6
0.20	v / m^3/kg	0.8857	0.9025	1.025	1.144	1.262	1.379	1.495	1.612	1.728
393.4	h / kJ/kg	2706.2	2720.5	2824.0	2924.8	3025.3	3126.6	3229.1	3332.9	3438.3
	s / kJ/(kg · K)	7.1269	7.1629	7.4068	7.6191	7.8107	7.9870	8.1510	8.3050	8.4503
	u / kJ/kg	2529.1	2540.0	2619.0	2695.9	2772.9	2850.8	2930.0	3010.6	3092.8
0.40	v / m^3/kg	0.4624		0.5054	0.5672	0.6274	0.6867	0.7455	0.8040	0.8624
416.8	h / kJ/kg	2738.1		2811.9	2916.9	3019.6	3122.3	3225.6	3330.1	3436.0
	s / kJ/(kg · K)	6.8954		7.0659	7.2872	7.4832	7.6618	7.8272	7.9821	8.1281
	u / kJ/kg	2553.1		2609.7	2690.0	2768.7	2847.6	2927.4	3008.5	3091.0
0.70	v / m^3/kg	0.2728		0.2823	0.3198	0.3553	0.3899	0.4241	0.4579	0.4915
438.1	h / kJ/kg	2762.7		2791.9	2904.4	3010.9	3115.7	3220.4	3325.9	3432.5
	s / kJ/(kg · K)	6.7070		6.7727	7.0100	7.2130	7.3953	7.5630	7.7193	7.8664
	u / kJ/kg	2571.8		2594.3	2680.6	2762.2	2842.7	2923.6	3005.4	3088.4
1.0	v / m^3/kg	0.1943			0.2206	0.2464	0.2712	0.2955	0.3194	0.3432
453.0	h / kJ/kg	2777.1			2891.3	3001.9	3109.0	3215.2	3321.6	3428.9
	s / kJ/(kg · K)	6.5850			6.8251	7.0360	7.2224	7.3924	7.5502	7.6982
	u / kJ/kg	2582.8			2670.6	2755.5	2837.8	2919.7	3002.2	3085.8
2.0	v / m^3/kg	0.09958			0.1044	0.1192	0.1326	0.1454	0.1579	0.1701
485.5	h / kJ/kg	2798.4			2841.4	2969.7	3085.7	3197.2	3307.2	3417.1
	s / kJ/(kg · K)	6.3392			6.4265	6.6713	6.8733	7.0518	7.2149	7.3665
	u / kJ/kg	2599.2			2632.6	2731.3	2820.5	2906.4	2991.5	3076.9

Table A.5 (cont.)

P / MPa (T^{sat} / K)		800	850	900	950	1000	1050	1100	1150	1200
0.070	v / m^3/kg	5.272	5.602	5.932	6.262	6.592	6.922	7.252	7.582	7.911
363.1	h / kJ/kg	3546.6	3655.0	3765.2	3877.1	3990.8	4106.3	4223.6	4342.5	4463.2
	s / kJ/(kg · K)	9.0741	9.2056	9.3316	9.4526	9.5692	9.6819	9.7910	9.8967	9.9994
	u / kJ/kg	3177.5	3262.9	3349.9	3438.8	3529.4	3621.8	3715.9	3811.8	3909.4
0.101325	v / m^3/kg	3.641	3.870	4.098	4.326	4.554	4.782	5.010	5.237	5.465
373.1	h / kJ/kg	3546.3	3654.7	3765.0	3876.9	3990.7	4106.1	4223.4	4342.4	4463.0
	s / kJ/(kg · K)	8.9032	9.0347	9.1607	9.2818	9.3984	9.5111	9.6202	9.7260	9.8287
	u / kJ/kg	3177.3	3262.7	3349.8	3438.6	3529.3	3621.6	3715.8	3811.7	3909.3
0.20	v / m^3/kg	1.844	1.959	2.075	2.191	2.306	2.422	2.538	2.653	2.769
393.4	h / kJ/kg	3545.3	3653.9	3764.3	3876.3	3990.1	4105.7	4223.0	4342.0	4462.7
	s / kJ/(kg · K)	8.5884	8.7201	8.8463	8.9674	9.0842	9.1969	9.3061	9.4119	9.5146
	u / kJ/kg	3176.6	3262.0	3349.2	3438.2	3528.8	3621.3	3715.5	3811.4	3909.0
0.40	v / m^3/kg	0.921	0.979	1.037	1.095	1.153	1.211	1.268	1.326	1.384
416.8	h / kJ/kg	3543.3	3652.2	3762.8	3875.1	3989.0	4104.7	4222.2	4341.3	4462.0
	s / kJ/(kg · K)	8.2667	8.3987	8.5251	8.6465	8.7634	8.8763	8.9855	9.0914	9.1942
	u / kJ/kg	3175.1	3260.8	3348.1	3437.2	3528.0	3620.5	3714.8	3810.8	3908.4
0.70	v / m^3/kg	0.5250	0.5584	0.5917	0.6249	0.6581	0.6913	0.7244	0.7576	0.7906
438.1	h / kJ/kg	3540.4	3649.7	3760.7	3873.2	3987.4	4103.3	4220.9	4340.1	4461.0
	s / kJ/(kg · K)	8.0056	8.1382	8.2650	8.3867	8.5039	8.6169	8.7263	8.8323	8.9352
	u / kJ/kg	3172.9	3258.9	3346.5	3435.8	3526.7	3619.4	3713.8	3809.8	3907.5
1.0	v / m^3/kg	0.3668	0.3903	0.4137	0.4370	0.4603	0.4836	0.5069	0.5301	0.5533
453.0	h / kJ/kg	3537.4	3647.2	3758.5	3871.3	3985.8	4101.8	4219.6	4339.0	4460.0
	s / kJ/(kg · K)	7.8382	7.9714	8.0986	8.2206	8.3380	8.4512	8.5608	8.6669	8.7699
	u / kJ/kg	3170.6	3257.0	3344.8	3434.3	3525.4	3618.2	3712.7	3808.9	3906.7
2.0	v / m^3/kg	0.1821	0.1941	0.2060	0.2178	0.2296	0.2413	0.2530	0.2647	0.2764
485.5	h / kJ/kg	3527.5	3638.8	3751.3	3865.1	3980.3	4097.0	4215.3	4335.2	4456.6
	s / kJ/(kg · K)	7.5090	7.6439	7.7725	7.8956	8.0138	8.1277	8.2377	8.3442	8.4476
	u / kJ/kg	3163.2	3250.6	3339.3	3429.5	3521.2	3614.4	3709.3	3805.7	3903.8

Table A.5 (cont.)

P / MPa		T / K								
(T^{sat} / K)		sat	600	650	700	750	800	900	1000	1100
3.0	v / m^3/kg	0.06666	0.08631	0.09534	0.1040	0.1124	0.1206	0.1368	0.1527	0.1642
507.0	h / kJ/kg	2803.3	3060.9	3178.5	3292.5	3405.0	3517.4	3744.0	3974.8	4147.0
	s / kJ/(kg · K)	6.1858	6.6546	6.8431	7.0120	7.1673	7.3124	7.5792	7.8223	7.9885
	u / kJ/kg	2603.3	2801.9	2892.5	2980.5	3067.9	3155.6	3333.7	3516.8	3654.5
4.0	v / m^3/kg	0.04978	0.06306	0.07027	0.07701	0.08350	0.08982	0.1021	0.1142	0.1229
523.5	h / kJ/kg	2800.9	3034.3	3159.1	3277.3	3392.8	3507.3	3736.7	3969.3	4142.5
	s / kJ/(kg · K)	6.0697	6.4877	6.6877	6.8629	7.0222	7.1700	7.4402	7.6852	7.8523
	u / kJ/kg	2601.8	2782.0	2878.1	2969.3	3058.8	3148.0	3328.1	3512.5	3650.8
6.0	v / m^3/kg	0.03245	0.03961	0.04510	0.05001	0.05461	0.05902	0.0675	0.0757	0.0816
548.7	h / kJ/kg	2784.6	2975.2	3117.7	3245.9	3367.7	3486.6	3721.9	3958.2	4133.3
	s / kJ/(kg · K)	5.8901	6.2232	6.4517	6.6416	6.8097	6.9633	7.2404	7.4894	7.6583
	u / kJ/kg	2589.9	2737.5	2847.1	2945.8	3040.0	3132.5	3316.8	3503.8	3643.4
8.0	v / m^3/kg	0.02353	0.02761	0.03241	0.03646	0.04014	0.04361	0.05020	0.05650	0.06101
568.2	h / kJ/kg	2758.6	2905.5	3072.4	3212.6	3341.6	3465.4	3706.9	3947.0	4124.0
	s / kJ/(kg · K)	5.7448	5.9969	6.2645	6.4724	6.6505	6.8103	7.0948	7.3478	7.5186
	u / kJ/kg	2570.4	2684.6	2813.1	2920.9	3020.5	3116.5	3305.3	3495.0	3636.0
10	v / m^3/kg	0.01803	0.02009	0.02471	0.02829	0.03144	0.03436	0.03980	0.04496	0.04862
584.1	h / kJ/kg	2725.5	2819.8	3022.5	3177.3	3314.6	3443.7	3691.7	3935.8	4114.7
	s / kJ/(kg · K)	5.6159	5.7754	6.1007	6.3304	6.5199	6.6866	6.9788	7.2359	7.4087
	u / kJ/kg	2545.1	2618.9	2775.4	2894.4	3000.2	3100.1	3293.7	3486.2	3628.5
	v / m^3/kg	0.01034		0.01405	0.01726	0.01978	0.02199	0.02594	0.02957	0.03212
615.3	h / kJ/kg	2610.9		2868.6	3078.8	3242.2	3386.9	3653.0	3907.3	4091.3
	s / kJ/(kg · K)	5.3108		5.7198	6.0322	6.2579	6.4447	6.7583	7.0263	7.2039
	u / kJ/kg	2455.8		2657.8	2820.0	2945.6	3057.1	3263.9	3463.7	3609.6
20	v / m^3/kg	0.00586		0.00790	0.01158	0.01388	0.01577	0.01901	0.02188	0.02387
638.9	h / kJ/kg	2411.4		2624.9	2961.6	3162.4	3326.5	3613.1	3878.4	4067.7
	s / kJ/(kg · K)	4.9299		5.2618	5.7635	6.0410	6.2530	6.5910	6.8706	7.0534
	u / kJ/kg	2294.2		2466.8	2730.1	2884.7	3011.0	3232.9	3440.7	3590.4

Table A.5 (cont.)

P / MPa (T^{sat} / K)		T / K								
		700	750	800	850	900	950	1000	1050	1100
30	v / m^3/kg	0.00543	0.00787	0.00952	0.01088	0.01207	0.01317	0.01420	0.01519	0.01563
	h / kJ/kg	2631.5	2976.2	3194.4	3371.8	3530.2	3678.1	3819.7	3957.4	4020.2
	s / kJ/(kg · K)	5.1754	5.6529	5.9350	6.1503	6.3315	6.4914	6.6366	6.7710	6.8303
	u / kJ/kg	2468.6	2739.9	2908.7	3045.6	3168.1	3283.0	3393.6	3501.8	3551.4
40	v / m^3/kg	0.00261	0.00483	0.00640	0.00760	0.00862	0.00953	0.01038	0.01117	0.01152
	h / kJ/kg	2222.5	2754.1	3049.3	3263.9	3444.6	3607.7	3760.4	3906.6	3972.8
	s / kJ/(kg · K)	4.5379	5.2741	5.6560	5.9165	6.1232	6.2997	6.4564	6.5990	6.6614
	u / kJ/kg	2118.1	2560.8	2793.3	2959.9	3099.8	3226.4	3345.4	3459.9	3511.9
50	v / m^3/kg	0.00204	0.00323	0.00459	0.00567	0.00658	0.00737	0.00810	0.00877	0.00907
	h / kJ/kg	2075.5	2536.4	2898.8	3153.7	3358.2	3537.3	3701.5	3856.4	3926.0
	s / kJ/(kg · K)	4.2956	4.9314	5.4003	5.7097	5.9437	6.1374	6.3059	6.4571	6.5226
	u / kJ/kg	1973.7	2375.1	2669.5	2870.0	3029.3	3168.6	3296.5	3417.7	3472.3
60	v / m^3/kg	0.00183	0.00250	0.00350	0.00444	0.00525	0.00596	0.00660	0.00720	0.00746
	h / kJ/kg	2014.0	2390.0	2759.7	3046.3	3273.6	3468.3	3643.8	3807.2	3880.2
	s / kJ/(kg · K)	4.1804	4.6986	5.1764	5.5244	5.7845	5.9951	6.1752	6.3347	6.4034
	u / kJ/kg	1904.1	2239.8	2549.7	2779.6	2958.5	3110.6	3247.6	3375.5	3432.7
70	v / m^3/kg	0.00172	0.00216	0.00286	0.00364	0.00435	0.00498	0.00555	0.00608	0.00632
	h / kJ/kg	1978.2	2304.5	2648.9	2948.2	3193.5	3402.1	3588.1	3759.7	3835.8
	s / kJ/(kg · K)	4.1040	4.5539	4.9985	5.3617	5.6424	5.8681	6.0590	6.2265	6.2982
	u / kJ/kg	1858.0	2153.0	2448.5	2693.7	2889.2	3053.5	3199.3	3333.9	3393.6
80	v / m^3/kg	0.00164	0.00197	0.00248	0.00310	0.00371	0.00427	0.00479	0.00527	0.00548
	h / kJ/kg	1954.3	2251.1	2567.9	2864.7	3120.2	3340.0	3535.3	3714.2	3793.3
	s / kJ/(kg · K)	4.0459	4.4552	4.8640	5.2241	5.5164	5.7543	5.9547	6.1293	6.2039
	u / kJ/kg	1823.2	2093.2	2369.4	2617.0	2823.6	2998.3	3152.2	3293.0	3355.2
100	v / m^3/kg	0.00153	0.00176	0.00207	0.00247	0.00291	0.00335	0.00377	0.00416	0.00434
	h / kJ/kg	1924.9	2188.4	2466.7	2743.0	3000.5	3231.5	3440.4	3631.3	3715.2
	s / kJ/(kg · K)	3.9586	4.3221	4.6813	5.0163	5.3109	5.5607	5.7751	5.9614	6.0405
	u / kJ/kg	1771.5	2012.5	2259.3	2495.8	2709.5	2896.6	3063.6	3215.2	3281.6

Table A.6 Properties of liquid water.

P / MPa		T / K								
		400	425	450	475	500	525	550	575	600
	(P^{sat} / MPa)	0.2458	0.5002	0.9320	1.616	2.639	4.102	6.117	8.814	12.34
sat	ρ / kg/m^3	937.48	915.27	890.35	862.50	831.32	796.13	755.81	708.30	649.41
	h / kJ/kg	532.95	640.24	749.29	860.74	975.46	1094.70	1220.27	1355.18	1505.22
	s / kJ/(kg · K)	1.60122	1.86074	2.10895	2.34826	2.58113	2.81029	3.03906	3.27231	3.51877
	u / kJ/kg	532.68	639.70	748.25	858.87	972.29	1089.55	1212.18	1342.73	1486.21
0.50	ρ / kg/m^3	937.61								
	h / kJ/kg	533.12								
	s / kJ/(kg · K)	1.60098								
	u / kJ/kg	532.59								
0.70	ρ / kg/m^3	937.72	915.39							
	h / kJ/kg	533.26	640.37							
	s / kJ/(kg · K)	1.60079	1.86051							
	u / kJ/kg	532.51	639.60							
1.0	ρ / kg/m^3	937.87	915.56	890.39						
	h / kJ/kg	533.46	640.55	749.33						
	s / kJ/(kg · K)	1.60051	1.86017	2.10885						
	u / kJ/kg	532.40	639.46	748.21						
1.4	ρ / kg/m^3	938.08	915.79	890.65						
	h / kJ/kg	533.74	640.80	749.54						
	s / kJ/(kg · K)	1.60013	1.85972	2.10832						
	u / kJ/kg	532.25	639.27	747.96						
2.0	ρ / kg/m^3	938.38	916.13	891.05	862.80					
	h / kJ/kg	534.15	641.16	749.85	860.89					
	s / kJ/(kg · K)	1.59956	1.85904	2.10751	2.34764					
	u / kJ/kg	532.02	638.98	747.60	858.57					
3.0	ρ / kg/m^3	938.89	916.70	891.70	863.57	831.66				
	h / kJ/kg	534.84	641.78	750.37	861.29	975.54				
	s / kJ/(kg · K)	1.59861	1.85792	2.10618	2.34603	2.58042				
	u / kJ/kg	531.64	638.50	747.00	857.81	971.93				
5.0	ρ / kg/m^3	939.91	917.83	892.99	865.09	833.52	797.19			
	h / kJ/kg	536.21	643.01	751.42	862.09	975.99	1094.66			
	s / kJ/(kg · K)	1.59672	1.85569	2.10352	2.34284	2.57651	2.80805			
	u / kJ/kg	530.89	637.56	745.82	856.31	969.99	1088.38			
7.0	ρ / kg/m^3	940.91	918.95	894.27	866.59	835.35	799.53	757.21		

Table A.6 (cont.)

P / MPa		T / K								
		400	425	450	475	500	525	550	575	600
	h / kJ/kg	537.59	644.24	752.47	862.90	976.46	1094.59	1219.84		
	s / kJ/(kg · K)	1.59485	1.85347	2.10090	2.33970	2.57265	2.80317	3.03616		
	u / kJ/kg	530.15	636.63	744.64	854.82	968.08	1085.84	1210.60		
10	ρ / kg/m^3	942.41	920.62	896.17	868.81	838.03	802.92	761.81	711.05	
	h / kJ/kg	539.66	646.11	754.07	864.15	977.21	1094.59	1218.55	1353.66	
	s / kJ/(kg · K)	1.59207	1.85018	2.09701	2.33505	2.56699	2.79603	3.02663	3.26678	
	u / kJ/kg	529.05	635.24	742.91	852.64	965.28	1082.14	1205.42	1339.60	
14	ρ / kg/m^3	944.39	922.81	898.66	871.71	841.51	807.28	767.61	719.65	655.80
	h / kJ/kg	542.43	648.61	756.24	865.87	978.30	1094.74	1217.15	1349.22	1500.24
	s / kJ/(kg · K)	1.58840	1.84586	2.09191	2.32899	2.55964	2.78685	3.01457	3.24931	3.50625
	u / kJ/kg	527.61	633.44	740.66	849.81	961.67	1077.40	1198.91	1329.76	1478.89
20	ρ / kg/m^3	947.30	926.03	902.30	875.93	846.54	813.50	775.70	731.10	675.12
	h / kJ/kg	546.61	652.39	759.54	868.53	980.10	1095.26	1215.64	1344.01	1486.27
	s / kJ/(kg · K)	1.58298	1.83949	2.08444	2.32015	2.54901	2.77373	2.99769	3.22588	3.46794
	u / kJ/kg	525.50	630.80	737.37	845.70	956.47	1070.67	1189.85	1316.65	1456.64
30	ρ / kg/m^3	952.05	931.25	908.15	882.65	854.45	823.11	787.86	747.43	699.48
	h / kJ/kg	553.62	658.78	765.17	873.20	983.46	1096.78	1214.36	1338.09	1471.22
	s / kJ/(kg · K)	1.57417	1.82919	2.07241	2.30603	2.53223	2.75335	2.97211	3.19206	3.41862
	u / kJ/kg	522.10	626.57	732.14	839.21	948.35	1060.33	1176.28	1297.95	1428.33
50	ρ / kg/m^3	961.12	941.15	919.17	895.12	868.88	840.19	808.67	773.74	734.55
	h / kJ/kg	567.77	671.83	776.88	883.23	991.32	1101.70	1215.12	1332.60	1455.55
	s / kJ/(kg · K)	1.55730	1.80962	2.04979	2.27979	2.50153	2.71692	2.92795	3.13682	3.34608
	u / kJ/kg	515.75	618.70	722.48	827.38	933.77	1042.18	1153.29	1267.98	1387.48
100	ρ / kg/m^3	981.82	963.43	943.51	922.10	899.20	874.79	848.78	821.07	791.51
	h / kJ/kg	603.78	705.56	807.98	911.18	1015.39	1120.89	1227.99	1337.07	1448.46
	s / kJ/(kg · K)	1.51866	1.76548	1.99963	2.22282	2.43662	2.64250	2.84178	3.03571	3.22533
	u / kJ/kg	501.93	601.77	701.99	802.73	904.18	1006.57	1110.18	1215.27	1322.12

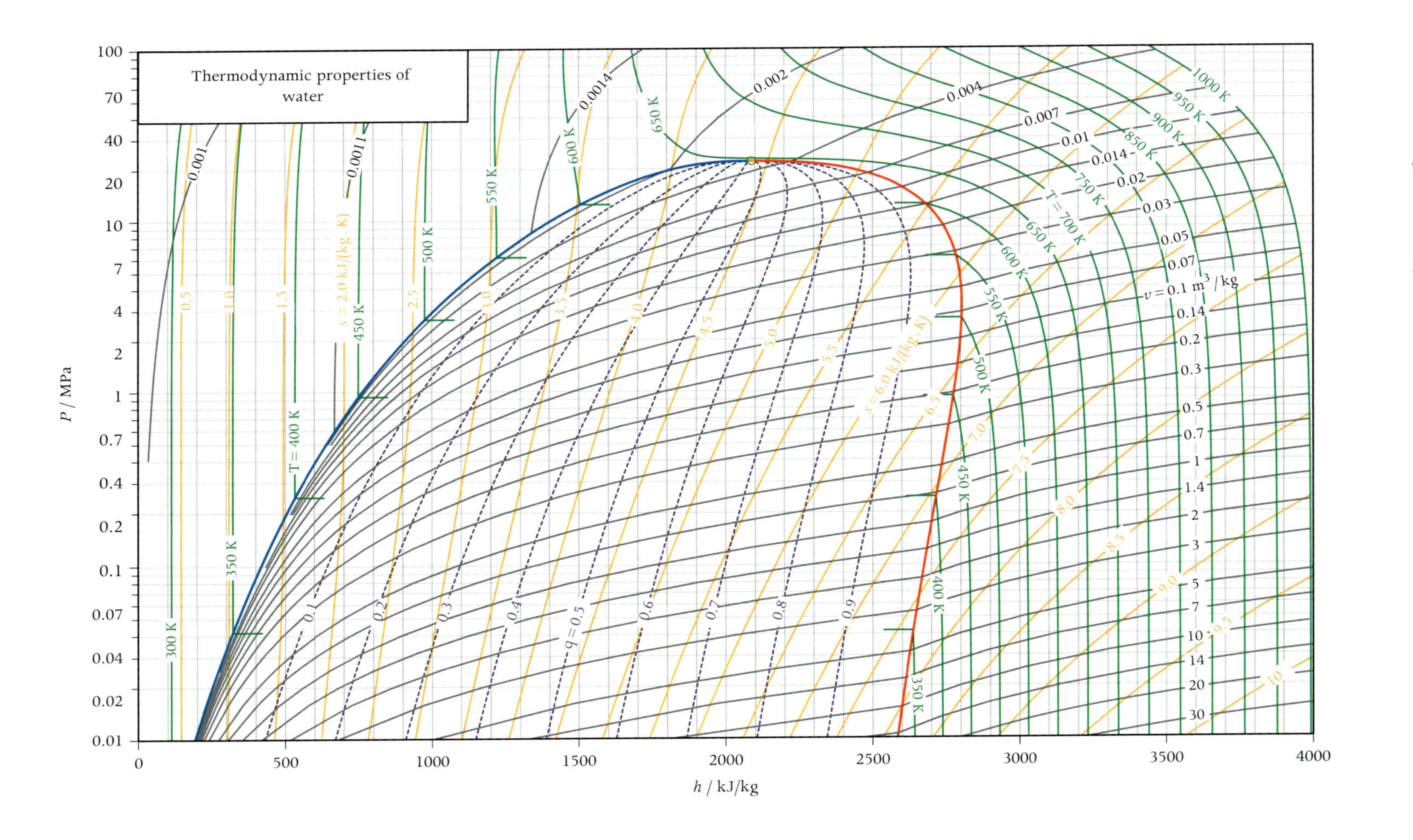
Thermodynamic properties of water
P / MPa
h / kJ/kg
100
70
40
20
10
7
4
2
1
0.7
0.4
0.2
0.1
0.07
0.04
0.02
0.01
0
500
1000
1500
2000
2500
3000
3500
4000
300 K
350 K
T = 400 K
450 K
500 K
550 K
600 K
650 K
T = 700 K
750 K
850 K
900 K
950 K
1000 K
s = 2.0 kJ/(kg · K)
s = 6.0 kJ/(kg · K)
q = 0.5
v = 0.1 m³/kg

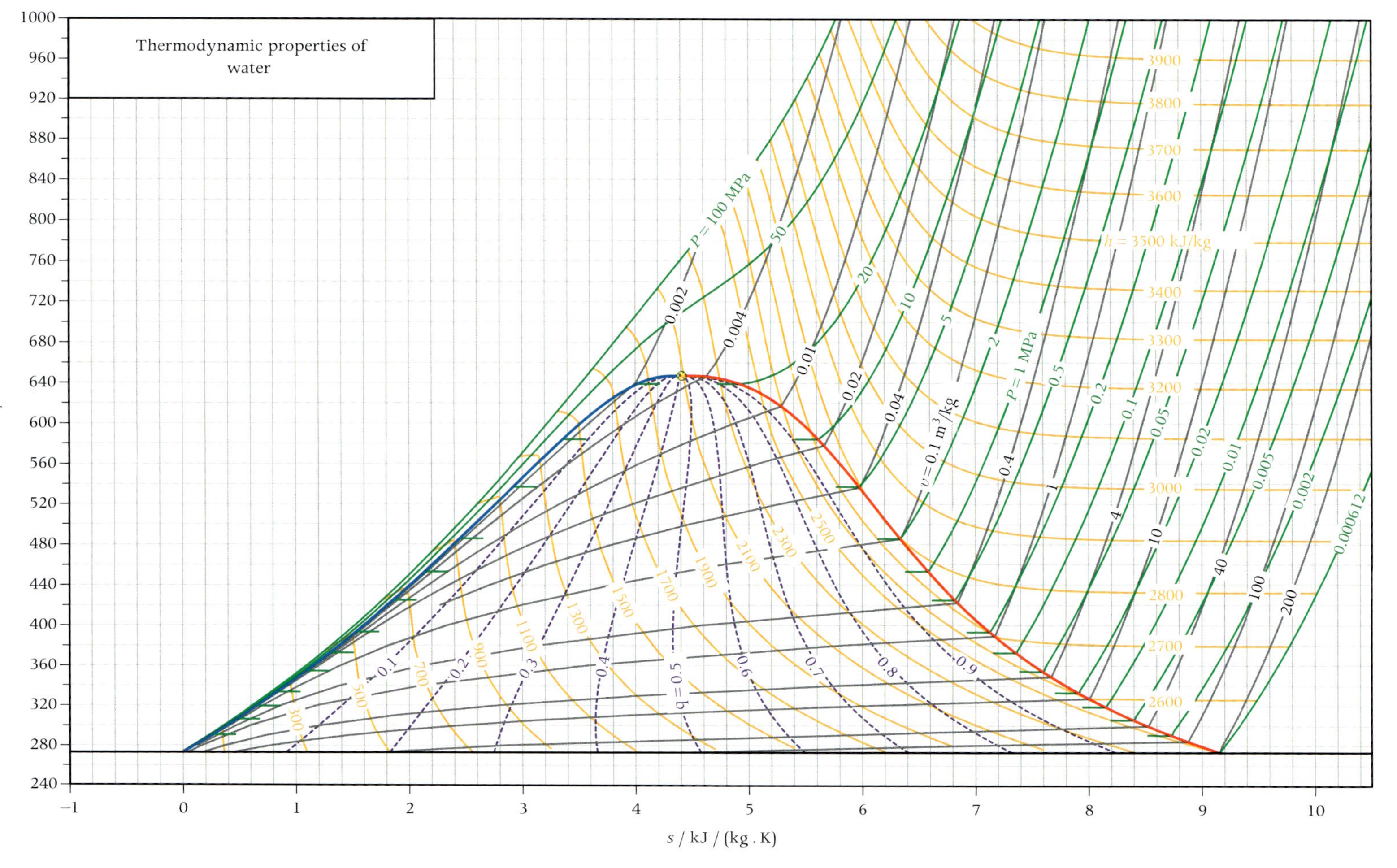
Thermodynamic properties of water
T / K
s / kJ / (kg . K)
P = 100 MPa
50
20
10
5
2
P = 1 MPa
0.5
0.2
0.1
0.05
0.02
0.01
0.005
0.002
0.000612
v = 0.1 m3/kg
0.002
0.004
0.01
0.02
0.04
0.4
1
4
10
40
100
200
h = 3500 kJ/kg
3900
3800
3700
3600
3400
3300
3200
3000
2800
2700
2600
2500
2300
2100
1900
1700
1500
1300
1100
900
700
500
300
q = 0.5
0.1
0.2
0.3
0.4
0.6
0.7
0.8
0.9
1000
960
920
880
840
800
760
720
680
640
600
560
520
480
440
400
360
320
280
240
−1
0
1
2
3
4
5
6
7
8
9
10

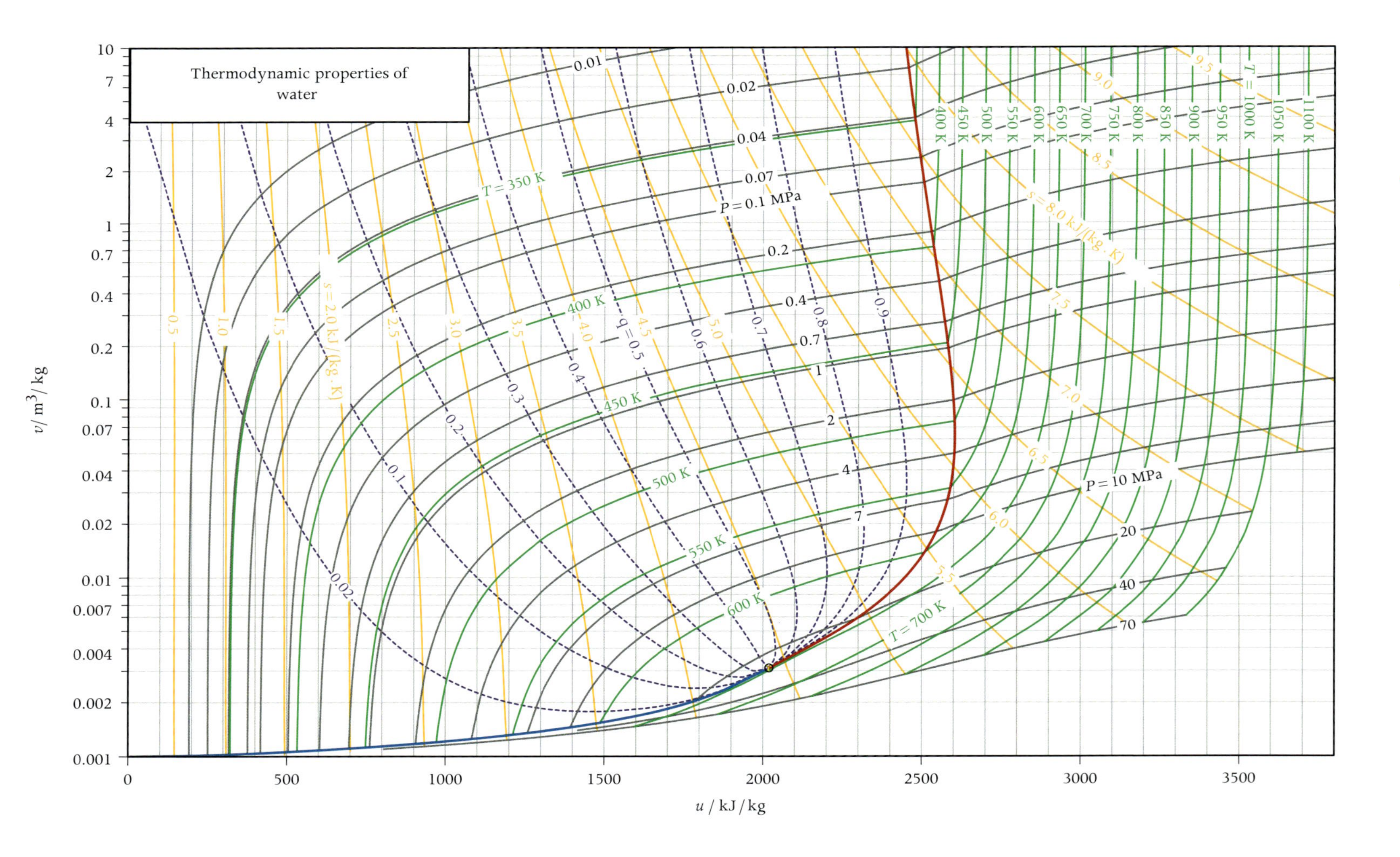
Thermodynamic properties of water
u / kJ / kg
v/ m³/ kg
P = 0.1 MPa
P = 10 MPa
T = 1000 K
T = 350 K
T = 700 K
s = 8.0 kJ/(kg . K)
s = 2.0 kJ/(kg . K)
q = 0.5

A.4 Refrigerant R134a

Properties in tables and charts are calculated with StanMix, which implements the thermodynamic model for R134a documented in T. P. van der Stelt, N. R. Nannan, and P. Colonna, "The iPRSV equation of state," *Fluid Phase Equilib.*, **330**, 24–35, 2012.
Reference state: $\hat{h}_0 = \hat{h}^{IG}(P = 1 \text{ bar}, T = 25\ °C) = 0$ J/mol (with $\hat{h} \equiv \hat{h}^{IG} + \hat{h}^{R}$), $\hat{s}_0 = \hat{s}^{IG}(P = 1 \text{ bar}, T = 25\ °C) = 0$ J/ (mol · K) (with $\hat{s} \equiv \hat{s}^{IG} + \hat{s}^{R}$).

Table A.7 Properties of saturated R134a as function of temperature.

T / K	P^{sat} / MPa	v^L / m^3/kg	v^V / m^3/kg	h^L / kJ/kg	h^{LV} / kJ/kg	h^V / kJ/kg	s^L / kJ/(kg · K)	s^{LV} / kJ/(kg · K)	s^V / kJ/(kg · K)
200	0.006566	0.000679	2.473	−314.59	241.67	−72.92	−1.3	1.2083	−0.0721
204	0.008784	0.000683	1.884	−310.16	239.79	−70.37	−1.2585	1.1754	−0.0831
208	0.01160	0.000687	1.453	−305.68	237.89	−67.79	−1.2368	1.1437	−0.0931
212	0.01513	0.000691	1.134	−301.17	235.97	−65.20	−1.2153	1.1131	−0.1023
216	0.01950	0.000696	0.8944	−296.62	234.02	−62.59	−1.1941	1.0834	−0.1106
220	0.02488	0.000700	0.7127	−292.02	232.05	−59.97	−1.1730	1.0548	−0.1182
224	0.03141	0.000705	0.5734	−287.37	230.04	−57.33	−1.1521	1.0270	−0.1251
228	0.03928	0.000710	0.4655	−282.68	228.00	−54.68	−1.1314	1.0000	−0.1313
232	0.04869	0.000715	0.3810	−277.95	225.92	−52.02	−1.1108	0.9738	−0.1370
236	0.05984	0.000720	0.3143	−273.16	223.80	−49.36	−1.0904	0.9483	−0.1421
240	0.07296	0.000726	0.2612	−268.32	221.62	−46.70	−1.0701	0.9234	−0.1467
244	0.08829	0.000732	0.2185	−263.44	219.40	−44.04	−1.0499	0.8992	−0.1508
246.99	0.101325	0.000737	0.1920	−259.76	217.70	−42.05	−1.0350	0.8814	−0.1536
250	0.1160	0.000741	0.1692	−256.01	215.95	−40.06	−1.0199	0.8638	−0.1561
254	0.1380	0.000748	0.1437	−250.99	213.58	−37.41	−1.0001	0.8409	−0.1592
258	0.1632	0.000755	0.1227	−245.92	211.14	−34.78	−0.9804	0.8184	−0.1620
262	0.1918	0.000762	0.1053	−240.79	208.63	−32.16	−0.9607	0.7963	−0.1644
266	0.2242	0.000770	0.09078	−235.60	206.03	−29.56	−0.9411	0.7746	−0.1666
270	0.2607	0.000778	0.07861	−230.35	203.36	−26.99	−0.9217	0.7532	−0.1685
274	0.3017	0.000787	0.06834	−225.03	200.59	−24.44	−0.9022	0.7321	−0.1702
278	0.3476	0.000796	0.05963	−219.65	197.71	−21.93	−0.8829	0.7112	−0.1717
282	0.3986	0.000806	0.05222	−214.19	194.73	−19.46	−0.8635	0.6905	−0.1730
286	0.4551	0.000816	0.04587	−208.66	191.63	−17.03	−0.8442	0.6700	−0.1742
290	0.5177	0.000827	0.04041	−203.04	188.40	−14.64	−0.8249	0.6497	−0.1752
294	0.5866	0.000838	0.03570	−197.34	185.03	−12.31	−0.8055	0.6294	−0.1762
298	0.6623	0.000851	0.03161	−191.55	181.51	−10.04	−0.7862	0.6091	−0.1771
302	0.7451	0.000864	0.02805	−185.66	177.82	−7.84	−0.7668	0.5888	−0.1780

Table A.7 (cont.)

T / K	P^{sat} / MPa	v^L / m³/kg	v^V / m³/kg	h^L / kJ/kg	h^{LV} / kJ/kg	h^V / kJ/kg	s^L / kJ/(kg · K)	s^{LV} / kJ/(kg · K)	s^V / kJ/(kg · K)
306	0.8357	0.000879	0.02494	−179.66	173.95	−5.71	−0.7473	0.5685	−0.1789
310	0.9343	0.000894	0.02221	−173.55	169.88	−3.67	−0.7278	0.5480	−0.1798
314	1.0416	0.000911	0.01980	−167.32	165.60	−1.72	−0.7081	0.5274	−0.1807
316	1.0985	0.000920	0.01871	−164.15	163.37	−0.78	−0.6982	0.5170	−0.1812
318	1.1578	0.000929	0.01768	−160.96	161.08	0.12	−0.6883	0.5065	−0.1818
320	1.2195	0.000939	0.01671	−157.72	158.72	1.00	−0.6783	0.4960	−0.1823
322	1.2836	0.000949	0.01580	−154.45	156.29	1.85	−0.6683	0.4854	−0.1830
324	1.3503	0.000960	0.01494	−151.13	153.79	2.66	−0.6583	0.4747	−0.1836
326	1.4195	0.000971	0.01413	−147.78	151.22	3.44	−0.6482	0.4639	−0.1843
328	1.4913	0.000983	0.013358	−144.38	148.55	4.18	−0.6380	0.4529	−0.1851
330	1.5659	0.000995	0.012632	−140.93	145.81	4.87	−0.6277	0.4418	−0.1859
332	1.6432	0.001009	0.011945	−137.44	142.96	5.53	−0.6174	0.4306	−0.1868
334	1.7234	0.001022	0.011295	−133.89	140.02	6.13	−0.6070	0.4192	−0.1878
336	1.8065	0.001037	0.010678	−130.28	136.97	6.69	−0.5965	0.4077	−0.1888
338	1.8926	0.001053	0.010092	−126.62	133.81	7.19	−0.5859	0.3959	−0.1900
340	1.9817	0.001069	0.009536	−122.89	130.52	7.63	−0.5751	0.3839	−0.1913
342	2.074	0.001087	0.009007	−119.09	127.09	8.01	−0.5643	0.3716	−0.1927
344	2.169	0.001106	0.008503	−115.21	123.52	8.31	−0.5533	0.3591	−0.1942
346	2.268	0.001126	0.008023	−111.25	119.79	8.54	−0.5421	0.3462	−0.1959
348	2.370	0.001148	0.007565	−107.20	115.88	8.68	−0.5308	0.3330	−0.1978
350	2.476	0.001172	0.007127	−103.05	111.77	8.72	−0.5193	0.3193	−0.1999
352	2.585	0.001198	0.006708	−98.78	107.44	8.66	−0.5075	0.3052	−0.2022
354	2.697	0.001227	0.006305	−94.39	102.86	8.47	−0.4954	0.2906	−0.2048
356	2.814	0.001258	0.005918	−89.85	98.00	8.15	−0.4830	0.2753	−0.2078
358	2.934	0.001293	0.005545	−85.14	92.80	7.65	−0.4703	0.2592	−0.2111
360	3.058	0.001333	0.005183	−80.24	87.20	6.97	−0.4571	0.2422	−0.2148
362	3.185	0.001378	0.004831	−75.09	81.13	6.04	−0.4433	0.2241	−0.2192
364	3.317	0.001430	0.004487	−69.64	74.47	4.82	−0.4288	0.2046	−0.2242
366	3.453	0.001492	0.004146	−63.81	67.02	3.22	−0.4134	0.1831	−0.2302
368	3.593	0.001568	0.003804	−57.43	58.51	1.08	−0.3966	0.1590	−0.2376
370	3.738	0.001668	0.003451	−50.24	48.36	−1.88	−0.3777	0.1307	−0.2470
372	3.886	0.001812	0.003064	−41.52	35.17	−6.35	−0.3549	0.0945	−0.2604
374.21	4.056	0.002339	0.002339	−20.78	0.0	−20.78	−0.3002	0.0	−0.3002

Table A.8 Properties of saturated R134a as function of pressure.

P / MPa	T^{sat} / K	v^L / m^3/kg	v^V / m^3/kg	h^L / kJ/kg	h^{LV} / kJ/kg	h^V / kJ/kg	s^L / kJ/(kg · K)	s^{LV} / kJ/(kg · K)	s^V / kJ/(kg · K)
0.010	205.84	0.000685	1.6686	−308.10	238.92	−69.18	−1.2485	1.1607	−0.0878
0.011	207.35	0.000686	1.5146	−306.42	238.20	−68.22	−1.2403	1.1488	−0.0915
0.012	208.87	0.000688	1.3749	−304.70	237.47	−67.23	−1.2321	1.1369	−0.0952
0.014	210.43	0.000690	1.2481	−302.95	236.73	−66.22	−1.2237	1.1250	−0.0988
0.015	212.01	0.000691	1.1331	−301.16	235.96	−65.20	−1.2153	1.1130	−0.1023
0.017	213.62	0.000693	1.0286	−299.33	235.18	−64.15	−1.2067	1.1010	−0.1057
0.019	215.26	0.000695	0.9338	−297.46	234.39	−63.08	−1.1980	1.0889	−0.1091
0.021	216.93	0.000697	0.8478	−295.55	233.57	−61.99	−1.1892	1.0767	−0.1124
0.023	218.63	0.000699	0.7697	−293.60	232.73	−60.87	−1.1802	1.0645	−0.1157
0.025	220.36	0.000701	0.6988	−291.61	231.87	−59.73	−1.1711	1.0523	−0.1188
0.028	222.12	0.000703	0.6344	−289.56	230.99	−58.57	−1.1619	1.0399	−0.1219
0.031	223.92	0.000705	0.5760	−287.47	230.09	−57.38	−1.1525	1.0276	−0.1250
0.035	225.75	0.000707	0.5230	−285.33	229.16	−56.17	−1.1430	1.0151	−0.1279
0.038	227.61	0.000709	0.4748	−283.14	228.20	−54.94	−1.1334	1.0026	−0.1308
0.043	229.52	0.000712	0.4311	−280.89	227.22	−53.68	−1.1235	0.9900	−0.1336
0.047	231.46	0.000714	0.3913	−278.59	226.21	−52.39	−1.1136	0.9773	−0.1363
0.052	233.44	0.000717	0.3553	−276.23	225.16	−51.07	−1.1034	0.9646	−0.1389
0.058	235.45	0.000720	0.3225	−273.82	224.09	−49.73	−1.0931	0.9517	−0.1414
0.065	237.51	0.000723	0.2928	−271.34	222.98	−48.36	−1.0827	0.9388	−0.1439
0.072	239.62	0.000726	0.2658	−268.79	221.83	−46.96	−1.0720	0.9258	−0.1462
0.079	241.76	0.000729	0.2412	−266.18	220.65	−45.53	−1.0612	0.9127	−0.1485
0.088	243.95	0.000732	0.2189	−263.49	219.42	−44.07	−1.0502	0.8994	−0.1507
0.098	246.19	0.000735	0.1987	−260.74	218.15	−42.58	−1.0389	0.8861	−0.1528
0.101325	246.99	0.000737	0.1920	−259.76	217.70	−42.05	−1.0350	0.8814	−0.1536
0.11	248.48	0.000739	0.1803	−257.90	216.84	−41.06	−1.0275	0.8727	−0.1549
0.12	250.81	0.000743	0.1636	−254.99	215.48	−39.52	−1.0159	0.8591	−0.1568
0.13	253.20	0.000747	0.14838	−252.00	214.06	−37.94	−1.0041	0.8454	−0.1586
0.15	255.64	0.000751	0.13458	−248.92	212.59	−36.33	−0.9920	0.8316	−0.1604
0.16	258.14	0.000755	0.12205	−245.74	211.06	−34.69	−0.9797	0.8176	−0.1621
0.18	260.69	0.000760	0.11066	−242.48	209.46	−33.02	−0.9672	0.8035	−0.1637
0.20	263.29	0.000765	0.10031	−239.11	207.80	−31.32	−0.9544	0.7892	−0.1651
0.22	265.96	0.000770	0.09091	−235.65	206.06	−29.59	−0.9413	0.7748	−0.1666
0.25	268.69	0.000776	0.08236	−232.07	204.24	−27.83	−0.9280	0.7601	−0.1679
0.28	271.49	0.000781	0.07459	−228.38	202.34	−26.04	−0.9144	0.7453	−0.1691

Table A.8 (cont.)

P / MPa	T^{sat} / K	v^L / m^3/kg	v^V / m^3/kg	h^L / kJ/kg	h^{LV} / kJ/kg	h^V / kJ/kg	s^L / kJ/(kg · K)	s^{LV} / kJ/(kg · K)	s^V / kJ/(kg · K)
0.31	274.34	0.000788	0.06753	−224.57	200.34	−24.23	−0.9006	0.7303	−0.1703
0.34	277.27	0.000794	0.06112	−220.63	198.24	−22.39	−0.8864	0.7150	−0.1714
0.38	280.27	0.000801	0.05528	−216.56	196.04	−20.52	−0.8719	0.6995	−0.1724
0.42	283.34	0.000809	0.04998	−212.34	193.71	−18.64	−0.8570	0.6837	−0.1734
0.46	286.48	0.000817	0.04517	−207.98	191.25	−16.73	−0.8418	0.6676	−0.1743
0.51	289.71	0.000826	0.04079	−203.46	188.64	−14.81	−0.8263	0.6512	−0.1751
0.57	293.01	0.000835	0.03680	−198.76	185.88	−12.88	−0.8103	0.6344	−0.1760
0.63	296.39	0.000846	0.03318	−193.89	182.94	−10.95	−0.7940	0.6172	−0.1767
0.70	299.86	0.000857	0.02989	−188.82	179.81	−9.01	−0.7772	0.5997	−0.1775
0.78	303.41	0.000869	0.02690	−183.55	176.47	−7.08	−0.7599	0.5816	−0.1783
0.86	307.06	0.000883	0.02418	−178.05	172.89	−5.16	−0.7422	0.5631	−0.1791
1.0	310.80	0.000897	0.02170	−172.32	169.05	−3.27	−0.7239	0.5439	−0.1799
1.1	314.63	0.000914	0.01945	−166.33	164.90	−1.42	−0.7050	0.5241	−0.1809
1.2	318.56	0.000932	0.01740	−160.05	160.42	0.37	−0.6855	0.5036	−0.1819
1.3	322.59	0.000952	0.015538	−153.47	155.56	2.09	−0.6654	0.4822	−0.1831
1.4	326.73	0.000975	0.013840	−146.54	150.25	3.71	−0.6444	0.4599	−0.1846
1.6	330.98	0.001002	0.012293	−139.23	144.43	5.20	−0.6227	0.4364	−0.1863
1.8	335.33	0.001032	0.010881	−131.50	138.01	6.51	−0.6000	0.4116	−0.1885
2.0	339.80	0.001068	0.009592	−123.27	130.86	7.59	−0.5763	0.3851	−0.1911
2.2	344.38	0.001110	0.008411	−114.47	122.83	8.36	−0.5512	0.3567	−0.1945
2.4	349.08	0.001161	0.007327	−104.98	113.69	8.72	−0.5246	0.3257	−0.1989
2.7	353.90	0.001225	0.006325	−94.62	103.10	8.49	−0.4960	0.2913	−0.2047
3.0	358.84	0.001309	0.005391	−83.10	90.49	7.39	−0.4648	0.2522	−0.2126
3.3	363.91	0.001428	0.004502	−69.89	74.77	4.88	−0.4295	0.2055	−0.2240
3.7	369.11	0.001620	0.003610	−53.57	53.13	−0.44	−0.3864	0.1439	−0.2425
4.056	374.21	0.002339	0.002339	−20.78	0.00	−20.78	−0.3002	0.0000	−0.3002

Table A.9 Properties of gaseous R134a.

P / MPa (T^{sat} / K)		sat	220	230	240	250	260	270	280	290
0.020	v / m^3/kg	0.8736	0.888	0.930	0.971	1.012	1.053	1.095	1.136	1.177
(216.4)	h / kJ/kg	−62.33	−59.86	−52.88	−45.70	−38.33	−30.78	−23.04	−15.11	−7.00
	s / kJ/(kg · K)	−0.1114	−0.1001	−0.0691	−0.0385	−0.0085	0.0212	0.0504	0.0792	0.1076
0.030	v / m^3/kg	0.5986		0.6174	0.6451	0.6727	0.7003	0.7279	0.7554	0.7829
(223.2)	h / kJ/kg	−57.86		−53.08	−45.89	−38.51	−30.94	−23.19	−15.26	−7.14
	s / kJ/(kg · K)	−0.1238		−0.1027	−0.0721	−0.0420	−0.0123	0.0170	0.0458	0.0743
0.040	v / m^3/kg	0.4577		0.4612	0.4821	0.5029	0.5238	0.5445	0.5652	0.5859
(228.3)	h / kJ/kg	−54.46		−53.28	−46.08	−38.69	−31.11	−23.35	−15.41	−7.28
	s / kJ/(kg · K)	−0.1318		−0.1267	−0.0960	−0.0658	−0.0361	−0.0068	0.0220	0.0505
0.050	v / m^3/kg	0.3717			0.3843	0.4011	0.4178	0.4345	0.4512	0.4678
(232.5)	h / kJ/kg	−51.69			−46.26	−38.86	−31.27	−23.50	−15.55	−7.42
	s / kJ/(kg · K)	−0.1377			−0.1147	−0.0845	−0.0547	−0.0254	0.0035	0.0320
0.070	v / m^3/kg	0.2715			0.2725	0.2846	0.2967	0.3088	0.3208	0.3327
(239.2)	h / kJ/kg	−47.27			−46.64	−39.22	−31.61	−23.82	−15.85	−7.71
	s / kJ/(kg · K)	−0.1457			−0.1431	−0.1128	−0.0830	−0.0536	−0.0246	0.0040
0.101325	v / m^3/kg	0.1920				0.1946	0.2031	0.2116	0.2200	0.2284
(247.0)	h / kJ/kg	−42.05				−39.79	−32.14	−24.32	−16.32	−8.15
	s / kJ/(kg · K)	−0.1536				−0.1444	−0.1144	−0.0849	−0.0558	−0.0272
0.14	v / m^3/kg	0.1417					0.1453	0.1515	0.1577	0.1639
(254.3)	h / kJ/kg	−37.19					−32.81	−24.95	−16.91	−8.71
	s / kJ/(kg · K)	−0.1595					−0.1425	−0.1128	−0.0836	−0.0548
0.20	v / m^3/kg	0.1012						0.1043	0.1088	0.1132
(263.1)	h / kJ/kg	−31.47						−25.95	−17.85	−9.59
	s / kJ/(kg · K)	−0.1650						−0.1443	−0.1149	−0.0859
0.30	v / m^3/kg	0.06872							0.07065	0.07374
(273.8)	h / kJ/kg	−24.55							−19.47	−11.10
	s / kJ/(kg · K)	−0.1701							−0.1518	−0.1224

Table A.9 (cont.)

P / MPa		T / K								
(T^{sat} / K)		300	310	320	330	340	350	360	370	380
0.020	v / m^3/kg	1.218	1.259	1.300	1.341	1.382	1.423	1.464	1.505	1.545
(216.4)	h / kJ/kg	1.28	9.74	18.37	27.18	36.15	45.30	54.60	64.07	73.70
	s / kJ/(kg · K)	0.1357	0.1635	0.1909	0.2180	0.2448	0.2713	0.2975	0.3234	0.3491
0.030	v / m^3/kg	0.8103	0.8378	0.865	0.893	0.920	0.947	0.975	1.002	1.029
(223.2)	h / kJ/kg	1.15	9.62	18.26	27.07	36.05	45.19	54.50	63.98	73.61
	s / kJ/(kg · K)	0.1024	0.1302	0.1576	0.1847	0.2115	0.2380	0.2642	0.2902	0.3159
0.040	v / m^3/kg	0.6066	0.6273	0.6479	0.6685	0.6891	0.7097	0.7303	0.7508	0.7714
(228.3)	h / kJ/kg	1.02	9.49	18.14	26.95	35.94	45.09	54.40	63.88	73.52
	s / kJ/(kg · K)	0.0787	0.1065	0.1339	0.1610	0.1879	0.2144	0.2406	0.2666	0.2923
0.050	v / m^3/kg	0.4844	0.5010	0.5175	0.5341	0.5506	0.5671	0.5836	0.6001	0.6165
(232.5)	h / kJ/kg	0.88	9.36	18.02	26.84	35.83	44.99	54.31	63.79	73.43
	s / kJ/(kg · K)	0.0602	0.0880	0.1155	0.1426	0.1695	0.1960	0.2222	0.2482	0.2739
0.070	v / m^3/kg	0.3447	0.3566	0.3685	0.3804	0.3923	0.4041	0.4160	0.4278	0.4396
(239.2)	h / kJ/kg	0.62	9.11	17.77	26.61	35.61	44.78	54.11	63.60	73.25
	s / kJ/(kg · K)	0.0322	0.0600	0.0875	0.1147	0.1416	0.1682	0.1945	0.2205	0.2462
0.101325	v / m^3/kg	0.2367	0.2450	0.2533	0.2616	0.2699	0.2782	0.2864	0.2946	0.3028
(247.0)	h / kJ/kg	0.19	8.71	17.40	26.25	35.27	44.45	53.80	63.30	72.97
	s / kJ/(kg · K)	0.0011	0.0290	0.0566	0.0839	0.1108	0.1374	0.1637	0.1898	0.2155
0.14	v / m^3/kg	0.1700	0.1761	0.1822	0.1883	0.1944	0.2004	0.2064	0.2124	0.2184
(254.3)	h / kJ/kg	−0.33	8.21	16.92	25.80	34.84	44.05	53.41	62.93	72.61
	s / kJ/(kg · K)	−0.0264	0.0016	0.0293	0.0566	0.0836	0.1103	0.1366	0.1627	0.1885
0.20	v / m^3/kg	0.1176	0.1220	0.1263	0.1307	0.1350	0.1393	0.1435	0.1478	0.1520
(263.1)	h / kJ/kg	−1.16	7.43	16.18	25.10	34.17	43.41	52.81	62.36	72.06
	s / kJ/(kg · K)	−0.0573	−0.0291	−0.0013	0.0261	0.0532	0.0800	0.1064	0.1326	0.1585
0.30	v / m^3/kg	0.07680	0.07983	0.08283	0.0858	0.0888	0.0917	0.0946	0.0975	0.1004
(273.8)	h / kJ/kg	−2.59	6.09	14.92	23.90	33.04	42.34	51.78	61.39	71.14
	s / kJ/(kg · K)	−0.0935	−0.0651	−0.0371	−0.0094	0.0179	0.0448	0.0714	0.0977	0.1237

Table A.9 (cont.)

P / MPa (T^{sat} / K)		T / K 400	420	440	460	480	500	520	540	560
0.020	v / m^3/kg	1.627	1.709	1.791	1.872	1.954	2.036	2.117	2.199	2.281
(216.4)	h / kJ/kg	93.44	113.79	134.76	156.32	178.46	201.18	224.46	248.29	272.67
	s / kJ/(kg · K)	0.3997	0.4494	0.4981	0.5460	0.5931	0.6395	0.6852	0.7301	0.7745
0.030	v / m^3/kg	1.084	1.139	1.193	1.248	1.302	1.357	1.411	1.466	1.520
(223.2)	h / kJ/kg	93.35	113.72	134.69	156.25	178.40	201.12	224.40	248.24	272.63
	s / kJ/(kg · K)	0.3665	0.4162	0.4650	0.5129	0.5600	0.6064	0.6520	0.6970	0.7414
0.040	v / m^3/kg	0.8124	0.853	0.894	0.935	0.976	1.017	1.058	1.099	1.140
(228.3)	h / kJ/kg	93.27	113.64	134.62	156.19	178.34	201.07	224.35	248.20	272.59
	s / kJ/(kg · K)	0.3429	0.3926	0.4414	0.4893	0.5365	0.5829	0.6285	0.6735	0.7179
0.050	v / m^3/kg	0.6494	0.6823	0.7151	0.7479	0.7807	0.8135	0.8463	0.879	0.912
(232.5)	h / kJ/kg	93.19	113.57	134.55	156.12	178.28	201.01	224.30	248.15	272.54
	s / kJ/(kg · K)	0.3246	0.3743	0.4231	0.4711	0.5182	0.5646	0.6103	0.6553	0.6996
0.070	v / m^3/kg	0.4632	0.4867	0.5103	0.5338	0.5572	0.5807	0.6041	0.6276	0.6510
(239.2)	h / kJ/kg	93.02	113.41	134.41	155.99	178.16	200.90	224.20	248.06	272.46
	s / kJ/(kg · K)	0.2969	0.3466	0.3955	0.4434	0.4906	0.5370	0.5827	0.6277	0.6721
0.101325	v / m^3/kg	0.3192	0.3356	0.3519	0.3682	0.3845	0.4008	0.4170	0.4332	0.4494
(247.0)	h / kJ/kg	92.76	113.18	134.19	155.79	177.98	200.73	224.04	247.91	272.32
	s / kJ/(kg · K)	0.2663	0.3161	0.3650	0.4130	0.4602	0.5066	0.5523	0.5974	0.6418
0.14	v / m^3/kg	0.2304	0.2423	0.2542	0.2660	0.2779	0.2897	0.3015	0.3133	0.3250
(254.3)	h / kJ/kg	92.44	112.88	133.92	155.54	177.75	200.52	223.85	247.73	272.15
	s / kJ/(kg · K)	0.2394	0.2892	0.3382	0.3862	0.4335	0.4800	0.5257	0.5708	0.6152
0.20	v / m^3/kg	0.1605	0.1689	0.1773	0.1857	0.1940	0.2024	0.2107	0.2190	0.2272
(263.1)	h / kJ/kg	91.94	112.42	133.50	155.16	177.39	200.19	223.54	247.44	271.89
	s / kJ/(kg · K)	0.2094	0.2594	0.3084	0.3566	0.4039	0.4504	0.4962	0.5413	0.5857
0.30	v / m^3/kg	0.1062	0.1119	0.1176	0.1232	0.1289	0.1345	0.1401	0.1456	0.1512
(273.8)	h / kJ/kg	91.10	111.65	132.79	154.51	176.79	199.64	223.03	246.97	271.45
	s / kJ/(kg · K)	0.1749	0.2251	0.2742	0.3225	0.3699	0.4165	0.4624	0.5076	0.5521

Table A.9 (cont.)

P / MPa		T / K								
(T^{sat} / K)		sat	290	300	310	320	330	340	350	360
0.40	v / m^3/kg	0.05204	0.05396	0.05636	0.05872	0.06105	0.06336	0.06565	0.06791	0.07016
(282.1)	h / kJ/kg	−19.39	−12.68	−4.06	4.70	13.62	22.68	31.88	41.24	50.75
	s / kJ/(kg · K)	−0.1730	−0.1496	−0.1203	−0.0916	−0.0633	−0.0354	−0.0079	0.0192	0.0460
0.50	v / m^3/kg	0.04182	0.04204	0.04405	0.04603	0.04797	0.04988	0.05176	0.05363	0.05547
(288.9)	h / kJ/kg	−15.29	−14.34	−5.60	3.27	12.28	21.42	30.70	40.12	49.69
	s / kJ/(kg · K)	−0.1749	−0.1717	−0.1420	−0.1129	−0.0844	−0.0562	−0.0285	−0.0012	0.0257
0.70	v / m^3/kg	0.02989		0.02991	0.03145	0.03296	0.03443	0.03586	0.03728	0.03867
(299.9)	h / kJ/kg	−9.00		−8.88	0.24	9.47	18.80	28.25	37.82	47.52
	s / kJ/(kg · K)	−0.1775		−0.1771	−0.1472	−0.1179	−0.0892	−0.0610	−0.0332	−0.0059
1.0	v / m^3/kg	0.02068				0.02159	0.02275	0.02387	0.02496	0.02603
(312.5)	h / kJ/kg	−2.44				4.83	14.54	24.31	34.15	44.10
	s / kJ/(kg · K)	−0.1803				−0.1574	−0.1275	−0.0983	−0.0698	−0.0418
1.4	v / m^3/kg	0.01435					0.01481	0.01577	0.01667	0.01754
(325.4)	h / kJ/kg	3.23					8.01	18.42	28.79	39.16
	s / kJ/(kg · K)	−0.1841					−0.1695	−0.1384	−0.1084	−0.0792
2.0	v / m^3/kg	0.00943							0.01027	0.01105
(340.4)	h / kJ/kg	7.71							19.14	30.61
	s / kJ/(kg · K)	−0.1915							−0.1584	−0.1261
2.5	v / m^3/kg	0.00703								0.00786
(350.5)	h / kJ/kg	8.72								21.73
	s / kJ/(kg · K)	−0.2004								−0.1638
3.0	v / m^3/kg	0.00535								0.00545
(359.1)	h / kJ/kg	7.31								9.02
	s / kJ/(kg · K)	−0.2130								−0.2083

Table A.9 (cont.)

P / MPa (T^{sat} / K)		370	380	390	400	410	420	430	440	450
0.40	v / m^3/kg	0.07239	0.07461	0.07682	0.07901	0.08119	0.08337	0.0855	0.0877	0.0898
(282.1)	h / kJ/kg	60.40	70.20	80.15	90.25	100.49	110.88	121.41	132.08	142.90
	s / kJ/(kg · K)	0.0724	0.0986	0.1244	0.1500	0.1753	0.2003	0.2251	0.2496	0.2739
0.50	v / m^3/kg	0.05730	0.05912	0.06092	0.06271	0.06448	0.06625	0.06801	0.06977	0.07151
(288.9)	h / kJ/kg	59.40	69.25	79.25	89.39	99.67	110.10	120.66	131.37	142.22
	s / kJ/(kg · K)	0.0523	0.0786	0.1046	0.1303	0.1556	0.1808	0.2056	0.2302	0.2546
0.70	v / m^3/kg	0.04004	0.04140	0.04274	0.04407	0.04538	0.04669	0.04799	0.04928	0.05056
(299.9)	h / kJ/kg	57.35	67.31	77.41	87.64	98.01	108.51	119.15	129.93	140.84
	s / kJ/(kg · K)	0.0210	0.0476	0.0738	0.0997	0.1253	0.1506	0.1757	0.2004	0.2250
1.0	v / m^3/kg	0.02707	0.02809	0.02909	0.03007	0.03105	0.03201	0.03296	0.03391	0.03484
(312.5)	h / kJ/kg	54.14	64.29	74.56	84.95	95.46	106.09	116.85	127.73	138.75
	s / kJ/(kg · K)	−0.0142	0.0128	0.0395	0.0658	0.0917	0.1174	0.1427	0.1677	0.1925
1.4	v / m^3/kg	0.01837	0.01918	0.01996	0.02073	0.02148	0.02222	0.02294	0.02366	0.02436
(325.4)	h / kJ/kg	49.56	60.03	70.57	81.20	91.93	102.76	113.69	124.74	135.91
	s / kJ/(kg · K)	−0.0507	−0.0227	0.0046	0.0316	0.0580	0.0841	0.1099	0.1353	0.1604
2.0	v / m^3/kg	0.01176	0.01244	0.01308	0.01369	0.01428	0.01486	0.01542	0.01597	0.01651
(340.4)	h / kJ/kg	41.86	53.00	64.10	75.20	86.34	97.52	108.78	120.11	131.53
	s / kJ/(kg · K)	−0.0953	−0.0656	−0.0368	−0.0086	0.0189	0.0458	0.0723	0.0983	0.1240
2.5	v / m^3/kg	0.00858	0.00923	0.00982	0.01038	0.01091	0.01142	0.01191	0.01238	0.01284
(350.5)	h / kJ/kg	34.30	46.35	58.14	69.78	81.36	92.92	104.49	116.10	127.77
	s / kJ/(kg · K)	−0.1293	−0.0972	−0.0666	−0.0371	−0.0085	0.0194	0.0466	0.0733	0.0995
3.0	v / m^3/kg	0.00634	0.00703	0.00762	0.00815	0.00865	0.00912	0.00956	0.00999	0.01040
(359.1)	h / kJ/kg	24.91	38.62	51.47	63.88	76.04	88.06	100.02	111.96	123.90
	s / kJ/(kg · K)	−0.1647	−0.1281	−0.0948	−0.0634	−0.0333	−0.0043	0.0238	0.0512	0.0781
4.0	v / m^3/kg		0.00394	0.00472	0.00530	0.00579	0.00622	0.00662	0.00700	0.00735
(373.5)	h / kJ/kg		15.74	34.66	50.04	64.09	77.47	90.47	103.23	115.86
	s / kJ/(kg · K)		−0.2026	−0.1534	−0.1145	−0.0798	−0.0475	−0.0169	0.0124	0.0408

Table A.9 (cont.)

P / MPa (T^{sat} / K)		T / K 460	480	500	520	540	560	580	600	620
0.40	v / m^3/kg	0.0920	0.0963	0.1005	0.1047	0.1090	0.1132	0.1174	0.1215	0.1257
(282.1)	h / kJ/kg	153.86	176.19	199.08	222.52	246.50	271.02	296.06	321.63	347.73
	s / kJ/(kg · K)	0.2980	0.3455	0.3922	0.4382	0.4834	0.5280	0.5720	0.6153	0.6581
0.50	v / m^3/kg	0.07325	0.07671	0.08014	0.08356	0.0870	0.0904	0.0937	0.0971	0.1005
(288.9)	h / kJ/kg	153.20	175.59	198.53	222.01	246.03	270.58	295.66	321.26	347.38
	s / kJ/(kg · K)	0.2788	0.3264	0.3732	0.4193	0.4646	0.5092	0.5532	0.5966	0.6394
0.70	v / m^3/kg	0.05183	0.05436	0.05687	0.05935	0.06182	0.06428	0.06673	0.06916	0.07159
(299.9)	h / kJ/kg	151.89	174.39	197.42	220.99	245.09	269.71	294.85	320.51	346.69
	s / kJ/(kg · K)	0.2492	0.2971	0.3441	0.3903	0.4358	0.4806	0.5247	0.5682	0.6111
1.0	v / m^3/kg	0.03577	0.03760	0.03941	0.04120	0.04298	0.04473	0.04648	0.04822	0.04995
(312.5)	h / kJ/kg	149.89	172.56	195.75	219.45	243.67	268.40	293.64	319.39	345.65
	s / kJ/(kg · K)	0.2170	0.2652	0.3125	0.3590	0.4047	0.4497	0.4939	0.5376	0.5806
1.4	v / m^3/kg	0.02506	0.02643	0.02778	0.02911	0.03041	0.03171	0.03299	0.03426	0.03552
(325.4)	h / kJ/kg	147.19	170.10	193.51	217.40	241.78	266.66	292.04	317.91	344.28
	s / kJ/(kg · K)	0.1851	0.2339	0.2817	0.3285	0.3745	0.4198	0.4643	0.5081	0.5514
2.0	v / m^3/kg	0.01703	0.01806	0.01906	0.02004	0.02100	0.02195	0.02288	0.02380	0.02471
(340.4)	h / kJ/kg	143.04	166.36	190.11	214.30	238.95	264.06	289.64	315.70	342.24
	s / kJ/(kg · K)	0.1493	0.1989	0.2474	0.2948	0.3413	0.3870	0.4319	0.4761	0.5196
2.5	v / m^3/kg	0.01329	0.01416	0.01500	0.01582	0.01662	0.01740	0.01817	0.01893	0.01967
(350.5)	h / kJ/kg	139.50	163.20	187.26	211.71	236.59	261.90	287.66	313.88	340.56
	s / kJ/(kg · K)	0.1253	0.1757	0.2248	0.2728	0.3197	0.3657	0.4109	0.4554	0.4991
3.0	v / m^3/kg	0.01080	0.01157	0.01230	0.01301	0.01370	0.01438	0.01504	0.01568	0.01632
(359.1)	h / kJ/kg	135.88	159.99	184.39	209.12	234.24	259.75	285.69	312.07	338.90
	s / kJ/(kg · K)	0.1044	0.1557	0.2055	0.2540	0.3014	0.3478	0.3933	0.4380	0.4820
4.0	v / m^3/kg	0.00769	0.00833	0.00894	0.00952	0.01007	0.01061	0.01113	0.01164	0.01214
(373.5)	h / kJ/kg	128.42	153.48	178.61	203.94	229.56	255.50	281.82	308.52	335.64
	s / kJ/(kg · K)	0.0684	0.1217	0.1730	0.2227	0.2710	0.3182	0.3644	0.4096	0.4541

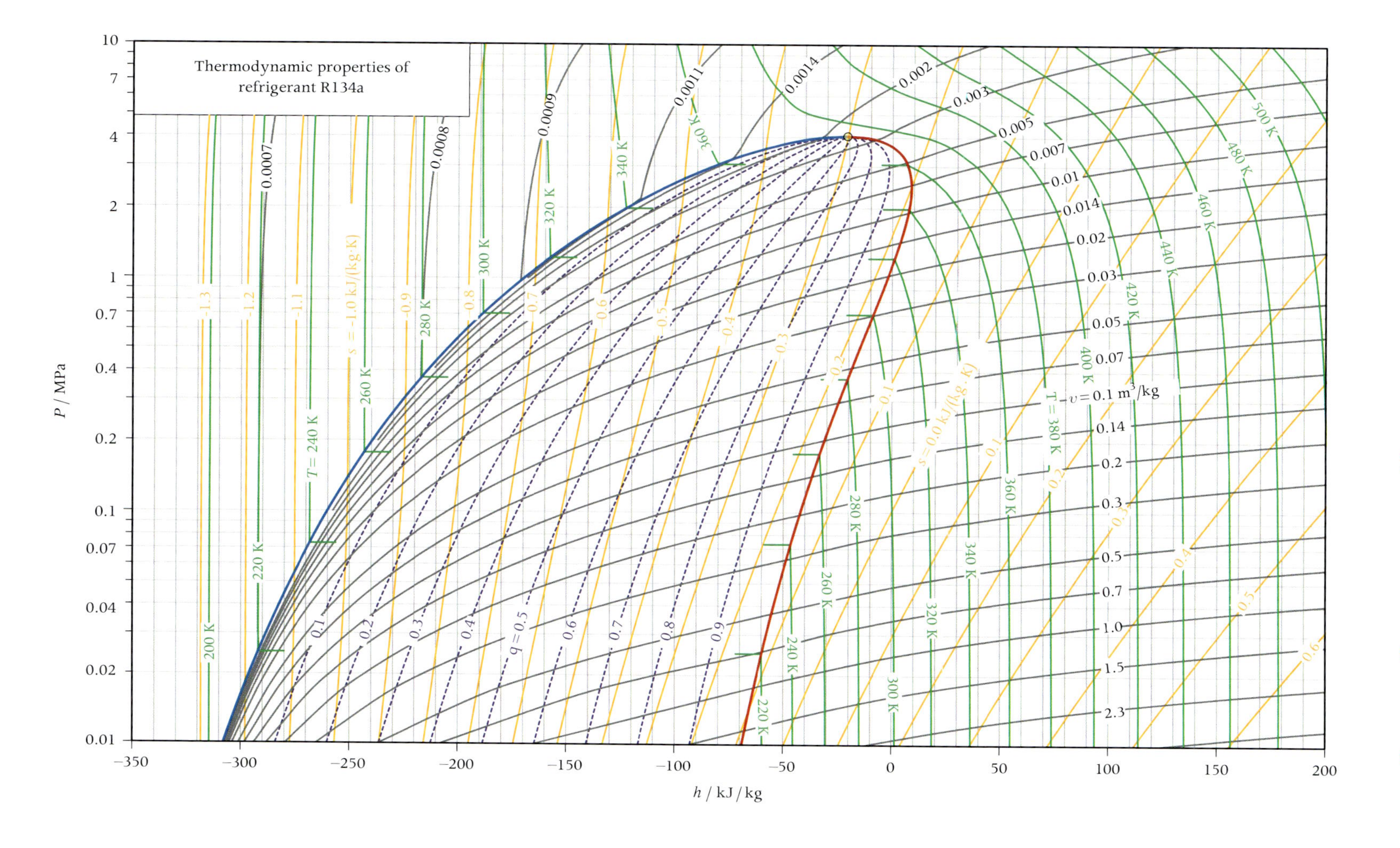
Thermodynamic properties of refrigerant R134a
P / MPa
h / kJ / kg
$v = 0.1$ m^3/kg
$s = 0.0$ kJ/(kg·K)
$s = -1.0$ kJ/(kg·K)
$T = 380$ K
$T = 240$ K
$q = 0.5$

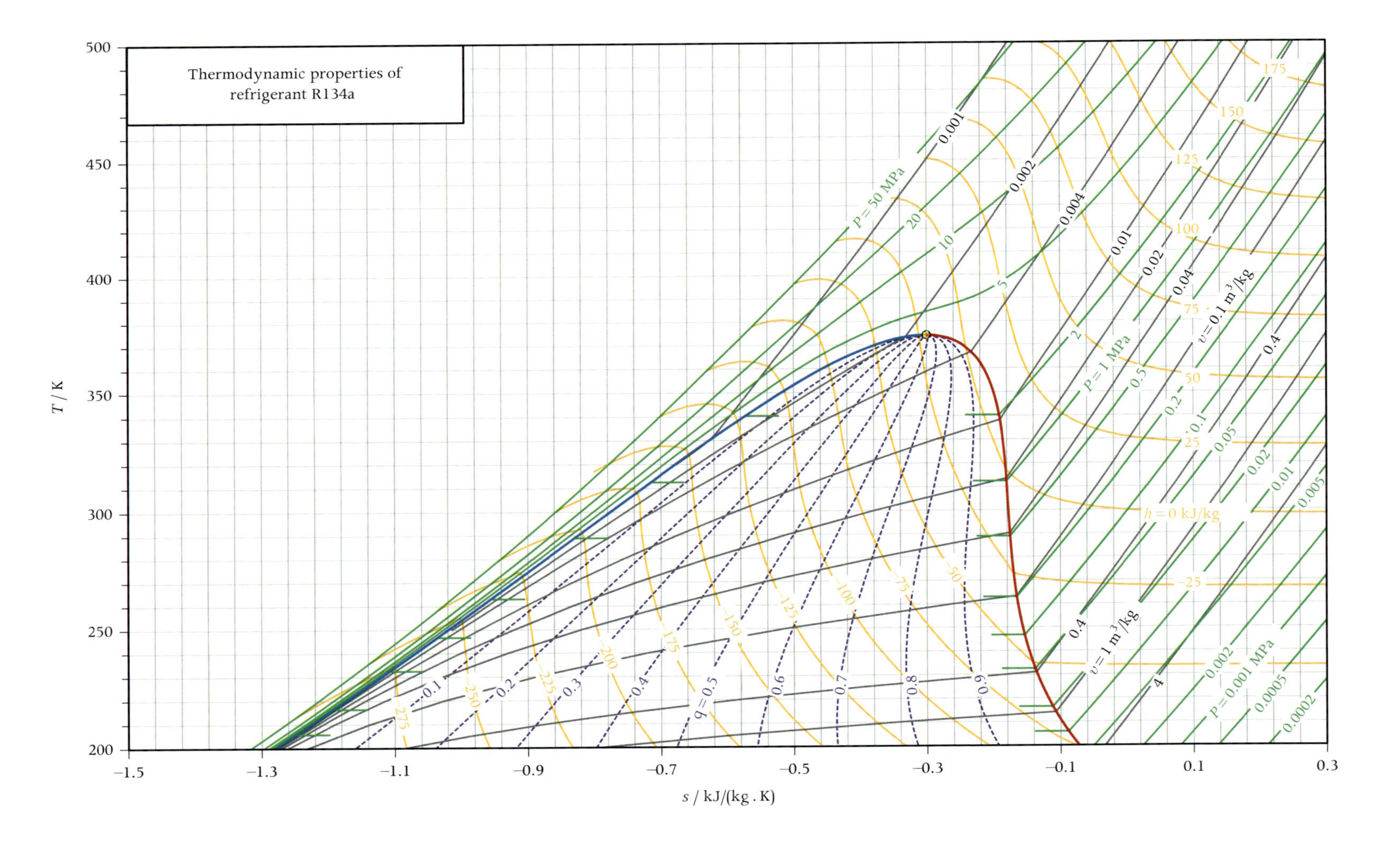
Thermodynamic properties of refrigerant R134a
T / K
s / kJ/(kg . K)
P = 50 MPa
P = 1 MPa
P = 0.001 MPa
υ = 0.1 m³/kg
υ = 1 m³/kg
h = 0 kJ/kg
q = 0.5

A.5 Methane

Properties in tables and charts are calculated with RefProp, which implements the thermodynamic model for methane documented in U. Setzmann and W. Wagner, "A new equation of state and tables of thermodynamic properties for methane covering the range from the melting line to 625 K at pressures up to 1000 MPa," *J. Phys. Chem. Ref. Data*, **20**, (6), 1061–1151, 1991.
Reference state:

$$h_0 = h^{sat,L}(T^{NBP} = 111.7\ \text{K}) = 0\ \text{kJ/kg},$$
$$s_0 = s^{sat,L}(T^{NBP} = 111.7\ \text{K}) = 0\ \text{kJ/}\left(\text{kg}\cdot\text{K}\right).$$

Table A.10 Properties of saturated methane as function of temperature.

T / K	P^{sat} / MPa	v^L / m^3/kg	v^V / m^3/kg	h^L / kJ/kg	h^{LV} / kJ/kg	h^V / kJ/kg	s^L / kJ/(kg · K)	s^{LV} / kJ/(kg · K)	s^V / kJ/(kg · K)
90.6941	0.01170	0.002215	3.988	−71.82	544.26	472.44	−0.7	6.0011	5.2911
95	0.01981	0.002244	2.457	−57.27	538.06	480.78	−0.5534	5.6637	5.1103
100	0.03438	0.002278	1.482	−40.27	530.48	490.21	−0.3793	5.3048	4.9255
105	0.05638	0.002315	0.942	−23.12	522.44	499.31	−0.2125	4.9756	4.7631
110	0.08813	0.002354	0.6257	−5.81	513.84	508.02	−0.0522	4.6712	4.6191
111.6672	0.101325	0.002368	0.5505	0.00	510.83	510.83	0.0000	4.5746	4.5746
115	0.13221	0.002396	0.4312	11.69	504.59	516.28	0.1025	4.3877	4.4902
120	0.19143	0.002440	0.3066	29.41	494.61	524.02	0.2521	4.1218	4.3738
125	0.26876	0.002487	0.2239	47.37	483.80	531.17	0.3972	3.8704	4.2676
130	0.36732	0.002538	0.1672	65.63	472.04	537.67	0.5385	3.6310	4.1695
135	0.49035	0.002593	0.1273	84.22	459.20	543.42	0.6764	3.4015	4.0779
140	0.64118	0.002653	0.0985	103.20	445.14	548.34	0.8116	3.1796	3.9912
145	0.82322	0.002720	0.0772	122.65	429.68	552.32	0.9446	2.9633	3.9079
150	1.0400	0.002794	0.0612	142.64	412.58	555.23	1.0761	2.7506	3.8267
155	1.2950	0.002878	0.0490	163.31	393.58	556.89	1.2069	2.5392	3.7461
160	1.5921	0.002973	0.0394	184.80	372.27	557.07	1.3378	2.3267	3.6645
165	1.9351	0.003086	0.0318	207.33	348.12	555.45	1.4701	2.1098	3.5799
170	2.3283	0.003221	0.02566	231.24	320.30	551.54	1.6054	1.8841	3.4895
175	2.7765	0.003391	0.02059	257.09	287.44	544.52	1.7466	1.6425	3.3891
180	3.2852	0.003620	0.01629	285.94	246.89	532.83	1.8991	1.3716	3.2707
185	3.8617	0.003978	0.01243	320.51	191.97	512.49	2.0765	1.0377	3.1142
190.56	4.599	0.006148	0.006148	415.59	0.0	415.59	2.5624	0.0	2.5624

Table A.11 Properties of gaseous methane.

P / MPa (T^{sat} / K)		T / K: sat	150	175	200	225	250	300	350	400
0.101325	v / m^3/kg	0.5505	0.7558	0.8866	1.016	1.146	1.275	1.532	1.789	2.045
(111.7)	h / kJ/kg	510.83	593.32	646.08	698.70	751.49	804.76	914.09	1029.08	1151.60
	s / kJ/(kg · K)	4.5746	5.2103	5.5357	5.8167	6.0654	6.2899	6.6883	7.0426	7.3695
0.20	v / m^3/kg	0.2944	0.3771	0.4449	0.5117	0.5779	0.6437	0.7748	0.9053	1.0356
(120.6)	h / kJ/kg	524.94	589.98	643.61	696.75	749.91	803.45	913.13	1028.36	1151.04
	s / kJ/(kg · K)	4.3601	4.8432	5.1739	5.4578	5.7082	5.9338	6.3336	6.6886	7.0159
0.50	v / m^3/kg	0.1250	0.1434	0.1726	0.2007	0.2280	0.2550	0.3083	0.3611	0.4137
(135.4)	h / kJ/kg	543.79	579.07	635.79	690.72	745.05	799.42	910.22	1026.17	1149.34
	s / kJ/(kg · K)	4.0716	4.3192	4.6692	4.9627	5.2186	5.4478	5.8516	6.2088	6.5375
1.0	v / m^3/kg	0.06370	0.06436	0.08154	0.09685	0.1114	0.1254	0.1529	0.1798	0.2064
(149.1)	h / kJ/kg	554.81	557.27	621.60	680.18	736.71	792.60	905.35	1022.51	1146.52
	s / kJ/(kg · K)	3.8406	3.8571	4.2545	4.5675	4.8340	5.0695	5.4805	5.8415	6.1724

Table A.12 Properties of liquid methane.

P / MPa		T / K: 100	110	120	130	140	150	160	170	180
	(P^{sat} / MPa)	0.03438	0.08813	0.1914	0.3673	0.6412	1.040	1.592	2.328	3.285
sat	ρ / kg/m^3	438.89	424.78	409.90	394.04	376.87	357.90	336.31	310.50	276.23
	h / kJ/kg	−40.27	−5.81	29.41	65.63	103.20	142.64	184.80	231.24	285.94
	s / kJ/(kg · K)	−0.3793	−0.0522	0.2521	0.5385	0.8116	1.0761	1.3378	1.6054	1.8991
0.101325	ρ / kg/m^3	438.94	424.79							
	h / kJ/kg	−40.16	−5.79							
	s / kJ/(kg · K)	−0.3798	−0.0523							
0.50	ρ / kg/m^3	439.24	425.15	410.25	394.22					
	h / kJ/kg	−39.54	−5.21	29.82	65.78					
	s / kJ/(kg · K)	−0.3827	−0.0555	0.2492	0.5370					
2.0	ρ / kg/m^3	440.37	426.50	411.88	396.27	379.27	360.21	337.78		
	h / kJ/kg	−37.19	−2.99	31.84	67.52	104.37	142.99	184.55		
	s / kJ/(kg · K)	−0.3933	−0.0673	0.2357	0.5212	0.7943	1.0606	1.3287		

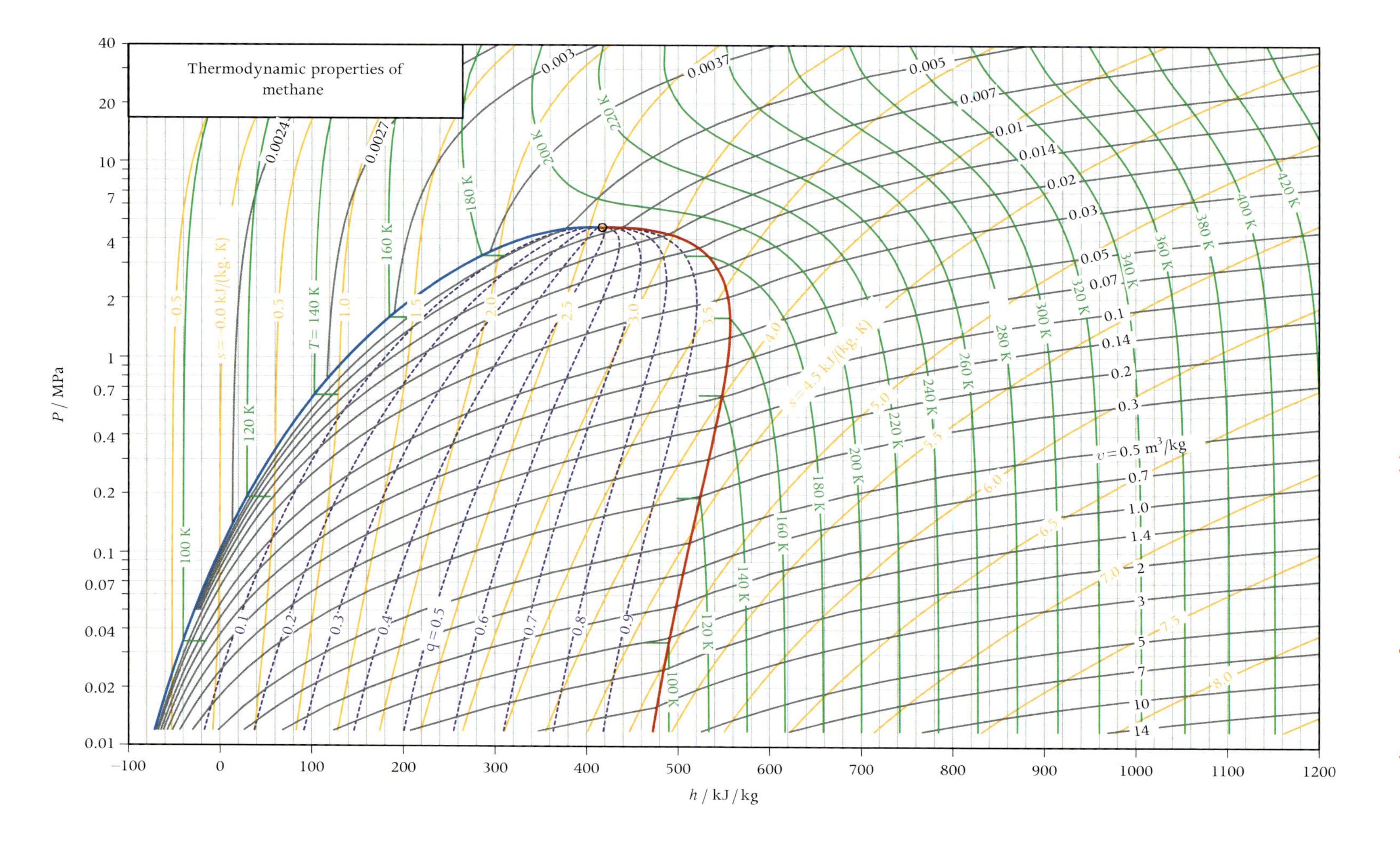
Thermodynamic properties of methane
P / MPa
h / kJ/kg
T = 140 K
100 K
120 K
160 K
180 K
200 K
220 K
240 K
260 K
280 K
300 K
320 K
340 K
360 K
380 K
400 K
420 K
s = 0.0 kJ/(kg. K)
s = 4.5 kJ/(kg. K)
v = 0.5 m³/kg
q = 0.5

A.6 Propane

Properties in tables and charts are calculated with RefProp, which implements the thermodynamic model for propane documented in E. Lemmon, M. McLinden, and W. Wagner, "Thermodynamic properties of propane III. A reference equation of state for temperatures from the melting line to 650 K and pressures up to 1000 MPa," *J. Chem. Eng. Data*, **54**, (*12*), 3141–3180, 2009.
Reference state: $h_0 = h^{\text{sat,L}}(T = 0\,°\text{C}) = 200\ \text{kJ/kg}$, $s_0 = s^{\text{sat,L}}(T = 0\,°\text{C}) = 1\ \text{kJ/}(\text{kg}\cdot\text{K})$.

Table A.13 Properties of saturated propane as function of temperature.

T / K	P^{sat} / MPa	v^{L} / m³/kg	v^{V} / m³/kg	h^{L} / kJ/kg	h^{LV} / kJ/kg	h^{V} / kJ/kg	s^{L} / kJ/(kg · K)	s^{LV} / kJ/(kg · K)	s^{V} / kJ/(kg · K)
200	0.02019	0.001625	1.846	32.5	456.10	488.63	0.3	2.2805	2.5731
210	0.03603	0.001654	1.079	53.98	446.66	500.64	0.3971	2.1270	2.5241
220	0.06057	0.001685	0.6671	75.80	436.89	512.69	0.4984	1.9859	2.4843
230	0.09678	0.001718	0.4319	98.03	426.68	524.71	0.5970	1.8551	2.4521
231.04	0.101325	0.001722	0.4139	100.36	425.59	525.95	0.6070	1.8421	2.4491
240	0.1480	0.001753	0.2908	120.72	415.93	536.65	0.6932	1.7330	2.4262
250	0.2180	0.001791	0.2024	143.93	404.52	548.45	0.7874	1.6181	2.4055
260	0.3107	0.001832	0.1449	167.72	392.33	560.05	0.8801	1.5090	2.3890
270	0.4304	0.001877	0.1061	192.16	379.21	571.37	0.9714	1.4045	2.3759
280	0.5817	0.001926	0.07926	217.31	365.01	582.32	1.0619	1.3036	2.3655
290	0.7691	0.001981	0.06016	243.27	349.51	592.78	1.1517	1.2052	2.3569
300	0.9977	0.002043	0.04623	270.15	332.45	602.60	1.2412	1.1082	2.3494
310	1.272	0.002114	0.03586	298.10	313.50	611.60	1.3310	1.0113	2.3423
320	1.599	0.002198	0.02798	327.30	292.17	619.47	1.4214	0.9130	2.3345
330	1.983	0.002300	0.02185	358.06	267.74	625.80	1.5134	0.8113	2.3247
340	2.431	0.002429	0.01698	390.86	238.94	629.80	1.6082	0.7028	2.3109
350	2.951	0.002606	0.01298	426.70	203.26	629.95	1.7082	0.5807	2.2890
360	3.555	0.002894	0.009490	468.18	154.19	622.36	1.8204	0.4283	2.2487
369.89	4.251	0.004536	0.004536	555.24	0.00	555.24	2.0516	0.0000	2.0516

Table A.14 Properties of gaseous propane.

P / MPa		T / K								
(T^{sat} / K)		sat	250	300	350	400	450	500	550	600
0.050	v / m^3/kg	0.7969	0.9296	1.123	1.314	1.504	1.6934	1.883	2.072	2.261
(216.2)	h / kJ/kg	508.09	556.04	634.90	724.49	825.25	937.02	1059.28	1191.35	1332.57
	s / kJ/(kg · K)	2.4985	2.7044	2.9914	3.2671	3.5359	3.7990	4.0564	4.3080	4.5536
0.101325	v / m^3/kg	0.4139	0.4519	0.5494	0.6450	0.7397	0.8338	0.9277	1.021	1.115
(231.0)	h / kJ/kg	525.95	553.81	633.46	723.46	824.47	936.40	1058.76	1190.92	1332.20
	s / kJ/(kg · K)	2.4491	2.5650	2.8549	3.1319	3.4014	3.6648	3.9225	4.1742	4.4200
0.20	v / m^3/kg	0.2194	0.2219	0.2738	0.3236	0.3724	0.4207	0.4686	0.5164	0.5641
(247.7)	h / kJ/kg	545.75	549.31	630.61	721.45	822.95	935.19	1057.78	1190.10	1331.50
	s / kJ/(kg · K)	2.4099	2.4241	2.7201	2.9998	3.2705	3.5347	3.7928	4.0449	4.2909
0.40	v / m^3/kg	0.1139		0.1321	0.1585	0.1838	0.2085	0.2329	0.2572	0.2812
(267.7)	h / kJ/kg	568.78		624.53	717.27	819.83	932.73	1055.77	1188.42	1330.07
	s / kJ/(kg · K)	2.3787		2.5752	2.8608	3.1344	3.4001	3.6592	3.9119	4.1583
0.70	v / m^3/kg	0.06606		0.07102	0.08766	0.10293	0.1176	0.1319	0.1460	0.1600
(286.5)	h / kJ/kg	589.23		614.40	710.68	815.02	928.99	1052.73	1185.89	1327.93
	s / kJ/(kg · K)	2.3597		2.4455	2.7421	3.0205	3.2887	3.5493	3.8030	4.0500
1.0	v / m^3/kg	0.04612			0.05920	0.07055	0.08121	0.09152	0.1016	0.1116
(300.1)	h / kJ/kg	602.69			703.65	810.05	925.17	1049.66	1183.34	1325.79
	s / kJ/(kg · K)	2.3493			2.6605	2.9444	3.2154	3.4775	3.7322	3.9800
2.0	v / m^3/kg	0.02163			0.02538	0.03262	0.03873	0.04438	0.04977	0.05502
(330.4)	h / kJ/kg	626.02			675.36	792.07	911.93	1039.22	1174.79	1318.62
	s / kJ/(kg · K)	2.3243			2.4694	2.7812	3.0634	3.3315	3.5898	3.8400
4.0	v / m^3/kg	0.00714				0.01315	0.01740	0.02082	0.02390	0.02680
(366.5)	h / kJ/kg	605.46				745.91	882.53	1017.31	1157.38	1304.28
	s / kJ/(kg · K)	2.1920				2.5620	2.8840	3.1680	3.4349	3.6904
7.0	v / m^3/kg					0.00409	0.00822	0.01080	0.01291	0.01480
	h / kJ/kg					621.03	829.87	982.31	1130.87	1283.02
	s / kJ/(kg · K)					2.1922	2.6873	3.0088	3.2919	3.5566
10	v / m^3/kg					0.00299	0.00493	0.00698	0.00864	0.01010
	h / kJ/kg					576.05	775.57	947.14	1105.07	1262.71
	s / kJ/(kg · K)					2.0545	2.5246	2.8866	3.1877	3.4620
20	v / m^3/kg					0.00243	0.00291	0.00357	0.00432	0.00507
	h / kJ/kg					551.77	709.10	875.05	1042.61	1210.44
	s / kJ/(kg · K)					1.9279	2.2983	2.6479	2.9673	3.2593

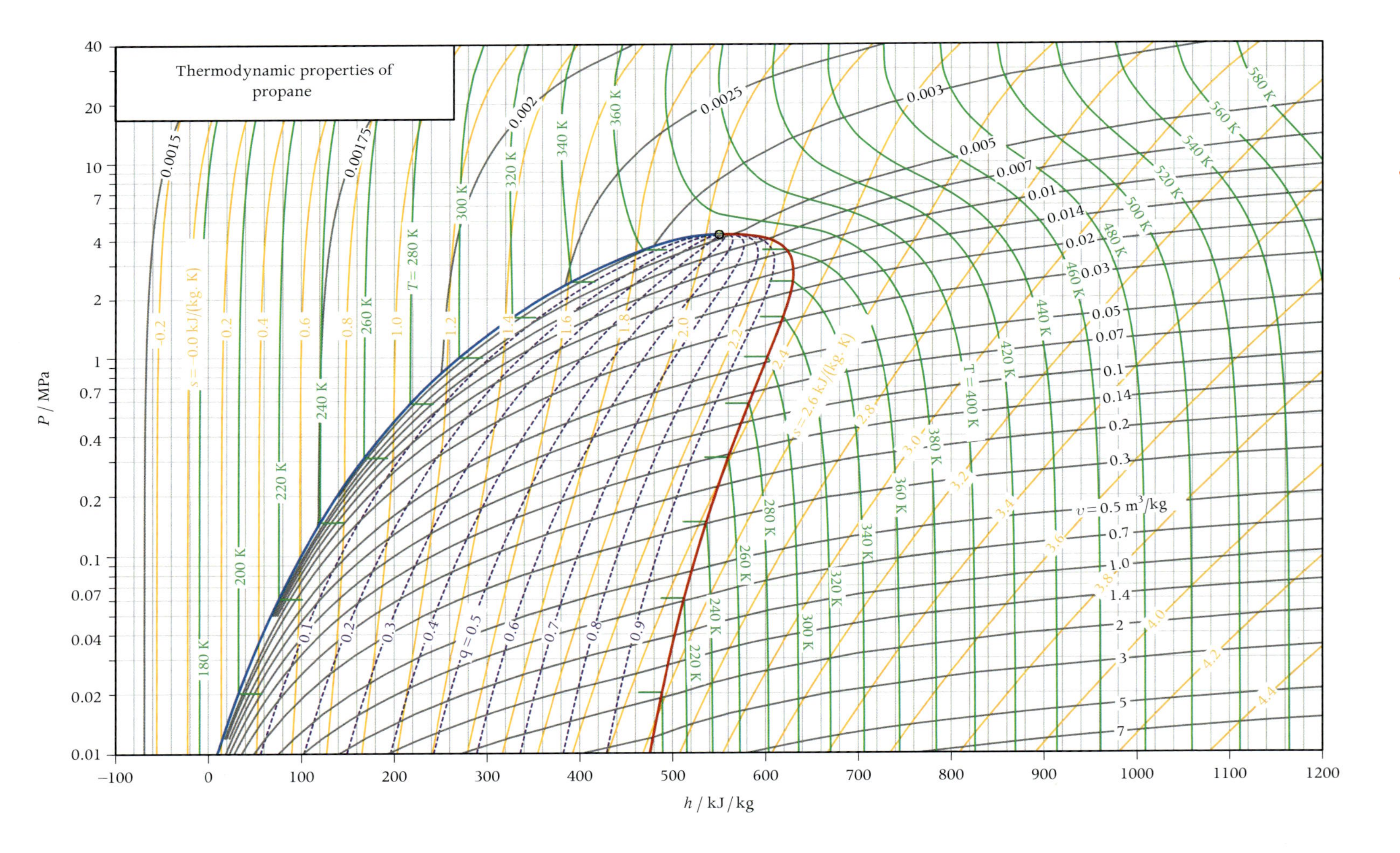
Thermodynamic properties of propane
P / MPa
h / kJ / kg
$s = 2.6$ kJ/(kg · K)
$v = 0.5$ m^3/kg
$T = 400$ K
$T = 280$ K
$q = 0.5$
$s = -0.0$ kJ/(kg · K)

A.7 Ammonia

Properties in tables and charts are calculated with RefProp, which implements the thermodynamic model for ammonia documented in R. Tillner-Roth, F. Harms-Watzenberg, and H. Baehr, "Eine neue Fundamentalgleichung fuer Ammoniak," *DKV-Tagungsbericht*, **20**, 167–181, 1993.

Reference state: $u_0 = u^{sat,L}(T_{triple} = 195.495\ \text{K}) = 0\ \text{kJ/kg}$, $s_0 = s^{sat,L}(T_{triple} = 195.495\ \text{K}) = 0\ \text{kJ/(kg} \cdot \text{K)}$.

Table A.15 Properties of saturated ammonia as function of temperature.

T / K	p^{sat} / MPa	v^L / m^3/kg	v^V / m^3/kg	h^L / kJ/kg	h^{LV} / kJ/kg	h^V / kJ/kg	s^L / kJ/(kg · K)	s^{LV} / kJ/(kg · K)	s^V / kJ/(kg · K)
200	0.008651	0.001373	11.23	19.00	1473.86	1492.85	0.1	7.3693	7.4653
204	0.01164	0.001382	8.500	35.95	1464.33	1500.28	0.1799	7.1781	7.3580
208	0.01546	0.001390	6.515	53.00	1454.60	1507.61	0.2627	6.9933	7.2560
212	0.02029	0.001399	5.052	70.15	1444.69	1514.83	0.3443	6.8146	7.1588
216	0.02633	0.001408	3.961	87.38	1434.56	1521.94	0.4248	6.6415	7.0663
220	0.03379	0.001417	3.136	104.71	1424.22	1528.93	0.5042	6.4737	6.9780
224	0.04293	0.001426	2.508	122.14	1413.65	1535.79	0.5827	6.3109	6.8936
228	0.05403	0.001436	2.023	139.65	1402.85	1542.50	0.6601	6.1529	6.8129
232	0.06739	0.001446	1.646	157.25	1391.81	1549.06	0.7365	5.9992	6.7357
236	0.08333	0.001456	1.350	174.93	1380.53	1555.46	0.8120	5.8497	6.6617
239.82	0.101325	0.001466	1.124	191.92	1369.50	1561.42	0.8833	5.7104	6.5937
240	0.1022	0.001467	1.115	192.70	1368.98	1561.69	0.8866	5.7041	6.5907
244	0.1245	0.001478	0.9275	210.56	1357.18	1567.74	0.9602	5.5622	6.5224
248	0.1504	0.001489	0.7767	228.49	1345.10	1573.59	1.0329	5.4238	6.4567
252	0.1806	0.001500	0.6545	246.51	1332.73	1579.24	1.1048	5.2886	6.3935
256	0.2154	0.001512	0.5547	264.61	1320.08	1584.69	1.1759	5.1566	6.3324
260	0.2553	0.001524	0.4727	282.79	1307.12	1589.91	1.2461	5.0274	6.2735
262	0.2774	0.001530	0.4372	291.91	1300.52	1592.43	1.2809	4.9638	6.2448
264	0.3009	0.001536	0.4049	301.05	1293.85	1594.90	1.3156	4.9009	6.2165
266	0.3260	0.001543	0.3755	310.22	1287.09	1597.31	1.3500	4.8387	6.1887
268	0.3527	0.001549	0.3486	319.40	1280.25	1599.65	1.3842	4.7770	6.1613
270	0.3811	0.001556	0.3239	328.61	1273.32	1601.93	1.4183	4.7160	6.1343
272	0.4112	0.001562	0.3014	337.84	1266.31	1604.15	1.4522	4.6556	6.1077
274	0.4432	0.001569	0.2807	347.09	1259.21	1606.30	1.4859	4.5957	6.0815

Table A.15 (cont.)

T / K	P^{sat} / MPa	v^L / m^3/kg	v^V / m^3/kg	h^L / kJ/kg	h^{LV} / kJ/kg	h^V / kJ/kg	s^L / kJ/(kg · K)	s^{LV} / kJ/(kg · K)	s^V / kJ/(kg · K)
276	0.4771	0.001576	0.2617	356.36	1252.02	1608.38	1.5194	4.5363	6.0557
278	0.5130	0.001583	0.2443	365.66	1244.74	1610.40	1.5528	4.4775	6.0303
280	0.5509	0.001590	0.2282	374.99	1237.36	1612.35	1.5860	4.4191	6.0051
282	0.5910	0.001597	0.2134	384.33	1229.89	1614.22	1.6190	4.3613	5.9803
284	0.6333	0.001604	0.1998	393.71	1222.32	1616.03	1.6519	4.3039	5.9559
286	0.6779	0.001611	0.1871	403.11	1214.65	1617.76	1.6846	4.2470	5.9317
288	0.7249	0.001619	0.1755	412.53	1206.88	1619.41	1.7172	4.1905	5.9078
290	0.7744	0.001627	0.1646	421.99	1199.00	1620.99	1.7497	4.1345	5.8841
292	0.8264	0.001634	0.1546	431.47	1191.02	1622.49	1.7820	4.0788	5.8608
294	0.8810	0.001642	0.1453	440.98	1182.92	1623.90	1.8141	4.0235	5.8377
296	0.9384	0.001650	0.1367	450.53	1174.71	1625.24	1.8461	3.9686	5.8148
298	0.9986	0.001658	0.1287	460.10	1166.39	1626.49	1.8781	3.9141	5.7921
300	1.062	0.001667	0.1212	469.71	1157.95	1627.66	1.9098	3.8598	5.7697
304	1.197	0.001684	0.1078	489.03	1140.70	1629.73	1.9731	3.7523	5.7254
308	1.345	0.001702	0.09604	508.49	1122.94	1631.44	2.0358	3.6459	5.6818
312	1.506	0.001720	0.08580	528.12	1104.64	1632.76	2.0983	3.5405	5.6388
316	1.682	0.001740	0.07682	547.91	1085.76	1633.67	2.1603	3.4359	5.5963
320	1.873	0.001760	0.06892	567.90	1066.26	1634.16	2.2221	3.3321	5.5542
324	2.079	0.001781	0.06194	588.08	1046.12	1634.20	2.2837	3.2288	5.5124
328	2.302	0.001804	0.05576	608.49	1025.26	1633.75	2.3450	3.1258	5.4709
332	2.543	0.001827	0.05026	629.14	1003.66	1632.80	2.4063	3.0231	5.4294
336	2.802	0.001852	0.04537	650.05	981.24	1631.29	2.4675	2.9203	5.3878
340	3.080	0.001878	0.04099	671.27	957.93	1629.19	2.5287	2.8174	5.3462
345	3.456	0.001913	0.03615	698.24	927.42	1625.67	2.6054	2.6882	5.2936
350	3.866	0.001952	0.03191	725.80	895.22	1621.02	2.6824	2.5578	5.2402
355	4.311	0.001994	0.02818	754.03	861.10	1615.12	2.7600	2.4256	5.1856
360	4.793	0.002040	0.02488	783.03	824.76	1607.79	2.8384	2.2910	5.1294
365	5.314	0.002091	0.02195	812.95	785.86	1598.80	2.9180	2.1530	5.0710
370	5.878	0.002149	0.01933	843.96	743.90	1587.86	2.9991	2.0106	5.0096
375	6.485	0.002216	0.01698	876.30	698.26	1574.56	3.0824	1.8620	4.9444
380	7.140	0.002293	0.01484	910.32	648.01	1558.33	3.1686	1.7053	4.8739
385	7.845	0.002386	0.01289	946.55	591.76	1538.31	3.2590	1.5370	4.7960
390	8.604	0.002502	0.01108	985.90	527.17	1513.08	3.3557	1.3517	4.7075
395	9.422	0.002658	0.009362	1030.29	449.62	1479.91	3.4634	1.1383	4.6017
400	10.30	0.002902	0.007628	1085.02	346.89	1431.91	3.5949	0.8672	4.4622
405.40	11.34	0.004444	0.004444	1262.38	0.0	1262.38	4.0257	0.0	4.0257

Table A.16 Properties of saturated ammonia as function of pressure.

P / MPa	T^{sat} / K	v^L / m^3/kg	v^V / m^3/kg	h^L / kJ/kg	h^{LV} / kJ/kg	h^V / kJ/kg	s^L / kJ/(kg · K)	s^{LV} / kJ/(kg · K)	s^V / kJ/(kg · K)
0.010	201.93	0.001377	9.800	27.17	1469.28	1496.45	0.1367	7.2762	7.4128
0.015	207.56	0.001389	6.703	51.14	1455.67	1506.81	0.2537	7.0131	7.2669
0.020	211.78	0.001398	5.121	69.21	1445.23	1514.44	0.3399	6.8241	7.1640
0.025	215.19	0.001406	4.157	83.90	1436.62	1520.52	0.4086	6.6760	7.0846
0.030	218.07	0.001412	3.505	96.35	1429.23	1525.58	0.4661	6.5539	7.0200
0.035	220.58	0.001418	3.035	107.22	1422.70	1529.93	0.5156	6.4499	6.9655
0.040	222.80	0.001423	2.679	116.90	1416.84	1533.75	0.5593	6.3592	6.9185
0.045	224.80	0.001428	2.400	125.65	1411.49	1537.15	0.5983	6.2788	6.8771
0.050	226.63	0.001433	2.175	133.65	1406.57	1540.22	0.6337	6.2064	6.8401
0.055	228.32	0.001437	1.990	141.04	1401.99	1543.02	0.6662	6.1406	6.8067
0.060	229.88	0.001441	1.834	147.90	1397.70	1545.60	0.6961	6.0802	6.7763
0.065	231.34	0.001444	1.702	154.32	1393.66	1547.98	0.7239	6.0244	6.7483
0.070	232.70	0.001448	1.588	160.36	1389.84	1550.20	0.7499	5.9725	6.7224
0.080	235.22	0.001454	1.402	171.47	1382.75	1554.22	0.7973	5.8786	6.6759
0.101325	239.82	0.001466	1.124	191.92	1369.50	1561.42	0.8833	5.7104	6.5937
0.12	243.25	0.001476	0.9597	207.20	1359.42	1566.61	0.9464	5.5886	6.5350
0.14	246.47	0.001484	0.8307	221.61	1349.76	1571.37	1.0052	5.4764	6.4816
0.16	249.34	0.001492	0.7331	234.50	1341.00	1575.50	1.0570	5.3783	6.4353
0.18	251.93	0.001500	0.6564	246.19	1332.96	1579.15	1.1036	5.2910	6.3946
0.20	254.30	0.001507	0.5946	256.92	1325.49	1582.40	1.1458	5.2123	6.3581
0.24	258.53	0.001519	0.5010	276.10	1311.92	1588.02	1.2204	5.0745	6.2949
0.28	262.23	0.001531	0.4334	292.96	1299.76	1592.72	1.2849	4.9566	6.2415
0.32	265.53	0.001541	0.3821	308.08	1288.67	1596.75	1.3420	4.8531	6.1951
0.36	268.53	0.001551	0.3419	321.83	1278.43	1600.26	1.3932	4.7609	6.1541
0.40	271.27	0.001560	0.3094	334.46	1268.88	1603.34	1.4398	4.6776	6.1174
0.44	273.80	0.001568	0.2827	346.18	1259.91	1606.09	1.4826	4.6015	6.0841
0.48	276.17	0.001576	0.2602	357.14	1251.42	1608.55	1.5222	4.5314	6.0536
0.52	278.38	0.001584	0.2411	367.43	1243.35	1610.77	1.5591	4.4664	6.0255
0.56	280.46	0.001591	0.2247	377.15	1235.64	1612.79	1.5937	4.4057	5.9994
0.60	282.43	0.001598	0.2104	386.37	1228.25	1614.62	1.6262	4.3488	5.9750
0.64	284.31	0.001605	0.1977	395.15	1221.15	1616.30	1.6569	4.2952	5.9521
0.68	286.09	0.001612	0.1866	403.54	1214.30	1617.83	1.6861	4.2444	5.9306
0.72	287.80	0.001618	0.1766	411.57	1207.67	1619.24	1.7139	4.1963	5.9102

Table A.16 (cont.)

P / MPa	T^{sat} / K	v^L / m^3/kg	v^V / m^3/kg	h^L / kJ/kg	h^{LV} / kJ/kg	h^V / kJ/kg	s^L / kJ/(kg · K)	s^{LV} / kJ/(kg · K)	s^V / kJ/(kg · K)
0.76	289.43	0.001624	0.1676	419.29	1201.26	1620.55	1.7404	4.1504	5.8909
0.80	291.00	0.001630	0.1596	426.72	1195.03	1621.75	1.7658	4.1067	5.8724
0.84	292.51	0.001636	0.1522	433.88	1188.97	1622.85	1.7901	4.0648	5.8549
0.88	293.96	0.001642	0.1455	440.81	1183.07	1623.88	1.8135	4.0246	5.8381
0.92	295.37	0.001648	0.1394	447.51	1177.32	1624.83	1.8360	3.9859	5.8220
0.96	296.73	0.001653	0.1337	454.01	1171.70	1625.71	1.8578	3.9487	5.8065
1.0	298.05	0.001659	0.1285	460.32	1166.20	1626.52	1.8788	3.9128	5.7916
1.2	304.09	0.001684	0.1075	489.44	1140.33	1629.77	1.9744	3.7500	5.7244
* 1.4	309.40	0.001708	0.09230	515.36	1116.59	1631.95	2.0578	3.6089	5.6666
* 1.6	314.17	0.001731	0.08078	538.85	1094.46	1633.31	2.1320	3.4836	5.6156
* 1.8	318.51	0.001752	0.07174	560.44	1073.59	1634.03	2.1992	3.3707	5.5698
* 2.0	322.50	0.001773	0.06445	580.49	1053.75	1634.24	2.2606	3.2674	5.5281
* 2.4	329.66	0.001813	0.05340	617.02	1016.40	1633.42	2.3704	3.0832	5.4537
2.8	335.97	0.001852	0.04540	649.89	981.41	1631.30	2.4670	2.9211	5.3882
3.2	341.64	0.001889	0.03933	680.05	948.10	1628.15	2.5538	2.7752	5.3290
3.6	346.80	0.001927	0.03456	708.09	916.03	1624.13	2.6331	2.6414	5.2745
4.0	351.55	0.001964	0.03071	734.47	884.86	1619.34	2.7064	2.5170	5.2234
4.4	355.96	0.002002	0.02752	759.51	854.33	1613.84	2.7749	2.4001	5.1750
4.8	360.07	0.002040	0.02484	783.45	824.23	1607.68	2.8395	2.2891	5.1286
5.2	363.94	0.002080	0.02255	806.49	794.38	1600.87	2.9009	2.1827	5.0837
5.6	367.58	0.002120	0.02056	828.81	764.61	1593.42	2.9596	2.0801	5.0397
6.0	371.04	0.002162	0.01882	850.54	734.77	1585.31	3.0162	1.9803	4.9965
6.5	375.12	0.002217	0.01693	877.06	697.16	1574.22	3.0843	1.8585	4.9428
7.0	378.96	0.002276	0.01527	903.08	658.90	1561.98	3.1503	1.7387	4.8890
8.0	386.05	0.002408	0.01250	954.50	579.01	1533.51	3.2786	1.4998	4.7785
10	398.32	0.002804	0.008225	1064.72	385.87	1450.60	3.5463	0.9688	4.5150
11.333	405.40	0.004444	0.004444	1262.38	0.0	1262.38	4.0257	0.0	4.0257

Table A.17 Properties of gaseous ammonia.

P / MPa (T^{sat} / K)		T / K sat	220	230	240	250	260	270	280	290
0.020	v / m^3/kg	5.121	5.329	5.579	5.828	6.076	6.324	6.571	6.817	7.063
(211.8)	h / kJ/kg	1514.44	1531.77	1552.68	1573.48	1594.22	1614.97	1635.75	1656.60	1677.53
	s / kJ/(kg · K)	7.1640	7.2443	7.3372	7.4257	7.5104	7.5918	7.6702	7.7460	7.8195
0.030	v / m^3/kg	3.505	3.538	3.707	3.875	4.042	4.208	4.374	4.539	4.703
(218.1)	h / kJ/kg	1525.58	1529.72	1551.01	1572.09	1593.06	1613.98	1634.90	1655.85	1676.87
	s / kJ/(kg · K)	7.0200	7.0389	7.1335	7.2233	7.3089	7.3909	7.4699	7.5461	7.6198
0.040	v / m^3/kg	2.679		2.772	2.899	3.025	3.150	3.275	3.399	3.523
(222.8)	h / kJ/kg	1533.7		1549.3	1570.7	1591.9	1613.0	1634.0	1655.1	1676.2
	s / kJ/(kg · K)	6.9185		6.9873	7.0782	7.1648	7.2475	7.3270	7.4036	7.4777
0.050	v / m^3/kg	2.175		2.210	2.313	2.415	2.516	2.616	2.716	2.815
(226.6)	h / kJ/kg	1540.22		1547.60	1569.28	1590.70	1611.97	1633.17	1654.35	1675.55
	s / kJ/(kg · K)	6.8401		6.8724	6.9647	7.0521	7.1356	7.2156	7.2926	7.3670
0.070	v / m^3/kg	1.588			1.643	1.717	1.790	1.863	1.935	2.006
(232.7)	h / kJ/kg	1550.20			1566.42	1588.31	1609.95	1631.43	1652.84	1674.22
	s / kJ/(kg · K)	6.7224			6.7911	6.8804	6.9653	7.0464	7.1242	7.1993
0.101325	v / m^3/kg	1.124			1.125	1.178	1.229	1.280	1.331	1.381
(239.8)	h / kJ/kg	1561.42			1561.82	1584.49	1606.72	1628.67	1650.44	1672.12
	s / kJ/(kg · K)	6.5937			6.5954	6.6880	6.7752	6.8580	6.9372	7.0132
0.14	v / m^3/kg	0.831				0.84	0.88	0.92	0.96	0.99
(246.5)	h / kJ/kg	1571.37				1579.65	1602.65	1625.20	1647.44	1669.50
	s / kJ/(kg · K)	6.4816				6.5149	6.6052	6.6903	6.7712	6.8486
0.20	v / m^3/kg	0.595					0.611	0.638	0.665	0.691
(254.3)	h / kJ/kg	1582.40					1596.14	1619.69	1642.70	1665.37
	s / kJ/(kg · K)	6.3581					6.4115	6.5004	6.5841	6.6636
0.30	v / m^3/kg	0.406						0.418	0.437	0.455
(263.9)	h / kJ/kg	1594.81						1610.10	1634.53	1658.31
	s / kJ/(kg · K)	6.2175						6.2748	6.3637	6.4471

Table A.17 (cont.)

P / MPa (T^{sat} / K)		T / K 300	310	320	330	340	350	360	370	380
0.020	v / m^3/kg	7.309	7.554	7.799	8.044	8.289	8.534	8.779	9.024	9.268
(211.8)	h / kJ/kg	1698.56	1719.71	1740.98	1762.40	1783.96	1805.68	1827.57	1849.63	1871.86
	s / kJ/(kg · K)	7.8908	7.9601	8.0277	8.0936	8.1579	8.2209	8.2826	8.3430	8.4023
0.030	v / m^3/kg	4.867	5.032	5.195	5.359	5.523	5.686	5.850	6.013	6.176
(218.1)	h / kJ/kg	1697.97	1719.18	1740.51	1761.97	1783.57	1805.33	1827.24	1849.33	1871.58
	s / kJ/(kg · K)	7.6914	7.7609	7.8286	7.8947	7.9591	8.0222	8.0839	8.1444	8.2038
0.040	v / m^3/kg	3.647	3.770	3.893	4.017	4.140	4.262	4.385	4.508	4.630
(222.8)	h / kJ/kg	1697.39	1718.66	1740.04	1761.54	1783.19	1804.97	1826.92	1849.03	1871.30
	s / kJ/(kg · K)	7.5495	7.6192	7.6871	7.7532	7.8179	7.8810	7.9428	8.0034	8.0628
0.050	v / m^3/kg	2.915	3.014	3.112	3.211	3.310	3.408	3.506	3.605	3.703
(226.6)	h / kJ/kg	1696.80	1718.14	1739.57	1761.12	1782.80	1804.62	1826.59	1848.72	1871.02
	s / kJ/(kg · K)	7.4390	7.5090	7.5770	7.6433	7.7081	7.7713	7.8332	7.8939	7.9533
0.070	v / m^3/kg	2.078	2.149	2.220	2.290	2.361	2.432	2.502	2.572	2.643
(232.7)	h / kJ/kg	1695.63	1717.08	1738.62	1760.26	1782.02	1803.90	1825.93	1848.12	1870.46
	s / kJ/(kg · K)	7.2718	7.3422	7.4106	7.4772	7.5421	7.6056	7.6676	7.7284	7.7880
0.101325	v / m^3/kg	1.431	1.480	1.530	1.579	1.628	1.677	1.726	1.775	1.823
(239.8)	h / kJ/kg	1693.77	1715.43	1737.13	1758.91	1780.79	1802.78	1824.90	1847.17	1869.58
	s / kJ/(kg · K)	7.0866	7.1577	7.2266	7.2936	7.3589	7.4226	7.4850	7.5460	7.6057
0.14	v / m^3/kg	1.031	1.068	1.104	1.140	1.175	1.211	1.247	1.282	1.317
(246.5)	h / kJ/kg	1691.46	1713.37	1735.28	1757.24	1779.27	1801.39	1823.63	1845.99	1868.49
	s / kJ/(kg · K)	6.9230	6.9949	7.0644	7.1320	7.1978	7.2619	7.3245	7.3858	7.4458
0.20	v / m^3/kg	0.7173	0.7431	0.7688	0.7942	0.8195	0.8448	0.8699	0.8949	0.9199
(254.3)	h / kJ/kg	1687.82	1710.14	1732.39	1754.63	1776.90	1799.23	1821.64	1844.16	1866.80
	s / kJ/(kg · K)	6.7397	6.8129	6.8836	6.9520	7.0185	7.0832	7.1463	7.2080	7.2684
0.30	v / m^3/kg	0.4730	0.4908	0.5083	0.5256	0.5428	0.5599	0.5769	0.5939	0.6107
(263.9)	h / kJ/kg	1681.63	1704.66	1727.50	1750.23	1772.91	1795.59	1818.30	1841.08	1863.95
	s / kJ/(kg · K)	6.5262	6.6017	6.6742	6.7442	6.8119	6.8776	6.9416	7.0040	7.0650

Table A.17 (cont.)

P / MPa (T^{sat} / K)		T / K 400	420	440	460	480	500	520	540	560
0.020	v / m^3/kg	9.758	10.25	10.74	11.22	11.71	12.20	12.69	13.18	13.67
(211.8)	h / kJ/kg	1916.89	1962.67	2009.25	2056.66	2104.90	2154.01	2204.00	2254.87	2306.65
	s / kJ/(kg · K)	8.5177	8.6294	8.7378	8.8431	8.9458	9.0460	9.1440	9.2400	9.3342
0.030	v / m^3/kg	6.503	6.829	7.155	7.481	7.807	8.133	8.459	8.785	9.111
(218.1)	h / kJ/kg	1916.64	1962.46	2009.06	2056.49	2104.75	2153.87	2203.87	2254.76	2306.55
	s / kJ/(kg · K)	8.3193	8.4311	8.5395	8.6449	8.7476	8.8479	8.9459	9.0419	9.1361
0.040	v / m^3/kg	4.876	5.120	5.365	5.610	5.854	6.099	6.344	6.588	6.832
(222.8)	h / kJ/kg	1916.40	1962.24	2008.88	2056.32	2104.60	2153.74	2203.75	2254.65	2306.44
	s / kJ/(kg · K)	8.1785	8.2903	8.3987	8.5042	8.6069	8.7072	8.8053	8.9013	8.9955
0.050	v / m^3/kg	3.899	4.095	4.291	4.487	4.683	4.878	5.074	5.270	5.465
(226.6)	h / kJ/kg	1916.16	1962.03	2008.69	2056.15	2104.45	2153.60	2203.62	2254.53	2306.34
	s / kJ/(kg · K)	8.0691	8.1810	8.2895	8.3950	8.4977	8.5981	8.6962	8.7922	8.8864
0.070	v / m^3/kg	2.783	2.923	3.064	3.204	3.344	3.484	3.623	3.763	3.903
(232.7)	h / kJ/kg	1915.67	1961.60	2008.31	2055.81	2104.15	2153.33	2203.37	2254.30	2306.13
	s / kJ/(kg · K)	7.9039	8.0160	8.1246	8.2302	8.3330	8.4334	8.5315	8.6276	8.7219
0.101325	v / m^3/kg	1.921	2.018	2.115	2.212	2.309	2.405	2.502	2.599	2.696
(239.8)	h / kJ/kg	1914.91	1960.94	2007.72	2055.28	2103.67	2152.90	2202.98	2253.95	2305.80
	s / kJ/(kg · K)	7.7220	7.8343	7.9431	8.0488	8.1517	8.2522	8.3504	8.4466	8.5409
0.14	v / m^3/kg	1.388	1.459	1.529	1.600	1.670	1.740	1.810	1.880	1.950
(246.5)	h / kJ/kg	1913.96	1960.11	2006.98	2054.63	2103.08	2152.37	2202.50	2253.51	2305.40
	s / kJ/(kg · K)	7.5624	7.6750	7.7840	7.8899	7.9930	8.0936	8.1919	8.2881	8.3825
0.20	v / m^3/kg	0.9697	1.019	1.069	1.118	1.168	1.217	1.266	1.315	1.364
(254.3)	h / kJ/kg	1912.49	1958.82	2005.84	2053.62	2102.17	2151.54	2201.75	2252.82	2304.77
	s / kJ/(kg · K)	7.3856	7.4986	7.6080	7.7142	7.8175	7.9182	8.0167	8.1131	8.2075
0.30	v / m^3/kg	0.6442	0.6776	0.7108	0.7440	0.7770	0.8100	0.8429	0.8758	0.9086
(263.9)	h / kJ/kg	1910.03	1956.67	2003.94	2051.92	2100.65	2150.16	2200.50	2251.68	2303.73
	s / kJ/(kg · K)	7.1832	7.2969	7.4069	7.5135	7.6172	7.7183	7.8170	7.9136	8.0082

Table A.17 (cont.)

P / MPa (T^{sat} / K)		sat	300	310	320	330	340	350	360	370
0.40	v / m^3/kg	0.3094	0.3508	0.3645	0.3780	0.3913	0.4045	0.4175	0.4304	0.4433
(271.3)	h / kJ/kg	1603.34	1675.27	1699.06	1722.51	1745.75	1768.86	1791.90	1814.93	1837.98
	s / kJ/(kg · K)	6.1174	6.3696	6.4476	6.5221	6.5936	6.6626	6.7294	6.7943	6.8574
0.50	v / m^3/kg	0.2503	0.2774	0.2887	0.2998	0.3107	0.3214	0.3320	0.3425	0.3529
(277.3)	h / kJ/kg	1609.69	1668.72	1693.32	1717.43	1741.21	1764.76	1788.18	1811.53	1834.86
	s / kJ/(kg · K)	6.0393	6.2440	6.3247	6.4012	6.4744	6.5447	6.6126	6.6784	6.7423
0.70	v / m^3/kg	0.1815	0.1933	0.2019	0.2103	0.2184	0.2264	0.2343	0.2420	0.2497
(287.0)	h / kJ/kg	1618.6	1655.01	1681.42	1706.96	1731.89	1756.39	1780.61	1804.63	1828.53
	s / kJ/(kg · K)	5.9202	6.0445	6.1311	6.2122	6.2889	6.3621	6.4323	6.4999	6.5654
1.0	v / m^3/kg	0.1285	0.1299	0.1366	0.1430	0.1491	0.1551	0.1609	0.1666	0.1721
(298.0)	h / kJ/kg	1626.52	1632.60	1662.33	1690.39	1717.29	1743.38	1768.89	1794.00	1818.83
	s / kJ/(kg · K)	5.7916	5.8119	5.9095	5.9985	6.0813	6.1592	6.2332	6.3039	6.3719
1.4	v / m^3/kg	0.09230		0.09262	0.09785	0.1027	0.1074	0.1119	0.1162	0.1204
(309.4)	h / kJ/kg	1631.95		1633.97	1666.36	1696.48	1725.06	1752.56	1779.29	1805.47
	s / kJ/(kg · K)	5.6666		5.6732	5.7760	5.8687	5.9540	6.0338	6.1091	6.1808
2.0	v / m^3/kg	0.06445				0.06753	0.07132	0.07487	0.07825	0.08148
(322.5)	h / kJ/kg	1634.24				1661.53	1695.02	1726.23	1755.88	1784.41
	s / kJ/(kg · K)	5.5281				5.6117	5.7117	5.8022	5.8857	5.9639
3.0	v / m^3/kg	0.04217					0.04255	0.04563	0.04840	0.05096
(338.9)	h / kJ/kg	1629.85					1634.83	1675.72	1712.26	1746.00
	s / kJ/(kg · K)	5.3579					5.3726	5.4912	5.5941	5.6866
4.0	v / m^3/kg	0.03071							0.03301	0.03539
(351.5)	h / kJ/kg	1619.34							1660.05	1701.88
	s / kJ/(kg · K)	5.2234							5.3379	5.4526
5.0	v / m^3/kg	0.02365								0.02565
(362.0)	h / kJ/kg	1604.35								1648.70
	s / kJ/(kg · K)	5.1060								5.2272

Table A.17 (cont.)

P / MPa (T^{sat} / K)		T / K 380	390	400	410	420	430	440	460	480
0.40	v / m^3/kg	0.4561	0.4688	0.4815	0.4941	0.5067	0.5193	0.5318	0.5568	0.5817
(271.3)	h / kJ/kg	1861.09	1884.28	1907.56	1930.97	1954.51	1978.19	2002.03	2050.22	2099.12
	s / kJ/(kg · K)	6.9190	6.9793	7.0382	7.0960	7.1528	7.2085	7.2633	7.3704	7.4744
0.50	v / m^3/kg	0.3633	0.3736	0.3838	0.3940	0.4042	0.4143	0.4244	0.4445	0.4646
(277.3)	h / kJ/kg	1858.21	1881.60	1905.08	1928.65	1952.34	1976.16	2000.12	2048.51	2097.59
	s / kJ/(kg · K)	6.8046	6.8653	6.9248	6.9830	7.0400	7.0961	7.1512	7.2587	7.3632
0.70	v / m^3/kg	0.2572	0.2647	0.2722	0.2796	0.2870	0.2943	0.3016	0.3162	0.3306
(287.0)	h / kJ/kg	1852.37	1876.21	1900.06	1923.98	1947.97	1972.06	1996.27	2045.09	2094.52
	s / kJ/(kg · K)	6.6290	6.6909	6.7513	6.8104	6.8682	6.9249	6.9805	7.0890	7.1942
1.0	v / m^3/kg	0.1776	0.1831	0.1885	0.1938	0.1991	0.2043	0.2096	0.2199	0.2302
(298.0)	h / kJ/kg	1843.46	1867.98	1892.44	1916.88	1941.34	1965.85	1990.44	2039.91	2089.88
	s / kJ/(kg · K)	6.4376	6.5013	6.5632	6.6236	6.6825	6.7402	6.7967	6.9067	7.0130
1.4	v / m^3/kg	0.1245	0.1286	0.1326	0.1365	0.1404	0.1443	0.1481	0.1557	0.1632
(309.4)	h / kJ/kg	1831.25	1856.75	1882.05	1907.24	1932.36	1957.45	1982.56	2032.93	2083.65
	s / kJ/(kg · K)	6.2496	6.3158	6.3799	6.4421	6.5026	6.5616	6.6194	6.7313	6.8392
2.0	v / m^3/kg	0.08461	0.08766	0.09064	0.09356	0.09644	0.09927	0.1021	0.1076	0.1130
(322.5)	h / kJ/kg	1812.14	1839.28	1865.99	1892.39	1918.56	1944.59	1970.53	2022.31	2074.19
	s / kJ/(kg · K)	6.0379	6.1084	6.1760	6.2412	6.3043	6.3655	6.4251	6.5402	6.6506
3.0	v / m^3/kg	0.05338	0.05568	0.05790	0.06005	0.06214	0.06418	0.06618	0.07008	0.07388
(338.9)	h / kJ/kg	1777.82	1808.28	1837.74	1866.46	1894.63	1922.39	1949.85	2004.18	2058.11
	s / kJ/(kg · K)	5.7714	5.8506	5.9251	5.9961	6.0640	6.1293	6.1924	6.3132	6.4279
4.0	v / m^3/kg	0.03754	0.03953	0.04141	0.04319	0.04491	0.04657	0.04819	0.05131	0.05432
(351.5)	h / kJ/kg	1739.50	1774.35	1807.28	1838.83	1869.36	1899.12	1928.30	1985.46	2041.62
	s / kJ/(kg · K)	5.5529	5.6434	5.7268	5.8047	5.8783	5.9483	6.0154	6.1425	6.2620
5.0	v / m^3/kg	0.02779	0.02967	0.03139	0.03299	0.03451	0.03596	0.03736	0.04002	0.04256
(362.0)	h / kJ/kg	1695.47	1736.55	1774.06	1809.16	1842.53	1874.64	1905.79	1966.11	2024.70
	s / kJ/(kg · K)	5.3520	5.4587	5.5537	5.6403	5.7208	5.7963	5.8680	6.0020	6.1267

Table A.17 (cont.)

P / MPa (T^{sat} / K)		T / K 500	520	540	560	580	600	620	640	660
0.40	v / m^3/kg	0.6066	0.6314	0.6561	0.6808	0.7055	0.7301	0.7548	0.7794	0.8039
(271.3)	h / kJ/kg	2148.79	2199.25	2250.54	2302.68	2355.69	2409.57	2464.35	2520.03	2576.61
	s / kJ/(kg · K)	7.5758	7.6748	7.7715	7.8663	7.9593	8.0507	8.1405	8.2289	8.3159
0.50	v / m^3/kg	0.4845	0.5044	0.5243	0.5441	0.5639	0.5837	0.6034	0.6231	0.6428
(277.3)	h / kJ/kg	2147.40	2198.00	2249.40	2301.63	2354.72	2408.69	2463.53	2519.27	2575.91
	s / kJ/(kg · K)	7.4648	7.5640	7.6610	7.7560	7.8492	7.9406	8.0305	8.1190	8.2062
0.70	v / m^3/kg	0.3450	0.3593	0.3736	0.3879	0.4021	0.4163	0.4304	0.4446	0.4587
(287.0)	h / kJ/kg	2144.63	2195.48	2247.11	2299.54	2352.80	2406.91	2461.89	2517.75	2574.49
	s / kJ/(kg · K)	7.2965	7.3962	7.4936	7.5890	7.6824	7.7741	7.8643	7.9529	8.0402
1.0	v / m^3/kg	0.2404	0.2505	0.2606	0.2707	0.2807	0.2907	0.3007	0.3107	0.3206
(298.0)	h / kJ/kg	2140.45	2191.69	2243.66	2296.38	2349.90	2404.24	2459.42	2515.46	2572.37
	s / kJ/(kg · K)	7.1162	7.2167	7.3148	7.4106	7.5045	7.5966	7.6871	7.7761	7.8636
1.4	v / m^3/kg	0.1706	0.1780	0.1853	0.1926	0.1998	0.2070	0.2142	0.2214	0.2285
(309.4)	h / kJ/kg	2134.85	2186.62	2239.03	2292.15	2346.02	2400.67	2456.13	2512.41	2569.54
	s / kJ/(kg · K)	6.9437	7.0452	7.1442	7.2407	7.3353	7.4279	7.5188	7.6081	7.6960
2.0	v / m^3/kg	0.1183	0.1236	0.1288	0.1340	0.1391	0.1443	0.1494	0.1544	0.1595
(322.5)	h / kJ/kg	2126.35	2178.94	2232.05	2285.78	2340.18	2395.30	2451.17	2507.82	2565.28
	s / kJ/(kg · K)	6.7571	6.8602	6.9604	7.0581	7.1536	7.2470	7.3386	7.4285	7.5169
3.0	v / m^3/kg	0.07760	0.08126	0.08486	0.08843	0.09195	0.09545	0.09892	0.1024	0.1058
(338.9)	h / kJ/kg	2111.96	2165.97	2220.30	2275.07	2330.37	2386.29	2442.86	2500.15	2558.17
	s / kJ/(kg · K)	6.5379	6.6438	6.7463	6.8459	6.9429	7.0377	7.1304	7.2214	7.3106
4.0	v / m^3/kg	0.05724	0.06009	0.06289	0.06564	0.06835	0.07104	0.07370	0.07633	0.07895
(351.5)	h / kJ/kg	2097.28	2152.79	2208.39	2264.24	2320.48	2377.22	2434.52	2492.44	2551.04
	s / kJ/(kg · K)	6.3756	6.4845	6.5894	6.6909	6.7896	6.8858	6.9797	7.0717	7.1618
5.0	v / m^3/kg	0.04501	0.04738	0.04970	0.05196	0.05419	0.05639	0.05857	0.06072	0.06285
(362.0)	h / kJ/kg	2082.30	2139.40	2196.32	2253.30	2310.51	2368.09	2426.12	2484.70	2543.88
	s / kJ/(kg · K)	6.2443	6.3563	6.4637	6.5673	6.6677	6.7653	6.8604	6.9534	7.0445

Table A.17 (cont.)

P / MPa (T^{sat} / K)		sat	370	380	390	400	420	440	460	480
5.0	v / m^3/kg	0.02365	0.02565	0.02779	0.02967	0.03139	0.03451	0.03736	0.04002	0.04256
(362.0)	h / kJ/kg	1604.35	1648.70	1695.47	1736.55	1774.06	1842.53	1905.79	1966.11	2024.70
	s / kJ/(kg · K)	5.1060	5.2272	5.3520	5.4587	5.5537	5.7208	5.8680	6.0020	6.1267
6.0	v / m^3/kg	0.01882		0.02097	0.02291	0.02459	0.02752	0.03010	0.03248	0.03471
(371.0)	h / kJ/kg	1585.31		1642.36	1693.29	1737.24	1813.85	1882.21	1946.08	2007.33
	s / kJ/(kg · K)	4.9965		5.1485	5.2808	5.3921	5.5791	5.7382	5.8802	6.0105
8.0	v / m^3/kg	0.01250			0.01360	0.01564	0.01859	0.02094	0.02300	0.02488
(386.0)	h / kJ/kg	1533.5			1573.44	1646.14	1749.24	1831.22	1903.77	1971.16
	s / kJ/(kg · K)	4.7785			4.8814	5.0656	5.3174	5.5082	5.6695	5.8130
10	v / m^3/kg	0.00823				0.00897	0.01296	0.01533	0.01725	0.01895
(398.3)	h / kJ/kg	1450.60				1487.29	1669.88	1773.79	1858.00	1932.94
	s / kJ/(kg · K)	4.5150				4.6070	5.0542	5.2962	5.4834	5.6430
12	v / m^3/kg						0.00878	0.01148	0.01338	0.01497
	h / kJ/kg						1561.02	1707.53	1808.15	1892.48
	s / kJ/(kg · K)						4.7436	5.0851	5.3090	5.4885
14	v / m^3/kg						0.00488	0.00860	0.01057	0.01212
	h / kJ/kg						1360.69	1628.57	1753.53	1849.67
	s / kJ/(kg · K)						4.2340	4.8604	5.1384	5.3432
16	v / m^3/kg						0.00320	0.00632	0.00845	0.00998
	h / kJ/kg						1205.04	1531.59	1693.47	1804.47
	s / kJ/(kg · K)						3.8454	4.6062	4.9667	5.2031
18	v / m^3/kg						0.00285	0.00464	0.00679	0.00832
	h / kJ/kg						1159.63	1423.28	1628.01	1757.03
	s / kJ/(kg · K)						3.7230	4.3354	4.7915	5.0663
20	v / m^3/kg						0.00268	0.00372	0.00551	0.00701
	h / kJ/kg						1135.59	1341.27	1559.71	1707.94
	s / kJ/(kg · K)						3.6526	4.1303	4.6164	4.9322

Table A.17 (cont.)

P / MPa		T / K								
(T^{sat} / K)		500	520	540	560	580	600	620	640	660
5.0	v / m^3/kg	0.04501	0.04738	0.04970	0.05196	0.05419	0.05639	0.05857	0.06072	0.06285
(362.0)	h / kJ/kg	2082.30	2139.40	2196.32	2253.30	2310.51	2368.09	2426.12	2484.70	2543.88
	s / kJ/(kg · K)	6.2443	6.3563	6.4637	6.5673	6.6677	6.7653	6.8604	6.9534	7.0445
6.0	v / m^3/kg	0.03685	0.03890	0.04090	0.04285	0.04476	0.04663	0.04848	0.05031	0.05211
(371.0)	h / kJ/kg	2067.01	2125.79	2184.10	2242.25	2300.46	2358.90	2417.70	2476.94	2536.71
	s / kJ/(kg · K)	6.1324	6.2476	6.3577	6.4634	6.5655	6.6646	6.7610	6.8550	6.9470
8.0	v / m^3/kg	0.02663	0.02830	0.02990	0.03145	0.03296	0.03443	0.03588	0.03730	0.03870
(386.0)	h / kJ/kg	2035.49	2097.93	2159.21	2219.84	2280.14	2340.38	2400.73	2461.35	2522.33
	s / kJ/(kg · K)	5.9443	6.0668	6.1824	6.2926	6.3985	6.5006	6.5995	6.6957	6.7896
10	v / m^3/kg	0.02049	0.02194	0.02331	0.02462	0.02589	0.02712	0.02833	0.02950	0.03066
(398.3)	h / kJ/kg	2002.66	2069.20	2133.75	2197.03	2259.56	2321.68	2383.66	2445.69	2507.91
	s / kJ/(kg · K)	5.7853	5.9158	6.0376	6.1527	6.2624	6.3677	6.4693	6.5678	6.6635
12	v / m^3/kg	0.01639	0.01769	0.01891	0.02007	0.02118	0.02225	0.02330	0.02431	0.02531
	h / kJ/kg	1968.50	2039.65	2107.74	2173.87	2238.75	2302.84	2366.50	2429.99	2493.49
	s / kJ/(kg · K)	5.6437	5.7833	5.9118	6.0320	6.1459	6.2545	6.3589	6.4597	6.5574
14	v / m^3/kg	0.01346	0.01466	0.01578	0.01683	0.01782	0.01878	0.01971	0.02061	0.02149
	h / kJ/kg	1933.03	2009.31	2081.26	2150.42	2217.76	2283.90	2349.30	2414.28	2479.08
	s / kJ/(kg · K)	5.5134	5.6630	5.7988	5.9246	6.0427	6.1548	6.2621	6.3652	6.4649
16	v / m^3/kg	0.01126	0.01240	0.01343	0.01440	0.01531	0.01619	0.01703	0.01784	0.01864
	h / kJ/kg	1896.32	1978.28	2054.38	2126.75	2196.65	2264.91	2332.09	2398.60	2464.72
	s / kJ/(kg · K)	5.3907	5.5515	5.6951	5.8267	5.9494	6.0651	6.1752	6.2808	6.3826
18	v / m^3/kg	0.00956	0.01064	0.01162	0.01252	0.01337	0.01418	0.01495	0.01570	0.01642
	h / kJ/kg	1858.53	1946.69	2027.21	2102.93	2175.49	2245.92	2314.92	2382.97	2450.42
	s / kJ/(kg · K)	5.2736	5.4466	5.5985	5.7363	5.8636	5.9830	6.0961	6.2041	6.3079
20	v / m^3/kg	0.00822	0.00925	0.01018	0.01103	0.01183	0.01258	0.01330	0.01399	0.01466
	h / kJ/kg	1819.95	1914.72	1999.87	2079.05	2154.34	2226.98	2297.82	2367.43	2436.23
	s / kJ/(kg · K)	5.1610	5.3469	5.5076	5.6517	5.7838	5.9069	6.0231	6.1336	6.2394

Table A.18 Properties of liquid ammonia.

P / MPa		T / K								
		200	225	250	275	300	325	350	375	400
	(P^{sat} / MPa)	0.008651	0.04551	0.1649	0.4599	1.062	2.133	3.866	6.485	10.30
sat	ρ / kg/m^3	728.12	699.91	669.18	636.03	599.97	559.69	512.38	451.34	344.56
	h / kJ/kg	19.00	126.51	237.49	351.72	469.71	593.16	725.80	876.30	1085.02
	s / kJ/(kg · K)	0.09601	0.60211	1.06900	1.50268	1.90983	2.29903	2.68242	3.08236	3.59493
0.50	ρ / kg/m^3	728.28	700.10	669.36	636.06					
	h / kJ/kg	19.47	126.91	237.75	351.75					
	s / kJ/(kg · K)	0.09502	0.60102	1.06804	1.50254					
0.70	ρ / kg/m^3	728.35	700.19	669.47	636.20					
	h / kJ/kg	19.67	127.09	237.91	351.87					
	s / kJ/(kg · K)	0.09461	0.60054	1.06747	1.50186					
1.0	ρ / kg/m^3	728.45	700.32	669.63	636.42					
	h / kJ/kg	19.96	127.35	238.14	352.06					
	s / kJ/(kg · K)	0.09401	0.59981	1.06662	1.50083					
1.4	ρ / kg/m^3	728.58	700.49	669.85	636.70	600.30				
	h / kJ/kg	20.35	127.71	238.46	352.32	469.84				
	s / kJ/(kg · K)	0.09320	0.59885	1.06548	1.49946	1.90838				
2.0	ρ / kg/m^3	728.78	700.75	670.18	637.13	600.88				
	h / kJ/kg	20.93	128.24	238.93	352.70	470.07				
	s / kJ/(kg · K)	0.09200	0.59742	1.06378	1.49742	1.90583				
4.0	ρ / kg/m^3	729.44	701.60	671.26	638.53	602.79	562.31	512.71		
	h / kJ/kg	22.87	130.02	240.50	353.98	470.87	593.06	725.65		
	s / kJ/(kg · K)	0.08800	0.59266	1.05815	1.49068	1.89743	2.28848	2.68125		

Table A.18 (cont.)

P / MPa		T / K								
		200	225	250	275	300	325	350	375	400
6.0	ρ / kg/m^3	730.10	702.44	672.33	639.91	604.64	565.01	517.31		
	h / kJ/kg	24.82	131.81	242.09	355.28	471.73	593.06	723.65		
	s / kJ/(kg · K)	0.08404	0.58794	1.05259	1.48405	1.88922	2.27757	2.66442		
8.0	ρ / kg/m^3	730.75	703.28	673.39	641.27	606.45	567.59	521.56	458.92	
	h / kJ/kg	26.77	133.60	243.69	356.61	472.62	593.17	722.01	870.35	
	s / kJ/(kg · K)	0.08011	0.58327	1.04710	1.47752	1.88120	2.26705	2.64874	3.05762	
10	ρ / kg/m^3	731.40	704.11	674.44	642.61	608.22	570.08	525.51	467.31	
	h / kJ/kg	28.73	135.40	245.30	357.96	473.55	593.38	720.66	864.17	
	s / kJ/(kg · K)	0.07622	0.57864	1.04166	1.47108	1.87333	2.25686	2.63399	3.02962	
12	ρ / kg/m^3	732.05	704.93	675.47	643.92	609.94	572.47	529.21	474.48	383.12
	h / kJ/kg	30.69	137.21	246.92	359.32	474.53	593.67	719.58	859.27	1043.74
	s / kJ/(kg · K)	0.07236	0.57405	1.03628	1.46474	1.86563	2.24699	2.62005	3.00524	3.48019
14	ρ / kg/m^3	732.69	705.74	676.49	645.21	611.63	574.79	532.70	480.78	403.88
	h / kJ/kg	32.66	139.02	248.54	360.70	475.54	594.05	718.71	855.29	1023.07
	s / kJ/(kg · K)	0.06853	0.56950	1.03096	1.45848	1.85808	2.23741	2.60680	2.98346	3.41583
16	ρ / kg/m^3	733.33	706.55	677.50	646.49	613.28	577.03	536.00	486.44	418.04
	h / kJ/kg	34.62	140.84	250.18	362.10	476.58	594.49	718.02	851.99	1009.86
	s / kJ/(kg · K)	0.06473	0.56500	1.02570	1.45231	1.85066	2.22810	2.59416	2.96364	3.37067
20	ρ / kg/m^3	734.60	708.15	679.49	648.98	616.48	581.32	542.15	496.34	438.26
	h / kJ/kg	38.57	144.49	253.49	364.95	478.75	595.58	717.14	846.91	992.65
	s / kJ/(kg · K)	0.05722	0.55610	1.01534	1.44021	1.83624	2.21020	2.57043	2.92840	3.30430

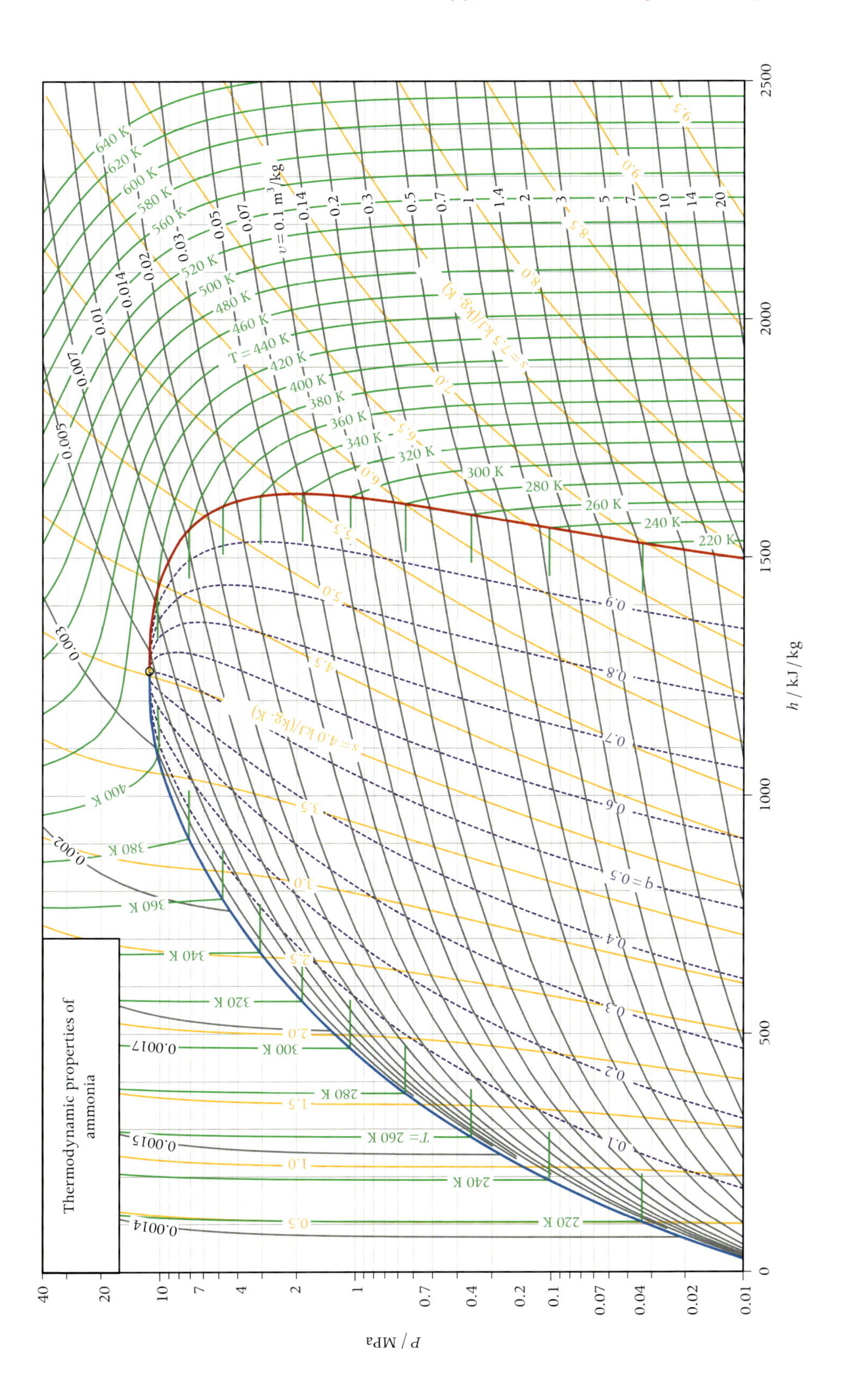
Thermodynamic properties of ammonia
h / kJ/kg
P / MPa
0
500
1000
1500
2000
2500
0.01
0.02
0.04
0.07
0.1
0.2
0.4
0.7
1
2
4
7
10
20
40
v = 0.1 m^3/kg
T = 440 K
T = 260 K
s = 4.0 kJ/(kg · K)
s = 7.5 kJ/(kg · K)
q = 0.5

A.8 Oxygen

Properties in tables and charts are calculated with RefProp, which implements the thermodynamic model for oxygen documented in R. Schmidt and W. Wagner, "A new form of the equation of state for pure substances and its application to oxygen," *Fluid Phase Equilib.*, **19**, (3), 175–200, 1985.
Reference state: $\hat{h}_0 = \hat{h}^{IG}(P = 1\text{ atm}, T = 25\text{ °C}) = -8680\text{ J/mol}$ (with $\hat{h} \equiv \hat{h}^{IG} + \hat{h}^{R}$), $\hat{s}_0 = \hat{s}^{IG}(P = 1\text{ atm}, T = 25\text{ °C}) = -205.0\text{ J/(mol}\cdot\text{K)}$ (with $\hat{s} \equiv \hat{s}^{IG} + \hat{s}^{R}$).

Table A.19 Properties of saturated oxygen as function of temperature.

T / K	P^{sat} / MPa	v^L / m³/kg	v^V / m³/kg	h^L / kJ/kg	h^{LV} / kJ/kg	h^V / kJ/kg	s^L / kJ/(kg · K)	s^{LV} / kJ/(kg · K)	s^V / kJ/(kg · K)
54.361	0.0001463	0.000766	96.54	−193.61	242.72	49.11	2.0921	4.4650	6.5571
60	0.0007258	0.000780	21.46	−184.19	238.38	54.19	2.2571	3.9729	6.2301
65	0.002335	0.000794	7.219	−175.81	234.47	58.66	2.3912	3.6072	5.9985
70	0.006262	0.000808	2.892	−167.42	230.51	63.09	2.5156	3.2930	5.8086
75	0.01455	0.000824	1.329	−159.02	226.48	67.45	2.6313	3.0197	5.6510
80	0.03012	0.000840	0.681	−150.61	222.31	71.69	2.7397	2.7788	5.5185
85	0.05683	0.000857	0.380	−142.18	217.92	75.75	2.8417	2.5638	5.4055
90	0.09935	0.000876	0.228	−133.69	213.24	79.55	2.9383	2.3693	5.3076
90.18781	0.101325	0.000876	0.224	−133.37	213.06	79.69	2.9419	2.3624	5.3042
95	0.1631	0.000895	0.145	−125.12	208.16	83.04	3.0303	2.1912	5.2215
100	0.2540	0.000917	0.09592	−116.45	202.60	86.16	3.1185	2.0260	5.1445
105	0.3785	0.000940	0.06612	−107.64	196.48	88.85	3.2033	1.8713	5.0746
110	0.5434	0.000966	0.04699	−98.64	189.70	91.05	3.2855	1.7245	5.0100
115	0.7556	0.000994	0.03424	−89.42	182.13	92.72	3.3657	1.5838	4.9495
120	1.022	0.001027	0.02544	−79.90	173.66	93.75	3.4444	1.4472	4.8915
125	1.351	0.001064	0.01919	−70.02	164.08	94.06	3.5222	1.3126	4.8349
130	1.749	0.001108	0.01463	−59.66	153.13	93.47	3.6001	1.1779	4.7780
135	2.225	0.001161	0.01120	−48.65	140.39	91.74	3.6791	1.0400	4.7191
140	2.788	0.001230	0.008565	−36.70	125.17	88.47	3.7612	0.8941	4.6552
145	3.448	0.001324	0.006455	−23.22	106.05	82.83	3.8498	0.7314	4.5812
150	4.219	0.001480	0.004653	−6.67	79.23	72.56	3.9546	0.5282	4.4828
154.58	5.043	0.002293	0.002293	32.42	0.0	32.42	4.2008	0.0	4.2008

Table A.20 Properties of gaseous oxygen.

P / MPa (T^{sat} / K)		T / K sat	200	300	400	500	600	700	800	1000
0.020	v / m^3/kg	0.9919	2.597	3.897	5.197	6.496	7.795	9.095	10.39	12.99
(77.10)	h / kJ/kg	69.25	181.54	272.91	365.78	461.39	560.15	661.88	766.18	980.84
	s / kJ/(kg · K)	5.5927	6.4646	6.8350	7.1020	7.3152	7.4952	7.6520	7.7912	8.0306
0.050	v / m^3/kg	0.4279	1.038	1.559	2.079	2.599	3.118	3.638	4.158	5.197
(83.9)	h / kJ/kg	74.91	181.39	272.84	365.74	461.37	560.14	661.88	766.18	980.85
	s / kJ/(kg · K)	5.4282	6.2260	6.5967	6.8638	7.0771	7.2571	7.4139	7.5531	7.7925
0.101325	v / m^3/kg	0.2239	0.5113	0.7688	1.026	1.282	1.539	1.796	2.052	2.565
(90.19)	h / kJ/kg	79.69	181.13	272.71	365.68	461.33	560.12	661.87	766.18	980.86
	s / kJ/(kg · K)	5.3042	6.0416	6.4129	6.6801	6.8935	7.0735	7.2303	7.3695	7.6089
0.20	v / m^3/kg	0.1197	0.2583	0.3893	0.5196	0.6498	0.7799	0.9099	1.040	1.300
(97.24)	h / kJ/kg	84.48	180.64	272.47	365.54	461.26	560.09	661.86	766.19	980.89
	s / kJ/(kg · K)	5.1861	5.8632	6.2355	6.5031	6.7166	6.8967	7.0535	7.1928	7.4322
0.50	v / m^3/kg	0.05086	0.1024	0.1554	0.2079	0.2601	0.3122	0.3643	0.4163	0.5203
(108.8)	h / kJ/kg	90.57	179.11	271.74	365.14	461.04	559.98	661.83	766.21	980.98
	s / kJ/(kg · K)	5.0250	5.6198	5.9955	6.2641	6.4779	6.6582	6.8152	6.9545	7.1940
1.0	v / m^3/kg	0.02600	0.05040	0.07748	0.1039	0.1302	0.1563	0.1824	0.2084	0.2605
(119.62)	h / kJ/kg	93.70	176.52	270.52	364.48	460.68	559.80	661.78	766.25	981.13
	s / kJ/(kg · K)	4.8958	5.4307	5.8121	6.0823	6.2969	6.4775	6.6346	6.7741	7.0137
2.0	v / m^3/kg	0.01263	0.02440	0.03851	0.05196	0.06520	0.07835	0.09144	0.1045	0.1306
(132.7)	h / kJ/kg	92.68	171.17	268.09	363.17	459.97	559.46	661.68	766.33	981.44
	s / kJ/(kg · K)	4.7461	5.2317	5.6255	5.8990	6.1149	6.2962	6.4537	6.5934	6.8333
5.0	v / m^3/kg	0.00280	0.00877	0.01516	0.02081	0.02624	0.03159	0.03688	0.04215	0.05264
(154.4)	h / kJ/kg	46.17	153.62	260.88	359.36	457.92	558.49	661.43	766.59	982.38
	s / kJ/(kg · K)	4.2905	4.9300	5.3679	5.6513	5.8712	6.0545	6.2131	6.3535	6.5941
10	v / m^3/kg		0.00360	0.00743	0.01045	0.01328	0.01601	0.01871	0.02137	0.02666
	h / kJ/kg		119.72	249.39	353.47	454.83	557.09	661.17	767.15	984.00
	s / kJ/(kg · K)		4.6211	5.1561	5.4560	5.6822	5.8686	6.0290	6.1704	6.4123
20	v / m^3/kg		0.00173	0.00370	0.00534	0.00683	0.00825	0.00964	0.01100	0.01368
	h / kJ/kg		75.21	229.99	343.68	449.89	555.14	661.24	768.67	987.49
	s / kJ/(kg · K)		4.2817	4.9208	5.2489	5.4860	5.6778	5.8414	5.9848	6.2289

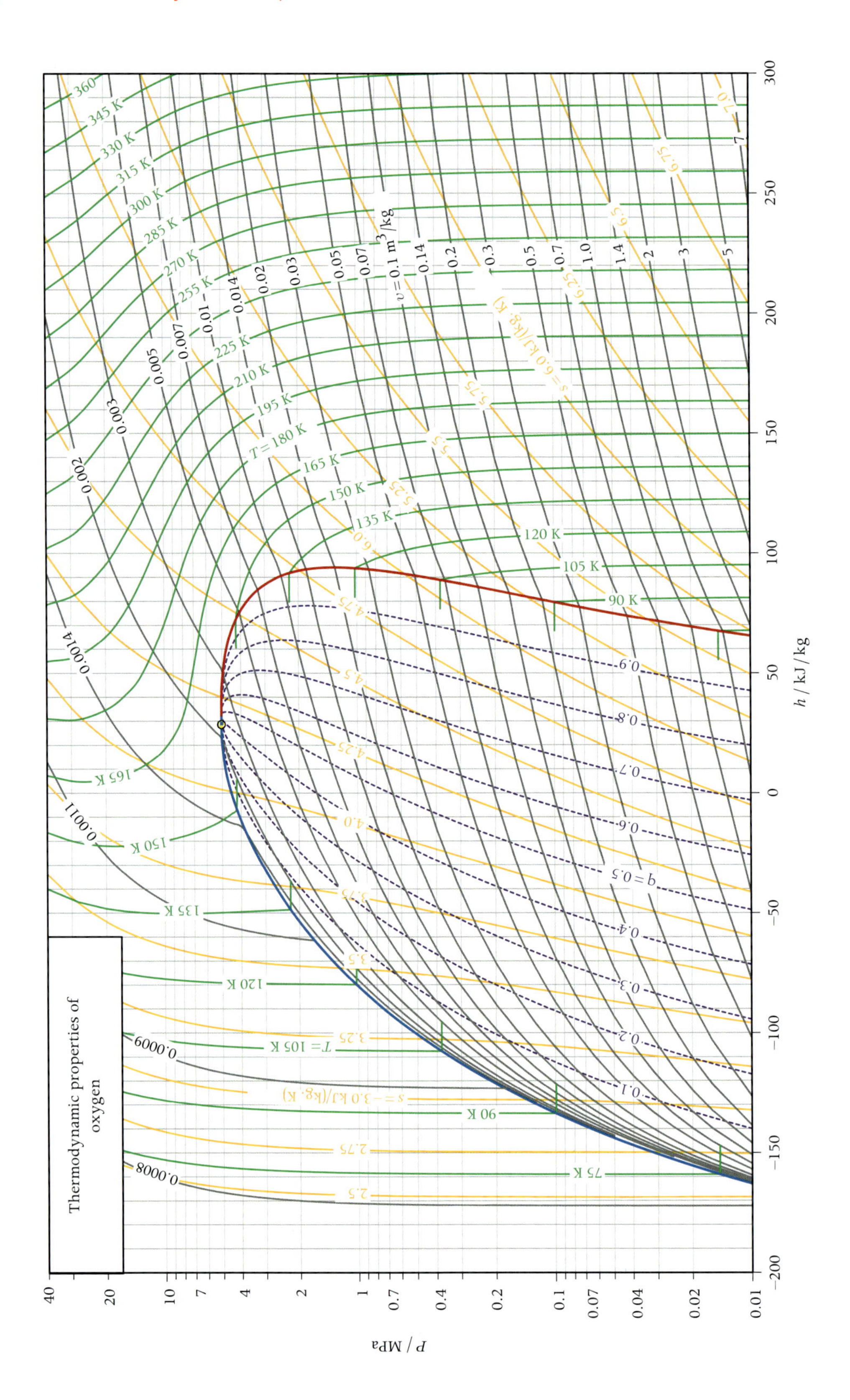
Thermodynamic properties of oxygen
P / MPa
h / kJ / kg
T = 105 K
T = 180 K
s = 3.0 kJ/(kg. K)
s = 6.0 kJ/(kg. K)
v = 0.1 m^3/kg
q = 0.5

A.9 Carbon Dioxide

Properties in tables and charts are calculated with RefProp, which implements the thermodynamic model for carbon dioxide documented in R. Span and W. Wagner, "A new equation of state for carbondioxide covering the fluid region from triple point temperature to 1100 K at pressures up to 800 MPa," *J. Phys. Chem. Ref. Data*, **25**, (6), 1509–1596, 1996.
Reference state: $h_0 = h^{\mathrm{sat,L}}(T = 0\ °\mathrm{C}) = 200\ \mathrm{kJ/kg}$, $s_0 = s^{\mathrm{sat,L}}(T = 0\ °\mathrm{C}) = 1\ \mathrm{kJ/(kg \cdot K)}$.

Table A.21 Properties of saturated carbon dioxide as function of temperature.

T / K	P^{sat} / MPa	v^L / m³/kg	v^V / m³/kg	h^L / kJ/kg	h^{LV} / kJ/kg	h^V / kJ/kg	s^L / kJ/(kg · K)	s^{LV} / kJ/(kg · K)	s^V / kJ/(kg · K)
220	0.5991	0.000858	0.063	86.73	344.91	431.64	0.6	1.5678	2.1194
222	0.6510	0.000863	0.058	90.67	341.64	432.31	0.5693	1.5389	2.1082
224	0.7062	0.000869	0.054	94.62	338.31	432.94	0.5868	1.5103	2.0972
226	0.7648	0.000874	0.050	98.59	334.94	433.53	0.6042	1.4820	2.0863
228	0.8270	0.000880	0.0463	102.57	331.51	434.09	0.6215	1.4540	2.0755
230	0.8929	0.000886	0.0430	106.57	328.03	434.60	0.6387	1.4262	2.0649
232	0.9626	0.000892	0.0399	110.59	324.49	435.07	0.6558	1.3986	2.0545
234	1.036	0.000898	0.0371	114.62	320.88	435.50	0.6729	1.3713	2.0441
236	1.114	0.000905	0.0346	118.67	317.21	435.88	0.6898	1.3441	2.0339
238	1.196	0.000912	0.0322	122.74	313.46	436.21	0.7067	1.3171	2.0237
240	1.282	0.000918	0.0300	126.84	309.65	436.49	0.7235	1.2902	2.0137
242	1.373	0.000925	0.0280	130.96	305.76	436.71	0.7402	1.2635	2.0037
244	1.469	0.000933	0.0262	135.10	301.78	436.89	0.7569	1.2368	1.9937
246	1.569	0.000940	0.0245	139.27	297.73	437.00	0.7736	1.2103	1.9838
248	1.675	0.000948	0.0229	143.48	293.58	437.05	0.7902	1.1838	1.9739
250	1.785	0.000956	0.0214	147.71	289.33	437.04	0.8068	1.1573	1.9641
252	1.901	0.000964	0.0201	151.98	284.99	436.97	0.8233	1.1309	1.9542
254	2.022	0.000973	0.0188	156.28	280.54	436.82	0.8399	1.1045	1.9443
256	2.148	0.000982	0.0176	160.63	275.97	436.60	0.8564	1.0780	1.9344
258	2.281	0.000991	0.0165	165.01	271.29	436.30	0.8730	1.0515	1.9245
260	2.419	0.001001	0.0155	169.44	266.48	435.92	0.8895	1.0249	1.9144

Table A.21 (cont.)

T / K	P^{sat} / MPa	v^L / m^3/kg	v^V / m^3/kg	h^L / kJ/kg	h^{LV} / kJ/kg	h^V / kJ/kg	s^L / kJ/(kg · K)	s^{LV} / kJ/(kg · K)	s^V / kJ/(kg · K)
262	2.563	0.001011	0.0146	173.92	261.53	435.45	0.9061	0.9982	1.9043
264	2.713	0.001022	0.0137	178.45	256.43	434.88	0.9228	0.9713	1.8941
266	2.870	0.001033	0.0128	183.04	251.18	434.22	0.9395	0.9443	1.8838
268	3.033	0.001045	0.0121	187.69	245.75	433.45	0.9563	0.9170	1.8733
270	3.203	0.001057	0.0113	192.41	240.14	432.56	0.9732	0.8894	1.8626
272	3.380	0.001070	0.0106	197.21	234.33	431.54	0.9902	0.8615	1.8517
274	3.564	0.001084	0.0100	202.08	228.30	430.38	1.0073	0.8332	1.8405
276	3.755	0.001099	0.0094	207.05	222.03	429.08	1.0246	0.8045	1.8291
278	3.954	0.001115	0.0088	212.12	215.49	427.60	1.0421	0.7751	1.8172
280	4.161	0.001132	0.0082	217.30	208.64	425.94	1.0598	0.7451	1.8050
282	4.375	0.001150	0.0077	222.61	201.46	424.07	1.0779	0.7144	1.7923
284	4.598	0.001170	0.0072	228.07	193.89	421.95	1.0963	0.6827	1.7789
286	4.829	0.001192	0.0067	233.70	185.86	419.57	1.1151	0.6499	1.7649
289	5.192	0.001229	0.0060	242.55	172.81	415.36	1.1443	0.5979	1.7423
292	5.576	0.001273	0.0054	252.02	158.16	410.18	1.1753	0.5416	1.7169
295	5.982	0.001329	0.0048	262.38	141.26	403.64	1.2087	0.4788	1.6876
298	6.412	0.001403	0.0042	274.14	120.80	394.95	1.2464	0.4054	1.6518
301	6.868	0.001518	0.0035	288.75	93.30	382.05	1.2930	0.3100	1.6029
304.13	7.377	0.002139	0.00214	332.25	0.00	332.25	1.4336	0.0000	1.4336

Table A.22 Properties of gaseous carbon dioxide.

P / MPa (T^{sat} / K)		sat	300	400	500	600	700	800	900	1000
1.0	v / m^3/kg	0.03845	0.0538	0.0742	0.0938	0.1131	0.1322	0.1513	0.1703	0.1893
(233.0)	h / kJ/kg	435.30	498.84	592.72	692.38	797.97	908.78	1024.04	1143.07	1265.28
	s / kJ/(kg · K)	2.0491	2.2894	2.5591	2.7813	2.9736	3.1443	3.2982	3.4383	3.5671
2.0	v / m^3/kg	0.01903	0.02537	0.03641	0.04655	0.05639	0.06609	0.07572	0.08530	0.09484
(253.6)	h / kJ/kg	436.85	488.36	587.56	689.19	795.82	907.27	1022.97	1142.32	1264.77
	s / kJ/(kg · K)	1.9461	2.1332	2.4187	2.6453	2.8396	3.0113	3.1657	3.3062	3.4352
5.0	v / m^3/kg	0.00638	0.00779	0.01374	0.01824	0.02241	0.02644	0.03039	0.03430	0.03817
(287.4)	h / kJ/kg	417.66	445.95	571.37	679.62	789.47	902.85	1019.85	1140.13	1263.29
	s / kJ/(kg · K)	1.7544	1.8509	2.2155	2.4570	2.6572	2.8319	2.9881	3.1298	3.2595
10	v / m^3/kg			0.00619	0.00884	0.01112	0.01325	0.01530	0.01731	0.01930
	h / kJ/kg			542.11	663.79	779.27	895.87	1014.98	1136.77	1261.06
	s / kJ/(kg · K)			2.0285	2.3007	2.5113	2.6910	2.8500	2.9934	3.1243
20	v / m^3/kg			0.00263	0.00425	0.00554	0.00670	0.00779	0.00885	0.00987
	h / kJ/kg			482.61	634.09	760.73	883.42	1006.45	1131.01	1257.38
	s / kJ/(kg · K)			1.7810	2.1212	2.3523	2.5414	2.7057	2.8524	2.9855

Table A.23 Properties of liquid carbon dioxide.

P / MPa		T / K								
		220	230	240	250	260	270	280	290	300
	(P^{sat} / MPa)	0.5991	0.8929	1.282	1.785	2.419	3.203	4.161	5.318	6.713
sat	ρ / kg/m^3	1166.1	1166.1	1166.1	1166.1	1166.1	1166.1	1166.1	1166.1	1166.1
	h / kJ/kg	86.73	106.57	126.84	147.71	169.44	192.41	217.30	245.63	283.38
	s / kJ/(kg · K)	0.5517	0.6387	0.7235	0.8068	0.8895	0.9732	1.0598	1.1544	1.2759
2.0	ρ / kg/m^3	1169.2	1131.6	1091.2	1046.9					
	h / kJ/kg	87.10	106.77	126.88	147.68					
	s / kJ/(kg · K)	0.5479	0.6353	0.7209	0.8058					
5.0	ρ / kg/m^3	1175.6	1139.3	1100.7	1058.9	1012.6	959.39	893.90		
	h / kJ/kg	87.93	107.37	127.16	147.47	168.60	191.05	215.90		
	s / kJ/(kg · K)	0.5401	0.6265	0.7107	0.7936	0.8764	0.9611	1.0515		
10	ρ / kg/m^3	1185.6	1151.2	1114.9	1076.4	1035.0	989.46	938.22	878.06	801.62
	h / kJ/kg	89.44	108.56	127.90	147.58	167.78	188.72	210.78	234.62	261.80
	s / kJ/(kg · K)	0.5277	0.6126	0.6949	0.7753	0.8545	0.9335	1.0137	1.0974	1.1894

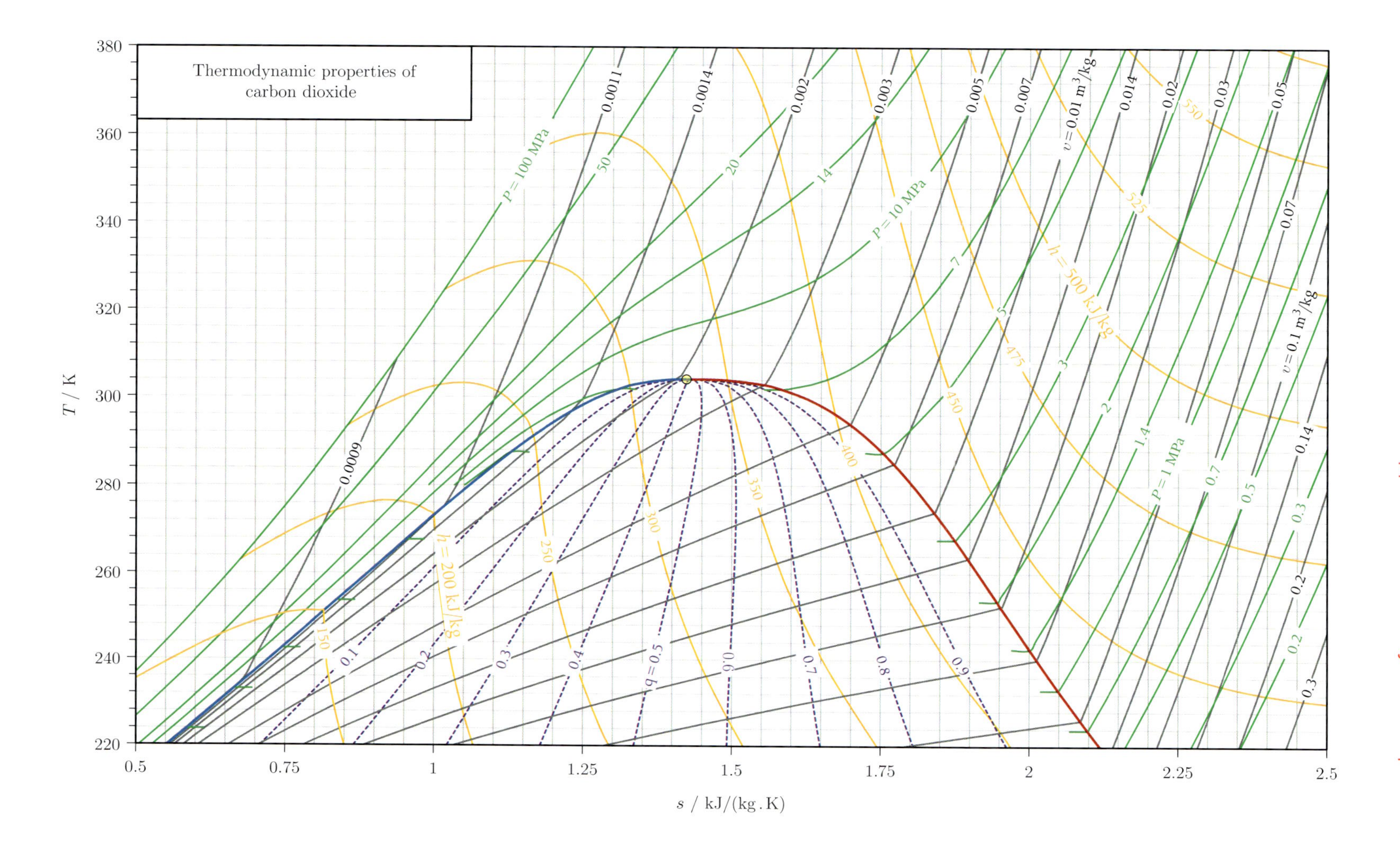

Thermodynamic properties of carbon dioxide
T / K
s / kJ/(kg.K)
380
360
340
320
300
280
260
240
220
0.5
0.75
1
1.25
1.5
1.75
2
2.25
2.5
P = 100 MPa
50
20
14
P = 10 MPa
7
5
3
2
1.4
P = 1 MPa
0.7
0.5
0.3
0.2
0.0009
0.0011
0.0014
0.002
0.003
0.005
0.007
v = 0.01 m³/kg
0.014
0.02
0.03
0.05
0.07
v = 0.1 m³/kg
0.14
0.2
0.3
h = 200 kJ/kg
150
250
300
350
400
450
475
h = 500 kJ/kg
525
550
0.1
0.2
0.3
0.4
q = 0.5
0.6
0.7
0.8
0.9

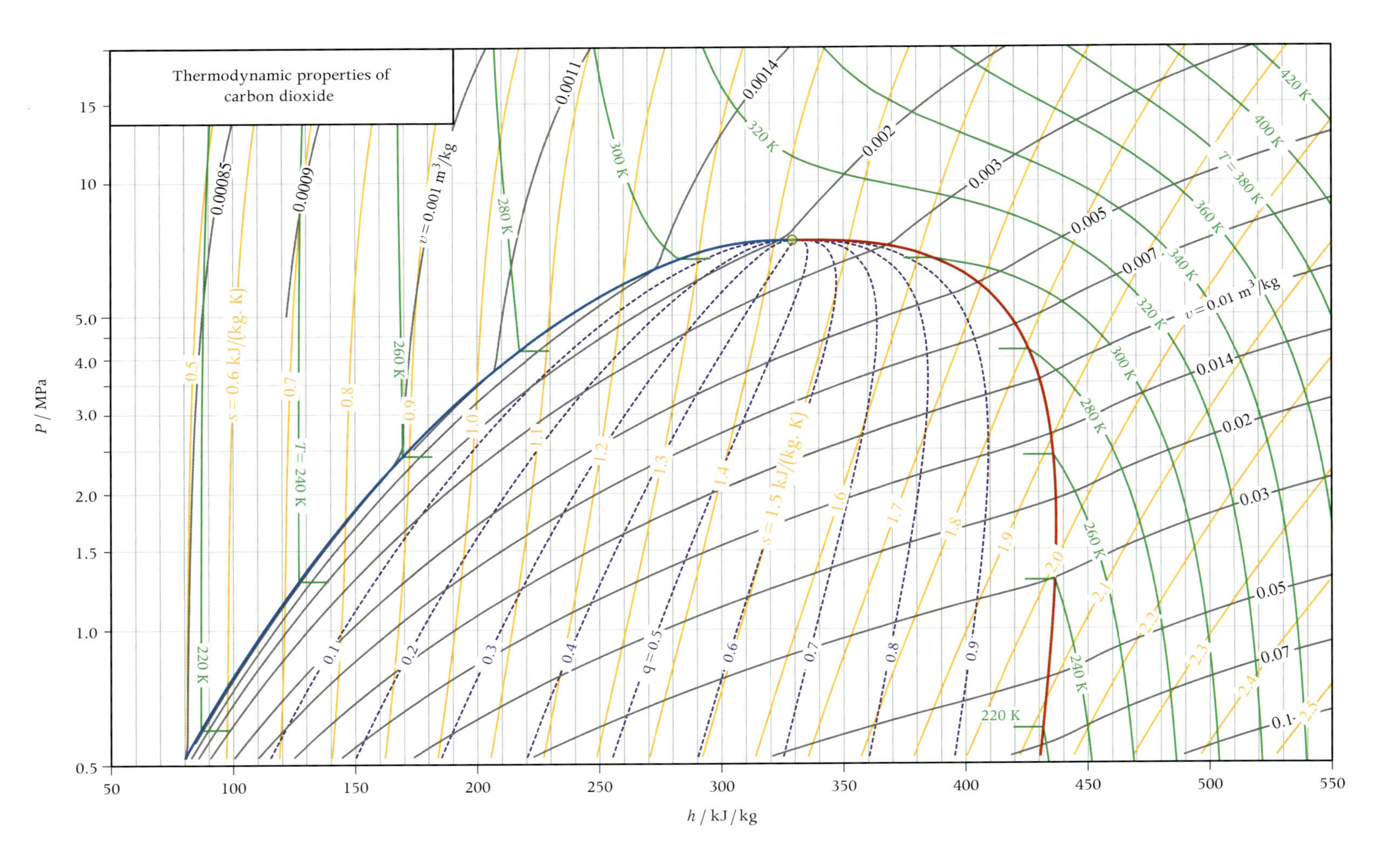

Thermodynamic properties of carbon dioxide
P / MPa
h / kJ / kg
15
10
5.0
4.0
3.0
2.0
1.5
1.0
0.5
50
100
150
200
250
300
350
400
450
500
550
T = 380 K
420 K
400 K
360 K
340 K
320 K
300 K
280 K
260 K
T = 240 K
240 K
220 K
v = 0.01 m^3/kg
v = 0.001 m^3/kg
0.014
0.02
0.03
0.05
0.07
0.1
0.007
0.005
0.003
0.002
0.0014
0.0011
0.0009
0.00085
s = 1.5 kJ/(kg. K)
s = 0.6 kJ/(kg. K)
q = 0.5

A.10 Siloxane MDM

Properties in tables and charts are calculated with RefProp, which implements the thermodynamic model for MDM documented in P. Colonna, N. R. Nannan, and A. Guardone, "Multiparameter equations of state for siloxanes: $[(CH_3)_3\text{-Si-}O_{1/2}]_2\text{-}[O\text{-Si-}(CH_3)_2]_{i=1\ldots3}$, and $[O\text{-Si-}(CH_3)_2]_6$," *Fluid Phase Equilib.*, **263**, (2), 115–130, 2008.

Reference state: $h_0 = h^{\text{sat,L}}(T^{\text{NBP}} = 425.66\ \text{K}) = 0\ \text{kJ/kg}$, $s_0 = s^{\text{sat,L}}(T^{\text{NBP}} = 425.66\ \text{K}) = 0\ \text{kJ/}(\text{kg}\cdot\text{K})$.

Table A.24 Properties of saturated MDM as function of temperature.

T / K	P^{sat} / MPa	v^{L} / m³/kg	v^{V} / m³/kg	h^{L} / kJ/kg	h^{LV} / kJ/kg	h^{V} / kJ/kg	s^{L} / kJ/(kg·K)	s^{LV} / kJ/(kg·K)	s^{V} / kJ/(kg·K)
250	0.00000932	0.001156	943.275	−335.10	226.95	−108.15	−1.0	0.9078	−0.0987
254	0.00001397	0.001162	639.200	−328.03	225.02	−103.01	−0.9785	0.8859	−0.0926
260	0.00002490	0.001170	367.070	−317.41	222.17	−95.25	−0.9372	0.8545	−0.0827
266	0.00004293	0.001178	217.751	−306.77	219.37	−87.40	−0.8967	0.8247	−0.0720
272	0.00007181	0.001187	133.1074	−296.10	216.63	−79.47	−0.8570	0.7964	−0.0606
278	0.0001168	0.001195	83.6574	−285.39	213.93	−71.45	−0.8181	0.7695	−0.0485
284	0.0001849	0.001204	53.9493	−274.63	211.28	−63.36	−0.7798	0.7439	−0.0359
290	0.0002859	0.001213	35.6321	−263.83	208.67	−55.17	−0.7422	0.7195	−0.0226
296	0.0004319	0.001222	24.0622	−252.98	206.09	−46.89	−0.7051	0.6963	−0.0089
302	0.0006390	0.001231	16.5879	−242.07	203.55	−38.52	−0.6686	0.6740	0.0054
308	0.0009269	0.001241	11.6571	−231.10	201.04	−30.07	−0.6327	0.6527	0.0200
314	0.001320	0.001250	8.3398	−220.07	198.55	−21.52	−0.5972	0.6323	0.0351
320	0.001848	0.001260	6.0669	−208.97	196.09	−12.88	−0.5622	0.6128	0.0506
326	0.002545	0.001270	4.4826	−197.80	193.65	−4.15	−0.5276	0.5940	0.0664
332	0.003453	0.001280	3.3604	−186.55	191.23	4.67	−0.4934	0.5760	0.0826
338	0.004620	0.001291	2.5535	−175.23	188.81	13.58	−0.4596	0.5586	0.0990
344	0.006099	0.001301	1.9650	−163.83	186.41	22.58	−0.4262	0.5419	0.1157
350	0.007953	0.001312	1.53013	−152.35	184.01	31.66	−0.3931	0.5258	0.1326
356	0.01025	0.001323	1.20468	−140.79	181.62	40.83	−0.3604	0.5102	0.1498
362	0.01307	0.001335	0.95827	−129.15	179.22	50.07	−0.3280	0.4951	0.1671
368	0.01649	0.001347	0.76963	−117.41	176.81	59.40	−0.2958	0.4805	0.1846
374	0.02060	0.001359	0.62370	−105.60	174.40	68.80	−0.2640	0.4663	0.2023
380	0.02551	0.001371	0.50969	−93.69	171.97	78.28	−0.2324	0.4526	0.2201
386	0.03132	0.001384	0.41979	−81.69	169.52	87.83	−0.2011	0.4392	0.2381
392	0.03815	0.001397	0.34828	−69.60	167.06	97.45	−0.1701	0.4262	0.2561

Table A.24 (cont.)

T / K	P^{sat} / MPa	v^L / m^3/kg	v^V / m^3/kg	h^L / kJ/kg	h^{LV} / kJ/kg	h^V / kJ/kg	s^L / kJ/(kg · K)	s^{LV} / kJ/(kg · K)	s^V / kJ/(kg · K)
398	0.04611	0.001411	0.29093	−57.42	164.56	107.14	−0.1393	0.4135	0.2742
404	0.05533	0.001425	0.24457	−45.15	162.03	116.88	−0.1087	0.4011	0.2924
410	0.06595	0.001440	0.20681	−32.78	159.47	126.69	−0.0783	0.3889	0.3106
415	0.07596	0.001453	0.18060	−22.40	157.30	134.90	−0.0532	0.3790	0.3258
420	0.08713	0.001466	0.15827	−11.95	155.10	143.15	−0.0282	0.3693	0.3411
425	0.09954	0.001479	0.13917	−1.43	152.86	151.43	−0.0034	0.3597	0.3563
425.68	0.101325	0.001481	0.13681	0.00	152.56	152.56	0.0000	0.3584	0.3584
430	0.1133	0.001493	0.12277	9.16	150.59	159.75	0.0214	0.3502	0.3716
435	0.1285	0.001508	0.10861	19.82	148.27	168.09	0.0460	0.3409	0.3868
440	0.1451	0.001523	0.09635	30.55	145.91	176.46	0.0704	0.3316	0.4020
445	0.1635	0.001538	0.08570	41.35	143.49	184.84	0.0948	0.3225	0.4172
450	0.1835	0.001554	0.076397	52.23	141.02	193.25	0.1190	0.3134	0.4324
455	0.2054	0.001571	0.068255	63.18	138.49	201.68	0.1431	0.3044	0.4475
460	0.2292	0.001589	0.061102	74.21	135.90	210.11	0.1672	0.2954	0.4626
465	0.2550	0.001607	0.054798	85.32	133.23	218.55	0.1911	0.2865	0.4776
470	0.2831	0.001627	0.049222	96.52	130.48	227.00	0.2150	0.2776	0.4926
475	0.3134	0.001647	0.044277	107.79	127.64	235.44	0.2387	0.2687	0.5074
480	0.3462	0.001669	0.039877	119.16	124.71	243.87	0.2624	0.2598	0.5222
485	0.3815	0.001692	0.035951	130.61	121.67	252.29	0.2860	0.2509	0.5369
490	0.4194	0.001716	0.032438	142.16	118.52	260.68	0.3096	0.2419	0.5514
495	0.4602	0.001742	0.029284	153.81	115.23	269.04	0.3331	0.2328	0.5659
500	0.5040	0.001770	0.026444	165.56	111.80	277.36	0.3566	0.2236	0.5801
505	0.5509	0.001800	0.023881	177.43	108.20	285.62	0.3800	0.2143	0.5943
510	0.6011	0.001833	0.021560	189.41	104.41	293.82	0.4034	0.2047	0.6082
515	0.6547	0.001868	0.019451	201.52	100.41	301.93	0.4269	0.1950	0.6218
520	0.7121	0.001907	0.017528	213.78	96.16	309.94	0.4503	0.1849	0.6353
525	0.7732	0.001951	0.015769	226.19	91.61	317.80	0.4739	0.1745	0.6484
530	0.8384	0.002000	0.014152	238.78	86.72	325.50	0.4975	0.1636	0.6611
535	0.9079	0.002056	0.012659	251.58	81.40	332.98	0.5213	0.1521	0.6734
540	0.9820	0.002121	0.011271	264.63	75.54	340.17	0.5453	0.1399	0.6851
545	1.061	0.002199	0.009967	277.99	68.97	346.96	0.5696	0.1266	0.6961
550	1.145	0.002297	0.008724	291.78	61.41	353.19	0.5944	0.1116	0.7061
555	1.235	0.002429	0.007503	306.24	52.24	358.48	0.6202	0.0941	0.7143
560	1.331	0.002640	0.006204	322.08	39.71	361.79	0.6482	0.0709	0.7191
564.09	1.415	0.003895	0.003895	350.05	0.0	350.05	0.6974	0.0	0.6974

Table A.25 Properties of gaseous MDM.

P / MPa (T^{sat} / K)		sat	250	300	350	400	450	500	550	600
						T / K				
1.00E–05	v / m^3/kg	881.2		1055	1230	1406	1582	1758	1933	2109
(250.7)	h / kJ/kg	–107.27		–41.27	32.15	112.53	199.66	293.03	392.09	496.33
	s / kJ/(kg · K)	–0.0977		0.1423	0.3684	0.5829	0.7880	0.9846	1.1733	1.3547
1.00E–04	v / m^3/kg	97.00		105.4	123.0	140.6	158.2	175.8	193.3	210.9
(276.0)	h / kJ/kg	–74.07		–41.28	32.14	112.53	199.66	293.03	392.08	496.33
	s / kJ/(kg · K)	–0.0525		0.0614	0.2874	0.5019	0.7070	0.9037	1.0924	1.2737
1.00E–03	v / m^3/kg	10.85			12.29	14.05	15.81	17.57	19.33	21.09
(309.3)	h / kJ/kg	–28.27			32.09	112.49	199.63	293.01	392.07	496.32
	s / kJ/(kg · K)	0.0232			0.2064	0.4209	0.6260	0.8227	1.0114	1.1928
0.01	v / m^3/kg	1.233				1.394	1.573	1.751	1.928	2.105
(355.4)	h / kJ/kg	39.91				112.11	199.35	292.79	391.90	496.18
	s / kJ/(kg · K)	0.1481				0.3393	0.5447	0.7414	0.9303	1.1117
0.10	v / m^3/kg	0.1386					0.1489	0.1688	0.1879	0.2067
(425.2)	h / kJ/kg	151.72					196.38	290.55	390.14	494.75
	s / kJ/(kg · K)	0.3569					0.4589	0.6573	0.8470	1.0290
0.101325	v / m^3/kg	0.1368					0.1468	0.1665	0.1854	0.2039
(425.7)	h / kJ/kg	152.56					196.33	290.52	390.11	494.73
	s / kJ/(kg · K)	0.3584					0.4584	0.6568	0.8465	1.0285
0.50	v / m^3/kg	0.02668						0.02675	0.03277	0.03776
(499.6)	h / kJ/kg	276.62						277.54	381.07	487.81
	s / kJ/(kg · K)	0.5789						0.5807	0.7780	0.9637
1.0	v / m^3/kg	0.01096							0.01216	0.01637
(541.2)	h / kJ/kg	341.80							363.44	477.23
	s / kJ/(kg · K)	0.6878							0.7275	0.9255

Table A.26 Properties of liquid MDM.

P / MPa (T^{sat} / K)		T / K 250	300	340	370	400	430	460	500	550
	(P^{sat} / MPa)	9.316E–6	5.621E–4	5.075E–3	0.01778	0.04903	0.1133	0.2292	0.5040	1.145
sat	ρ / kg/m^3	864.92	814.19	772.70	740.35	706.31	669.75	629.35	564.96	435.27
	h / kJ/kg	–335.10	–245.71	–171.44	–113.49	–53.34	9.16	74.21	165.56	291.78
	s / kJ/(kg · K)	–1.0065	–0.6807	–0.4485	–0.2852	–0.1290	0.0214	0.1672	0.3566	0.5944
0.101325	ρ / kg/m^3	865.06	814.38	772.93	740.61	706.52				
	h / kJ/kg	–335.02	–245.64	–171.37	–113.43	–53.32				
	s / kJ/(kg · K)	–1.0067	–0.6809	–0.4486	–0.2854	–0.1292				
0.5	ρ / kg/m^3	865.59	815.10	773.89	741.81	708.07	671.78	631.40		
	h / kJ/kg	–334.69	–245.33	–171.10	–113.19	–53.13	9.26	74.19		
	s / kJ/(kg · K)	–1.0072	–0.6815	–0.4493	–0.2862	–0.1301	0.0203	0.1662		
2	ρ / kg/m^3	867.55	817.78	777.39	746.17	713.64	679.15	641.73	584.14	482.86
	h / kJ/kg	–333.46	–244.17	–170.04	–112.25	–52.37	9.75	74.20	164.16	285.32
	s / kJ/(kg · K)	–1.0092	–0.6838	–0.4519	–0.2891	–0.1335	0.0162	0.1611	0.3485	0.5793
3	ρ / kg/m^3	868.84	819.53	779.66	748.96	717.15	683.70	647.86	594.28	509.83
	h / kJ/kg	–332.64	–243.40	–169.33	–111.61	–51.83	10.13	74.34	163.66	282.18
	s / kJ/(kg · K)	–1.0105	–0.6853	–0.4536	–0.2909	–0.1356	0.0137	0.1580	0.3441	0.5700

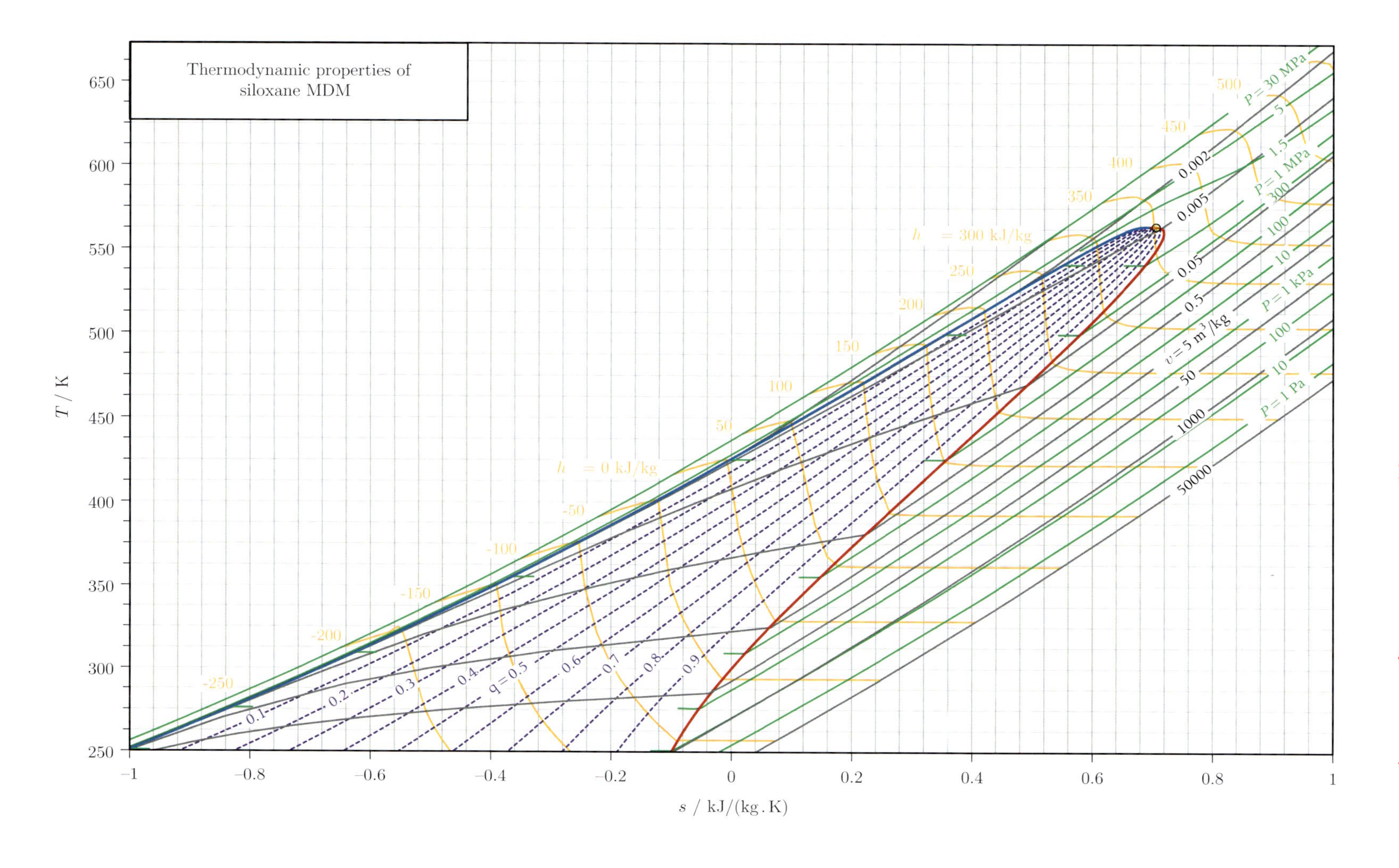
Thermodynamic properties of siloxane MDM
T / K
s / kJ/(kg.K)
P = 30 MPa
5
1.5
P = 1 MPa
300
100
10
P = 1 kPa
100
10
P = 1 Pa
0.002
0.005
0.05
0.5
v = 5 m³/kg
50
1000
50000
h = 300 kJ/kg
h = 0 kJ/kg
q = 0.5
0.1
0.2
0.3
0.4
0.6
0.7
0.8
0.9

A.11 Alkali Metal Potassium

Properties in tables and charts are calculated with TPSI , which implements the thermodynamic model for potassium documented in W. C. Reynolds, *Thermodynamic Properties in SI*. Dept. of Mechanical Engineering, Stanford University, 1979.
Reference state: $h_0 = u^{sat,L}(T = 800\ K) = 0\ kJ/kg$, $s_0 = s^{sat,L}(T = 800\ K) = 0\ kJ/(kg \cdot K)$.

Table A.27 Properties of saturated potassium as function of temperature.

T / K	P^{sat} / MPa	v^L / m^3/kg	v^V / m^3/kg	h^L / kJ/kg	h^{LV} / kJ/kg	h^V / kJ/kg	s^L / kJ/(kg · K)	s^{LV} / kJ/(kg · K)	s^V / kJ/(kg · K)
800	0.006363	0.001389	25.97	0.0	2060.7	2060.7	0.0	2.5758	2.5758
825	0.009283	0.001400	18.27	19.5	2048.6	2068.1	0.0240	2.4832	2.5071
850	0.01324	0.001412	13.14	39.2	2035.9	2075.1	0.0475	2.3951	2.4427
875	0.01849	0.001425	9.630	59.1	2022.5	2081.7	0.0706	2.3115	2.3821
900	0.02535	0.001437	7.187	79.2	2008.6	2087.8	0.0933	2.2318	2.3250
925	0.03414	0.001450	5.452	99.5	1994.1	2093.6	0.1155	2.1558	2.2713
950	0.04526	0.001463	4.199	119.9	1979.2	2099.1	0.1372	2.0833	2.2206
975	0.05911	0.001476	3.279	140.5	1963.8	2104.3	0.1586	2.0142	2.1727
1000	0.07614	0.001490	2.593	161.1	1948.1	2109.2	0.1794	1.9481	2.1275
1025	0.09686	0.001504	2.076	181.8	1932.1	2113.9	0.1998	1.8850	2.0848
1029.83	0.101325	0.001507	1.991	185.8	1929.0	2114.8	0.2037	1.8731	2.0769
1050	0.1218	0.001518	1.680	202.5	1916.0	2118.5	0.2198	1.8247	2.0445
1075	0.1514	0.001533	1.373	223.3	1899.6	2122.9	0.2393	1.7671	2.0064
1100	0.1864	0.001548	1.134	244.1	1883.2	2127.3	0.2583	1.7120	1.9703
1125	0.2272	0.001563	0.9444	264.8	1866.8	2131.6	0.2769	1.6594	1.9363
1150	0.2746	0.001579	0.7932	285.5	1850.5	2136.0	0.2951	1.6091	1.9042
1175	0.3290	0.001595	0.6715	306.2	1834.2	2140.4	0.3128	1.5610	1.8738
1200	0.3913	0.001611	0.5727	326.9	1818.1	2145.0	0.3301	1.5151	1.8452
1225	0.4619	0.001628	0.4918	347.5	1802.2	2149.7	0.3470	1.4712	1.8182
1250	0.5416	0.001645	0.4251	368.1	1786.4	2154.5	0.3636	1.4292	1.7927
1275	0.6309	0.001663	0.3697	388.7	1770.9	2159.6	0.3798	1.3889	1.7687
1300	0.7305	0.001681	0.3234	409.4	1755.5	2164.8	0.3957	1.3503	1.7460
1325	0.8409	0.001699	0.2844	430.1	1740.2	2170.3	0.4113	1.3133	1.7247
1350	0.9629	0.001718	0.2515	450.9	1724.9	2175.8	0.4268	1.2777	1.7045
1375	1.097	0.001737	0.2233	472.0	1709.6	2181.6	0.4420	1.2434	1.6854
1400	1.244	0.001757	0.1993	493.2	1694.2	2187.4	0.4572	1.2102	1.6673
1425	1.403	0.001778	0.1785	514.8	1678.6	2193.4	0.4722	1.1779	1.6502
1450	1.577	0.001798	0.1605	536.7	1662.5	2199.3	0.4873	1.1466	1.6338

Table A.27 (cont.)

T / K	P^{sat}/ MPa	v^L / m^3/kg	v^V / m^3/kg	h^L / kJ/kg	h^{LV} / kJ/kg	h^V / kJ/kg	s^L / kJ/(kg · K)	s^{LV} / kJ/(kg · K)	s^V / kJ/(kg · K)
1475	1.765	0.001820	0.1449	559.2	1645.9	2205.1	0.5024	1.1159	1.6182
1500	1.967	0.001842	0.1311	582.2	1628.5	2210.7	0.5176	1.0856	1.6033
1525	2.185	0.001864	0.1191	605.9	1610.0	2215.9	0.5330	1.0558	1.5888
1550	2.418	0.001888	0.1083	630.5	1590.3	2220.8	0.5487	1.0260	1.5747
1575	2.667	0.001911	0.09879	656.1	1568.8	2224.9	0.5648	0.9960	1.5609
1600	2.932	0.001936	0.09023	683.0	1545.1	2228.1	0.5815	0.9657	1.5471
1625	3.214	0.001961	0.08248	711.5	1518.5	2230.0	0.5988	0.9345	1.5333
1650	3.512	0.001987	0.07539	742.1	1488.0	2230.1	0.6171	0.9018	1.5189

Table A.28 Properties of gaseous potassium.

P / MPa		T / K								
(T^{sat} / K)		sat	1125	1200	1275	1350	1425	1500	1575	1650
0.101325	v / m^3/kg	1.991	2.251	2.435	2.611	2.781	2.947	3.111	3.274	3.435
(1030)	h / kJ/kg	2114.8	2209.1	2267.6	2319.2	2366.9	2412.1	2455.8	2498.5	2540.6
	s / kJ/(kg · K)	2.0769	2.1647	2.2150	2.2568	2.2931	2.3257	2.3556	2.3834	2.4095
0.2	v / m^3/kg	1.062	1.087	1.194	1.291	1.384	1.472	1.558	1.643	1.727
(1109)	h / kJ/kg	2128.8	2148.3	2225.1	2288.1	2343.2	2393.5	2440.9	2486.3	2530.3
	s / kJ/(kg · K)	1.9582	1.9756	2.0418	2.0927	2.1348	2.1711	2.2035	2.2330	2.2603
0.4	v / m^3/kg	0.5612			0.6154	0.6671	0.7155	0.7617	0.8064	0.8501
(1203)	h / kJ/kg	2145.6			2226.7	2296.4	2356.8	2411.3	2461.9	2510.0
	s / kJ/(kg · K)	1.8416			1.9071	1.9603	2.0038	2.0411	2.0741	2.1039
0.7	v / m^3/kg	0.3362				0.3616	0.3923	0.4210	0.4484	0.4747
(1293)	h / kJ/kg	2163.3				2229.9	2304.1	2368.6	2426.7	2480.3
	s / kJ/(kg · K)	1.7526				1.8030	1.8566	1.9007	1.9385	1.9718
1.0	v / m^3/kg	0.2429					0.2638	0.2853	0.3056	0.3249
(1357)	h / kJ/kg	2177.5					2254.8	2328.2	2393.0	2451.9
	s / kJ/(kg · K)	1.6989					1.7545	1.8048	1.8470	1.8835
2.0	v / m^3/kg	0.1292							0.1403	0.1512
(1504)	h / kJ/kg	2211.5							2291.2	2364.9
	s / kJ/(kg · K)	1.6010							1.6528	1.6985

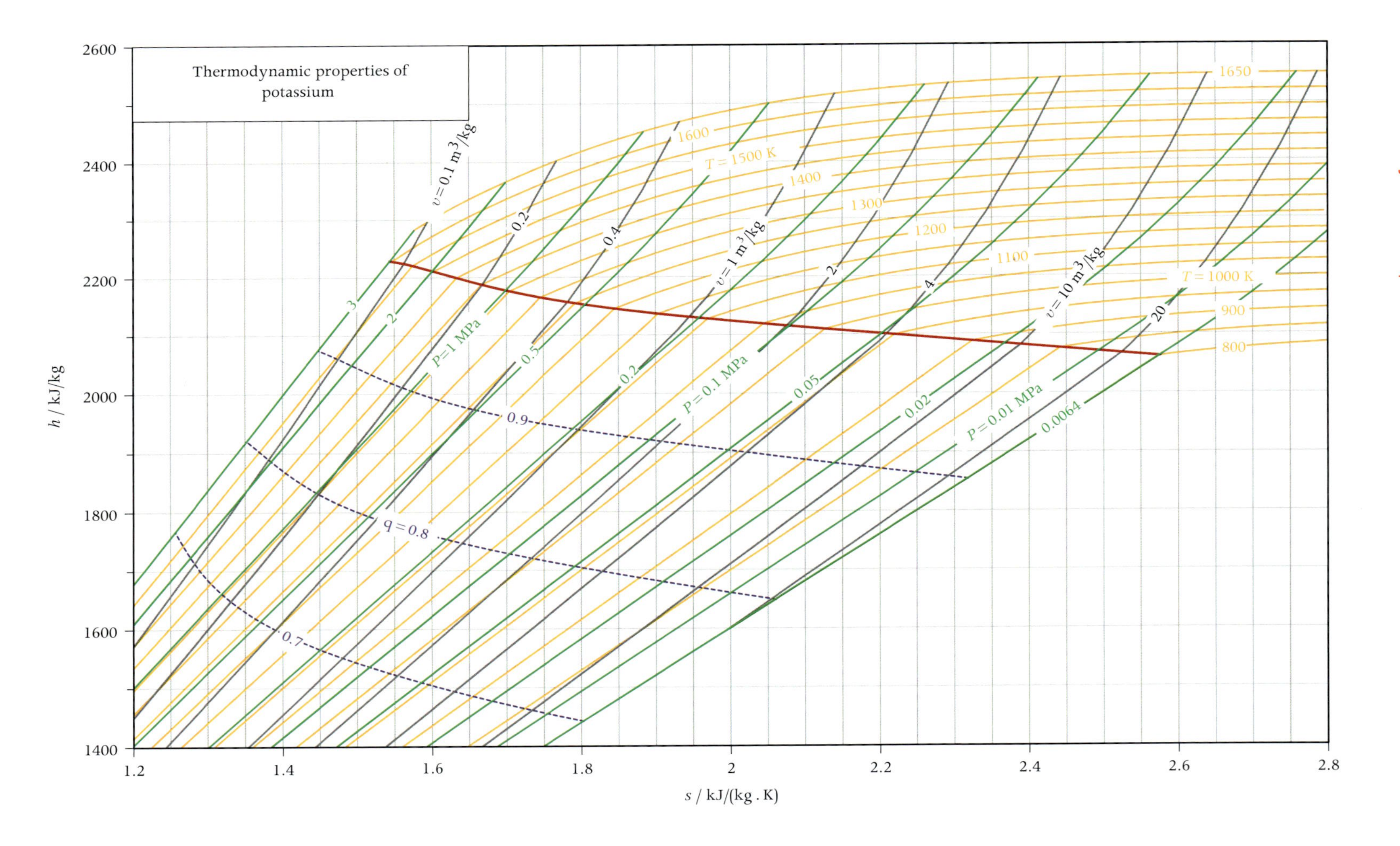
Thermodynamic properties of potassium
h / kJ/kg
s / kJ/(kg . K)
2600
2400
2200
2000
1800
1600
1400
1.2
1.4
1.6
1.8
2
2.2
2.4
2.6
2.8
1650
1600
T = 1500 K
1400
1300
1200
1100
T = 1000 K
900
800
v = 0.1 m3/kg
0.2
0.4
v = 1 m3/kg
2
4
v = 10 m3/kg
20
3
2
P = 1 MPa
0.5
0.2
P = 0.1 MPa
0.05
0.02
P = 0.01 MPa
0.0064
0.9
q = 0.8
0.7

A.12 The Complete iPRSV Thermodynamic Model

Here we see in detail how all the equations for the calculation of fluid thermodynamic properties are obtained from a volumetric equation of state. As an example, we derive all these properties from the pressure-explicit Peng–Robinson equation of state

$$P = \frac{\hat{R}T}{(\hat{v}-\mathsf{b})} - \frac{\mathsf{a}(T)}{\hat{v}(\hat{v}+\mathsf{b})+\mathsf{b}(\hat{v}-\mathsf{b})}.$$

In addition we consider the functional form for the attractive parameter $\mathsf{a}(T)$ proposed by Stryjek and Vera,[1] and improved by Van der Stelt, Nannan and Colonna,[2] in order to obtain more accurate estimations of saturation properties. This thermodynamic model is termed the iPRSV cubic equation of state, and it is implemented in StanMix. The following equations are therefore those coded in the software supporting this book, and you can thus use this tool to compute thermodynamic properties with a detailed understanding of what the program does.

In the iPRSV equation of state, $\mathsf{a}(T)$ is given by

$$\mathsf{a}(T) = \mathsf{a}_\mathrm{c} \cdot \alpha(T),$$

where

$$\mathsf{a}_\mathrm{c} = 0.457235\frac{\hat{R}^2 T_\mathrm{c}^2}{P_\mathrm{c}},$$

$$\alpha(T) = \left[1 + k\left(1-\sqrt{T_\mathrm{r}}\right)\right]^2,$$

$$k = k_0 + k_1\left\{\sqrt{[A - D(T_\mathrm{r}+B)]^2 + E} + A - D(T_\mathrm{r}+B)\right\}\sqrt{T_\mathrm{r}+C},$$

with $A = 1.1$, $B = 0.25$, $C = 0.2$, $D = 1.2$, $E = 0.01$, and

$$k_0 = 0.378893 + 1.4897153\omega - 0.17131848\,\omega^2 + 0.0196554\,\omega^3.$$

The covolume parameter b in the iPRSV equation of state is

$$\mathsf{b} = 0.077796\frac{\hat{R}T_\mathrm{c}}{P_\mathrm{c}}.$$

Where T_c is the critical temperature, T_r is the reduced temperature, and ω is the acentric factor.

The fluid-specific parameter κ_1 is obtained by fitting measured vapor pressures or by fitting calculated values to an accurate $P^\mathrm{sat} = P(T^\mathrm{sat})$ equation. Note that the dependence of α on the temperature is limited to subcritical temperatures. Note also that the acentric factor has been used to obtain the functional form of $\mathsf{a}(T)$ which best correlates experimental vapor pressure data for many different molecules (organic, inorganic, polar, and non-polar).

As shown in Section 6.3, in order to obtain a complete thermodynamic model, an equation for the ideal-gas specific heat capacity must complement the volumetric equation of state. A polynomial dependence on the temperature is sufficiently accurate for most engineering applications, therefore the equation

$$\hat{c}_P^\mathrm{IG}(T) = \alpha + \beta T + \gamma T^2 + \eta T^3$$

is the one used in StanMix .

Table A.29 Fluid input data for the iPRSV thermodynamic model. These data are stored for each fluid in the STANMIX database.

Critical temperature	T_c
Critical pressure	P_c
Acentric factor	ω
Saturation parameter	κ_1
Isobaric ideal-gas specific heat parameter 1	α
Isobaric ideal-gas specific heat parameter 2	β
Isobaric ideal-gas specific heat parameter 3	γ
Isobaric ideal-gas specific heat parameter 4	η

The fluid thermodynamic parameters which are the necessary input for the iPRSV thermodynamic model are given in Table A.29.

Depending on the problem to be solved, a certain thermodynamic property must be calculated for a state specified by two other properties. The calculation of the specific volume v given the pressure P and the temperature T, is the simplest calculation in the case of volumetric equations of state. If the equation is cubic in the specific volume, v can be obtained with Cardano's formula giving the roots of cubic equations. The value of the liquid specific volume can be greatly different from its vapor counterpart, and this can be the source of numerical and accuracy problems, therefore the computer calculation is often carried out in terms of compressibility factor Z, which varies much less in magnitude.

By recalling that the Peng–Robinson equation of state can be written as

$$Z^3 - (1 - B)\,Z^2 + \left(A - 3B^2 - 2B\right) Z - \left(AB - B^2 - B^3\right) = 0,$$

and by setting

$$D = B - 1,$$
$$E = A - 3B^2 - 2B,$$
$$F = B^3 + B^2 - AB,$$

so that the equation of state becomes

$$Z^3 + D \cdot Z^2 + E \cdot Z + F = 0, \tag{A.1}$$

the discriminant is

$$\Delta = \Delta_1^3 - \Delta_2^2,$$

where

$$\Delta_1 = \frac{D^2 - 3E}{9},$$

and

$$\Delta_2 = \frac{2D^3 - 9DE + 27F}{54}.$$

If, for the given values of P and T, the discriminant is positive, there exists a unique real root given by

$$Z = \left(\sqrt{\Delta_2^2 - \Delta_1^3} + |\Delta_2| \right)^{\frac{1}{3}}.$$

The corresponding specific volume is therefore

$$\hat{v} = Z \frac{\hat{R}T}{P}.$$

A single solution of the cubic equation means that the thermodynamic state is in the supercritical region. If the discriminant is negative, Equation (A.1) admits three real solutions (see Figure 6.17), and the specified state is subcritical. A check on the saturation state is necessary in order to establish which root represents a thermodynamic state. For example, if at the given pressure, the given temperature is greater than the saturation temperature at that pressure, the specified state is in the liquid phase, and the smallest root gives the specific volume for the specified state. In the case of two roots, these are usually computed with an iterative method, because this yields more accurate results and can be very fast, if the initial value or the bracketing values for the iteration are correctly chosen.

We have seen that saturation states can be found by equating the fugacity coefficients of the liquid and the vapor, that is

$$\ln \phi^{\mathrm{L}} - \ln \phi^{\mathrm{V}} = 0. \tag{A.2}$$

The fugacity coefficient of a volumetric equation of state is given by

$$\ln \phi = \frac{1}{\hat{R}T} \int_{\hat{v}=\infty}^{\hat{v}} \left[\frac{\hat{R}T}{\hat{v}} - P \right] d\hat{v} - \ln Z + (Z-1).$$

Substituting P with the right-hand side of the iPRSV equation of state and solving the integral gives

$$\ln \phi = (Z-1) - \ln\left(Z - \frac{bP}{\hat{R}T} \right) - \frac{a}{2\sqrt{2}b\hat{R}T} \ln \left[\frac{Z + \left(1+\sqrt{2}\right) \frac{bP}{\hat{R}T}}{Z + \left(1-\sqrt{2}\right) \frac{bP}{\hat{R}T}} \right].$$

Note that in order to obtain the analytic expression of the fugacity coefficient, one makes use of the indefinite integral $\int \frac{dx}{1-x^2} = \tanh^{-1} x + C$ and of the inverse hyperbolic tangent, defined as $\tanh^{-1} x = \frac{1}{2} [\ln(1+x) - \ln(1-x)]$.

Given a value for a subcritical temperature T, the value of the corresponding saturation pressure can be obtained by substituting T in (A.2) and iterating on P until a root is found. At each iteration step, the specific volume of the liquid and of the vapor phase must be calculated according to the procedure we have just illustrated.

Using (6.44) and by recalling that $\hat{h} \equiv \hat{u} + P\hat{v}$, the enthalpy as a function of T and P can be written as

$$\hat{h} = \hat{h}^{\mathrm{IG}} + \hat{R}T(Z-1) + \int_{\infty}^{\hat{v}} \left[T \left(\frac{\partial P}{\partial T} \right)_{\hat{v}} - P \right] d\hat{v}.$$

For the Peng–Robinson equation of state, one obtains

$$\left(\frac{\partial P}{\partial T} \right)_{\hat{v}} = \frac{\hat{R}}{\hat{v} - b} - \frac{\frac{da}{dT}}{\hat{v}\left(\hat{v}+b\right) + b\left(\hat{v}-b\right)}.$$

The derivative with respect to temperature of the attractive parameter a is

$$\frac{da}{dT} = 2a_c\left(1 + \tau_1 \cdot k\right)\left(\tau_1 \frac{dk}{dT} - \frac{k}{2\tau_2}\right),$$

with

$$T_r = \frac{T}{T_c},$$
$$\tau_1 = 1 - \sqrt{T_r},$$
$$\tau_2 = \sqrt{T_c \cdot T},$$
$$\frac{dk}{dT} = \frac{k_1\left(T_x + T_y\right)}{T_c}\left[\frac{1}{2T_z} - \frac{DT_z}{T_y}\right],$$

where

$$T_x = A - D(T_r + B),$$
$$T_y = \sqrt{T_x^2 + E},$$
$$T_z = \sqrt{T_r + C}.$$

Finally, the enthalpy is given by

$$\hat{h} = \hat{h}^{IG} + \hat{R}T\left(Z - 1\right) + \frac{T\frac{da}{dT} - a}{2\sqrt{2}b} ln\left[\frac{Z + \left(1 + \sqrt{2}\right)\frac{bP}{\hat{R}T}}{Z + \left(1 - \sqrt{2}\right)\frac{bP}{\hat{R}T}}\right].$$

The ideal-gas contribution is computed from the isobaric ideal-gas specific heat as

$$\hat{h}^{IG} = \int_{T_0}^{T} \hat{c}_P^{IG} dT = \alpha\left(T - T_0\right) + \beta\frac{\left(T^2 - T_0^2\right)}{2} + \gamma\frac{\left(T^3 - T_0^3\right)}{3} + \delta\frac{\left(T^4 - T_0^4\right)}{4}.$$

The internal energy can easily be obtained from the enthalpy as

$$\hat{u} = \hat{h} - P\hat{v}.$$

The expression of the entropy in terms of T and P is derived in a similar way. From (6.45), the entropy as a function of P, T, and $\hat{v}$ is

$$\hat{s} = \hat{s}^{IG} + \hat{R}\ln Z + \int_{\infty}^{\hat{v}}\left[\left(\frac{\partial P}{\partial T}\right)_{\hat{v}} - \frac{\hat{R}}{\hat{v}}\right]d\hat{v}.$$

For the Peng–Robinson equation of state the entropy is thus

$$\hat{s} = \hat{s}^{IG} + \hat{R}\ln\left(Z - \frac{bP}{\hat{R}T}\right) + \frac{\frac{da}{dT}}{2\sqrt{2}b}\ln\left[\frac{Z + \left(1 + \sqrt{2}\right)\frac{bP}{\hat{R}T}}{Z + \left(1 - \sqrt{2}\right)\frac{bP}{\hat{R}T}}\right],$$

with, cf. (6.11),

$$\hat{s}^{IG} = \int_{T_0}^{T} \frac{\hat{c}_P^{IG}}{T} dT - \hat{R}\ln\frac{P}{P_0}$$

$$= \alpha \ln \frac{T}{T_0} + \beta (T - T_0) + \gamma \frac{\left(T^2 - T_0^2\right)}{2} + \delta \frac{\left(T^3 - T_0^3\right)}{3} - \hat{R} \ln \frac{P}{P_0}.$$

As shown in Section 6.3, in the case of a volumetric equation of state, secondary properties can be computed as a function of a few partial derivatives, namely

$$\left(\frac{\partial P}{\partial \hat{v}}\right)_T, \quad \left(\frac{\partial P}{\partial T}\right)_{\hat{v}}, \quad \left(\frac{\partial^2 P}{\partial T^2}\right)_{\hat{v}},$$

and the integral

$$\int_{\infty}^{\hat{v}} \left(\frac{\partial^2 P}{\partial T^2}\right)_{\hat{v}} d\hat{v}.$$

For the Peng–Robinson equation of state these derivatives are

$$\left(\frac{\partial P}{\partial T}\right)_{\hat{v}} = \frac{\hat{R}}{\hat{v} - \mathsf{b}} - \frac{\dfrac{d\mathsf{a}}{dT}}{\hat{v}\left(\hat{v} + \mathsf{b}\right) + \mathsf{b}\left(\hat{v} - \mathsf{b}\right)},$$

$$\left(\frac{\partial P}{\partial \hat{v}}\right)_T = \frac{-\hat{R}T}{\left(\hat{v} - \mathsf{b}\right)^2} + \frac{2\mathsf{a}\left(\hat{v} + \mathsf{b}\right)}{\left[\hat{v}\left(\hat{v} + \mathsf{b}\right) + \mathsf{b}\left(\hat{v} - \mathsf{b}\right)\right]^2},$$

$$\left(\frac{\partial^2 P}{\partial T^2}\right)_{\hat{v}} = \frac{-\dfrac{d^2\mathsf{a}}{dT^2}}{\hat{v}\left(\hat{v} + \mathsf{b}\right) + \mathsf{b}\left(\hat{v} - \mathsf{b}\right)}.$$

The second derivative of the attractive parameter a with respect to temperature is

$$\frac{d^2\mathsf{a}}{dT^2} = 2\mathsf{a}_{\mathrm{c}} \left[\left(\tau_1 \frac{dk}{dT} - \frac{k}{2T_{\mathrm{c}}\sqrt{T_{\mathrm{r}}}} \right)^2 + (1 + \tau_1 k) \right.$$
$$\left. \times \left(\frac{k}{4T_{\mathrm{c}}^2 T_{\mathrm{r}}^{3/2}} - \frac{dk/dT}{T_{\mathrm{c}}\sqrt{T_{\mathrm{r}}}} + \tau_1 \frac{d^2k}{dT^2} \right) \right],$$

where

$$\frac{d^2k}{dT^2} = \frac{-k_1\left(T_x + T_y\right)}{T_{\mathrm{c}}^2} \left[\frac{D}{T_y T_z} + \frac{1}{4T_z^3} + \frac{D^2 T_z}{T_y^3}\left(T_x - T_y\right) \right].$$

The necessary integral is

$$\int_{\infty}^{\hat{v}} \left(\frac{\partial^2 P}{\partial T^2}\right)_{\hat{v}} d\hat{v} = \frac{\dfrac{d^2\mathsf{a}}{dT^2}}{2\sqrt{2}\mathsf{b}} \ln \left[\frac{Z + \left(1 + \sqrt{2}\right)\dfrac{\mathsf{b}P}{\hat{R}T}}{Z + \left(1 - \sqrt{2}\right)\dfrac{\mathsf{b}P}{\hat{R}T}} \right].$$

The isochoric specific heat capacity is

$$\hat{c}_v = \hat{c}_v^{\mathrm{IG}} + T \int_{\infty}^{\hat{v}} \left(\frac{\partial^2 P}{\partial T^2}\right)_{\hat{v}} d\hat{v}.$$

The isobaric specific heat capacity is

$$\hat{c}_P = \hat{c}_v - T \frac{\left(\dfrac{\partial P}{\partial T}\right)_{\hat{v}}^2}{\left(\dfrac{\partial P}{\partial \hat{v}}\right)_T}.$$

Their ratio is

$$\gamma = 1 - \frac{T}{\hat{c}_v} \frac{\left(\frac{\partial P}{\partial T}\right)_{\hat{v}}^2}{\left(\frac{\partial P}{\partial \hat{v}}\right)_T}.$$

Finally, the speed of sound is

$$c = \sqrt{-\hat{v}^2 \gamma \left(\frac{\partial P}{\partial \hat{v}}\right)_T}.$$

A.13 Extension of the iPRSV Model to Mixtures with the Wong–Sandler Mixing Rules

The iPRSV equation of state for mixtures is

$$P = \frac{\hat{R}T}{(\hat{v} - b_{mix})} - \frac{a_{mix}(T)}{\hat{v}(\hat{v} + b_{mix}) + b_{mix}(\hat{v} - b_{mix})}.$$

The parameters a_{mix} and b_{mix} can be computed with equations involving the parameters of the pure fluid components a_i and b_i and the mixture molar composition. This approach is common to many mixture thermodynamic models and such equations are called *mixing rules*. The Wong–Sandler mixing rules are implemented in StanMix and read

$$a_{mix} = \hat{R}T b_{mix} D,$$

$$b_{mix} = \frac{Q}{1 - D},$$

where

$$Q = \sum_{i=1}^{nc} \sum_{j=1}^{nc} z_i z_j \left(b_{ij} - \frac{a_{ij}}{\hat{R}T} \right),$$

$$D = \frac{\hat{g}^E}{\hat{R}T\sigma} + \sum_{i=1}^{nc} z_i \frac{a_i}{\hat{R}T b_i}.$$

For a Peng–Robinson-type equation of state

$$\sigma = \ln\left(\sqrt{2} - 1\right)/\sqrt{2}.$$

The activity coefficient model used in the mixing rules implemented in StanMix to compute $\hat{g}^E$ is the NRTL (non-random two liquid) model. According to this model

$$\frac{\hat{g}^E}{\hat{R}T} = \sum_{i=1}^{nc} x_i \left[\frac{\sum_{j=1}^{nc} x_j G_{ji} \tau_{ji}}{\sum_{j=1}^{nc} x_j G_{ji}} \right],$$

$$G_{ji} = e^{(-\alpha \tau_{ji})},$$

$$\ln \gamma_i = \frac{\sum_{j=1}^{nc} x_j G_{ji} \tau_{ji}}{\sum_{j=1}^{nc} x_j G_{ji}} + \sum_{j=1}^{nc} \left[\frac{x_j G_{ij}}{\sum_{k=1}^{nc} x_k G_{kj}} \left(\tau_{ij} - \frac{\sum_{k=1}^{nc} x_k G_{ki} \tau_{kj}}{\sum_{k=1}^{nc} x_k G_{kj}} \right) \right],$$

where $\alpha (= \alpha_{ij} = \alpha_{ji})$, τ_{ij}, and τ_{ji} are binary interaction parameters depending on the constituents of the mixture. They are usually calculated by fitting the model to experimental vapor–liquid equilibrium data at low pressure.

The fugacity coefficient of a component in the mixture is given by

$$\ln \tilde{\phi}_i = \frac{1}{\hat{R}T} \int_{\infty}^{\hat{v}} \left[\frac{\hat{R}T}{\hat{v}} - N \left(\frac{\partial P}{\partial N_i} \right)_{T,\hat{v},N_{j \neq i}} \right] d\hat{v} - \ln Z.$$

By substituting for the derivative of P with respect to the composition with the appropriate expression obtained from the iPRSV equation of state and by solving the integral, one obtains

$$\ln \tilde{\phi}_i = \hat{R} \ln \left(Z - \frac{P b_{\text{mix}}}{\hat{R}T} \right) + \frac{1}{b_{\text{mix}}} \left(\frac{\partial N b_{\text{mix}}}{\partial N_i} \right)_{T,\hat{v},N_{j \neq i}} \left(\frac{P \hat{v}}{\hat{R}T} - 1 \right)$$

$$- \frac{a_{\text{mix}}}{2\sqrt{2} b_{\text{mix}} \hat{R}T} \left[\frac{1}{a_{\text{mix}}} \left(\frac{1}{N^2} \frac{\partial N^2 a_{\text{mix}}}{\partial N_i} \right)_{T,\hat{v},N_{j \neq i}} \right.$$

$$\left. - \frac{1}{b_{\text{mix}}} \left(\frac{\partial N b_{\text{mix}}}{\partial N_i} \right)_{T,\hat{v},N_{j \neq i}} \right] \times \ln \frac{\hat{v} + \left(1 + \sqrt{2}\right) b_{\text{mix}}}{\hat{v} + \left(1 - \sqrt{2}\right) b_{\text{mix}}}.$$

The required derivatives of the mixing rules parameters with respect to composition are

$$\left(\frac{1}{N^2} \frac{\partial N^2 a_{\text{mix}}}{\partial N_i} \right)_{T,N_{j \neq i}} = \hat{R}T \left[D \left(\frac{\partial N b_{\text{mix}}}{\partial N_i} \right)_{T,N_{j \neq i}} + b_{\text{mix}} \left(\frac{\partial N D}{\partial N_i} \right)_{T,N_{j \neq i}} \right],$$

$$\left(\frac{\partial N b_{\text{mix}}}{\partial N_i} \right)_{T,N_{j \neq i}} = \frac{1}{1 - D} \left(\frac{1}{N} \frac{\partial N^2 Q}{\partial N_i} \right)_{T,N_{j \neq i}} - \frac{Q}{(1 - D)^2} \left(1 - \frac{\partial N D}{\partial N_i} \right)_{T,N_{j \neq i}},$$

$$\left(\frac{1}{N} \frac{\partial N^2 Q}{\partial N_i} \right)_{T,N_{j \neq i}} = \sum_{j=1}^{nc} 2 z_j \left(b_{ij} - \frac{a_{ij}}{\hat{R}T} \right),$$

$$\left(\frac{\partial N D}{\partial N_i} \right)_{T,N_{j \neq i}} = \frac{1}{\sigma} \left[\frac{\partial \left(N \hat{g}^E / \hat{R}T \right)}{\partial N_i} \right]_{T,N_{j \neq i}} + \frac{a_i}{\hat{R}T b_i} = \frac{\ln \gamma_i}{\sigma} + \frac{a_i}{\hat{R}T b_i}.$$

Again, the enthalpy is expressed with the help of (6.43) and (6.46) and it can be obtained as

$$\hat{h} = \hat{h}_{\text{mix}}^{\text{IG}} + \hat{R}T (Z - 1) + \int_{\infty}^{\hat{v}} \left[T \left(\frac{\partial P}{\partial T} \right)_v - P \right] d\hat{v}.$$

The ideal gas contribution to the enthalpy is given by, see (8.5),

$$\hat{h}_{\text{mix}}^{\text{IG}} = \sum_{i}^{nc} z_i \hat{h}_i^{\text{IG}},$$

where $\hat{h}_i^{\text{IG}}$ is the pure fluid ideal gas enthalpy of mixture component i.

Now we introduce two new variables, ν and ς, in order to make it easier to obtain the partial derivatives that are necessary to compute the enthalpy. These are given by

$$\nu = \hat{v} - b_{mix},$$
$$\varsigma = \hat{v}\left(\hat{v} + b_{mix}\right) + b_{mix}\left(\hat{v} - b_{mix}\right).$$

The iPRSV equation of state can therefore be written as

$$P = \frac{\hat{R}T}{\nu} - \frac{a_{mix}(T)}{\varsigma}.$$

The needed first order partial derivatives with respect to the temperature at constant specific molar volume and composition are

$$\left(\frac{\partial \nu}{\partial T}\right)_{\hat{v},N} = -\left(\frac{\partial b_{mix}}{\partial T}\right)_{\hat{v},N},$$

$$\left(\frac{\partial \varsigma}{\partial T}\right)_{\hat{v},N} = 2\nu\left(\frac{\partial b_{mix}}{\partial T}\right)_{\hat{v},N},$$

$$\left(\frac{\partial P}{\partial T}\right)_{\hat{v},N} = \frac{\hat{R}}{\nu} - \frac{\hat{R}T}{\nu^2}\left(\frac{\partial \nu}{\partial T}\right)_{\hat{v},N} - \frac{1}{\varsigma}\left(\frac{\partial a_{mix}}{\partial T}\right)_{\hat{v},N} + \frac{a_{mix}}{\varsigma^2}\left(\frac{\partial \varsigma}{\partial T}\right)_{\hat{v},N}$$

$$\left(\frac{\partial a_{mix}}{\partial T}\right)_N = \frac{\hat{R}}{1-D}\left[DQ + TQ\left(\frac{\partial D}{\partial T}\right)_N + TD\left(\frac{\partial Q}{\partial T}\right)_N + \frac{TDQ}{1-D}\left(\frac{\partial D}{\partial T}\right)_N\right],$$

$$\left(\frac{\partial b_{mix}}{\partial T}\right)_N = \frac{1}{1-D}\left(\frac{\partial Q}{\partial T}\right)_N + \frac{Q}{(1-D)^2}\left(\frac{\partial D}{\partial T}\right)_N,$$

$$\left(\frac{\partial Q}{\partial T}\right)_N = \sum_{i=1}^{nc}\sum_{j=1}^{nc} z_i z_j \frac{d\left(b_{ij} - \frac{a_{ij}}{\hat{R}T}\right)}{dT},$$

$$\left(\frac{\partial D}{\partial T}\right)_N = \sum_{i=1}^{nc} \frac{z_i}{\hat{R}Tb_i}\left(\frac{da_i}{dT} - \frac{a_i}{T}\right),$$

$$\frac{d\left(b_{ij} - \frac{a_{ij}}{\hat{R}T}\right)}{dT} = \frac{1}{2\hat{R}T}\left[\frac{a_i + a_j}{T} - \left(\frac{da_i}{dT} + \frac{da_j}{dT}\right)\right]\left(1 - k_{ij}\right).$$

At last, after solving the integral, the enthalpy relation reads

$$\hat{h} = \sum_i^{nc} z_i \hat{h}_i^{IG} + \hat{R}T(Z-1) + T\left(\frac{\partial b_{mix}}{\partial T}\right)_{\hat{v},N}\left(\frac{-b_{mix}\hat{R}T}{\hat{v} - b_{mix}} + \frac{a_{mix}\hat{v}}{\hat{v}\left(\hat{v} + b_{mix}\right) + b_{mix}\left(\hat{v} - b_{mix}\right)}\right)$$
$$+ \frac{1}{2\sqrt{2}b_{mix}}\left[T\left(\frac{\partial a_{mix}}{\partial T}\right)_{\hat{v},N} - a_{mix} - \frac{a_{mix}}{b_{mix}}\left(\frac{\partial b_{mix}}{\partial T}\right)_{\hat{v},N}\right] \times \ln\frac{\hat{v} + \left(1+\sqrt{2}\right)b_{mix}}{\hat{v} + \left(1-\sqrt{2}\right)b_{mix}}.$$

Similarly, using (6.45), the entropy for a mixture reads

$$\hat{s} = \hat{s}_{mix}^{IG} + \hat{R}\ln Z + \int_{\infty}^{\hat{v}}\left[\left(\frac{\partial P}{\partial T}\right)_{\hat{v}} - \frac{\hat{R}}{\hat{v}}\right]d\hat{v},$$

where, according to (8.6), the ideal gas contribution to the entropy is

$$\hat{s}^{\text{IG}}_{\text{mix}} = \sum_{i}^{nc} z_i \hat{s}^{\text{IG}}_i + \hat{R} \sum_{i}^{nc} z_i \ln z_i,$$

with $\hat{s}^{\text{IG}}_i$ being the pure fluid ideal gas entropy of mixture component i. Solving the integral in the expression for the entropy yields

$$\hat{s} = \sum_{i}^{nc} z_i \hat{s}^{\text{IG}}_i + \hat{R} \sum_{i}^{nc} z_i \ln z_i + \hat{R} \ln \left(Z - \frac{P \mathrm{b}_{\text{mix}}}{\hat{R} T} \right) + \frac{1}{\mathrm{b}_{\text{mix}}} \left(\frac{\partial \mathrm{b}_{\text{mix}}}{\partial T} \right)_{\hat{v},N}$$
$$\times \left(\frac{-\mathrm{b}_{\text{mix}} \hat{R} T}{\hat{v} - \mathrm{b}_{\text{mix}}} + \frac{\mathrm{a}_{\text{mix}} \hat{v}}{\hat{v} \left(\hat{v} + \mathrm{b}_{\text{mix}} \right) + \mathrm{b}_{\text{mix}} \left(\hat{v} - \mathrm{b}_{\text{mix}} \right)} \right)$$
$$+ \frac{1}{2\sqrt{2} \mathrm{b}_{\text{mix}}} \left[\left(\frac{\partial \mathrm{a}_{\text{mix}}}{\partial T} \right)_{\hat{v},N} - \frac{\mathrm{a}_{\text{mix}}}{\mathrm{b}_{\text{mix}}} \left(\frac{\partial \mathrm{b}_{\text{mix}}}{\partial T} \right)_{\hat{v},N} \right] \times \ln \frac{\hat{v} + \left(1 + \sqrt{2}\right) \mathrm{b}_{\text{mix}}}{\hat{v} + \left(1 - \sqrt{2}\right) \mathrm{b}_{\text{mix}}}.$$

Let us now see how the partial derivatives and the integral necessary for the calculation of all other thermodynamic properties are obtained, see Section 6.3. The first order partial derivative of pressure with respect to the specific volume at constant temperature and composition is the same as in the iPRSV equation for pure fluids, because the mixture parameters are not dependent on the specific volume, therefore

$$\left(\frac{\partial P}{\partial \hat{v}} \right)_{T,N} = \frac{-\hat{R} T}{\left(\hat{v} - \mathrm{b}_{\text{mix}} \right)^2} + \frac{2 \mathrm{a}_{\text{mix}} \left(\hat{v} + \mathrm{b}_{\text{mix}} \right)}{\hat{v} \left(\hat{v} + \mathrm{b}_{\text{mix}} \right) + \mathrm{b}_{\text{mix}} \left(\hat{v} - \mathrm{b}_{\text{mix}} \right)}.$$

The second order partial derivatives with respect to temperature at constant specific volume and compositions are

$$\left(\frac{\partial^2 \nu}{\partial T^2} \right)_{\hat{v},N} = -\left(\frac{\partial^2 \mathrm{b}_{\text{mix}}}{\partial T^2} \right)_{\hat{v},N},$$
$$\left(\frac{\partial^2 \varsigma}{\partial T^2} \right)_{\hat{v},N} = -2 \left(\frac{\partial \mathrm{b}_{\text{mix}}}{\partial T} \right)^2_{\hat{v},N} + 2\nu \left(\frac{\partial^2 \mathrm{b}_{\text{mix}}}{\partial T^2} \right)_{\hat{v},N},$$
$$\left(\frac{\partial^2 P}{\partial T^2} \right)_{v,N} = \frac{\hat{R}}{\nu^2} \left\{ 2 \left(\frac{\partial \nu}{\partial T} \right)_{\hat{v},N} \left[\frac{T}{\nu} \left(\frac{\partial \nu}{\partial T} \right)_{\hat{v},N} - 1 \right] - T \left(\frac{\partial^2 \nu}{\partial T^2} \right)_{\hat{v},N} \right\}$$
$$+ \frac{1}{\varsigma^2} \left\{ 2 \left(\frac{\partial \varsigma}{\partial T} \right)_{\hat{v},N} \left[\left(\frac{\partial \mathrm{a}_{\text{mix}}}{\partial T} \right)_{\hat{v},N} - \frac{\mathrm{a}_{\text{mix}}}{\varsigma} \left(\frac{\partial \varsigma}{\partial T} \right)_{\hat{v},N} \right] - \varsigma \left(\frac{\partial^2 \mathrm{a}_{\text{mix}}}{\partial T^2} \right)_{\hat{v},N} + \mathrm{a}_{\text{mix}} \left(\frac{\partial^2 \varsigma}{\partial T^2} \right)_{\hat{v},N} \right\},$$

$$\left(\frac{\partial^2 \mathrm{a}_{\text{mix}}}{\partial T^2} \right)_N = \hat{R} \left\{ \frac{2TDQ \left(\frac{\partial D}{\partial T} \right)^2_N}{(1-D)^3} + \frac{2DQ \left(\frac{\partial D}{\partial T} \right)_N + 2TQ \left(\frac{\partial D}{\partial T} \right)^2_N}{(1-D)^2} + \frac{2TD \left(\frac{\partial Q}{\partial T} \right)_N \left(\frac{\partial D}{\partial T} \right)_N + TDQ \left(\frac{\partial^2 D}{\partial T^2} \right)_N}{(1-D)^2} \right.$$
$$\left. + \frac{2Q \left(\frac{\partial D}{\partial T} \right)_N + 2D \left(\frac{\partial Q}{\partial T} \right)_N + 2T \left(\frac{\partial D}{\partial T} \right)_N \left(\frac{\partial Q}{\partial T} \right)_N}{(1-D)} + \frac{\left(\frac{\partial^2 D}{\partial T^2} \right)_N + TD \left(\frac{\partial^2 Q}{\partial T^2} \right)_N}{(1-D)} \right\},$$

$$\left(\frac{\partial^2 \mathrm{b}_{\text{mix}}}{\partial T^2} \right)_N = \frac{\left(\frac{\partial^2 Q}{\partial T^2} \right)_N}{(1-D)} + \frac{2 \left(\frac{\partial D}{\partial T} \right)_N \left(\frac{\partial Q}{\partial T} \right)_N + Q \left(\frac{\partial^2 D}{\partial T^2} \right)_N}{(1-D)^2} + \frac{2Q \left(\frac{\partial D}{\partial T} \right)^2_N}{(1-D)^3},$$

$$\left(\frac{\partial^2 Q}{\partial T^2}\right)_N = \sum_{i=1}^{nc}\sum_{j=1}^{nc} z_i z_j \frac{d^2\left(\mathsf{b}_{ij} - \frac{\mathsf{a}_{ij}}{\hat{R}T}\right)}{dT^2},$$

$$\left(\frac{\partial^2 D}{\partial T^2}\right)_N = \sum_{i=1}^{nc} \frac{z_i}{\hat{R}T\mathsf{b}_i}\left(\frac{d^2\mathsf{a}_i}{dT^2} - \frac{2}{T}\frac{d\mathsf{a}_i}{dT} + \frac{2\mathsf{a}_i}{T^2}\right),$$

and

$$\frac{d^2\left(\mathsf{b}_{ij} - \frac{\mathsf{a}_{ij}}{\hat{R}T}\right)}{dT^2} = \frac{1}{2\hat{R}T}\left[\frac{-2}{T^2}\left(\mathsf{a}_i + \mathsf{a}_j\right) + \frac{2}{T}\left(\frac{d\mathsf{a}_i}{dT} + \frac{d\mathsf{a}_j}{dT}\right) - \left(\frac{d^2\mathsf{a}_i}{dT^2} + \frac{d^2\mathsf{a}_j}{dT^2}\right)\right]\left(1 - k_{ij}\right).$$

Finally, the integral $\int_\infty^{\hat{v}}\left(\frac{\partial^2 P}{\partial T^2}\right)_{\hat{v},N} d\hat{v}$, which is required for the computation of the isochoric specific heat capacity, can be calculated as

$$\begin{aligned}
\int_\infty^{\hat{v}}\left(\frac{\partial^2 P}{\partial T^2}\right)_{\hat{v},N} d\hat{v} = {} & \frac{\hat{R}}{\hat{v} - \mathsf{b}_{\text{mix}}}\left\{\frac{T}{\hat{v} - \mathsf{b}_{\text{mix}}}\left(\frac{\partial \mathsf{b}_{\text{mix}}}{\partial T}\right)^2_{\hat{v},N} - 2\left(\frac{\partial \mathsf{b}_{\text{mix}}}{\partial T}\right)_{\hat{v},N} - T\left(\frac{\partial^2 \mathsf{b}_{\text{mix}}}{\partial T^2}\right)_{\hat{v},N}\right\} \\
& - \frac{2a\left(\mathsf{b}_{\text{mix}} - 3\hat{v}\right)}{\varsigma^2}\left(\frac{\partial \mathsf{b}_{\text{mix}}}{\partial T}\right)^2_{\hat{v},N} + \frac{1}{\mathsf{b}^2_{\text{mix}}\varsigma}\left\{2b\hat{v}\left(\frac{\partial \mathsf{a}_{\text{mix}}}{\partial T}\right)_{\hat{v},N}\left(\frac{\partial \mathsf{b}_{\text{mix}}}{\partial T}\right)_{\hat{v},N}\right. \\
& \left. + \mathsf{a}_{\text{mix}}\left[\mathsf{b}_{\text{mix}}\hat{v}\left(\frac{\partial^2 \mathsf{b}_{\text{mix}}}{\partial T^2}\right)_{\hat{v},N} - 2\left(\mathsf{b}_{\text{mix}} + \hat{v}\right)\left(\frac{\partial \mathsf{b}_{\text{mix}}}{\partial T}\right)^2_{\hat{v},N}\right]\right\} \\
& + \frac{1}{2\sqrt{2}\mathsf{b}^3_{\text{mix}}}\left\{\mathsf{b}^2_{\text{mix}}\left(\frac{\partial^2 \mathsf{a}_{\text{mix}}}{\partial T^2}\right)_{\hat{v},N} + 2\mathsf{a}_{\text{mix}}\left(\frac{\partial \mathsf{b}_{\text{mix}}}{\partial T}\right)^2_{\hat{v},N}\right. \\
& \left. - \mathsf{a}_{\text{mix}}\mathsf{b}_{\text{mix}}\left(\frac{\partial^2 \mathsf{b}_{\text{mix}}}{\partial T^2}\right)_{\hat{v},N} - 2\mathsf{b}_{\text{mix}}\left(\frac{\partial \mathsf{a}_{\text{mix}}}{\partial T}\right)_{\hat{v},N}\left(\frac{\partial \mathsf{b}_{\text{mix}}}{\partial T}\right)_{\hat{v},N}\right\} \times \ln\frac{\hat{v} + \left(1 + \sqrt{2}\right)\mathsf{b}_{\text{mix}}}{\hat{v} + \left(1 - \sqrt{2}\right)\mathsf{b}_{\text{mix}}}.
\end{aligned}$$

B Mathematical Relations between Partial Derivatives

The *derivative inversion rule* is

$$\left(\frac{\partial F}{\partial y}\right)_x = \frac{1}{(\partial y/\partial F)_x}. \tag{B.1}$$

The *triple product rule* (xyz-1 rule) is

$$\left(\frac{\partial F}{\partial x}\right)_y \left(\frac{\partial x}{\partial y}\right)_F \left(\frac{\partial y}{\partial F}\right)_x = -1. \tag{B.2}$$

The *chain rule*, often used to introduce another independent variable Φ, is

$$\left(\frac{\partial F}{\partial y}\right)_x = \frac{(\partial F/\partial \Phi)_x}{(\partial y/\partial \Phi)_x} = \left(\frac{\partial F}{\partial \Phi}\right)_x \left(\frac{\partial \Phi}{\partial y}\right)_x. \tag{B.3}$$

The *composite derivative rule* is

$$F(x, y) = G\left[\alpha(x, y), \beta(x, y)\right], \tag{B.4}$$

$$\left(\frac{\partial F}{\partial x}\right)_y = \left(\frac{\partial G}{\partial \alpha}\right)_\beta \left(\frac{\partial \alpha}{\partial x}\right)_y + \left(\frac{\partial G}{\partial \beta}\right)_\alpha \left(\frac{\partial \beta}{\partial x}\right)_y, \tag{B.5}$$

$$\left(\frac{\partial F}{\partial y}\right)_x = \left(\frac{\partial G}{\partial \alpha}\right)_\beta \left(\frac{\partial \alpha}{\partial y}\right)_x + \left(\frac{\partial G}{\partial \beta}\right)_\alpha \left(\frac{\partial \beta}{\partial y}\right)_x. \tag{B.6}$$

In particular, if $\alpha(x, y) = x$ then $(\partial \alpha/\partial x)_y = 1$ and $(\partial \alpha/\partial y)_x = 0$. In this case the composite derivative rule for $F(x, y) = G\left[x, z(x, y)\right]$ gives

$$\left(\frac{\partial F}{\partial x}\right)_y = \left(\frac{\partial G}{\partial x}\right)_z + \left(\frac{\partial G}{\partial z}\right)_x \left(\frac{\partial z}{\partial x}\right)_y, \tag{B.7}$$

and

$$\left(\frac{\partial F}{\partial y}\right)_x = \left(\frac{\partial F}{\partial z}\right)_x \left(\frac{\partial z}{\partial y}\right)_x. \tag{B.8}$$

C Numerical Schemes for Saturation Point and Flash Calculations

C.1 Numerical Scheme for Bubble and Dew Point Calculations

Figure C.1 shows, as an example, a simplified flowchart illustrating a computer procedure implementing a rather simple numerical scheme for the calculation of the bubble temperature of a mixture. Procedures for the calculations of the dew temperature and of the bubble and dew pressures are similar. Initial values for bubble and dew point temperatures and pressures can be determined by employing Wilson's approximation, which reads

$$\ln K_i = \ln\left(\frac{P_{c_i}}{P}\right) + 5.373\,(1+\omega_i)\left(1 - \frac{T_{c_i}}{T}\right), \tag{C.1}$$

where P_{c_i}, T_{c_i}, and ω_{c_i} are the critical pressure, the critical temperature, and the acentric factor of component i, respectively.
Such a simple approach is prone to convergence problems if the thermodynamic state is close to the vapor–liquid critical point. For such calculations, much more complex numerical methods must be adopted.

C.2 Numerical Scheme for Isothermal *PT*-flash Calculations

The isothermal flash point, or in short the *PT*-flash calculation allows us to compute the liquid (x_i) and vapor (y_i) composition of a fluid mixture in equilibrium, given P and T. Other types of flash calculations go by the name of *Ph*-, *Tv*-, *vs*-, and *vu*-flash calculation, depending on the specified variables. The *PT*-flash calculation benefits from the use of equilibrium factors K_i, auxiliary variables defined as

$$K_i \equiv \frac{\tilde{\phi}_i^{\mathrm{V}}}{\tilde{\phi}_i^{\mathrm{L}}}.$$

The equilibrium equations to be solved are

$$y_i \tilde{\phi}_i^{\mathrm{V}} = x_i \tilde{\phi}_i^{\mathrm{L}}.$$

In addition, the constraints coming from the mass balances and from the fact that the summation of the mole fractions equals 1, must also be satisfied. These are

$$q y_i + (1-q)\, x_i - z_i = 0,$$

$$\sum_i (y_i - x_i) = 0,$$

with q the vapor mole fraction. All equations can be combined into a single equation, the so-called Rachford–Rice equation, which reads

$$\sum_i \frac{z_i (K_i - 1)}{1 - q + qK_i} = 0.$$

A flowchart illustrating a computer procedure implementing a simple algorithm for the solution of the PT-flash problem is given in Figure C.2. Initial values for the equilibrium factors K_i are usually computed by employing Wilson's Equation (C.1).

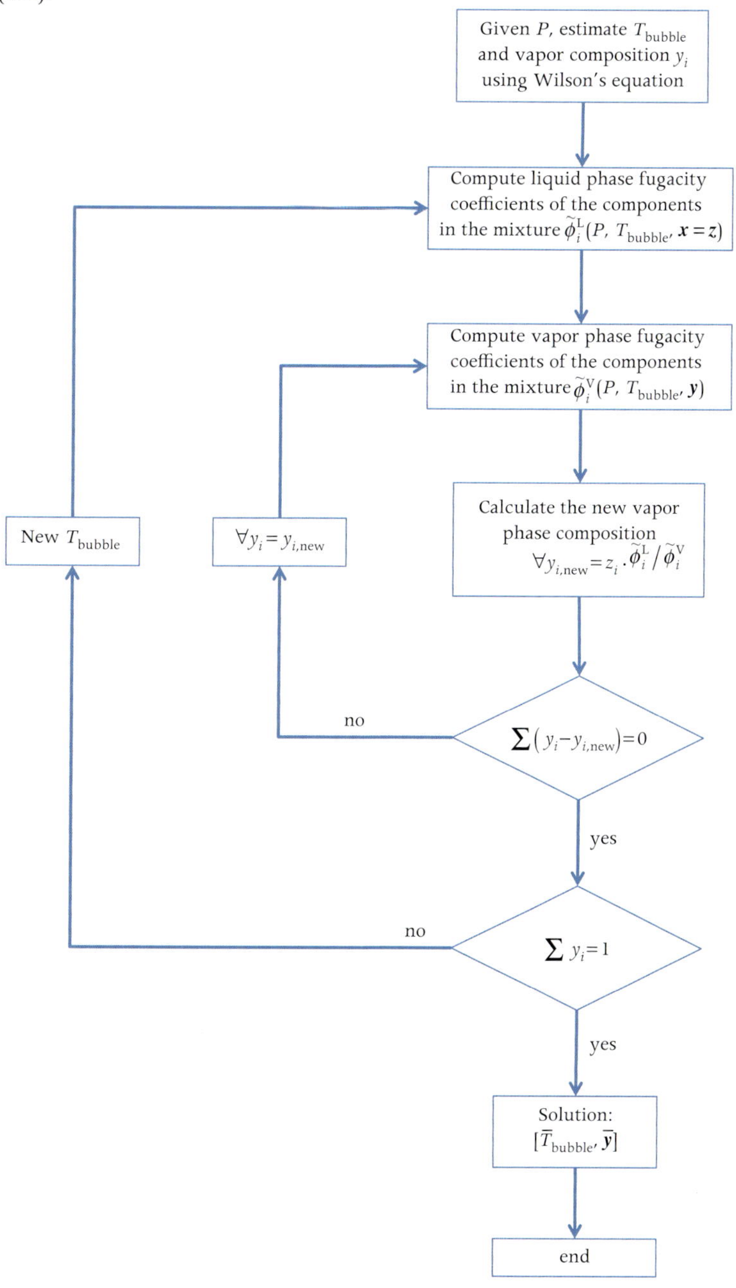

Figure C.1 Computational scheme of a simple algorithm for the calculation of the bubble temperature.

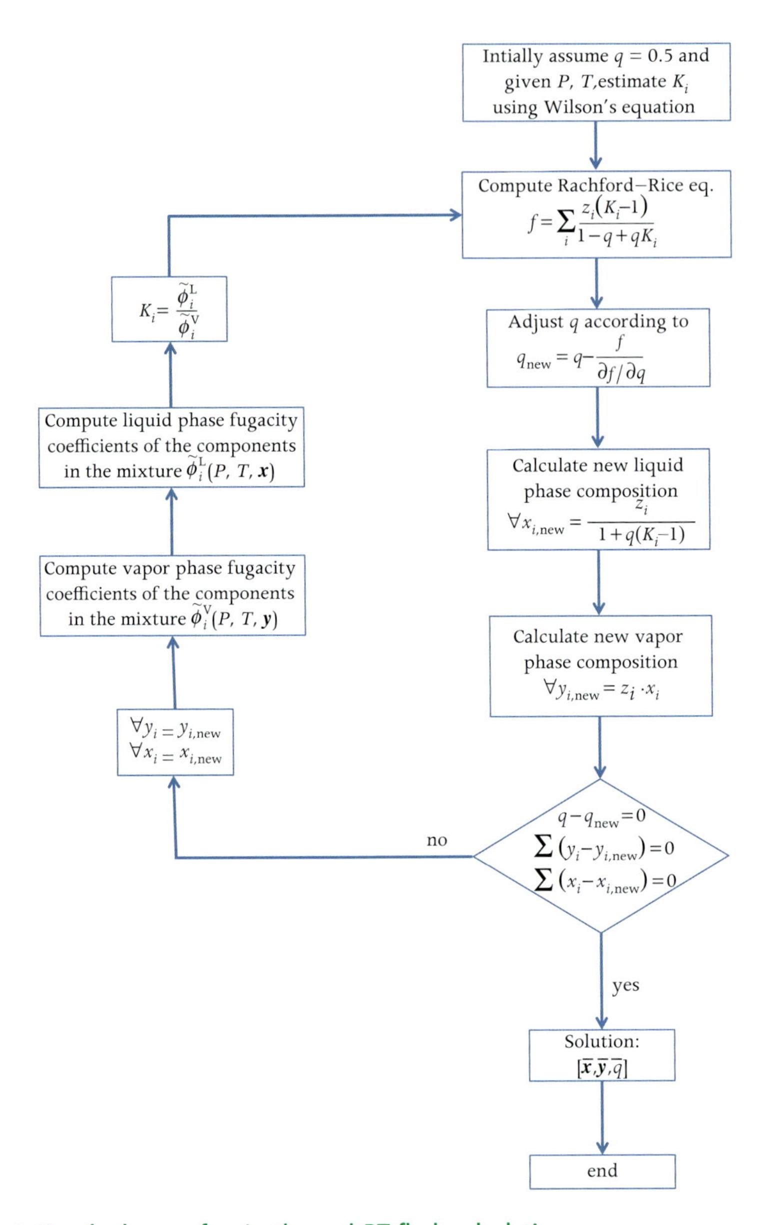

Figure C.2 Computational scheme of an isothermal *PT*-flash calculation.

D Chemical Equilibrium

D.1 Logarithms to the Base 10 of the Equilibrium Constant K

Given the generic chemical equation

$$\nu_1 C_1 + \nu_2 C_2 \rightleftharpoons \nu_3 C_3 + \nu_4 C_4,$$

the corresponding definition for the equilibrium constant is

$$K(T) \equiv \frac{x_3^{\nu_3} x_4^{\nu_4}}{x_1^{\nu_1} x_2^{\nu_2}} \left(\frac{P}{P_0} \right)^{\nu_3+\nu_4-\nu 1-\nu 2} \qquad (P_0 = 1.01325 \text{ bar}) .$$

Table D.1 lists temperature-dependent values 0f K for some example reactions.

Table D.1 Values of K as a function of temperature for several chemical reactions.

T/K	$H_2 \rightleftharpoons 2H$	$O_2 \rightleftharpoons 2O$	$H_2O \rightleftharpoons H_2 + \frac{1}{2}O_2$	$H_2O \rightleftharpoons OH + \frac{1}{2}H_2$	$CO_2 \rightleftharpoons CO + \frac{1}{2}O_2$	$N_2 \rightleftharpoons 2N$	$\frac{1}{2}N_2 + \frac{1}{2}O_2 \rightleftharpoons NO$	$Na \rightleftharpoons Na^+ + e^-$	$Cs \rightleftharpoons Cs^+ + e^-$
298	−71.210	−80.620	−40.047	−46.593	−45.059	−119.434	−15.187	−32.3	−25.1
400	−51.742	−58.513	−29.241	−33.910	−32.380	−87.473	−11.156	−24.3	−17.5
600	−32.667	−36.859	−18.663	−21.470	−20.041	−56.206	−7.219	−14.6	−10.0
800	−23.074	−25.985	−13.288	−15.214	−13.882	−40.521	−5.250	−9.58	−6.15
1000	−17.288	−19.440	−10.060	−11.444	−10.195	−31.084	−4.068	−6.54	−3.79
1200	−13.410	−15.062	−7.896	−8.922	−7.743	−24.619	−3.279	−4.47	−2.18
1400	−10.627	−11.932	−6.334	−7.116	−5.998	−20.262	−2.717	−2.97	−1.010
1600	−8.530	−9.575	−5.175	−5.758	−4.693	−16.869	−2.294	−1.819	−0.108
1800	−6.893	−7.740	−4.263	−4.700	−3.682	−14.225	−1.966	−0.913	+0.609
2000	−5.579	−6.269	−3.531	−3.852	−2.875	−12.016	−1.703	−0.175	+1.194
2200	−4.500	−5.064	−2.931	−3.158	−2.218	−10.370	−1.488	+0.438	+1.682
2400	−3.598	−4.055	−2.429	−2.578	−1.673	−8.992	−1.309	+0.956	+2.098
2600	−2.833	−3.206	−2.003	−2.087	−1.213	−7.694	−1.157	+1.404	+2.46
2800	−2.176	−2.475	−1.638	−1.670	−0.820	−6.640	−1.028	+1.792	+2.77
3000	−1.604	−1.840	−1.322	−1.302	−0.481	−5.726	−0.915	+2.13	+3.05
3200	−1.104	−1.285	−1.046	−0.983	−0.185	−4.925	−0.817	+2.44	+3.29
3500	−0.458	−0.571	−0.693	−0.557	+0.194	−3.893	−0.692	+2.84	3.62
4000	+0.406	+0.382	−0.221	−0.035	+0.695	−2.514	−0.526	+3.38	+4.07
4500	+1.078	+1.125	+0.153	+0.392	+1.082	−1.437	−0.345	+3.82	+4.43
5000	+1.619	+1.719	+0.450	+0.799	+1.388	−0.570	−0.298	+4.18	+4.73

D.2 StanJan

D.2.1 Dual problem

The objective is to numerically solve (10.52) and (10.53), which we give here for convenience,

$$\sum_{j=1}^{na} n_{ij}\bar{N}_{(j)}z_j = p_i, \qquad i = 1, \ldots, na, \tag{D.1}$$

and

$$\sum_{\substack{j=1 \\ \text{in } m}}^{nc} z_j = 1, \qquad m = 1, \ldots, ph, \tag{D.2}$$

in order to determine the na unknown element potentials and the p unknown phase moles. A convergent algorithm exists for this purpose, based on a related max–min problem: the *dual* problem.

We define three functions of the element potentials and phase moles as

$$W \equiv \sum_{m=1}^{p} \bar{N}_m (Z_m - 1) - \sum_{i=1}^{n} a\lambda_i p_i,$$

$$Z_m \equiv \sum_{\substack{j=1 \\ \text{in } m}}^{s} z_j,$$

and

$$H_i \equiv \sum_{j=1}^{s} \bar{N}_j n_{ij} z_j - p_i,$$

where z_j is defined in terms of the λ_i by (10.50), that is by

$$z_j = \exp\left(-\tilde{g}_j + \sum_{i=1}^{na} \lambda_i n_{ij}\right).$$

Note that $Z_m = 1$ for all phases present in the system when (D.2) is satisfied, and $H_i = 0$ for all atoms i when (D.1) is satisfied. Differentiating $W(\boldsymbol{\lambda}, \bar{N})$ gives

$$\frac{\partial W}{\partial \lambda_i} = H_i, \tag{D.3}$$

and

$$\frac{\partial W}{\partial \bar{N}_m} = Z_m - 1,$$

so that

$$dW = \sum_{i=1}^{a} H_i d\lambda_i + \sum_{m=1}^{p} (Z_m - 1)\, d\bar{N}_m. \tag{D.4}$$

Note that, at constant $\bar{N}_m$, W is stationary ($dW = 0$) with respect to arbitrary variations in the element potentials at any state for which the atomic constraints are satisfied ($H_i = 0$).

We further define

$$D_{im} \equiv \sum_{\substack{j=1 \\ \text{in } m}}^{s} n_{ij} z_j,$$

and

$$Q_{ik} \equiv \sum_{j=1}^{s} \bar{N}_j n_{ij} n_{kj} z_j. \tag{D.5}$$

Then we have that

$$dZ_m = \sum_{i=1}^{na} D_{im} d\lambda_i \tag{D.6}$$

and

$$dH_i = \sum_{k=1}^{a} Q_{ik} d\lambda_k + \sum_{m=1}^{p} D_{im} d\bar{N}_m. \tag{D.7}$$

The nature of the stationary point in W is revealed by

$$\frac{\partial^2 W}{\partial \lambda_i \partial \lambda_k} = \frac{\partial H_i}{\partial \lambda_k} = Q_{ik} > 0. \tag{D.8}$$

Since $Q_{ii} > 0$, W is a minimum at the extremum, and W is a concave function of the element potentials. This means that the minimum W point, where the population constraints are satisfied for fixed phase moles, can be found by the method of steepest descent, in which we move down the path in the λ space along which W decreases most rapidly, until we find the minimum point.

Now consider a path in the $(\boldsymbol{\lambda}, \bar{\mathbf{N}})$ space along which the H_i all vanish. From (D.4), we see that W on this path is also stationary with respect to arbitrary variations $\bar{N}_m$ when (D.2) is satisfied for all phases present. Along this path the λ_i are fixed by the $\bar{N}_m$, and we must consider this in the analysis.

Between any two states for which the H_i are zero, from (D.7),

$$\sum_{k=1}^{a} Q_{ik} d\lambda_k = -\sum_{m=1}^{p} D_{im} d\bar{N}_m.$$

This tells us how the W-minimizing λ_i will change when we change the $\bar{N}_m$. We define a matrix E_{im} such that

$$\sum_{k=1}^{a} Q_{ik} E_{km} = -D_{im}. \tag{D.9}$$

Then, between two nearby states where the H_i all vanish,

$$d\lambda_i = \sum_{m=1}^{p} E_{im} d\bar{N}_m.$$

Then, from (D.6), along the path of states where all $H_i = 0$ we have

$$dZ_m = \sum_{n=1}^{p} A_{mn} d\bar{N}_n, \tag{D.10}$$

where

$$dA_{mn} = \sum_{i=1}^{a} D_{im} E_{in}. \tag{D.11}$$

Equation (D.11) shows how the Z_m change when we change the $\bar{N}_m$ along the path where all $H_i = 0$. Let $W^*(\bar{N})$ denote the value of W along such a path. Then, from (D.4),

$$\frac{\partial W^*}{\partial \bar{N}_m} = V_m, \tag{D.12}$$

where

$$V_m = Z_m - 1.$$

Therefore, using (D.10),

$$\frac{\partial^2 W^*}{\partial \bar{N}_m \partial \bar{N}_n} = \frac{\partial Z_m}{\partial \bar{N}_n} = A_{mn}. \tag{D.13}$$

Hence, using (D.11), (D.9), and (D.5),

$$\begin{aligned}\frac{\partial^2 W^*}{\partial \bar{N}_m \partial \bar{N}_n} &= \sum_{i=1}^{a} D_{im} E_{im} \\ &= -\sum_{i=1}^{a}\sum_{k=1}^{a} Q_{ik} E_{km} E_{im} \\ &= -\sum_{j=1}^{a}\sum_{i=1}^{a}\sum_{k=1}^{a} \bar{N}_j z_j n_{ij} E_{im} E_{km} \\ &= -\sum_{j=1}^{s} \bar{N}_j z_j \left(\sum_{i=1}^{a} n_{ij} E_{im}\right)^2 < 0.\end{aligned}$$

Hence, W^*, is a maximum at the stationary point; moreover, W^* is a convex function of the $\bar{N}_n$. This means that the method of steepest ascent, in which we move up the $W^*(\bar{N})$ surface along the most rapidly rising path, can be used to find the maximum.

Summarizing, W is a minimum for given phase moles at any state for which the atomic constraints are satisfied. We denote such states by W^*. Where W^* is in turn a maximum with respect to the phase moles when the mole-fraction-sum constraints are satisfied. These facts form the basis for a convergent solution algorithm.

At the equilibrium solution,

$$W^*_{\max} = -\sum_{i=1}^{na} \lambda_i p_i, \tag{D.14a}$$

but, since λ_i is the G/RT per mole of i atoms,

$$W^*_{\max} = \frac{G}{RT}. \tag{D.14b}$$

The max–min problem for W is the *dual* of the Gibbs minimization problem, with the dual function W having physical significance only in the equilibrium state. To help the reader understand this max–min problem, consider the one-species case (10.50), that is

$$z_j = \exp\left(-\tilde{g}_j + \sum_{i=1}^{na} \lambda_i n_{ij}\right),$$

becomes

$$z = \exp\left(-\tilde{g} + \lambda n\right), \tag{D.15}$$

and the atomic constraint is

$$\bar{N} n z = p. \tag{D.16}$$

For this case

$$W = \bar{N}\left[\exp\left(-\tilde{g} + \lambda n\right) - 1\right] - \lambda p.$$

At fixed N, W is a concave function of λ, sketched in Figure D.1a. The atomic constraint (D.15) is satisfied when

$$\lambda = \frac{1}{n}\left[\tilde{g} + \ln\left(\frac{p}{n\bar{N}}\right)\right].$$

The path along which the constraints are satisfied is sketched in Figure D.1b. Thus,

$$W^* = \frac{p}{n} - \bar{N} - \frac{p}{n}\left[\tilde{g} + \ln\left(\frac{p}{n\bar{N}}\right)\right].$$

In Figure D.1c W^* is sketched as a function of $\bar{N}$. The maximum of W^* occurs when

$$\bar{N} = \frac{p}{n},$$

for which the element potential is, as expected,

$$\lambda = \frac{\tilde{g}}{n},$$

and the equilibrium value of W^* is

$$W^*_{\max} = -\bar{N}\tilde{g} = -\bar{N} n \lambda.$$

D.2.2 Detail of the numerical solution

The solution of the max–min problem proceeds in three modes. We shall first describe these graphically in terms of the previous example and its extension into more dimensions, and then present the analytical details. The surface $W(\lambda, \bar{N})$, for $n = 1$, $p = 1$, $\tilde{g} = 0$, is shown in Figure D.1d. One way to solve the problem is to march down the surface at constant N until the minimum point is reached. The loci of such minima define a road that leads up the valley to the saddle point, where the the equilibrium solution is located.

Imagine that this surface is a hillside. On the hillside, a hiker has the choice of going up or down, and he must go down in λ space to reach the minimum W at fixed $\bar{N}$. If there were two λ's instead of just one, his downhill path could be taken in many ways; the fastest way down is the "path of steepest descent." Once at the roadway, the hiker has a choice of up or down, and he must go up in N-space to reach the point of maximum W^* where the phase-mole-sums are all unity. But he must stay on the road, where the population constraints are satisfied. He would have a choice of several uphill roads if there were more than one N, and he should choose the road of steepest ascent in order to reach the summit most quickly. This is the basis for the numerical method for solution of the max–min problem for W; steepest descent variation of the element potentials at constant phase moles, followed by steepest-ascent variation of the phase mole while maintaining the atomic constraints.

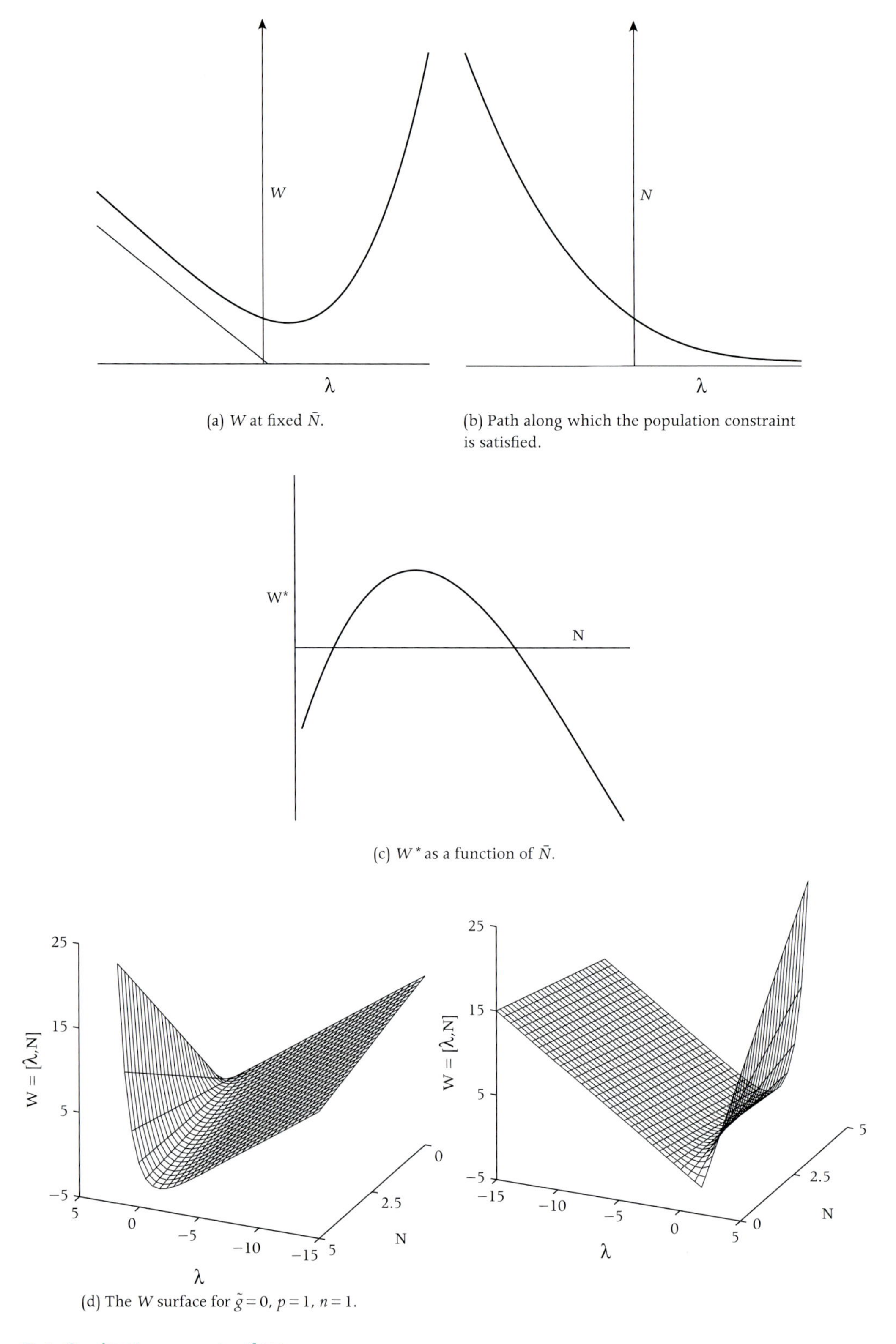

(a) W at fixed $\bar{N}$.

(b) Path along which the population constraint is satisfied.

(c) W^* as a function of $\bar{N}$.

(d) The W surface for $\tilde{g} = 0$, $p = 1$, $n = 1$.

Figure D.1 Qualitative aspects of W.

At the bottom of the valley, where the terrain is flat, it is hard to tell which way is down; here steepest descent methods often have difficulty, or are slow; but Newton–Raphson methods work well when we are close to a solution. So we use steepest descent to get close, and Newton–Raphson to zoom to the minimum. The same basic procedure is applied ascending towards the top. The phase moles are adjusted by steepest ascent until we are very near the solution, at which point a Newton–Raphson iteration is employed.

In a two-phase problem, the hiker could be going nicely uphill, reducing the moles of one phase, and run abruptly into a fence labeled, "No more moles – keep out of negative mole-land!" At this point our hiker must turn and follow the fence uphill, maintaining zero moles for this phase, until the path of steepest ascent leaves the fence. This is the basic method by which the solution process decides which phases are present. We shall now outline the mathematics of these processes.

Mode 1: steepest descent of W in λ space at fixed N.

If a segment of the descent path in λ space has length ds, then

$$d\lambda_i = f_i ds$$

and

$$\sum_{i=1}^{na} = f_i f_i = 1. \tag{D.17}$$

Then,

$$\frac{dW}{ds} = \sum\nolimits_{i=1}^{na} H_i f_i,$$

where the f_i are the direction cosines for the descent path. To find the path of steepest descent, we seek the f_i that maximize dW/ds, subject to (D.16). Thus, we put

$$d\left(\frac{dW}{ds} - \frac{\beta}{2}\sum\nolimits_{i=1}^{na} f_i f_i\right) = 0,$$

where $\beta/2$ is a Lagrange multiplier for the normalizing constraint (D.17). Hence, the steepest descent path is that for which

$$\sum_{i=1}^{na} \left(H_i - \beta f_i\right) df_i = 0$$

for arbitrary df_i. Thus, the direction cosines for the path of steepest descent are given by

$$f_i = \frac{H_i}{\beta}.$$

Then (D.17) gives

$$\sum_{i=1}^{na} H_i H_i = \beta^2, \tag{D.18}$$

so that

$$\frac{dW}{ds} = \sum_{i=1}^{na} \frac{H_i H_i}{\beta} = \beta, \tag{D.19}$$

which must he negative for descent. Hence, from (D.18)

$$\beta = -\sqrt{\sum_{i=1}^{na} H_i H_i}.$$

For a given set of phase moles and potentials, we calculate the H_i and determine the path of steepest descent towards the set of λ_i which, for the given phase moles, will render all of $H_i = 0$. The distance that we should go along this path is estimated using a Taylor series expansion of dW/ds,

$$\frac{dW}{ds} = \left.\frac{dW}{ds}\right|_0 + \left.\frac{d^2W}{ds^2}\right|_0 \Delta s + \cdots .$$

Since we seek $dW/ds = 0$, using (D.19) and (D.3),

$$0 = \beta + \sum_{k=1}^{na} \frac{\partial}{\partial \lambda_k}\left(\frac{dW}{ds}\right)\frac{d\lambda_k}{ds} = \beta + \sum_{k=1}^{na}\sum_{i=1}^{na} \frac{\partial^2 W}{\partial \lambda_k \lambda_i} f_i f_k \Delta s.$$

So, using (D.8)

$$\Delta s = \frac{-\beta}{\sum_{k=1}^{na}\sum_{i=1}^{na} Q_{ik} f_i f_k}.$$

Since the matrix Q_{ik} is determined by the current values of the phase moles and element potentials, the path of steepest descent and the trial step we should take down it are easily calculated. This method for adjusting the element potentials is denoted as mode 1 in STANJAN.

In mode 1, the first thing done after the various quantities have been evaluated for the new potentials is to examine the behavior of W on the old path of steepest descent. If the minimum point has been passed, then a "valley interpolation" is made using both the old state and the new state to estimate the location between them on the descent path where the minimum W point occurs. This interpolated state is then used to start a fresh descent. If the trial points are jumping back and forth across a valley, overshooting the minimum point, then "damping" is turned on, in which the step along the path is reduced from that estimated above. This allows the trial point to descend down a steep hillside and then turn to follow the gentler downflow of the valley towards the point of minimum W. Mode 1 is very robust and works effectively, even when the trial state is far from the state of minimum W. For example, convergence has been obtained in cases where the initial mole fraction sum Z was as large as 10^{17}! Improvements occur along the descent path at the rate of a decade or two per step, and it is not long before the errors in the H_i become quite small. At this point an adjustment is made in the phase moles, and the solution process continues.

Mode 2: Newton–Raphson adjustment of the λ at fixed $\bar{N}$

Near the point of minimum W, it becomes difficult to tell the direction of steepest descent, and instead a Newton–Raphson iteration is used to find the point where the H_i all vanish. Equation (D.7) becomes, for fixed phase moles,

$$\sum_{k=1}^{na} Q_{ik} \Delta \lambda_k = -H_i. \tag{D.20}$$

This is solved to obtain the desired changes. This procedure is denoted as mode 2 in STANJAN. It is adopted whenever the H errors are suitably small, and abandoned in favor of mode 1 if the computed changes are too large.

Phase moles adjustments in modes 1 and 2

After a set of λ_i rendering the H_i all zero (approximately) has been found for a given set of $\bar{N}_m$, we must adjust the N for maximum W^*. When the solution is not close, a steepest ascent method is used to adjust the $\bar{N}_m$. Let r_m represent the direction cosines in $\bar{\mathbf{N}}$ space, and ds^* an element of the path along which we seek the maximum of W^*. Then

$$d\bar{N}_m = r_m ds^*,$$

and

$$\sum_{m=1}^{p} r_m r_m = 1. \tag{D.21}$$

From (D.12),

$$\frac{dW^*}{ds^*} = \sum_{m=1}^{p} V_m r_m.$$

To find the path of steepest ascent, we seek the r_n that maximize dW^*/ds, subject to (D.21). Thus we set

$$d\left(\frac{dW^*}{ds^*} - \frac{\alpha}{2}\sum_{m=1}^{p} r_m r_m\right) = 0,$$

where $\alpha/2$ is a Lagrange multiplier for the normalizing constraint (D.17). Hence, the steepest ascent path is that for which

$$\sum_{m=1}^{p} (V_m - \alpha r_m)\, df_m = 0$$

for arbitrary dr_m. Thus, the direction cosines for the path of steepest ascent are given by

$$r_m = \frac{V_m}{\alpha}.$$

Then (D.21) gives

$$\sum_{m=1}^{p} V_m V_m = \alpha^2, \tag{D.22}$$

so that

$$\frac{W^*}{ds^*} = \sum_{m=1}^{p} \frac{V_m V_m}{\alpha} = \alpha, \tag{D.23}$$

which must be positive for ascent. Hence, from (D.22)

$$\alpha = \sqrt{\sum_{m=1}^{p} V_m V_m}.$$

The distance that we must go along the path is estimated from a Taylor's series expansion of dW^*/ds^*,

$$\frac{W^*}{ds^*} = \left.\frac{W^*}{ds^*}\right|_0 + \left.\frac{W^{*2}}{ds^*2}\right|_0 + \cdots.$$

Since we scale the point $dW^*/ds = 0$, using (D.23),

$$0 = \alpha + \sum_{n=1}^{p} \frac{\partial}{\partial \bar{N}_m}\left(\frac{dW^*}{ds^*}\right)\frac{d\bar{N}_m}{ds^*}$$
$$= \alpha + \sum_{n=1}^{p}\sum_{m=1}^{p} \frac{\partial^2 W^*}{\partial \bar{N}_n \partial \bar{N}_m} r_n r_m \cdot \Delta s^*.$$

So, using (D.13),

$$\Delta s^* = -\frac{\alpha}{\sum_{n=1}^{p}\sum_{m=1}^{p} A_m r_n r_m}.$$

This is used to estimate the distance along the path to the maximum W^* point. If this distance would produce negative phase moles, then the changes are reduced to prevent this occurrence. Limits placed on the changes, "ridge interpolation," oscillation damping, and other numerical tricks add to the robustness of the program. If the path of steepest ascent reaches a state where one $\bar{N}_m$ is zero, then that phase will be absent in the system, and the phase is dropped from the phase sums. A new path of steepest ascent is then computed, and a step towards the maximum of W^* is taken. The possibility of inactive phases becoming active is considered in each phase adjustment.

Mode 3: Newton–Raphson adjustment of the λ and $\bar{N}$

When the atomic population errors are small and the mole fraction sums for all active phases are very nearly unity, a Newton–Raphson scheme is used to adjust the λ_i and $\bar{N}_m$ simultaneously. The equation system (D.6), (D.7) is used to calculate the changes necessary to bring all H_i to zero and all Z_m to unity. This mode is adopted whenever the solution is very near and abandoned whenever the changes it requests are too large. This mode is always the last mode used before a converged solution; it is called Mode 3 in STANJAN.

Accuracy and convergence

The element potentials and phase moles are adjusted to an accuracy of 1 part in 10^8, and the mole fraction sums are made unity to 1 part in 10^{10}. This accuracy is maintained, even in nearly singular cases, with the help of the matrix conditioning procedure described in the following Section. On a PC, mole fractions as small as 10^{-300} are displayed.

D.3 Independent Atoms, Basis Species, and Matrix Conditioning

In some systems the atoms are not independently variable; for example, in a system consisting of CO, COS, and S, the atoms C and O are not independently variable. In order to avoid singular matrices in the solution, we must work only with the independent atoms. These are identified in the STANJAN initializer, described in the following. The atom sums above are then carried out only over the independent atoms, and only the element potentials of independent atoms are computed.

In any system there will be a small set of base species which together could contain all of the atoms. While there is one base species per independent atom, there need not be a one-to-one correspondence between the independent atoms and base species. Usually there are many possible sets of base species; the most useful are those that dominate the system, and the STANJAN initializer identifies these. For example, in a system containing a mixture of CO, CO_2, O_2, and C(S) at 3000 K and 1 atm, with a C:O ratio of 2 : 1, the base species will be C(S) and CO (see example in Section 10.7, page 281).

The base species play a key role in obtaining accurate solutions when the matrices are nearly singular. This is accomplished by a process we call *matrix conditioning*. For example, consider a system containing CO, CO_2, and O_2, with a C:O ratio of 1 : 2, at a low temperature. The system will consist almost entirely of CO_2, and the population equations will be

$$\begin{aligned} N_{CO_2} + N_{CO} &= 1, \\ 2N_{CO_2} + N_{CO} + 2N_{O_2} &= 2. \end{aligned} \tag{D.24}$$

Since N_{CO} and N_{O_2} are both very small, these two equations are very nearly the same, i.e., the system is very nearly linearly dependent. Conditioning removes this difficulty.

We need to solve equations of the form

$$\sum_{i=1}^{na} Q_{ik} X_k = Y_i, \tag{D.25}$$

where X_k denotes the solution vector and the Y_i the right-hand side (see (D.13) and (D.14)). The Q_{ik} matrix associated with this system is

$$Q_{ik} = \bar{N} \begin{bmatrix} z_{CO_2} + z_{CO} & 2z_{CO_2} + z_{CO} \\ 2z_{CO_2} + z_{CO} & 4z_{CO_2} + z_{CO} + 4z_{O_2} \end{bmatrix}, \tag{D.26}$$

because $z_{CO} = 1$, and the other z_j are very small, this matrix is very nearly singular, and hence the solutions of (D.9) and (D.20) are very hard to construct accurately.

The idea of matrix conditioning is to form linear combinations of the equations that remove all but one base species from each equation. This is equivalent to multiplying the (D.26) by a conditioning matrix C_{ni}; this produces

$$\sum_{i=1}^{na} \sum_{k=1}^{na} \sum_{j=1}^{na} C_{ni} \bar{N}_{(j)} n_{ij} n_{kj} z_j X_k = \sum_{i=1}^{na} C_{ni} Y_i.$$

Now, for the nth equation, we select C_{ni} such that the only base species retained is the nth;

$$\sum_{i=1}^{na} C_{ni} n_{ij} = \begin{cases} 1 \text{ if } j \text{ is the } n \text{ th base species,} \\ 0 \text{ if } j \text{ is any other species.} \end{cases} \tag{D.27}$$

Then, instead of solving (D.26), we solve the conditioned equations

$$\sum_{k=1}^{na} Q^*_{ik} X_k = Y^*_i,$$

where the conditioned matrix and right-hand side are

$$Q^*_{ik} = \sum_{n=1}^{na} C_{in} Q_{nk},$$

and

$$Y^*_i = \sum_{k=1}^{na} C_{ik} Y_k.$$

In the C—O example above, we take CO_2 and O_2 as the base species. The first equation does not contain the base O_2, and hence is already conditioned. The second conditioned equation is formed by subtracting the first of (D.25) from half of the second. Thus, the two conditioned equations are

$$\left(N_{CO_2} + N_{CO}\right) X_1 + \left(2N_{CO_2} + N_{CO}\right) X_2 = Y_1,$$

and

$$\left(-N_{CO}\right) X_1 + \left(2N_{O_2} - N_{CO}\right) X_2 = -Y_1 + \frac{1}{2} Y_2.$$

This pair of equations will be linearly independent and will yield accurate numerical results, even when the moles of CO and O_2 are very small.

STANJAN computes Q_{ik} *exactly*, so that the base species vanish completely from other than their own equations. The conditioned versions of Equations (D.13) and (D.14) are solved, rather than the primitive equations. The bases are reviewed and changed if necessary whenever a phase appears or disappears during the solution, and a new conditioning matrix is calculated.

This matrix conditioning allows STANJAN to solve accurately, even when two original equations differ by as little as one part in 10^{20} or more! The matrix-conditioning process is also used in the phase-redistribution process in the initializer (Section D.4) to help maintain high accuracy in nearly singular systems.

The population equations may also be conditioned, producing

$$\sum_{k=1}^{na} \sum_{j=1}^{na} C_{ik} \bar{N}_{(j)} n_{kj} z_j = \sum_{i=1}^{na} C_{ik} p_k = p_i^*.$$

Note that, by (D.27), all but one of the base species drop out from each of these conditioned population equations. For example, the conditioning of the system of Equations (D.24) produces

$$N_{CO_2} + N_{CO} = 1, \qquad \text{(D.28a)}$$

$$-\frac{1}{2} N_{CO} + N_{O_2} = 0. \qquad \text{(D.28b)}$$

The right-hand sides of these equations (the "conditioned populations") are just the moles of the base species, and this fact is used to compute the conditioned populations. The balancing procedure described in Section 10.7, page 281 is required whenever one of the conditioned populations p_i^* is zero, and the conditioned population equation provides this balance. Note that (D.28b) is the balance equation (10.59).

D.4 Initialization

The solution requires an initial guess, and a good guess leads to a fast solution. The STANJAN initializer is one of the most important reasons for its success in treating general problems. Problems that could not be initialized by early STANJAN versions now run nicely, and problems that took dozens of iterations are now initialized so well that only a few iterations are required. Thus, the STANJAN initializer may be of considerable interest to those who prefer to use other methods for equilibrium solution. The basic idea of the initializer is to create an approximate distribution of the atoms from which the phase moles and mole fractions of key species can be estimated. These estimated mole fractions are then used to estimate the element potentials, in much the same way as the examples in Section 10.7. The initializer does this in a way that works for an arbitrary problem.

The initializer begins by distributing the system atoms to a set of base species in a way that makes an approximate Gibbs function as small as possible. This approximate Gibbs function is that obtained by neglecting the $\ln x_j$ corrections to the Gibbs functions. The initializer minimizes

$$\frac{G^*}{RT} = \sum_{j=1}^{na} \tilde{g}_j N_j, \tag{D.29}$$

subject to the atomic constraints

$$\sum_{j=1}^{nc} n_{ij} N_j = p_i, \qquad i = 1, \cdots, na,$$

and to the constraints

$$N_j \geq 0, j = 1, 2, \ldots, s.$$

This minimization problem is a classic problem in linear programming, and is solved by the simplex method. The simplex method is a widely adopted algorithm for the solution of maximization/minimization problems. The maximum or minimum is searched by using a set of operations that, with some approximation, can be visualized as the construction of simplices, the extension to arbitrary dimensions of the concept of triangle or tetrahedron.

The theorems of linear programming show that the solution will be one where only a small set of species have non-zero moles – in the approximation (D.29). At each step in the simplex process one has identified a set of "base species" that contain the atoms, with all other species having zero moles. The simplex process is a base-species-replacement process in which the function to be minimized is continually reduced by changing the base species set until no further reductions are possible. There are always as many base species as there are independent atoms in the system, and the base species moles together contain all of the atoms.

The process begins by placing all atoms in "false" mono-atomic species. The atoms are transferred to the real species by a simplex minimization of the total number of false moles. Important conclusions are drawn at the end of this simplex process. If it is impossible to eliminate the false species, then the assigned populations were impossible. If a false species remains as a base with zero moles, then that atom is not linearly independent in the system. This is the process by which independent atoms are identified.

Once the atoms are placed in real species, the simplex minimization process continues until no further reduction in G^* can be achieved. The dominant species are then identified as the base species; the moles of all other species are zero at this point. The conditioning matrix is calculated for this set of base species.

The initial distribution of atoms to the base species allows estimation of the phase moles and mole fractions of the base species, with one base species per independent atom. In order to estimate the element potentials, one linear equation in the element potentials must be obtained for each independent atom in the system, i.e., for each base species. For each base there are two primary possibilities:

I) If the estimated mole fraction is greater than zero, this value is used to derive one linear equation relating the element potentials (see examples in Section 10.7).

II) If the base species was estimated to have zero moles, then the dominant balancing species is identified. The balancing species is a secondary species that appears in the conditioned population equation with a negative coefficient; the one with the largest expected mole fraction is chosen as the dominant balancing species. Four possibilities exist:
   a) If no balancing species can be found, then the species is excluded and the initialization is repeated.
   b) If the balancing species is in the same phase as the zero-moles base species, then the balance equation is used to derive a linear equation relating the potentials (see example in Section 10.7). If the phase has zero moles, the "phase-redistribution" flag is set.

c) If the balancing species is in another phase, then the other zero-moles base species are examined and the bases are reordered, so that the dominant zero-moles base is considered first.
d) If the zero-moles base is the dominant zero-moles base in a phase containing zero moles, then its mole fraction is set to unity and this value is used to derive a linear equation relating the element potentials, and the phase-redistribution flag is set.

The conditioning matrix is recomputed whenever the bases are changed. These processes produce a set of linear algebraic equations, which is solved for the estimated element potentials.

In determining the balancing species, estimates of the element potentials are used to estimate the species mole fractions. The simplex Lagrange multipliers themselves provide the first estimates of the element potentials for purposes of selecting the balancing species. Then, after a set of element potentials has been obtained from the process described above, the selection of balancing species is repeated using these potentials, and different balancers are chosen if appropriate. Thus, the final element potentials are consistent with the choices of balancing species used to generate them.

If the phase redistribution flag was set by the element potential estimating process described above, then it is necessary to redistribute the atoms to populate an empty phase. The idea is to redistribute so that the balancing species are present in about the right amount, which will force the species that they balance to be present. The first step is to estimate the mole fractions of the balancing species using the estimated element potentials. Then the atoms are redistributed amongst the set of species consisting of the original base species plus the balancing species, seeking to bring the mole fractions of the balancing species as close to their targets as possible. This is accomplished by a second simplex calculation in which the sum of the differences between the target mole fractions and actual mole fractions is minimized, subject to the atomic constraints, to the constraint that all moles must be non-negative, and to the constraint that the target mole fraction cannot be exceeded. Usually these targets are met precisely. The net effect is that approximately the right number of atoms are put into the phases which, on first estimation, had zero moles. This intricate simplex process is described in more detail in the following.

If a phase redistribution is required, the element potentials are re-estimated using the revised mole distribution, following the procedure described above. On each pass, the base set is checked to see that they are the dominant species, and if necessary bases are changed. Thus, at the end the phase moles and element potentials are all based on a consistent set of dominant species. These initial estimates of the phase moles and element potentials will generate approximately the correct mole fractions, and so the equilibrium solution by the method described in Section D.2.2 usually converges to high accuracy in just a few iterations.

When running a sequence of calculations involving the same species and atomic populations at nearby states, the full initialization process is avoided. The mole fractions of the base species from the previous run are instead used to estimate the element potentials, and the phase moles of the previous run are used. STANJAN also provides the option of freezing the composition, and these runs do not require initialization or equilibrium solution.

The phase-redistribution simplex process uses the variables

$$y_j = -N_j + \bar{N}_{(j)} x_j^* \text{ (balancers)}, \tag{D.30a}$$

for the balancing species assigned target mole fractions

$$y_j = N_j \text{ (bases)}, \tag{D.30b}$$

for the base species. Equation (D.30a) is rewritten as

$$y_j = x_j^* \left(\sum_{j'o \text{ in } (j)} y_{j'} + \sum_{j''b \text{ in } (j)} N_{j''} \right) - N_j, \tag{D.31}$$

where the sum over $j'o$ denotes a sum over the base species, and $j''b$ denotes a sum over the balancing species, in the phase of balancing species j. The set of (D.31) for the balancing species (there may be more than one) is inverted to give, for the balancing species,

$$N_j = \sum_{j''b} T_{jj''} y_{j''} + \sum_{j'o} B_{jj'} y_{j'}, \tag{D.32a}$$

where $j''b$ denotes a sum over balancing species and $j'o$ denotes a sum over base species and **T** and **B** result from the inversion. Then, for the base species,

$$N_j = y_j \text{ (bases)}. \tag{D.32b}$$

Equations (D.32) allow expression of the N_j in terms of the simplex variables Y_j. The atomic constraints (10.44) can then be expressed in the form

$$\sum_{jbo} R_{ij} y_j = p_i, \qquad i = 1, \ldots, na, \tag{D.33}$$

where jbo denotes a sum over all variables in this simplex problem (the base and balancing species); R_{ij} is computed from T_{jj}, B_{jj}, and n_{ij} Finally, the simplex variables also satisfy

$$y_j \geq 0 \qquad \text{for all } j. \tag{D.34}$$

Note that (D.34) keeps the balancing mole fractions no greater than their targets.

The simplex process is initiated by establishing a feasible set using additional false mono-atomic species having $y_j = N_j$ and minimizing the sum of the false moles, subject to the constraints. Then the problem described above is solved.

In order to improve the accuracy in nearly singular problems, the atomic constraint equations (D.33) are multiplied by the conditioning matrix C_{ik} discussed in Section D.3. In some problems with very rare balancing species, the constraint equations may not be very well satisfied after some simplex base change. A correction is then made. The approach is to select the constraint equation best satisfied, and then to treat that simplex base as a known quantity in a smaller set of equations for the other simplex bases. The net result is a remarkable increase in accuracy.

In summary, the initializer is a sophisticated program that makes a very good guess as to the distribution of atoms to dominant species and phases in the system. It possesses remarkable accuracy in very-nearly-singular situations. The importance of the initializer cannot be overemphasized.

E The Method of Lagrange Multipliers

Lagrange multipliers are used in the theoretical solution of problems of the form

$$F\{x) = \min,$$

subject to the constraints

$$C_k\{x) = \text{constant}, \qquad k = 1, ..., c. \tag{E.1}$$

In general, F and C_k may be non-linear functions of the solution vector $\mathbf{x} = \mathbf{x_1}, \mathbf{x_2}, ..., \mathbf{x_n}$. The differential of F is

$$dF = \sum_{i=1}^{n} H_i dx_i, \tag{E.2}$$

where

$$H_i = \frac{\partial F}{\partial x_i}. \tag{E.3}$$

Now, for F to be a minimum with respect to arbitrary variations, $dF = 0$ for arbitrary dx_i that satisfy the constraints

$$dC_k = 0 = \sum_{i=1}^{n} A_{ik} dx_i, \tag{E.4}$$

where

$$A_{ik} = \frac{\partial C_k}{\partial x_i}.$$

If we have n variables and c constraints, only $n - c$ of the variables may be freely varied. Before examining the conditions under which dF is zero for arbitrary variations of the free x_i, we need to represent the changes in the restricted x_i in terms of the change in the free ones, and then substitute for the changes in the restricted variables in E.2. This substitution is equivalent to subtracting a linear combination of the Equations E.4 from E.2, such that the restricted dx_i drop out of the result. This subtraction yields

$$dF = \sum_{k=1}^{c} \lambda_k dC_k = \sum_{i=1}^{n} \left(H_i - \sum_{k=1}^{c} \lambda_k A_{ik} \right) dx_i, \tag{E.5}$$

where the coefficients λ_i must be chosen such that they drop out the restricted dx_i. In order for restricted dx_i drop out, the coefficient of each one of them must be zero, so for these is,

$$H_i = \sum_{k=1}^{c} \lambda_k A_{ik} = 0. \tag{E.6}$$

For the remaining freely varied x_i, there must be no variation that changes F (to the first order) which requires that the coefficients of these dx_i also vanish in E.5. Hence, E.6 must hold for all is. Equation E.6 represents a set of n simultaneous equations for the solution vector x_i. The constraints E.1 provide c additional equations for the λ_k, called the *Lagrange multipliers*. If F and C_i are quadratic functions of dx_i, then E.6 will be a linear equation system; this is the case in the application to finding the paths of steepest descent described in Appendix D.2.1. In the element potential theory, F is the Gibbs function, and the resulting equations are nonlinear.

NOTATION

Remarks

- Molar properties are indicated with ˆ, e.g., $\hat{v}$ is the molar volume.
- Partial molar properties are indicated with ¯, e.g., $\bar{g}_i$ is the partial molar Gibbs energy of mixture component i.
- Properties of component i in a mixture are indicated with ˜, e.g., $\tilde{f}_i$ is fugacity of component i in a mixture.
- In order to distinguish between the parenthesis which implies a multiplication of factors, as in $(1 - x)v$, and the parenthesis that is used to indicate that a variable is a function of other variables, like in $T^{sat} = T^{sat}(P^{sat})$, we indicate the latter with $\{$, such that $T^{sat} = T^{sat}\{P^{sat})$ cannot be equivocated.

Roman symbols

$\hat{h}_f^o$	Molar enthalpy of formation
A	Area
A	Helmholtz energy
a	Specific Helmholtz energy
B	Second virial coefficient
C	Third virial coefficient
c	Speed of light
c	Speed of sound
c_P	Specific heat at constant pressure (isobaric)
c_V	Specific heat at constant volume (isochoric)
$đ$	Inexact differential
E	Total energy
e	Total specific energy
E_k	Kinetic energy
E_p	Potential energy
f	Fugacity
G	Gibbs energy
g	Gravitational acceleration
g	Specific Gibbs energy
H	Enthalpy
h	Specific enthalpy
K	Equilibrium constant of a reaction
k	Boltzmann constant
k	Henry's constant
k_G	Gravitational constant
k_{ij}	Binary interaction parameter
M	Mass
m	Mass of a molecule
M_0	Rest mass
N	Number of moles
n	Polytropic exponent
n'	Number of molecules per unit volume
N_A	Avogadro's number
nc	Number of components in a mixture
P	Pressure
P_i	Partial pressure of component i in an ideal gas mixture
p_i	probability of realizing quantum state i at time t
Q	Energy transfer as heat
$\dot{Q}$	Thermal power (rate of energy transfer as heat)
R	Gas constant
S	Entropy
s	Specific entropy
T	Temperature
U	Internal energy
u	Specific internal energy
V	Volume
v	Specific volume
W	Work
w	Specific work
w	Weight
$\dot{W}$	Shaft power, or mechanical power (rate of energy transfer as work)
w_i	Mole fraction of species i in the solid phase
x	Distance
x	Quality or vapor fraction
x_i	Mole fraction of species i in the liquid phase

y_i	Mole fraction of species i in the vapor phase
Z	Compressibility factor
z	Elevation
z_i	Mole fraction of species i, in a single unspecified phase (liquid, vapor, or solid)

Calligraphic symbols

a	Acceleration
$\mathcal{F}$	Force
$\mathcal{H}$	Magnetic field
$\mathcal{M}$	Dipole moment
$\mathcal{I}$	Momentum
$\mathcal{V}$	Velocity

Greek symbols

α	Isentropic compressibility
α	Parameter in the temperature-dependent polynomial equation for c_P^{IG}
β	Isobaric compressibility
β	Parameter in the temperature-dependent polynomial equation for c_P^{IG}
β_v	Volumetric compression ratio
β_P	Pressure ratio
χ_i	Mole fraction of species i in an ideal gas mixture
δ	Dimensionless density in Helmholtz equations of state
η	Energy conversion efficiency
η	Parameter in the temperature-dependent polynomial equation for c_P^{IG}
η_s	Isentropic efficiency
η_{II}	II-law efficiency
η_{max}	Carnot efficiency
Ξ	Exergy
ξ	Specific exergy
Ξ_d	Exergy destruction
ξ_f	Specific flow exergy
γ	Parameter in the temperature-dependent polynomial equation for c_P^{IG}
γ	Specific heat ratio
γ_i	Activity coefficient of component i in a mixture
κ	Isothermal compressibility
λ_i	Element potential of the atom i
ν	Frequency
ν_i	Number of moles of species i
ω	Acentric factor
ϕ	Fugacity coefficient
ρ	Density
ρ_{cutoff}	Cutoff ratio
τ	Inverse dimensionless temperature in Helmholtz equations of state
ξ_i	Mass fraction of species i in an ideal gas mixture

Acronyms

COP	Coefficient of performance
HHV	Higher heating value
LHV	Lower heating value
TIT	Turbine inlet temperature

Superscripts

boil	Boiling point
freeze	Freezing point
BPE	Boiling point elevation
bubble	At the bubble point state
dew	At the dew point state
E	Excess (or departure from ideal solution)
FPD	Freezing point depression
IG	Ideal gas
IS	Ideal solution
L	Liquid
NBP	Normal boiling point
o	Standard reference state (1 atm, 25 °C = 298.15 K)
R	Residual (or departure from ideal gas)
S	Solid
sat	At the saturation state
V	Vapor

Subscripts

0	At a reference state (datum state)
∞	At infinite dilution

s	Isentropic
b	Boundary
c	At the critical point
comp	Compressor
cond	Condenser
CV	Control volume
env	Environment
evap	Evaporator
gen	Generator
mix	Mixture
P	Products
phys	Physical
R	Reactants
r	Reduced
react	Reaction
sol	Solar
sys	System
turb	Turbine

Special symbols

a	Attractive parameter in cubic equations of state
b	Repulsive parameter in cubic equations of state
c	Parameter for the general formulation of cubic equations of state
d	Parameter for the general formulation of cubic equations of state
$\mathcal{P}$	Production
h	Planck's constant

INDEX